Scott Swinton

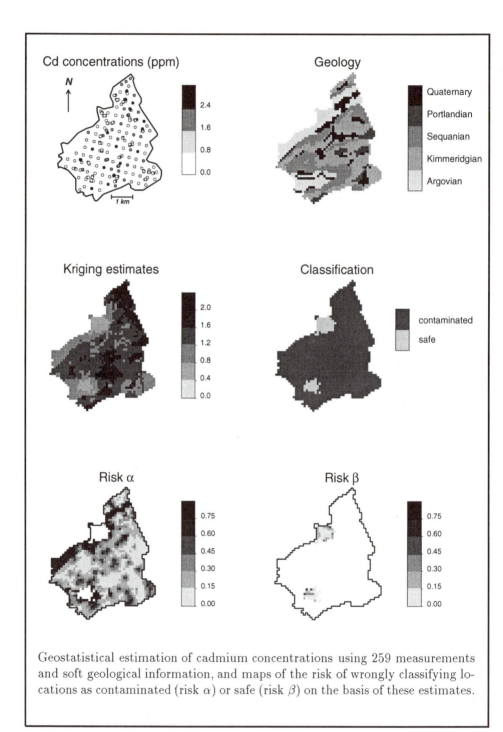

Geostatistical estimation of cadmium concentrations using 259 measurements and soft geological information, and maps of the risk of wrongly classifying locations as contaminated (risk α) or safe (risk β) on the basis of these estimates.

Geostatistics for Natural Resources Evaluation

PIERRE GOOVAERTS

Department of Civil and Environmental Engineering
The University of Michigan, Ann Arbor

New York   Oxford
Oxford University Press
1997

Oxford University Press

Oxford   New York
Athens   Auckland   Bangkok   Bogota   Bombay   Buenos Aires
Calcutta   Cape Town   Dar es Salaam   Delhi   Florence   Hong Kong
Istanbul   Karachi   Kuala Lumpur   Madras   Madrid   Melbourne
Mexico City   Nairobi   Paris   Singapore   Taipei   Tokyo   Toronto   Warsaw

and associated companies in
Berlin   Ibadan

Copyright © 1997 by Oxford University Press, Inc.

Published by Oxford University Press, Inc.
198 Madison Avenue, New York, New York 10016

Oxford is a registered trademark of Oxford University Press

Library of Congress Cataloging-in-Publication Data
Goovaerts, Pierre.
Geostatistics for natural resources evaluation / Pierre Goovaerts.
p.   cm. — (Applied geostatistics series)
Includes bibliographical references and index.
ISBN 0-19-511538-4
1. Geology—Statistical methods.   I. Title.   II. Series.
QE33.2.M3G66   1997
550'.72—dc21   97-2521

1 3 5 7 9 8 6 4 2

Printed in the United States of America
on acid-free paper

To

Nathalie,

Maxime,

and Xavier

Foreword

> As a last remark, beware that uncertainty . . . arises from our imperfect knowledge
> of that phenomenon, it is data-dependent and most importantly model-dependent,
> that model specifying our prior concept (decisions) about the phenomenon. No
> model, hence no uncertainty measure, can ever be objective.

Pierre Goovaerts closes an authoritative work with a comment that many mas-
ters could envy, showing his commanding distance from a subject he otherwise
meticulously articulated over 400 pages. This is the signature of the very best.
This (still) young man is neither French nor South African. He is neither mining
engineer nor petroleum engineer. He is an agronomist with a trademark no-non-
sense practicality that, from the beginning, made geostatistics different and
hence, I dare say, successful. An author must have rare insight to present such a
compendium of methods and tools without letting them take command of the
book. Pierre built his book around the study goals, giving to data analysis and
data integration the lead they deserve. Never mistake the tools for the goals—
this is the recipe for a good geostatistical study, and Pierre has demonstrated that
it also works for a good book. His illustration of every concept, every kriging
system, by an example from the same data set (heavy metal contamination in the
Swiss Jura) is an enchantment and a necessary breather after some tough reading.
   Characteristically, exploratory data analysis (chapter 2) precedes the introduction of the random function (chapter 3). The chapter on inference and modeling of a multivariate model (chapter 4) is, in my opinion, the best ever written on
the subject. Pierre could have been bolder by giving precedence to uncertainty

assessment over estimation, but he chose to present first the estimation tools (chapters 5 and 6). This is the most complete yet cohesive exposé of all the various flavors of kriging and cokriging. The emphasis is not on the illusory kriging variance but rather on building estimators that can account for the large diversity of information types characteristic of earth sciences. The very reason for geostatistics and the future of the discipline lie in the modeling of uncertainty, at each node through conditional distributions (chapter 7) and globally through stochastic images (conditional simulations, chapter 8). In modern geostatistics, which is driven by conditional simulations, kriging is an engine, and not the only one, to build models of conditional probability distributions. Kriging estimates and kriging variances have lost their original luster: the former because of their uneven smoothing and the latter because of their data independence.

The practice of geostatistics has always been ahead of academic publications. This book finally may have caught up with the use of random function models in the earth sciences, but geostatistics has already freed itself from the frame of such models. It befits that the door of an era be closed by a man of the future.

Stanford University   *A.G. Journel*
May 1996

Acknowledgments

Most of this book was written in 1994 during my second year as a postdoc at the Department of Geological and Environmental Sciences (Stanford University). The idea of a geostatistical book was launched by André Journel on a flight back from the forum "Geostatistics for the Next Century," held in Montreal in June 1993. In addition to initiating the project, André was an enthusiastic adviser and a tireless reviewer, tracking any sentence that failed to follow his golden rules of clarity and practicality. Without his ceaseless support, this book probably would never have appeared on the shelves.

The 20 months of writing were followed by a 6-month peer review. I wish to thank the following persons for their time and patience in reading all or part of the 550-page draft manuscript: Clayton Deutsch, Jennifer Dungan, Guy Gérard, Jaime Gómez-Hernández, Ricardo Olea, Philippe Sonnet, Mohan Srivastava, Hans Wackernagel, and Richard Webster. Special thanks to Mohan Srivastava for his thorough review of the book and his 106 comments, sometimes irritating but always pertinent. The draft manuscript also was used as a textbook for pre-qualifying exams of Stanford University Ph.D. students and benefited from the careful reading of Phaedon Kyriakidis, Srinivas Rao, and Tingting Yao.

Most of my knowledge about indicator geostatistics and stochastic simulation was gained during my 2 years at Stanford, and I am grateful to Gilles Bourgault and many graduate students for their stimulating discussions and seminars. This time at Stanford would not have been possible without the financial support of Stanford University, the Belgian National Fund for Scientific Research, the Belgian American Educational Foundation, NATO, and a Fulbright-Hays grant. I am especially indebted to the Belgian National Fund for Scientific Research for supporting my research for 7 years.

Geostatistics cannot exist without data. Throughout the book, case studies have proved valuable complements to the theoretical introduction of concepts and algorithms. I thank Jean-Pascal Dubois of the Swiss Federal Institute of Technology at Lausanne for providing me with a priceless environmental data set and for authorizing the publication of these data.

Louvain-la-Neuve, Belgium   P.G.
December 1996

Contents

Chapter 1

Introduction

Earth sciences data are typically distributed in space and/or in time. Knowledge of an attribute value, say, a mineral grade or a pollutant concentration, is thus of little interest unless location and/or time of measurement are known and accounted for in the data analysis. Geostatistics provides a set of statistical tools for incorporating the spatial and temporal coordinates of observations in data processing.

The development of geostatistics in the 1960s resulted from the need for a methodology to evaluate the recoverable reserves in mining deposits. Priority was given to practicality, a current trademark of geostatistics that explains its success and application in such diverse fields as mining, petroleum, soil science, oceanography, hydrogeology, remote sensing, and environmental sciences. Until the late 1980s, geostatistics was essentially viewed as a means to describe spatial patterns and interpolate the value of the attribute of interest at unsampled locations. Geostatistics is now increasingly used to model the uncertainty about unknown values through the generation of alternative images (realizations) that all honor the data and reproduce aspects of the patterns of spatial dependence or other statistics deemed consequential for the problem at hand. A given scenario or transfer function (remediation process, flow simulator) can be applied to the set of realizations, allowing the uncertainty of the response (remediation efficiency, flow properties) to be assessed. Stochastic imaging is one of the most vibrant and promising areas of research in geostatistics.

Often there are only a few measurements of the attribute of interest; the resultant predicted maps thus provide poor resolution, and the corresponding uncertainty may be very large. In such situations, it is critical to account for information that is more densely sampled. For example, insufficient pollutant data can be supplemented with indirect, yet exhaustive, information provided by the calibration of a soil or land use map. Integration of secondary data in prediction and simulation algorithms is another active avenue of research in geostatistics.

Theoretical developments and applications of geostatistical tools are being published in an ever-increasing variety of journals and congress proceedings. Yet, as of 1997, there was no textbook to provide students and practitioners with a comprehensive as well as practical overview of the ever-growing palette of geostatistical concepts and algorithms. Ten years after David's (1977) and Journel and Huijbregts' (1978) books on mining geostatistics, Isaaks and Srivastava (1989) wrote a remarkable introduction to applied geostatistics at the undergraduate level. Their focus was on exploratory data analysis and spatial prediction, with little development of more advanced topics, such as assessment of uncertainty and stochastic imaging. The guidebook associated with the geostatistical software library GSLIB (Deutsch and Journel, 1992a) gave a complete presentation of recent geostatistical developments, particularly in the area of conditional simulation, yet it was not intended to be a theoretical reference textbook. This book aims at bridging the gap between Isaaks and Srivastava's introductory book and GSLIB's more complete user guide.

The main text of the book is divided into seven chapters, covering the most important areas of geostatistical methodology. The presentation follows the typical steps of a geostatistical analysis, introducing tools for description, quantitative modeling of spatial continuity, spatial prediction and uncertainty assessment. To facilitate reading and as an attempt at standardization, this book uses the notation of the GSLIB guidebook.

The various tools are illustrated using a multivariate soil data set related to heavy metal contamination of a 14.5 km^2 region in the Swiss Jura. The geostatistical analysis was carried out using the GSLIB software. Although this data set gives the book a definite environmental flavor, presentation of the algorithms is general and intended for students and practitioners desiring to gain an understanding of the methodology. Mathematical developments underlying most interpolation algorithms are given; therefore, the reader should have some prior notions of linear algebra, in addition to an undergraduate knowledge of statistics. These theoretical developments may be skipped, however, on first reading, without altering comprehension of the case studies.

1.1 The Jura Data Set

The data used throughout this book were collected by the Swiss Federal Institute of Technology at Lausanne. A detailed description of the sampling, field, and laboratory procedures is given in Atteia et al. (1994) and Webster et al. (1994). Data were recorded at 359 locations scattered in space; see Figure 1.1. Concentrations of seven heavy metals (cadmium, cobalt, chromium, copper, nickel, lead, and zinc) in the topsoil were measured at each location. Geologic and land use maps provide exhaustive, albeit soft (indirect), categorical information related to metal content. This data set shares three features common to most earth science data sets: (1) data are auto- and cross-correlated in space; (2) several attributes are involved jointly; and (3)

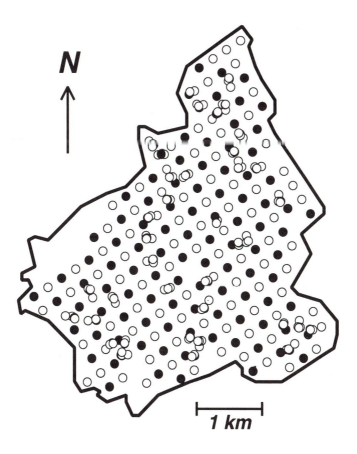

Figure 1.1: Map showing the split of the 359 data locations into a test set (closed circles) and a prediction set (open circles).

a few precise analytical measurements are supplemented by more numerous categorical data (soft information).

The large sample size allows the data to be divided into a validation set (100 test locations) and a prediction set (259 locations). The prediction set includes geology and land use, nickel and zinc concentrations at all 359 locations, and the concentrations of other heavy metals at 259 locations. The prediction set is considered to be the only information available for characterizing the entire study area. The validation set is used to check results provided by the various interpolation and simulation algorithms proposed.

A typical geostatistical analysis is conducted on the Jura data with the following objectives:

1. Describe the patterns of spatial dependence of heavy metals, and relate them to the distribution of potential sources, such as rock types and human activities (land use).

2. Build a probabilistic model of the spatial distribution of heavy metals in the region.

3. Estimate the metal concentrations at test locations.

4. Model the probability distributions of metal concentrations at test locations, and assess the risk of exceeding critical thresholds.

5. Identify test locations where remedial measures should be taken.

6. Model joint spatial uncertainty of metal concentrations through a set of alternative numerical models (stochastic imaging), and assess the risk involved in declaring the study area safe.

Prediction and classification of test locations are checked against the true values from the validation set.

1.2   Plan of The Book

The book starts with an exploratory analysis of the Jura data set in Chapter 2. The univariate distribution of categorical and continuous attributes and their relationships are first described ignoring data locations. Data locations are then considered for modeling spatial continuity and cross dependence between attributes.

Chapter 3 introduces the random function concept, which allows a probabilistic presentation of the various tools introduced in the previous chapter. Chapter 4 addresses the problem of inferring statistics that are representative of the study area and not only of the available sample. Theoretical and practical issues related to modeling experimental (cross) semivariograms are discussed.

The subsequent two chapters introduce the problem of interpolation and the multiple versions of the kriging (interpolation) paradigm. Chapter 5 presents kriging techniques that utilize only values of the attribute under study. Algorithms for incorporating secondary information are discussed in Chapter 6.

Chapter 7 introduces the Gaussian and indicator algorithms for modeling local probability distributions of either continuous or categorical attributes. The use of these models to assess the uncertainty about unknown values and determination of optimum estimates is discussed.

Chapter 8 presents algorithms to generate multiple realizations distributed in space of either continuous or categorical attributes. The simulation techniques presented include sequential Gaussian and indicator algorithms, the

LU decomposition algorithm, probability field simulation, and simulated annealing. Different ways of summarizing and visualizing the spatial uncertainty model provided by the series of alternative realizations are reviewed.

1.3 Terminology

Although all statistical concepts and notations are defined in the text and are summarized in Appendix B, a few terms used extensively throughout the book are now introduced.

- *Attribute.* Physical properties are called "attributes" and denoted by lowercase letters, such as z or s. *Continuous* attributes such as metal concentrations are measured on a continuous quantitative scale, whereas *categorical* attributes take only a limited number of states, usually non-ordered, e.g., rock types or land uses.

- *Variable.* The variable Z or S, denoted by capital letters, is defined as the set of possible values or states that the attribute z or s can take over the study area or at a location with coordinates vector $\mathbf{u}$. In the latter case, the variable is denoted $Z(\mathbf{u})$ or $S(\mathbf{u})$.

- *Individual.* The attribute value is measured on a physical sample, such as a piece of rock or a core of soil taken from the field. In sections 2.1 and 2.2 where no account is taken of data locations, that physical sample is referred to as an *individual*. In subsequent chapters, each physical sample is associated to a precise *location* $\mathbf{u}_\alpha$ in the study area.

- *Population.* The population is defined as the set of all measurements of the attribute of interest that could be made over the study area. The finite collection of measurements available is referred to as a *sample* or sample set.

- *Parameter.* Parameters are constant (not random) quantities of a model, for example, the range parameter of a semivariogram model or the mean parameter of a lognormal probability distribution function modeling a histogram.

- *Statistics.* Statistics are quantities summarizing a distribution, which may involve several attributes and/or several locations in space. *Univariate*, *bivariate*, and *multivariate* statistics relate, respectively, to one, two, and multiple attributes. The terminology *one-point*, *two-point*, and *multiple-point* statistics is used when the variables relate to the same attribute at one, two, and multiple locations. For example, the correlation coefficient is a bivariate statistic, whereas the semivariogram is a two-point statistic. The cross semivariogram is a bivariate two-point statistic because it involves two different attributes at two different locations.

Chapter 2

Exploratory Data Analysis

The objective of this chapter is to introduce the Jura prediction data set through its most salient features. This purely descriptive part is a preliminary step toward building a numerical and probabilistic model for uncertainty in spatial prediction. Although the ultimate goal of the study is characterization of the whole area, the set of measurements at all 359 sites is, at this time, considered to be an exhaustive population. Thus, all statistics computed from these data are considered to be exhaustive statistics or sample population parameters; hence the traditional ^ superscript (for estimation) is intentionally not used. In Chapter 4, the population is expanded to the entire study area, and the issue of inferring unknown population parameters from sample statistics is addressed.

The univariate (one attribute at a time) distributions of categorical and continuous variables are described in section 2.1. Section 2.2 looks at the joint relations between pairs of colocated metal concentrations. In section 2.3, the patterns of variation of metal concentrations are described and related to those of potential sources, such as rock types and land uses. Spatial relations between concentrations of different metals are analyzed in section 2.4. The main features of the Jura data set are summarized in section 2.5.

2.1   Univariate Description

This section begins with a straightforward description of the two categorical variables, land use and rock type. Subsequently, continuous variables (metal concentrations) are described, pooled over the entire data set, and then split according to rock type and land use.

2.1.1   Categorical variables

Let $\{s(\alpha),\ \alpha = 1, \ldots, n\}$ be the set of observations of the categorical attribute s measured on n individuals α. The set of K possible states s_k that any value $s(\alpha)$ can take is denoted by $\{s_1, \ldots, s_K\}$. For example, $s(\alpha) = s_k$ if the state s_k is observed on the αth individual. The K states are exhaustive and mutually exclusive in the sense that each individual belongs to one and only one state s_k. The distribution (histogram) of categorical data is completely described by a frequency table, which lists the K states and their frequency of occurrence. The frequency of occurrence of state s_k, denoted $f(s_k)$ or simply p_k, can be expressed as the arithmetic average of n indicator data:

$$f(s_k) = \frac{1}{n} \sum_{\alpha=1}^{n} i(\alpha; s_k) \tag{2.1}$$

where the indicator datum $i(\alpha; s_k)$ associated with the αth individual is set to 1 if the state s_k is observed and zero otherwise, that is,

$$i(\alpha; s_k) = \left\{ \begin{array}{ll} 1 & \text{if } s(\alpha) = s_k \\ 0 & \text{otherwise} \end{array} \right. \tag{2.2}$$

Table 2.1 gives the list of land uses and rock types[1] in the study area and the corresponding sample proportions. Most of the sampled locations are under permanent grass, which is either grazed (64.1%) or cut for hay twice a year (21.2%); only 1.9% of these locations are cultivated and the remaining 12.8% is forest, mainly spruce (*Picea abies*). Apart from the Portlandian formation, which represents only 1.2% of the sites, the four other geologic formations are in fairly equal proportions.

In addition to the frequency of each state, it is worth knowing how often two states s_k and $v_{k'}$ corresponding to two different categorical attributes jointly occur. For example, what is the proportion of sampled locations that are simultaneously under forest and in the Quaternary formation? This information is provided by the joint frequency of occurrence $f(s_k, v_{k'})$, which can be expressed as the average of an indicator product:

$$f(s_k, v_{k'}) = \frac{1}{n} \sum_{\alpha=1}^{n} i(\alpha; s_k) \cdot i(\alpha; v_{k'}) \tag{2.3}$$

where $i(\alpha; v_{k'})$ is defined as in equation (2.2).

Table 2.2 gives the joint frequencies of observations for all possible pairings of rock type and land use. Forests are preferentially located on Kimmeridgian rocks, whereas most pastures are on Kimmeridgian and Sequanian rocks. Meadow and tillage are equally represented on each geologic formation except Portlandian. Note that 13 out of 20 bivariate categories contain less than 10 individuals (frequency $\leq 4\%$).

---

[1] The geologic classification is actually stratigraphic, not lithological. Throughout the text, the term *rock type* should be understood as *stratigraphic class*.

Table 2.1: Frequency of occurrence of different land uses and rock types ($n=259$).

Land use	Frequency		Rock type	Frequency
Forest	12.8%		Argovian	20.5%
Pasture	21.2%		Kimmeridgian	32.8%
Meadow	64.1%		Sequanian	24.3%
Tillage	1.9%		Portlandian	1.2%
			Quaternary	21.2%

2.1.2 Continuous variables

Frequency distribution

Let $\{z(\alpha), \ \alpha = 1, \ldots, n\}$ be the set of measurements of the continuous attribute z on the n individuals α. Once again, the actual location of these data is ignored for now. The distribution of continuous values is typically depicted by a histogram with the range of data values discretized into a specific number of classes of equal width and the relative proportion of data within each class expressed by the height of bars. These relative proportions define the class frequencies, hence the histogram depicts the frequency distribution of z-values for a given definition of classes.

Figure 2.1 shows the seven histograms of metal concentrations expressed in parts per million (ppm; S.I. units $= \mathrm{mg\ kg^{-1}}$). The long upper tails of the histograms of Cd, Cu, Pb, and Zn values indicate the presence of a few large concentrations. The other histograms are fairly symmetric, with the distribution of cobalt values being somewhat bimodal.

Cumulative frequency distribution

Sometimes it is important to know the proportion of data values that are below or above critical thresholds, for example, grades below which mining is not profitable or concentrations above which remedial action should be taken. The proportion of data $z(\alpha)$ that do not exceed a certain threshold value z_k,

Table 2.2: Joint frequencies of occurrence of land uses and rock types ($n=259$).

	Forest	Pasture	Meadow	Tillage
Argovian	2.7%	2.3%	15.1%	0.4%
Kimmeridgian	8.5%	6.9%	17.0%	0.4%
Sequanian	1.2%	9.3%	13.1%	0.7%
Portlandian	0.4%	0.4%	0.4%	0.0%
Quaternary	0.0%	2.3%	18.5%	0.4%

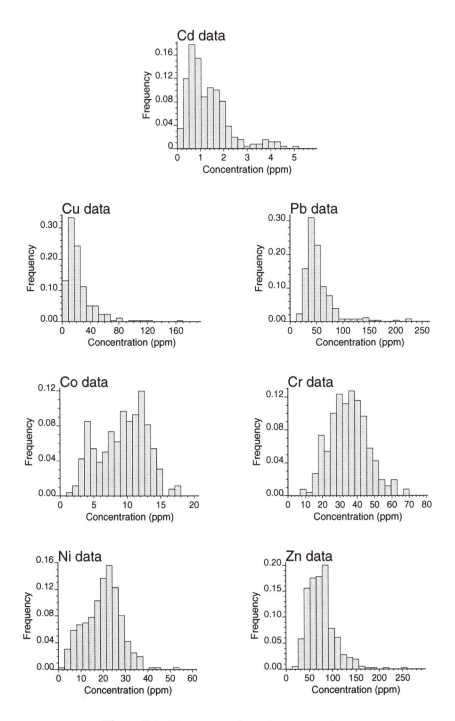

Figure 2.1: Histograms of metal concentrations.

denoted $F(z_k)$, is given by the indicator average:

$$F(z_k) = \frac{1}{n} \sum_{\alpha=1}^{n} i(\alpha; z_k) \qquad (2.4)$$

where the indicator transform of a datum $z(\alpha)$ is defined as

$$i(\alpha; z_k) = \begin{cases} 1 & \text{if } z(\alpha) \leq z_k \\ 0 & \text{otherwise} \end{cases} \qquad (2.5)$$

These proportions can be computed for a series of threshold values z_k, resulting in the cumulative frequency distribution function $F(z_k)$ and its graphical representation, the cumulative histogram. Most often, the range of data values is not discretized, but the n values are ordered from smallest to largest, and each value $z(\alpha)$ is plotted versus the proportion of data that are less than it, $F(z(\alpha))$.

Figure 2.2 shows the cumulative distributions of metal concentrations. The vertical dashed line indicates, for each metal, the tolerable maximum for healthy soils, as defined by the Swiss Federal Office of Environment, Forests and Landcape (FOEFL, 1987); see Table 2.3 (page 15) for exact values. The percentage of data exceeding[2] these critical thresholds is given at the top of each graph. These proportions are large for Cd (65.3%) and Pb (42.1%), and they are smaller for Cu (8.5%), Ni (0.3%), and Zn (0.6%). The concentrations of the other metals are below their tolerable maxima. The extent of the pollution is most important for those variables with an asymmetric distribution of values.

Summary statistics

Important features of a distribution are its central value and measures of its spread and symmetry.

The central value of a distribution is usually taken as the arithmetic mean, defined as

$$m = \frac{1}{n} \sum_{\alpha=1}^{n} z(\alpha) \qquad (2.6)$$

For highly asymmetric distributions, a more appropriate central value is the median, M, which is the value corresponding to a cumulative frequency of 0.5, i.e., the value that splits the distribution into two halves: lower-valued data and higher-valued data.

The p-quantile value of the distribution, noted q_p, is the value that a proportion p of the data does not exceed; i.e., q_p is such that $F(q_p) = p$. The median is the 0.5 quantile of the distribution, $M = q_{0.5}$. Other quantiles of

---

[2]Throughout the book, these individuals or sites will be referred to as "contaminated" or "polluted," although the critical threshold may be exceeded as a result of rocks that are naturally rich in that metal, in the absence of any man-made pollution.

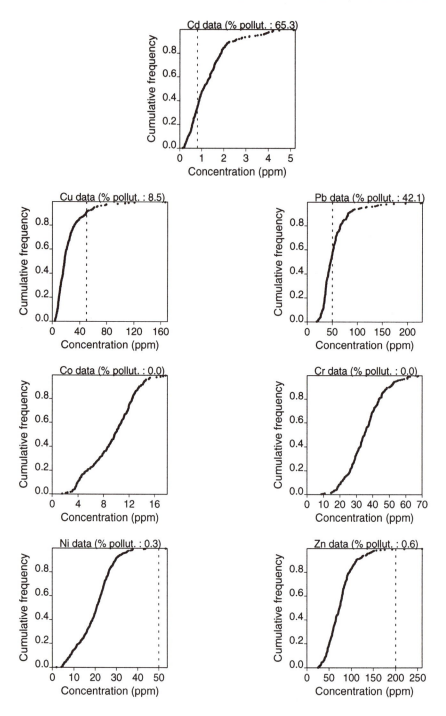

Figure 2.2: Cumulative distributions of metal concentrations and proportions of data that exceed the tolerable maxima represented by the vertical dashed lines. For cobalt and chromium, the critical threshold is larger than the maximum observed.

widespread use are the nine deciles that correspond to cumulative frequencies
0.1, 0.2, 0.3, ..., 0.9. The minimum and maximum values of the distribution
define the range of variation of the variable.

A measure of spread around the mean is the variance, defined as

$$\sigma^2 = \frac{1}{n} \sum_{\alpha=1}^{n} (z(\alpha) - m)^2 \tag{2.7}$$

The square root of the variance, σ, is called the standard deviation, and its
ratio to the mean, σ/m, for non-negative variables, is the unit-free coefficient
of variation.

A distribution is said to be symmetric if

$$F(m - z) = 1 - F(m + z) \qquad \forall z \tag{2.8}$$

Relation (2.8) entails that the mean and median of the distribution are equal,
$m = M$. A measure of asymmetry is the coefficient of skewness, defined as

$$\varphi = \frac{1}{n} \sum_{\alpha=1}^{n} (z(\alpha) - m)^3 / \sigma^3 \tag{2.9}$$

For symmetric distributions, φ is zero. If a distribution has a long tail of
large values, then φ is positive and the distribution is said to be positively
skewed. On the other hand, a distribution with a long tail of small values
has a negative skewness. For most practical purposes, only the skewness sign
is of interest. A simpler measure of skewness would then be the difference
between the mean and median of the distribution, $\varphi' = m - M$.

Table 2.3 lists, for each metal, the values of the main statistics. Note the
departure between mean and median for the distributions of Cd, Cu, and Pb
values, which are strongly positively skewed.

Table 2.3: Summary statistics for the prediction data set (units = ppm).

	Cd	Cu	Pb	Co	Cr	Ni	Zn
n	259	259	259	259	259	359	359
Mean	1.31	23.7	53.9	9.30	35.1	20.0	75.9
Median	1.07	17.6	46.4	9.80	34.8	20.7	73.6
Minimum	0.14	3.96	18.9	1.55	8.72	1.98	25.0
Maximum	5.13	166.4	229.6	17.7	67.6	53.2	259.8
Std. deviation	0.91	20.7	29.7	3.57	10.9	8.08	30.8
Coef. of var.	0.70	0.87	0.55	0.38	0.31	0.40	0.41
Skewness	1.50	2.86	2.89	-0.18	0.29	0.11	1.49
Tolerable max.	0.8	50.0	50.0	25.0	75.0	50.0	200.0

Extreme values and data transformation

Most variables in earth science have an asymmetric distribution, such as the concentrations of cadmium or copper in Figure 2.1 (page 12). A few very small or very large values may strongly affect summary statistics like the mean or variance of the data, the linear correlation coefficient, or measures of spatial continuity (the covariance or semivariogram, to be introduced later). Such extreme values can be handled as follows:

1. Declare the extreme values erroneous and remove them.

2. Classify the extreme values into a separate statistical population.

3. Use robust statistics, which are less sensitive to extreme values.

4. Transform the data to reduce the influence of extreme values.

The decision to discard extreme values must be made with particular care because such values are typically of greater interest. In environmental applications, large concentrations indicate potentially critical areas, whereas in mining the focus in on the selection of high grades. Data should be dismissed only if they are clearly wrong. The spatial location of extreme values is helpful in detecting erroneous data. An isolated very small or very large value may be suspicious, yet this may not be sufficient to remove it.

Spatial clusters or strings of small or large values might reflect the presence of a fracture, a local source of pollution, or a particular lithologic formation. One may then consider treating the extreme values as belonging to another statistical population. Although the decision to split the data into more homogeneous subsets is best based on physical considerations, that decision is largely conditioned by the density of data available. Each subset should have enough data to allow reliable inference of statistics for each population. This may not be the case for the typically small population of extreme values.

If there is no physical reason for discarding extreme values or treating them separately, one may want to reduce their influence. Statistics such as the mean, correlation coefficient, and semivariogram have more robust counterparts: the median, rank correlation coefficient, and madogram or various relative variograms (Srivastava and Parker, 1989), respectively. Robust statistics should be used in conjunction with "traditional" measures (see discussion in section 2.3.3).

Data transformation is commonly used to reduce the influence of extreme values. One such transform consists of taking the square root or logarithm of strictly positive measurements. Both transformations are monotonic increasing, hence they preserve the rank of the original data in the cumulative distribution. This means that order relations between any pair of original values, say $z(\alpha) \leq z(\beta)$, hold true for their transforms: $y(\alpha) \leq y(\beta)$, with $y(\alpha) = \ln(z(\alpha))$ or $y(\alpha) = \sqrt{z(\alpha)}$. The data ranks $r(z(\alpha))$, with $r(z(\alpha)) = 1$ to n, can also be used as transformed scores, in which case the influence of a datum depends only on its rank not on its actual value. Rather than the ranks $r(z(\alpha))$, one may prefer the standardized ranks $r(z(\alpha))/n$, which are valued between 0 and 1. These are referred to as the uniform transforms of the original data $z(\alpha)$. The concept of rank transform is the cornerstone of the indicator approach introduced subsequently.

Unfortunately, the user may not be fully aware of the consequences of data transformation. Taking the logarithm of data may appear natural: for pH, the logarithmic transformation of concentrations is an integral part of the measurement process. One may ask why an a priori transform implicit to the measurement taken is deemed more appropriate than a transform performed a posteriori by the geostatistician. In fact, the problem does not lie in the transform itself, but rather in the back-transform of results, such as interpolated values. For square root or logarithmic transforms, the back-transform through squaring or exponentiation tends to exaggerate any error associated with the interpolation. Such exaggeration of errors is most dramatic for extreme values, which are often of greatest interest. The back-transform may erase most of the benefits of having more robust statistics calculated on the transformed scores. Therefore, one must question the appropriateness of any data transform. Such decisions should be made carefully and call for much more than a quick look at the shape of the sample histogram and the desire to make that sample histogram symmetric.

The Jura data were validated earlier (Webster et al., 1994) and are hereafter considered error-free. For the reasons stated previously, no transformation was performed a priori. However, the sensitivity of statistics to extreme values is evaluated throughout the exploratory data analysis.

The impact of land use and rock type

The data are now split into several subsets according to rock type and land use, and the corresponding distributions of metal concentrations are retrieved. The objective is to better understand the relation between metal concentrations and environmental factors.

Conditional distributions
Let $\{(z(\alpha), s(\alpha)), \ \alpha = 1, \ldots, n\}$ be the set of joint measurements of the continuous attribute z and the categorical attribute s on n individuals. The distribution of the z-values, given that a particular state s_k is observed, is said to be *conditional* to s_k.

Table 2.4 lists, for each metal, the average concentrations for each rock type and land use. For all metals, the largest concentrations are measured on soils under pasture. The smallest concentrations are generally observed in forest soils or on Argovian rocks. One may think that there is a preferential location of forest on Argovian rocks. However, Table 2.2 (page 11) shows these two categories to be weakly related: only 21% $(2.7/12.8 = 0.21)$ of the forest soil is located on the Argovian formation. The discrepancy between forest and other land uses is particularly important for copper concentrations. The larger concentrations on agricultural soil may originate from animal feed,

Table 2.4: Average concentrations of seven heavy metals for each
land use and rock type (units=ppm).

	Cd	Cu	Pb	Co	Cr	Ni	Zn
Land use							
Forest	1.40	9.19	50.6	7.69	30.1	18.2	57.7
Pasture	2.0	27.1	61.8	10.0	45.1	23.4	87.3
Meadow	1.06	25.5	52.0	9.40	32.9	19.2	75.9
Tillage	0.96	22.6	52.6	8.37	27.1	18.5	76.0
Rock type							
Argovian	1.14	16.4	41.3	5.40	28.6	12.3	62.2
Kimmeridgian	1.35	22.2	56.4	11.0	35.4	25.0	77.6
Sequanian	1.51	27.9	62.9	9.97	40.0	20.4	84.2
Portlandian	1.85	17.3	48.0	9.40	39.3	22.9	72.6
Quaternary	1.16	28.7	52.3	9.60	34.9	18.8	77.6

which the forest does not receive. The average concentrations of cobalt and
nickel in the soil on Argovian formations are half those measured on the other
formations.

Conditional cumulative frequencies
In addition to the average metal concentration within a category $s_{k'}$, it is
worth knowing the corresponding proportion of data that are above or below
the critical threshold. The proportion of data no greater than z_k is given
by the conditional cumulative frequency $F(z_k|s_{k'})$, which is computed as the
average of an indicator product:

$$F(z_k|s_{k'}) \;=\; \frac{1}{n_{k'}} \sum_{\alpha=1}^{n} i(\alpha; s_{k'}) \cdot i(\alpha; z_k) \qquad (2.10)$$

where $n_{k'} = \sum_{\alpha=1}^{n} i(\alpha; s_{k'})$ is the number of individuals belonging to category $s_{k'}$.

Table 2.5 gives the conditional percentages of data that exceed tolerable
maxima for the three metals (Cd, Cu, Pb) with widespread contamination.
There is no pollution by copper under forest and tillage, and it is of small
extent on Argovian and Portlandian rocks. For cadmium and lead, the effect
of land use is not as clear. The proportion of Cd and Pb data exceeding the
critical threshold is smaller for the soils on Argovian rocks.

Subdivision of the data set
The previous analysis of conditional distributions reveals that statistics of
certain categories differ in terms of average metal concentrations and proportions above critical thresholds. This raises questions about the prior decision
of pooling all the data together and raises the possibility of splitting the data
into several subsets.

Table 2.5: Percentage of individuals that exceed tolerable maxima (Cd, Cu, or Pb) within each land use and rock type. The results for tillage and Portlandian rocks relate to less than 10 individuals (see Table 2.1).

	Cd	Cu	Pb
Land use			
Forest	78.8	0.0	51.5
Pasture	89.1	14.9	43.0
Meadow	54.8	8.4	39.8
Tillage	60.0	0.0	40.0
Rock type			
Argovian	33.9	1.9	22.6
Kimmeridgian	76.5	7.1	49.4
Sequanian	71.4	12.7	50.8
Portlandian	100.0	0.0	33.3
Quaternary	69.1	12.7	40.0

It would make sense to consider the concentrations for each land use or rock type as a separate population. The physical processes that control the spatial continuity of metal concentrations in forest soil are likely to be different from those in fertilized soils. Unfortunately, too few data are available to compute reliable statistics for each separate combination of rock types and land uses (see Table 2.2). An alternative would be to regroup categories with similar properties. I decided, however, to pool the whole set of measurements to cover the common situation where the lack of data prevents one from splitting the data.

2.2 Bivariate Description

The next step in the exploration of the Jura data consists of looking at the relation between pairs of metal concentrations measured at the same locations. Figure 2.2 (page 14) shows that a large proportion of data exceed the tolerable maxima for Cd, Cu, or Pb, so the bivariate description focuses on the relations between these three metals and the two more densely sampled metals, nickel and zinc.

2.2.1 The scattergram

Let $\{(z_i(\alpha), z_j(\alpha)), \alpha = 1, \ldots, n\}$ be the set of measurements of the two continuous attributes z_i and z_j on the same n individuals. This can be displayed in a scattergram in which the components of each data pair are plotted against one another. Figure 2.3 shows the scattergrams of nickel and zinc values versus the concentrations of Cd, Cu, and Pb. The latter three metals are related to zinc: larger concentrations of each metal tend to be associated with larger concentrations of zinc. Cadmium and nickel concentrations are also positively correlated.

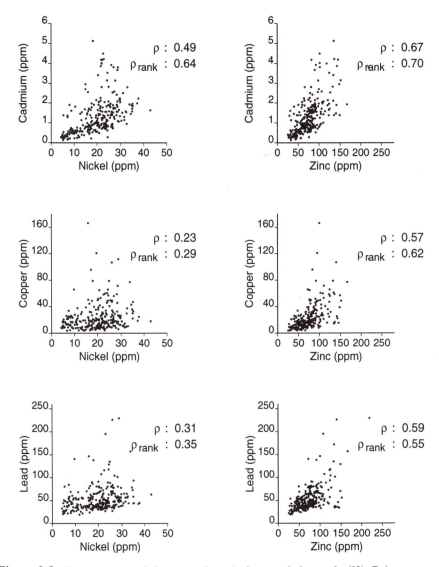

Figure 2.3: Scattergrams of the two exhaustively sampled metals (Ni, Zn) versus the three metals (Cd, Cu, Pb) with widespread contamination.

2.2.2   Measures of bivariate relation

As for the univariate case, one is interested in statistics that summarize the main features of the bivariate relation. The most frequently used statistics are the covariance and its standardized form, the linear correlation coefficient.

The covariance σ_{ij} is a measure of the joint variation of Z_i and Z_j around their means. It is computed as

$$\sigma_{ij} \;=\; \frac{1}{n} \sum_{\alpha=1}^{n} (z_i(\alpha) \;-\; m_i) \cdot (z_j(\alpha) \;-\; m_j) \tag{2.11}$$

where m_i and m_j are the arithmetic means of variables Z_i and Z_j, respectively. The covariance becomes the variance if $i = j$, see relation (2.7). The correlation coefficient ρ_{ij} is readily deduced as

$$\rho_{ij} \;=\; \frac{\sigma_{ij}}{\sigma_i \cdot \sigma_j} \quad \in [-1, 1] \tag{2.12}$$

where σ_i and σ_j are the standard deviations of Z_i and Z_j, respectively. The unit-free correlation coefficient is easier to interpret than the covariance, which depends on the measurement scales of the two variables.

The following are guidelines for using the linear correlation coefficient to quantify dependence between two variables:

1. The quantity ρ_{ij} provides a measure only of linear relation between two variables. It can be misleading if interpreted otherwise (see discussion in section 3.2.2). Two variables may be highly dependent and yet have zero linear correlation: a classic example is $Z_i = [Z_j]^2$, where Z_j has a symmetric distribution.

2. Like the variance, the correlation coefficient is strongly affected by extreme values. A more robust measure is the rank correlation coefficient ρ_{ij}^R, which considers the ranks of the data, $r(z_i(\alpha))$ and $r(z_j(\alpha))$, rather than the original values:

$$\rho_{ij}^R \;=\; \frac{1}{n} \frac{\sum_{\alpha=1}^{n} [r(z_i(\alpha)) \;-\; m_{Ri}] \cdot [r(z_j(\alpha)) \;-\; m_{Rj}]}{\sigma_{Ri} \cdot \sigma_{Rj}} \tag{2.13}$$

where m_{Ri} and σ_{Ri} are the mean and standard deviation of the n ranks $r(z_i(\alpha))$. A large deviation between ρ_{ij} and ρ_{ij}^R reflects either a non-linear relation between the two variables Z_i and Z_j or the presence of pairs of extreme values.

3. When dealing with N_v variables, $N_v(N_v - 1)/2$ scattergrams can be drawn. The user may be tempted to bypass the cumbersome plotting and description of too many scattergrams and focus on the linear correlation coefficients alone. However, remember that the correlation coefficient extracts only a small part of the information provided by the scattergram.

Table 2.6: Matrix of linear correlation coefficients.

	Cd	Cu	Pb	Co	Cr	Ni	Zn
Cd	1.00						
Cu	0.12	1.00					
Pb	0.22	0.78	1.00				
Co	0.25	0.22	0.19	1.00			
Cr	0.61	0.21	0.30	0.45	1.00		
Ni	0.49	0.23	0.31	0.75	0.69	1.00	
Zn	0.67	0.57	0.59	0.47	0.67	0.63	1.00

Table 2.6 gives the linear correlation coefficients computed among the seven heavy metals from the same set of 259 individuals. The strongest correlations ($\rho > 0.70$) are for the pairs Cu-Pb and Co-Ni. Figure 2.4 is a scattergram of the linear versus the rank correlation coefficient for the 21 pairs of heavy metals. Both measures are similar, which indicates that extreme values do not greatly affect the linear correlation coefficients.

2.3   Univariate Spatial Description

To account for data locations, measurements of continuous and categorical attributes are denoted $z(\mathbf{u}_\alpha)$ and $s(\mathbf{u}_\alpha)$, where $\mathbf{u}_\alpha$ is the vector of spatial coordinates of the αth individual. For example, $s(\mathbf{u}_\alpha) = s_k$ if category s_k is observed at location $\mathbf{u}_\alpha$. The objective is to describe and quantify the relation between measurements of the same attribute at any two data locations $\mathbf{u}_\alpha$ and $\mathbf{u}_\beta$.

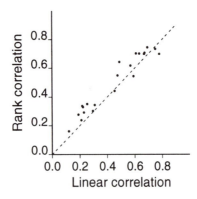

Figure 2.4: Scattergram of the rank correlation coefficient versus the linear correlation coefficient for the 21 pairs of heavy metals.

2.3.1 Location maps

Any spatial analysis should start with a posting of data values. Figures 2.5 and 2.6 show the location maps of the sample data, with darker shading indicating larger metal concentrations. All metal concentrations were measured at 259 locations, with Ni and Zn concentrations being recorded at 100 additional locations (Figure 2.6, bottom maps).

The data configuration resulted from a combination of regular and nested sampling schemes (Webster et al., 1994). The basic grid is a square mesh with 107 grid nodes at intervals of 250 m. Out of these 107 grid nodes, 38 were randomly selected so that the proportion of 38 nodes belonging to rock type s_k corresponds to the proportion of the study area covered by that rock

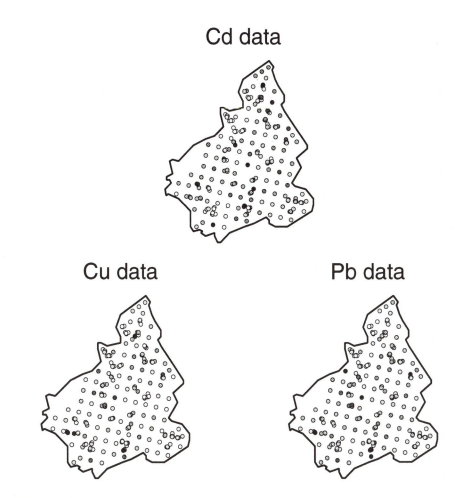

Figure 2.5: Location maps of Cd, Cu, and Pb data. Darker shading indicates larger concentrations.

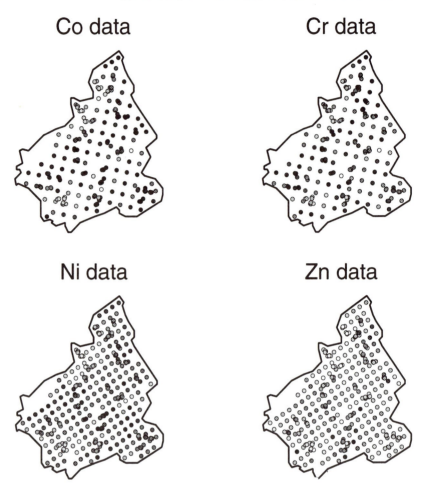

Figure 2.6: Location maps of Co, Cr, Ni, and Zn data. Darker shading indicates larger concentrations.

type. Starting from each of these 38 nodes, the surveyors chose a first location 100 m away. From that location they chose a second location 40 m away; from that second location, a third location was chosen 16 m away; and finally from that third location, a fourth location was chosen 6 m away. The distances were fixed but the directions were random. The objective of this nested sampling was to cover a wide range of spatial scales between 0 and 250 m.

In section 2.1.2, many sites were found to exceed the tolerable maximum for Cd, Cu, and Pb. The indicator maps in Figure 2.7, which show the contaminated locations in black, complete the description. The density of black dots reflects the greater extent of contamination by cadmium and lead. The pattern of spatial distribution of contaminated sites is also informative. The black dots in Figure 2.7 are not randomly distributed—they tend to

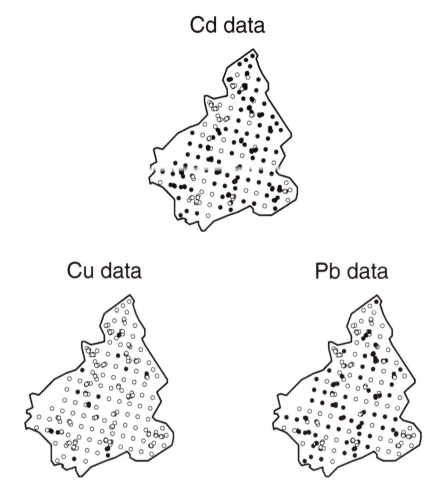

Figure 2.7: Indicator maps showing in black locations that exceed the tolerable maximum for Cd, Cu, and Pb.

cluster. Such clustering could reflect potential sources, such as originating from the rock or man-made pollution. In general, observations that are close to each other on the ground are also alike in metal concentrations.

2.3.2 The h-scattergram

Spatial relations between data can be displayed using an h-scattergram. Just as the scattergram is a plot of all pairs of values related to two different attributes measured at the same location, the h-scattergram is a plot of all pairs of measurements $(z(\mathbf{u}_\alpha), z(\mathbf{u}_\alpha + \mathbf{h}))$ on the same attribute z at locations separated by a given distance h in a particular direction θ. The vector notation $\mathbf{h}$ accounts for both distance and direction. By convention, the value at the start of the vector $\mathbf{h}$, $z(\mathbf{u}_\alpha)$, is called the tail value, whereas the value at the end, $z(\mathbf{u}_\alpha + \mathbf{h})$, is the head value.

Data pairs are typically grouped into classes of distances (lags) and angles, $[h \pm \Delta h]$ and $[\theta \pm \Delta \theta]$, so that each $\mathbf{h}$-scattergram is built on a sufficient number of pairs. For example, consider the relation between Cd values in the east-west (E-W) direction. Figure 2.8 shows the $\mathbf{h}$-scattergrams of Cd values in the easterly[3] direction ($\Delta \theta = 22.5°$) for six different classes of distances ($\Delta h = 100$ m). The abscissa corresponds to tail values and the ordinate to head values. Because the two axes relate to the same variable, a perfect correlation would entail that all points lie on the first bisector (45° dashed line). The spread of the cloud of points around this 45° line reflects the variability between data values. The increasing inflation of the cloud with increasing separation distance h reflects the increasing dissimilarity between measurements farther apart.

The $\mathbf{h}$-scattergram also draws attention to possible outliers or misrecorded values. Isolated pairs on the $\mathbf{h}$-scattergram should be identified, located on the map, and then analyzed more carefully. For small separation distances, outlier pairs involving the same location, say, $\mathbf{u}_\alpha$, indicate that datum $z(\mathbf{u}_\alpha)$ is very unlike its neighbors. For example, in Figure 2.8, a few pairs fall far from the 45° line on the $\mathbf{h}$-scattergram for $|\mathbf{h}| = 214$ m. Three of these pairs involve the same location with Cd concentration of 4.50 ppm; the second measurements of these pairs are 0.40, 0.77, and 1.43 ppm. If there is a sound physical reason for considering the extreme value (4.50 ppm) as erroneous, then it should be removed from the data set. Such a decision should not, however, be based on a single lag h, since the same datum may not yield outlier pairs for other lags h or in other directions.

Distinct clouds of points on a $\mathbf{h}$-scattergram may indicate the presence of separate populations with different spatial continuity. The data could be split into more homogeneous subsets if sufficient data are available within each subset.

2.3.3 Measures of spatial continuity and variability

Each $\mathbf{h}$-scattergram displays the relation between pairs of z-values for a given class of distance and direction. The similarity or dissimilarity between data separated by a vector $\mathbf{h}$ can be quantified by several measures (Deutsch and Journel, 1992a, p. 40).

Covariance function

The covariance and correlation coefficient introduced in section 2.2.2 can be extended to measure similarity between non-colocated data. The covariance

---

[3] The $\mathbf{h}$-scattergram in the westerly direction is the symmetrical projection about the first bisector of the cloud of points of the easterly $\mathbf{h}$-scattergram.

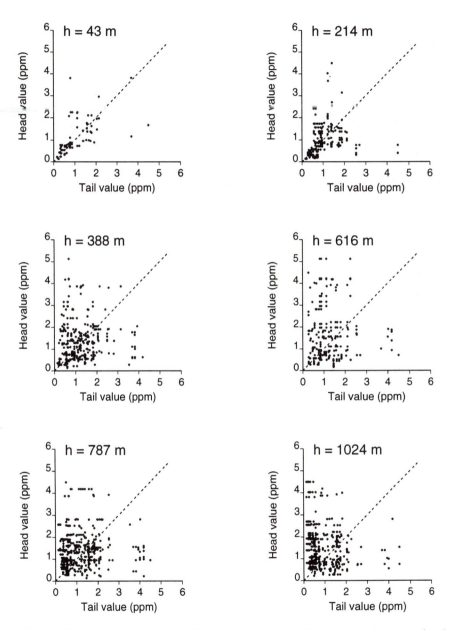

Figure 2.8: h-scattergrams of Cd concentrations in the easterly direction (angle tolerance = 22.5°) for six classes of distances h (distance tolerance=100 m). The spread of points increases as the separation distance increases.

between data values separated by a vector $\mathbf{h}$ is computed as follows:

$$C(\mathbf{h}) = \frac{1}{N(\mathbf{h})} \sum_{\alpha=1}^{N(\mathbf{h})} z(\mathbf{u}_\alpha) \cdot z(\mathbf{u}_\alpha + \mathbf{h}) - m_{-\mathbf{h}} \cdot m_{+\mathbf{h}} \qquad (2.14)$$

with

$$m_{-\mathbf{h}} = \frac{1}{N(\mathbf{h})} \sum_{\alpha=1}^{N(\mathbf{h})} z(\mathbf{u}_\alpha) \qquad m_{+\mathbf{h}} = \frac{1}{N(\mathbf{h})} \sum_{\alpha=1}^{N(\mathbf{h})} z(\mathbf{u}_\alpha + \mathbf{h})$$

where $N(\mathbf{h})$ is the number of data pairs within the class of distance and direction, and $m_{-\mathbf{h}}$ and $m_{+\mathbf{h}}$ are the means of the corresponding tail and head values (lag means). The covariance can be computed for different lags $\mathbf{h}_1$, $\mathbf{h}_2$, ... and the ordered set of covariances $C(\mathbf{h}_1)$, $C(\mathbf{h}_2)$, ... is called the experimental autocovariance function, or simply, the experimental covariance function.

Correlogram

A unit-free measure of similarity between data is the standardized form of the covariance:

$$\rho(\mathbf{h}) = \frac{C(\mathbf{h})}{\sqrt{\sigma^2_{-\mathbf{h}} \cdot \sigma^2_{+\mathbf{h}}}} \in [-1, +1] \qquad (2.15)$$

with

$$\sigma^2_{-\mathbf{h}} = \frac{1}{N(\mathbf{h})} \sum_{\alpha=1}^{N(\mathbf{h})} [z(\mathbf{u}_\alpha) - m_{-\mathbf{h}}]^2$$

$$\sigma^2_{+\mathbf{h}} = \frac{1}{N(\mathbf{h})} \sum_{\alpha=1}^{N(\mathbf{h})} [z(\mathbf{u}_\alpha + \mathbf{h}) - m_{+\mathbf{h}}]^2$$

where $\sigma^2_{-\mathbf{h}}$ and $\sigma^2_{+\mathbf{h}}$ are the variances of the tail and head values (lag variances). The ordered set of correlation coefficients $\rho(\mathbf{h}_1)$, $\rho(\mathbf{h}_2)$, ... is called the experimental autocorrelation function or correlogram.

Semivariogram

Unlike the covariance and correlation functions, which are measures of similarity, the experimental semivariogram $\gamma(\mathbf{h})$ measures the average dissimilarity between data separated by a vector $\mathbf{h}$. It is computed as half the average squared difference between the components of every data pair:

$$\gamma(\mathbf{h}) = \frac{1}{2N(\mathbf{h})} \sum_{\alpha=1}^{N(\mathbf{h})} [z(\mathbf{u}_\alpha) - z(\mathbf{u}_\alpha + \mathbf{h})]^2 \qquad (2.16)$$

where $[z(\mathbf{u}_\alpha) - z(\mathbf{u}_\alpha + \mathbf{h})]$ is an h-increment of attribute z.

The semivariogram value at a given lag $\mathbf{h}$, sometimes called semivariance, can be interpreted as the moment of inertia of the $\mathbf{h}$-scattergram about its first bisector (Figure 2.9). Indeed, the orthogonal distance of any point $(z(\mathbf{u}_\alpha), z(\mathbf{u}_\alpha + \mathbf{h}))$ to the first bisector is $d_\alpha = |z(\mathbf{u}_\alpha) - z(\mathbf{u}_\alpha + \mathbf{h})| \cdot \cos 45°$. The moment of inertia about the $45°$ line is the average of all such squared distances:

$$
\begin{aligned}
\frac{1}{N(\mathbf{h})} \sum_{\alpha=1}^{N(\mathbf{h})} d_\alpha^2 &= \frac{1}{N(\mathbf{h})} \sum_{\alpha=1}^{N(\mathbf{h})} [z(\mathbf{u}_\alpha) - z(\mathbf{u}_\alpha + \mathbf{h})]^2 \cdot [\cos 45°]^2 \\
&= \frac{1}{2N(\mathbf{h})} \sum_{\alpha=1}^{N(\mathbf{h})} [z(\mathbf{u}_\alpha) - z(\mathbf{u}_\alpha + \mathbf{h})]^2 = \gamma(\mathbf{h}) \quad (2.17)
\end{aligned}
$$

Hence, the semivariogram value increases as the points spread out farther from the first bisector of the $\mathbf{h}$-scattergram.

Example

In Figure 2.10, the three top graphs show, respectively, the experimental covariance function, correlogram, and semivariogram of cadmium computed in the easterly direction using an angular tolerance of $22.5°$ and a distance tolerance of 100 m. The decreasing behavior of the covariance function and correlogram reflects the decreasing similarity of data values as the separation distance h increases.

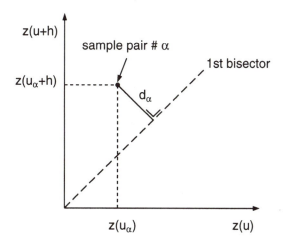

Figure 2.9: Interpretation of the semivariogram value $\gamma(\mathbf{h})$ as the moment of inertia of the $\mathbf{h}$-scattergram around the first bisector. $\gamma(\mathbf{h})$ is the average of all squared orthogonal distances d_α to that bisector.

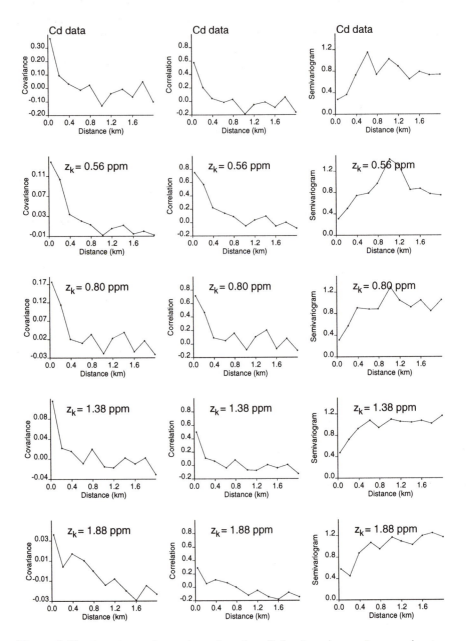

Figure 2.10: Experimental covariance functions (left column), correlograms (center column), and semivariograms (right column) of cadmium in the easterly direction. The three top graphs are computed from the original values, whereas the other graphs relate to the indicator data displayed in Figure 2.11, page 33. The indicator semivariogram values are standardized by the variance of the indicator data.

Both the covariance and the correlation between Cd data in the E-W direction decrease to approximately zero at a separation distance of about 400 m; this distance is called the E-W range. .he range distance, the experimental semivariogram stops increasing and 1. ictuates around a sill value. Such a semivariogram is said to be bounded.

The semivariogram value may not tend to zero when h tends to zero, although by definition $\gamma(0) = 0$. Such a discontinuity at the origin of the semivariogram, called the nugget effect, relates to measurement error and/or spatial sources of variation at distances smaller than the shortest sampling interval (Journel and Huijbregts, 1978, p. 39). The relative nugget effect is defined as the ratio of the nugget discontinuity to the sill value.

Remarks

1. The covariance and semivariogram value computed in opposite directions are identical, i.e., $C(\mathbf{h}) = C(-\mathbf{h})$ and $\gamma(\mathbf{h}) = \gamma(-\mathbf{h})$, since the corresponding $\mathbf{h}$-scattergram and $-\mathbf{h}$-scattergram are symmetric with respect to the first bisector.

2. Setting the angular tolerance $\Delta\theta$ to $90°$ amounts to pooling the data pairs in all directions. The resulting covariance function or semivariogram is called *omnidirectional*, whereas the term *directional* is used whenever $\Delta\theta < 90°$.

3. Like other variance-type statistics, the values of the covariance or semivariogram are sensitive to extreme data values. The following are ways to handle the problem of robustness:

   (a) Transform the data (see section 2.1.2) to reduce the skewness of their histograms.

   (b) Use other summary statistics of the $\mathbf{h}$-scattergram that are less sensitive to extreme values. The sensitivity of the semivariogram to extreme z-values comes from the squaring of $\mathbf{h}$-increments, see equation (2.16). A more general measure of the spatial variability is the variogram of order ω, defined as the mean absolute deviation to the power ω:

$$\gamma_\omega(\mathbf{h}) = \frac{1}{2N(\mathbf{h})} \sum_{\alpha=1}^{N(\mathbf{h})} |z(\mathbf{u}_\alpha) - z(\mathbf{u}_\alpha + \mathbf{h})|^\omega \quad \text{with} \quad \omega \in [0, 2]$$

(2.18)

   For $\omega = 2$, one retrieves the traditional semivariogram $\gamma(\mathbf{h})$. The smaller ω, the lesser the influence of extreme values on the measure $\gamma_\omega(\mathbf{h})$. Two commonly used measures are the madogram ($\omega = 1$) and rodogram ($\omega = 1/2$) (Deutsch and Journel, 1992a, p. 41). These measures are not substitutes for the traditional semivariogram; rather, they should be used to infer features, such as range and anisotropy (see related discussion in section 4.2.4).

(c) Remove data pairs that unduly influence the value of the summary statistics, say, $\gamma(\mathbf{h})$, for specific lags $\mathbf{h}$. The approach consists of discarding data pairs that appear isolated on the $\mathbf{h}$-scattergram and recalculating the corresponding semivariogram value. The contribution of a few pairs should not have an inordinate influence on the value of the summary statistics. Beware that removing a data pair from the calculation of the semivariogram for a specific lag $\mathbf{h}$ does not entail removing the corresponding two data values from the sample set; these two data values can still contribute to other pairs at the same lag or at other lags.

Such "cleansing" of the $\mathbf{h}$-scattergram calls for interactive visualization tools that provide a complete description of each extreme pair, including locations of both pair values in the study area, and allows a quick update of the semivariogram value after outlier pairs have been removed; for example, see Froidevaux (1990), Englund and Sparks (1991), Haslett et al. (1991).

2.3.4   Application to indicator transforms

The covariance function and semivariogram provide measures of spatial continuity or variability over the full range of attribute values. The pattern of spatial continuity or variability may, however, differ, depending on whether the attribute value is small, medium, or large. For example, in many environmental applications, random large concentrations coexist with a background of small values that vary more continuously in space. Depending on whether large concentrations are clustered or scattered in space, our interpretation of the physical processes controlling contamination and our decision for remediation may change.

Indicator transform

The characterization of the spatial distribution of z-values above or below a given threshold value z_k starts by coding each datum value $z(\mathbf{u}_\alpha)$ as an indicator datum $i(\mathbf{u}_\alpha; z_k)$, defined as

$$i(\mathbf{u}_\alpha; z_k) = \begin{cases} 1 & \text{if } z(\mathbf{u}_\alpha) \leq z_k \\ 0 & \text{otherwise} \end{cases} \qquad (2.19)$$

Consider, for example, the four threshold values $z_k = 0.56, 0.80, 1.38,$ and 1.88 ppm corresponding, respectively, to the 0.2, 0.35, 0.6, and 0.8 quantiles of the distribution of Cd concentrations. The corresponding four indicator maps in Figure 2.11 show in black the locations where these thresholds are exceeded, i.e., where $i(\mathbf{u}_\alpha; z_k) = 0$. This series of maps reveals a change

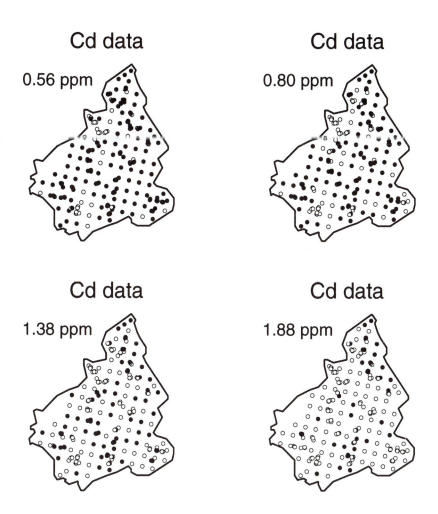

Figure 2.11: Indicator maps showing in black locations that exceed the second decile, the tolerable maximum, the sixth- and the eighth-decile threshold. The clustering of small values (white dots in top left graph) contrasts with the scatter of large values (black dots in bottom right graph).

in spatial continuity from small to large Cd values. Whereas most of the small concentrations ($z(\mathbf{u}_\alpha) \leq 0.56$ ppm) are confined to two distinct areas, large concentrations ($z(\mathbf{u}_\alpha) > 1.88$ ppm) are scattered over the entire area. The two intermediate maps show a transition from clustered small values to randomly distributed large values.

The measures previously introduced to quantify the spatial continuity or variability of z-values can also be applied to the indicator data $i(\mathbf{u}_\alpha; z_k)$.

Indicator covariance function

The experimental indicator covariance at a given lag $\mathbf{h}$ is computed as

$$
\begin{aligned}
C_I(\mathbf{h}; z_k) &= \frac{1}{N(\mathbf{h})} \sum_{\alpha=1}^{N(\mathbf{h})} i(\mathbf{u}_\alpha; z_k) \cdot i(\mathbf{u}_\alpha + \mathbf{h}; z_k) - F_{-\mathbf{h}}(z_k) \cdot F_{+\mathbf{h}}(z_k) \\
&= F(\mathbf{h}; z_k) - F_{-\mathbf{h}}(z_k) \cdot F_{+\mathbf{h}}(z_k) \qquad (2.20)
\end{aligned}
$$

with

$$
F_{-\mathbf{h}}(z_k) = \frac{1}{N(\mathbf{h})} \sum_{\alpha=1}^{N(\mathbf{h})} i(\mathbf{u}_\alpha; z_k) \qquad F_{+\mathbf{h}}(z_k) = \frac{1}{N(\mathbf{h})} \sum_{\alpha=1}^{N(\mathbf{h})} i(\mathbf{u}_\alpha + \mathbf{h}; z_k)
$$

where $F_{-\mathbf{h}}(z_k)$ and $F_{+\mathbf{h}}(z_k)$ are the proportions of tail and head values not exceeding the threshold value z_k.

The indicator covariance $C_I(\mathbf{h}; z_k)$ appears as the "centering" of the two-point cumulative frequency $F(\mathbf{h}; z_k)$. That frequency measures how often two values of the same attribute z separated by a vector $\mathbf{h}$ are jointly no greater than the threshold value[4] z_k. For a small threshold value z_k, $F(\mathbf{h}; z_k)$ measures the connectivity between small values separated by a vector $\mathbf{h}$: the larger $F(\mathbf{h}; z_k)$, the better connected in space are the small z-values. The spatial connectivity of large values (e.g., values greater than a large threshold $z_{k'}$) can be measured by $F(\mathbf{h}; z_{k'})$, where the indicator transform is now $j(\mathbf{u}_\alpha; z_{k'}) = 1 - i(\mathbf{u}_\alpha; z_{k'})$. Note that $C_J(\mathbf{h}; z_k) = C_I(\mathbf{h}; z_k)$, $\forall\, z_k$.

Indicator correlogram

The indicator correlogram is the standardized form of the previous indicator covariance function:

$$
\rho_I(\mathbf{h}; z_k) = \frac{C_I(\mathbf{h}; z_k)}{\sqrt{\sigma^2_{-\mathbf{h}}(z_k) \cdot \sigma^2_{+\mathbf{h}}(z_k)}} \in [-1, +1] \qquad (2.21)
$$

where $\sigma^2_{-\mathbf{h}}(z_k) = F_{-\mathbf{h}}(z_k)[1 - F_{-\mathbf{h}}(z_k)]$ is the variance of the tail indicator values and $\sigma^2_{+\mathbf{h}}(z_k) = F_{+\mathbf{h}}(z_k)[1 - F_{+\mathbf{h}}(z_k)]$ is the variance of the head indicator values.

Indicator semivariogram

The indicator semivariogram is computed as

$$
\gamma_I(\mathbf{h}; z_k) = \frac{1}{2N(\mathbf{h})} \sum_{\alpha=1}^{N(\mathbf{h})} [i(\mathbf{u}_\alpha; z_k) - i(\mathbf{u}_\alpha + \mathbf{h}; z_k)]^2 \qquad (2.22)
$$

---

[4] At this stage, the same threshold value z_k is considered for both values.

The indicator variogram value $2\gamma_I(\mathbf{h}; z_k)$ measures how often two z-values separated by a vector $\mathbf{h}$ are on opposite sides of the threshold value z_k. In other words, $2\gamma_I(\mathbf{h}; z_k)$ measures the transition frequency between two classes of z-values as a function of $\mathbf{h}$. Unlike the indicator covariance function, the greater $\gamma_I(\mathbf{h}; z_k)$ or $\gamma_I(\mathbf{h}; z_{k'})$, the less connected in space are the small $(z(\mathbf{u}_\alpha) \leq z_k)$ or large values $(z(\mathbf{u}_\alpha) > z_{k'})$. Similar to the indicator covariance, $\gamma_I(\mathbf{h}; z_k) = \gamma_J(\mathbf{h}; z_k)$, for $j(\mathbf{u}_\alpha; z_k) = 1 - i(\mathbf{u}_\alpha; z_k)$.

Graphical interpretation

Indicator covariance and semivariogram values can be graphically interpreted as proportions of points (data pairs) that fall in specific areas of the $\mathbf{h}$-scattergram.

The non-centered indicator covariance $F(\mathbf{h}; z_k)$ represents the proportion of data pairs $(z(\mathbf{u}_\alpha), z(\mathbf{u}_\alpha + \mathbf{h}))$ that are jointly no greater than the threshold value z_k (Figure 2.12, horizontal hatched area).

The only data pairs that contribute to the indicator variogram $2\gamma_I(\mathbf{h}; z_k)$ are those where the two values $z(\mathbf{u}_\alpha)$ and $z(\mathbf{u}_\alpha + \mathbf{h})$ are on opposite sides of the threshold value z_k. The indicator variogram value corresponds to the proportion of points that fall in the vertical hatched area of Figure 2.12.

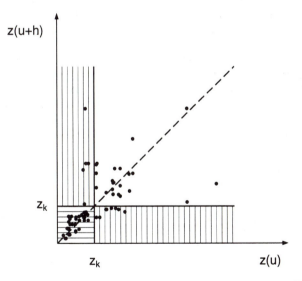

Figure 2.12: Graphical interpretation of the indicator spatial statistics. The non-centered indicator covariance $F(\mathbf{h}; z_k)$ and the indicator variogram value $2\gamma_I(\mathbf{h}; z_k)$ are the proportions of points falling in the horizontal and vertical hatched areas of the $\mathbf{h}$-scattergram, respectively.

Example

Figure 2.10 (page 30) shows the experimental indicator covariance functions, correlograms, and standardized semivariograms computed in the easterly direction for the four indicator maps displayed in Figure 2.11. These graphs confirm the visual impression from Figure 2.11 that small Cd concentrations are better correlated in space than large concentrations. The correlation for the first lag is 0.75 for $z_k=0.56$ ppm and only 0.3 for $z_k=1.88$ ppm.

Remarks

1. The terms *z-covariance* and *z-semivariogram* are used hereafter to emphasize that the measures of spatial continuity or spatial variability relate to values of the variable z instead of their indicator transforms.

2. Typically, experimental indicator semivariograms at extreme threshold values tend to be more erratic than those at the median threshold value. Indeed, for such extreme thresholds, the indicator semivariogram value depends on the spatial distribution of the few data pairs where the two z-values are on opposite sides of the threshold z_k.

3. Unlike the z-covariance or z-semivariogram, the indicator statistics are not affected by extreme values, since only the position of the data with respect to the threshold value z_k is considered. The indicator semivariogram at the median threshold value, in particular, may be used to detect patterns of spatial continuity whenever extreme-valued data render the traditional semivariogram $\gamma(\mathbf{h})$ erratic.

4. The sill of the experimental indicator semivariogram $\gamma_I(\mathbf{h}; z_k)$ is roughly equal to the indicator variance, $F(z_k)[1 - F(z_k)]$, where $F(z_k)$ is the mean of the indicator data $i(\mathbf{u}_\alpha; z_k)$. When comparing indicator semivariograms at different threshold values, it is good practice to standardize their sills to one by dividing the semivariogram values by the indicator variance.

2.3.5   Spatial continuity of metal concentrations

The spatial analysis is now extended to all directions and metals, and the corresponding patterns of spatial variation are interpreted in relation to geology and land use. The following discussion focuses on the semivariogram, which is most frequently used.

Spatial anisotropy

Figure 2.13 shows the experimental semivariograms of metal concentrations computed in four directions with an angular tolerance of $22.5°$. The pattern of increase generally does not vary with the direction; the discrepancies reflect experimental fluctuations that result from the small number of data pairs

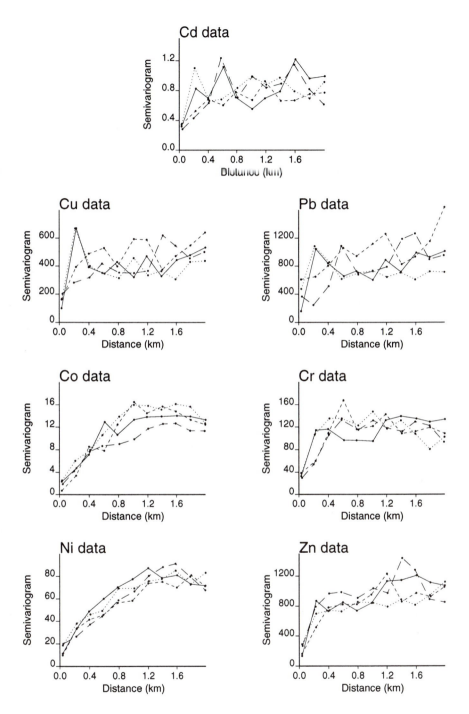

Figure 2.13: Experimental semivariograms for the seven heavy metals in four directions (— : 22.5°, − − : 67.5°, - - - : 112.5°, . . . : 157.5° ; $\Delta\theta = 22.5°$).

available for each lag. Cobalt concentration, however, appear to vary more continuously (smaller semivariogram values) in the 67.5° direction (SW-NE).

Sensitivity to extreme values

Figure 2.14 (solid line) shows the experimental omnidirectional semivariograms of metal concentrations. For the four positively skewed variables (Cd, Cu, Pb, Zn), the semivariogram is also computed on the logarithm transform and plotted on the same graph after rescaling to the z-variance (dashed line). The semivariograms of the logarithms appear slightly less erratic with a smaller nugget effect. The differences are, however, small and do not justify the log-transformation.

Interpreting patterns of spatial variation

The semivariograms of all attribute values have a small nugget effect, from 10 to 30% of the total variance (Figure 2.14, solid line). Two scales of spatial variation can be distinguished from inflexions of the experimental curves: a local scale (range $\approx$ 200 m) and a regional scale (range $\approx$ 1 km). The shape of Ni and Co semivariograms is dominated by the long-range structure, whereas the short-range structure is the major component for the other metals. The semivariogram of Zn combines the two structures, with variances in approximatively equal proportions.

The long-range structure of the semivariograms of Ni and Co concentrations are probably related to the control asserted by rock type, more precisely Argovian rocks (recall discussion related to Table 2.4). The concentrations of the other metals that are more influenced by land use show mainly the short-range variation. This suggests that the long-range structure relates to regional changes in geology (rock type) and the short-range structure to the distribution of sources of man-made contaminants.

Figure 2.15 shows the positions of the 359 sampling sites superimposed on the land use and geologic maps. These maps have been created by allocating unsampled locations to the land use or rock type of the nearest datum. The scatter of farmland and pastures contrasts with the greater continuity of geologic formations preferentially oriented SW-NE. Along that direction, note the grouping of forest soils in the eastern part of the study area.

Indicator semivariograms for rock types and land uses
Whenever the average value of an attribute z is very different from its average within a particular category s_k, the geometric layout of that category controls the shape and anisotropy of the z-semivariogram. The pattern of continuity (variability) of a category s_k can be characterized by semivariograms defined on an indicator coding of the presence/absence of that category. Define

$$i(\mathbf{u}_\alpha; s_k) = \begin{cases} 1 & \text{if } s(\mathbf{u}_\alpha) = s_k \\ 0 & \text{otherwise} \end{cases} \qquad (2.23)$$

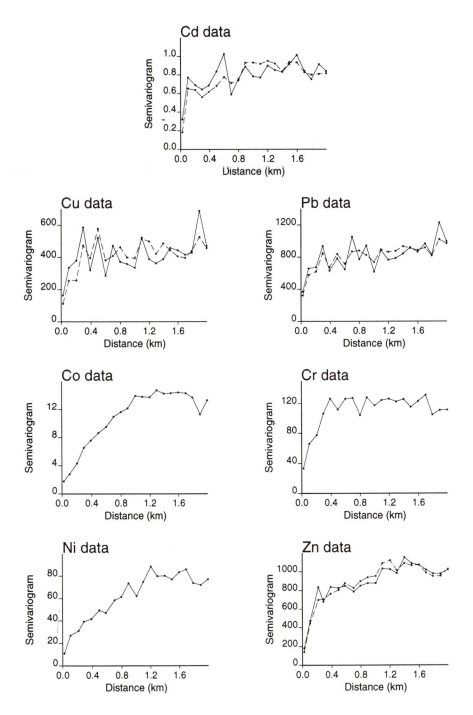

Figure 2.14: Experimental omnidirectional semivariograms of metal concentrations (solid line). For the positively skewed variables, the semivariogram of logarithms (dashed line) is rescaled to the original data variance.

Land use

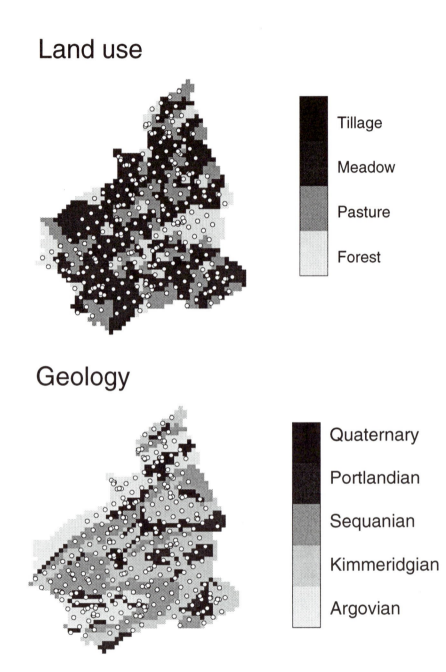

Figure 2.15: Locations of sampling sites superimposed on the land use and geologic maps.

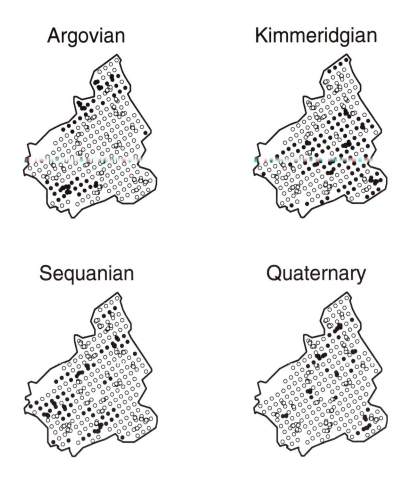

Figure 2.16: Indicator maps showing, in black, the locations belonging to a particular rock type.

The four indicator maps of Figure 2.16 show, in black, the locations where a particular rock type prevails, i.e., locations where $i(\mathbf{u}_\alpha; s_k) = 1$.

The indicator semivariogram for category s_k is then computed as

$$\gamma_I(\mathbf{h}; s_k) = \frac{1}{2N(\mathbf{h})} \sum_{\alpha=1}^{N(\mathbf{h})} [i(\mathbf{u}_\alpha; s_k) - i(\mathbf{u}_\alpha + \mathbf{h}; s_k)]^2 \qquad (2.24)$$

The indicator variogram value $2\gamma_I(\mathbf{h}; s_k)$ measures how often two locations a vector $\mathbf{h}$ apart belong to different categories $s_{k'} \neq s_k$. The smaller $2\gamma_I(\mathbf{h}; s_k)$, the better the spatial connectivity of category s_k. The ranges and shapes of the directional indicator semivariograms reflect the geometric patterns of category s_k.

Figure 2.17 shows the indicator semivariograms of the most common land uses and rock types computed in four directions with an angular tolerance of 22.5°, with the following results:

- For all rock types, the indicator semivariogram value equals zero at the first lag, which means that any two data locations less than 100 m apart belong to the same formation.

- For Argovian and Sequanian rocks, the longer SW-NE range (larger dashed line) reflects the corresponding preferential orientation of these two lithologic formations (Figure 2.16).

- The semivariograms for forest soils and Kimmeridgian rocks also show a better SW-NE continuity, though less pronounced.

- The indicator semivariograms for pastures and meadow are similar in all directions (isotropy) and have a shorter range than for forests.

The indicator structural analysis confirms the influence of Argovian rocks on the variability of cobalt concentrations. The indicator semivariogram of Argovian rocks and the semivariogram of cobalt concentrations share the same long-range structure and a better spatial continuity in the SW-NE direction (compare Figures 2.13 and 2.17). Such long-range anisotropy is not apparent for the Ni semivariogram, although, like cobalt, the average concentrations in nickel on Argovian rocks are half those measured on the other rocks. The direction-independent and mostly short-range variability of other metal concentrations matches that of meadow and pastures.

Semivariograms of residuals
When the pattern of variation of attribute z results from large differences in average z-values between categories s_k, filtering such differences should affect the shape of the z-semivariogram. Thus, one approach for interpreting the pattern of spatial variation of attribute z consists of

1. subtracting from each datum value $z(\mathbf{u}_\alpha)$ belonging to category $s_k = s(\mathbf{u}_\alpha)$ the average z-value within s_k, that is, the conditional mean $m_{|s_k}$,

2. computing the semivariogram $\gamma_R(\mathbf{h})$ of the residuals $r(\mathbf{u}_\alpha) = z(\mathbf{u}_\alpha) - m_{|s_k}$, and

3. comparing the sill and shape of semivariograms for original attribute values and for residuals.

Figure 2.18 shows the semivariograms of Ni and Co concentrations before and after filtering the conditional means for land uses and rock types given in Table 2.4 (page 18). For both metals, the conditional means are fairly constant from one land use to another. Hence, subtracting the conditional means does not affect Co and Ni semivariograms: the semivariogram of residuals (large dashed line) is close to the original semivariogram (solid line). In

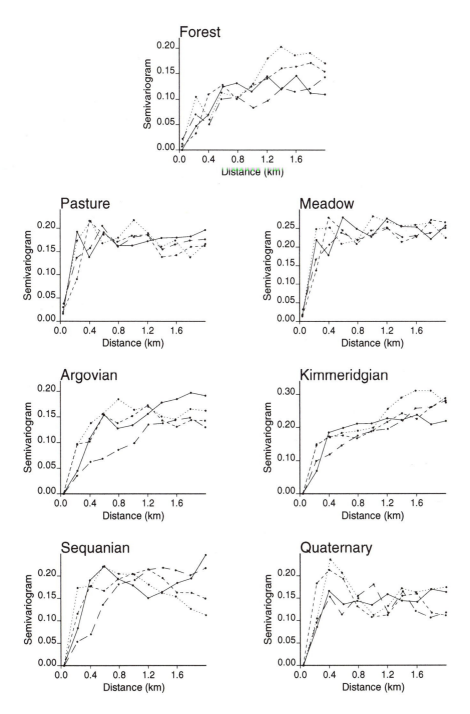

Figure 2.17: Experimental indicator semivariograms of the most common land uses and rock types in four directions (— : 22.5°, - - : 67.5°, - - - : 112.5°, ... : 157.5° ; $\Delta\theta = 22.5°$).

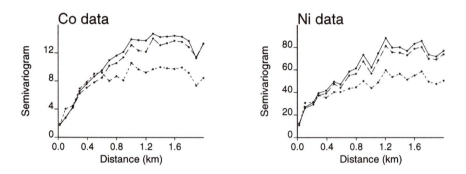

Figure 2.18: Experimental omnidirectional Co and Ni semivariograms before (solid line) and after subtracting from data the conditional means for land uses (large dashed line) and rock types (small dashed line).

contrast, filtering rock type means (small dashed line) drastically reduces the contribution of the long-range components to both semivariograms. This confirms the large-scale influence of geology on Co and Ni concentrations.

Indicator semivariograms for metal concentrations

Figure 2.19 shows, for each metal, the experimental omnidirectional standardized indicator semivariograms for three threshold values. The solid line refers to the second-decile threshold, whereas the large and small dashed lines denote, respectively, the fifth- and eighth-decile threshold. Three patterns can be distinguished:

1. Cu and Cr semivariograms appear to be similar across thresholds.

2. Indicator semivariograms for small concentrations of Cd, Ni, and Co have lesser nugget effect than those for larger concentrations. This suggests that homogeneous areas of small concentrations coexist within larger zones where high and medium concentrations are intermingled.

3. Small lead concentrations are less connected in space (greater nugget effect) than larger concentrations.

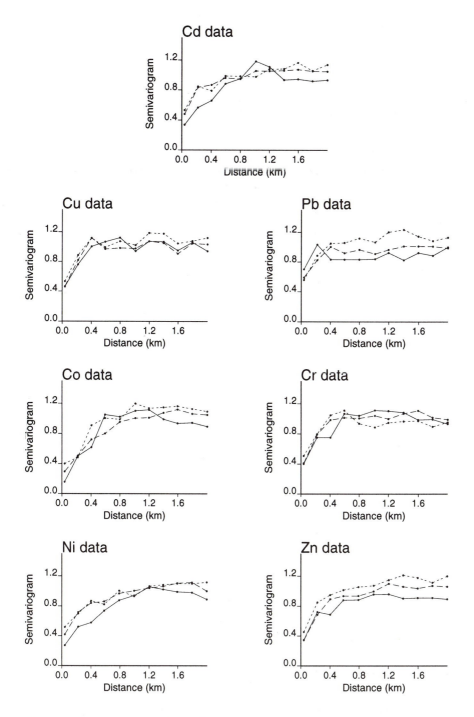

Figure 2.19: Experimental omnidirectional standardized indicator semivariograms for the second- (—), fifth- (– –) and eighth-decile (- - -) thresholds.

2.4   Bivariate Spatial Description

The next step in the spatial exploration of data consists of looking at the cross dependence between measurements of different attributes.

2.4.1   The cross h-scattergram

A cross h-scattergram is an h-scattergram where the tail and head values relate to two different attributes z_i and z_j. Figure 2.20 shows the cross h-scattergram between Ni and Cd concentrations in the E-W direction for four different classes of distances. The lag tolerance is $\Delta h = 100$ m.

At $|\mathbf{h}|=0$, the cross h-scattergram is the traditional scattergram of colocated values (Figure 2.3, page 20, top left graph). For each lag $|\mathbf{h}| \neq 0$, two cross h-scattergrams can be drawn, depending on what the head and tail attributes are: (1) cadmium and nickel or (2) nickel and cadmium. The left cross h-scattergrams show the similarity between a Ni value and the Cd value east of it, whereas the right cross h-scattergrams show the similarity between a Ni value and the Cd value west of it.

As in the univariate case, the increasing inflation of the cloud with increasing lag h indicates that the relation between Ni and Cd concentrations weakens as the separation distance increases. However, because the two variables are different, the inflation cannot be measured around the 45° line.

2.4.2   Measures of spatial cross continuity/variability

The covariance and correlogram measures are readily extended to the case where head and tail values relate to two different attributes.

Cross covariance function

The covariance between z_i- and z_j-values separated by a vector $\mathbf{h}$ is computed as

$$C_{ij}(\mathbf{h}) = \frac{1}{N(\mathbf{h})} \sum_{\alpha=1}^{N(\mathbf{h})} z_i(\mathbf{u}_\alpha) \cdot z_j(\mathbf{u}_\alpha + \mathbf{h}) - m_{i_{-\mathbf{h}}} \cdot m_{j_{+\mathbf{h}}} \qquad (2.25)$$

with

$$m_{i_{-\mathbf{h}}} = \frac{1}{N(\mathbf{h})} \sum_{\alpha=1}^{N(\mathbf{h})} z_i(\mathbf{u}_\alpha) \qquad m_{j_{+\mathbf{h}}} = \frac{1}{N(\mathbf{h})} \sum_{\alpha=1}^{N(\mathbf{h})} z_j(\mathbf{u}_\alpha + \mathbf{h})$$

where $N(\mathbf{h})$ is the number of pairs of data locations a vector $\mathbf{h}$ apart, and $m_{i_{-\mathbf{h}}}$ and $m_{j_{+\mathbf{h}}}$ are the means of tail z_i-values and head z_j-values. The ordered set of cross covariances $C_{ij}(\mathbf{h}_1), C_{ij}(\mathbf{h}_2), \ldots$ is called the experimental cross covariance function. In general, $C_{ij}(\mathbf{h}) \neq C_{ij}(-\mathbf{h})$, although $C_{ij}(\mathbf{h}) = C_{ji}(-\mathbf{h})$, as discussed subsequently.

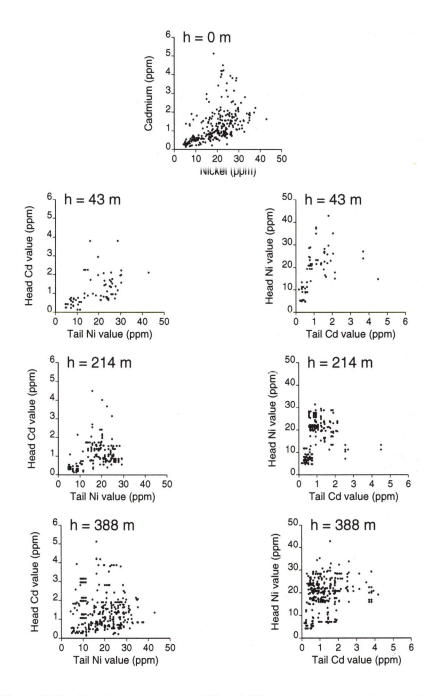

Figure 2.20: Cross h-scattergrams of Ni and Cd concentrations in the westerly (left column) and easterly (right column) directions (angle tolerance = 22.5^0) for four classes of distances.

Cross correlogram

The cross correlogram $\rho_{ij}(\mathbf{h})$ is given by

$$\rho_{ij}(\mathbf{h}) = \frac{C_{ij}(\mathbf{h})}{\sqrt{\sigma_{i\,-\mathbf{h}}^2 \cdot \sigma_{j\,+\mathbf{h}}^2}} \in [-1, +1] \tag{2.26}$$

with

$$\sigma_{i\,-\mathbf{h}}^2 = \frac{1}{N(\mathbf{h})} \sum_{\alpha=1}^{N(\mathbf{h})} [z_i(\mathbf{u}_\alpha) - m_{i\,-\mathbf{h}}]^2$$

$$\sigma_{j\,+\mathbf{h}}^2 = \frac{1}{N(\mathbf{h})} \sum_{\alpha=1}^{N(\mathbf{h})} [z_j(\mathbf{u}_\alpha + \mathbf{h}) - m_{j\,+\mathbf{h}}]^2$$

where $\sigma_{i\,-\mathbf{h}}^2$ and $\sigma_{j\,+\mathbf{h}}^2$ are the variances of tail z_i-values and head z_j-values.

Pseudo cross semivariogram

Some authors have introduced the pseudo cross semivariogram (Clark et al., 1989; Papritz et al., 1993), computed as

$$\gamma_{ij}^p(\mathbf{h}) = \frac{1}{2N(\mathbf{h})} \sum_{\alpha=1}^{N(\mathbf{h})} [z_i(\mathbf{u}_\alpha) - z_j(\mathbf{u}_\alpha + \mathbf{h})]^2 \tag{2.27}$$

where the difference $[z_i(\mathbf{u}_\alpha) - z_j(\mathbf{u}_\alpha + \mathbf{h})]$ is called a cross $\mathbf{h}$-increment.

Unlike in the $\mathbf{h}$-scattergram case, the first bisector of a cross $\mathbf{h}$-scattergram does not represent perfect correlation, hence the spread of the cloud around the 45° line, as measured by the pseudo cross semivariogram (2.27), is meaningless. Another shortcoming of the pseudo cross semivariogram is that its value tends to be overinfluenced by the variable with the largest values. Typically, the data should be transformed, for example, standardized to zero mean and unit variance, before computing the measure $\gamma_{ij}^p(\mathbf{h})$ (Myers, 1991).

The statistic $\gamma_{ij}^p(\mathbf{h})$ may be used when the two variables Z_i and Z_j relate to the same attribute measured at two different times (e.g., see Papritz and Flühler, 1994). In this particular case, the data need not be transformed since the two attributes are measured in the same unit and their variations are likely of the same magnitude.

The lag effect

Cross covariances computed in opposite directions are in general different: $C_{ij}(\mathbf{h})$ is different from $C_{ij}(-\mathbf{h})$. For example, the cross dependence between Cd and Ni concentrations is asymmetric in the E-W direction (Figure 2.21, top graphs).

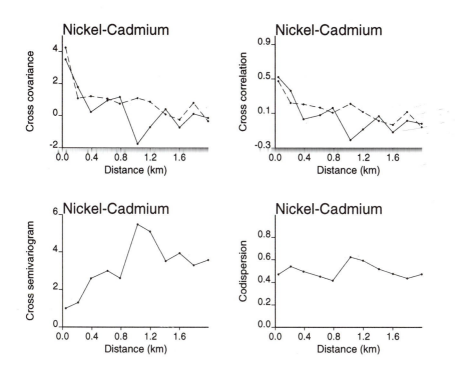

Figure 2.21: Four measures of spatial relationship between Ni and Cd concentrations in the E-W direction. The solid and dashed lines refer to the westerly and easterly direction, respectively. The experimental cross semivariogram and codispersion function are identical in both directions.

A substantial difference between $C_{ij}(\mathbf{h})$ and $C_{ij}(-\mathbf{h})$ would mean that one variable is lagging behind the other, an effect referred to as a *lag effect* (Journel and Huijbregts, 1978, p. 41). For example, this effect is observed in geochemistry, where different rates of precipitation may cause enrichment in some minerals to lag behind that of others along the direction of hydrothermal flow. A lag effect entails that the cross covariance function or the cross correlogram reaches its maximum at $|\mathbf{h}| \neq 0$.

A lag effect not backed by any physical interpretation is better ignored. Most often, deviations between opposite directions reflect experimental fluctuations that result from the small number of data pairs available. In what follows, all cross covariances $C_{ij}(\mathbf{h})$ refer to the average of $C_{ij}(-\mathbf{h})$ and $C_{ij}(\mathbf{h})$.

2.4.3 The scattergram of h-increments

The cross covariance measures how the value of an attribute z_i at one location is related to the value of another attribute z_j a vector $\mathbf{h}$ apart. Rather than the data pair $(z_i(\mathbf{u}_\alpha), z_j(\mathbf{u}_\alpha+\mathbf{h}))$, one may consider the pair of $\mathbf{h}$-increments, $([z_i(\mathbf{u}_\alpha)-z_i(\mathbf{u}_\alpha+\mathbf{h})], [z_j(\mathbf{u}_\alpha)-z_j(\mathbf{u}_\alpha+\mathbf{h})])$, which expresses the joint variation of gradients of z_i- and z_j-values from one location to another a vector $\mathbf{h}$ away. If both attributes are positively related, an increase (decrease) in z_i-value from $\mathbf{u}_\alpha$ to $\mathbf{u}_\alpha + \mathbf{h}$ tends to be associated with an increase (decrease) in z_j-value. Conversely, a negative correlation between attributes would entail that an increase (decrease) in z_i-value tends to be associated with a decrease (increase) in z_j-value.

Figure 2.22 shows the scattergrams between the $\mathbf{h}$-increments of Ni and Cd in the E-W direction for six different classes of distances (lag tolerance is $\Delta h = 100\ m$). The increasing spread of the cloud with separation distance reflects the increasing dissimilarity of values farther apart. The positive relation between $\mathbf{h}$-increments results from the positive correlation between attribute values.

2.4.4 Measures of joint variability

The similarity between $\mathbf{h}$-increments can be measured using the covariance and correlation coefficient.

Cross semivariogram

The cross semivariogram $\gamma_{ij}(\mathbf{h})$ is defined as half the non-centered covariance between $\mathbf{h}$-increments:

$$\gamma_{ij}(\mathbf{h}) = \frac{1}{2N(\mathbf{h})} \sum_{\alpha=1}^{N(\mathbf{h})} [z_i(\mathbf{u}_\alpha) - z_i(\mathbf{u}_\alpha + \mathbf{h})] \cdot [z_j(\mathbf{u}_\alpha) - z_j(\mathbf{u}_\alpha + \mathbf{h})] \quad (2.28)$$

Unlike the cross covariance or cross correlogram, the cross semivariogram is symmetric in i, j and $(\mathbf{h}, -\mathbf{h})$; interchanging i and j or substituting $-\mathbf{h}$ for $\mathbf{h}$ makes no difference in expression (2.28). The cross semivariogram thus cannot detect lag effects. In addition, the cross semivariogram can be computed only from those locations where both attributes are measured.

It is good practice to compare the magnitude of $\gamma_{ij}(\mathbf{h})$ values with the corresponding attribute variances σ_i^2 and σ_j^2. This is easily done by standardizing the two variables to unit variance and plotting the cross semivariogram values on the scale -1 to 1. The sill of the experimental cross semivariogram then directly reflects the magnitude of the correlation between variables.

Codispersion function

The codispersion coefficient is computed as

$$\nu_{ij}(\mathbf{h}) = \frac{\gamma_{ij}(\mathbf{h})}{\sqrt{\gamma_{ii}(\mathbf{h}) \cdot \gamma_{jj}(\mathbf{h})}} \quad \in [-1, +1] \quad (2.29)$$

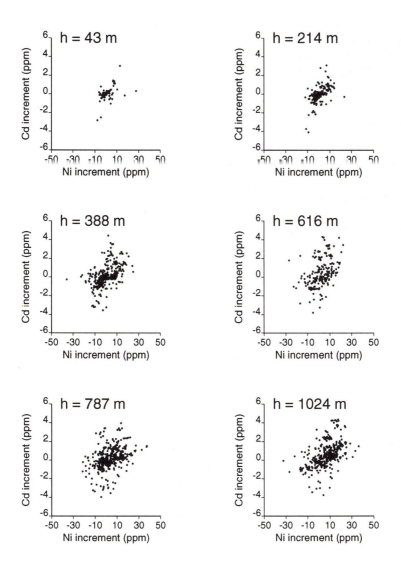

Figure 2.22: Scattergrams of h-increments of Ni and Cd concentrations in the easterly direction (angular tolerance $= 22.5°$) for six classes of distances (distance tolerance $= 100$ m).

The ordered set of codispersion coefficients $\nu_{ij}(\mathbf{h}_1)$, $\nu_{ij}(\mathbf{h}_2)$, ... is called the experimental codispersion function.

The codispersion coefficient can be interpreted as the correlation coefficient between h-increments when the scattergram is plotted in its symmetric form, that is, each pair of locations $(\mathbf{u}_\alpha, \mathbf{u}_\alpha + \mathbf{h})$ appears twice, once as the point of coordinates $([z_i(\mathbf{u}_\alpha) - z_i(\mathbf{u}_\alpha + \mathbf{h})], [z_j(\mathbf{u}_\alpha) - z_j(\mathbf{u}_\alpha + \mathbf{h})])$ and again as the point $([z_i(\mathbf{u}_\alpha + \mathbf{h}) - z_i(\mathbf{u}_\alpha)], [z_j(\mathbf{u}_\alpha + \mathbf{h}) - z_j(\mathbf{u}_\alpha)])$.

Beware that the relation $\nu_{ij}(\mathbf{h}) \in [-1, +1]$ may not be satisfied if the experimental cross semivariogram has not been calculated from the same data used to calculate the two semivariograms $\gamma_{ii}(\mathbf{h})$ and $\gamma_{jj}(\mathbf{h})$ (Journel and Huijbregts, 1978, p. 256).

Example

Figure 2.21 (page 49, bottom graphs) shows the experimental cross semivariogram and codispersion function between Ni and Cd values in the E-W direction. The increasing value of the cross semivariogram as a function of $\mathbf{h}$ reflects the gradual inflation of the cloud of points on the scattergram of $\mathbf{h}$-increments in Figure 2.22. The codispersion function indicates that the correlation between $\mathbf{h}$-increments does not change as the lag h increases.

2.4.5   Application to indicator transforms

As in the univariate case, the pattern of spatial cross dependence may depend on whether the values of either attribute are small, medium, or large. For example, large concentrations of two metals that share the same source of contamination are likely to be more strongly correlated than small or medium concentrations. The measures used to quantify spatial cross dependence between attribute values can be applied to indicator transforms.

Indicator cross covariance function

The indicator cross covariance is computed as

$$
\begin{aligned}
C_{ij}^{I}(\mathbf{h}; z_{ik}, z_{jk'}) &= \frac{1}{N(\mathbf{h})} \sum_{\alpha=1}^{N(\mathbf{h})} i(\mathbf{u}_\alpha; z_{ik}) \cdot i(\mathbf{u}_\alpha + \mathbf{h}; z_{jk'}) \\
&\quad - F_{i_{-\mathbf{h}}}(z_{ik}) \cdot F_{j_{+\mathbf{h}}}(z_{jk'}) \\
&= F_{ij}(\mathbf{h}; z_{ik}, z_{jk'}) - F_{i_{-\mathbf{h}}}(z_{ik}) \cdot F_{j_{+\mathbf{h}}}(z_{jk'}) \quad (2.30)
\end{aligned}
$$

where $F_{i_{-\mathbf{h}}}(z_{ik})$ and $F_{j_{+\mathbf{h}}}(z_{jk'})$ are the proportions of tail z_i-values and head z_j-values not exceeding the threshold values z_{ik} and $z_{jk'}$.

The cross covariance (2.30) is the centered version of the two-point joint cumulative frequency $F_{ij}(\mathbf{h}; z_{ik}, z_{jk'})$. The latter measures how often a z_i-value and the z_j-value a vector $\mathbf{h}$ apart are jointly no greater than their respective threshold values $(z_{ik}, z_{jk'})$, for example, how often two data locations separated by a vector $\mathbf{h}$ exceed the critical thresholds for two different metals.

Indicator cross correlogram

The standardized form of the indicator cross covariance function is the indicator cross correlogram:

$$\rho_{ij}^I(\mathbf{h}; z_{ik}, z_{jk'}) = \frac{C_{ij}^I(\mathbf{h}; z_{ik}, z_{jk'})}{\sqrt{\sigma_{i_{-\mathbf{h}}}^2(z_{ik}) \cdot \sigma_{j_{+\mathbf{h}}}^2(z_{jk'})}} \tag{2.31}$$

where the variance $\sigma_{i_{-\mathbf{h}}}^2(z_{ik})$ of tail indicator values $i(\mathbf{u}_\alpha; z_{ik})$ is equal to $F_{i_{-\mathbf{h}}}(z_{ik})[1 - F_{i_{-\mathbf{h}}}(z_{ik})]$.

Indicator cross semivariogram

The indicator cross semivariogram is computed as

$$\gamma_{ij}^I(\mathbf{h}; z_{ik}, z_{jk'}) = \frac{1}{2N(\mathbf{h})} \sum_{\alpha=1}^{N(\mathbf{h})} [i(\mathbf{u}_\alpha; z_{ik}) - i(\mathbf{u}_\alpha + \mathbf{h}; z_{ik})] \cdot$$
$$[i(\mathbf{u}_\alpha; z_{jk'}) - i(\mathbf{u}_\alpha + \mathbf{h}; z_{jk'})] \tag{2.32}$$

The only data pairs that have non-zero contributions to the indicator cross semivariogram are those where the values of both attributes z_i and z_j are on opposite sides of their threshold values $(z_{ik}, z_{jk'})$. The contribution of a data pair to $\gamma_{ij}^I(\mathbf{h}; z_{ik}, z_{jk'})$ can be positive $(+1)$ or negative (-1), depending on whether the z_i- and z_j-values jointly decrease (increase) from $\mathbf{u}_\alpha$ to $\mathbf{u}_\alpha + \mathbf{h}$, or vary in opposite ways. Therefore, the indicator cross semivariogram value cannot be interpreted as a joint frequency of transition, that is, it does not measure how often values of both attributes z_i and z_j are on opposite sides of threshold values.

Example

Figure 2.23 shows the experimental cross correlogram and cross semivariograms between Ni and Cd indicator data in the E-W direction; the threshold values are the second (left column), fifth (middle column), and eighth (right column) decile of the cumulative distribution. The cross correlation for the first lag drops from 0.78 for the second decile to 0.18 for the eighth decile threshold. Similarly, the relative nugget effect on the cross semivariogram increases with threshold value. Such strong dependence between Cd and Ni indicator data at small threshold values relates to the greater spatial continuity of small Cd and Ni concentrations observed in Figure 2.19 (page 45).

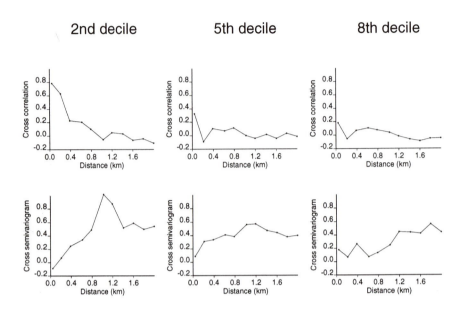

Figure 2.23: Two measures of spatial dependence between indicator transforms of Ni and Cd data: second- (left column), fifth- (middle column), and eighth-decile (right column) threshold. The cross semivariograms have been standardized by the covariance between indicator data.

Remark

Measures of spatial cross continuity and variability can be applied to other indicator data, such as these:

- $i(\mathbf{u}_\alpha; z_k)$ and $i(\mathbf{u}_\alpha; z_{k'})$ related to the same continuous attribute z but for two different threshold values z_k and $z_{k'}$

- $i(\mathbf{u}_\alpha; s_k)$ and $i(\mathbf{u}_\alpha; s_{k'})$ related to two different categories s_k and $s_{k'}$

- $i(\mathbf{u}_\alpha; z_k)$ and $i(\mathbf{u}_\alpha; s_k)$ related to a continuous and a categorical attribute

2.4.6   Spatial relations between metal concentrations

The previous spatial descriptive tools are now applied to the cross dependence between the two exhaustively sampled metals (Ni, Zn) and the three metals with widespread contamination (Cd, Cu, Pb).

Spatial anisotropy

Figure 2.24 shows the experimental cross semivariograms for the six pairs of variables considered in Figure 2.3 (page 20) and four directions. All variables

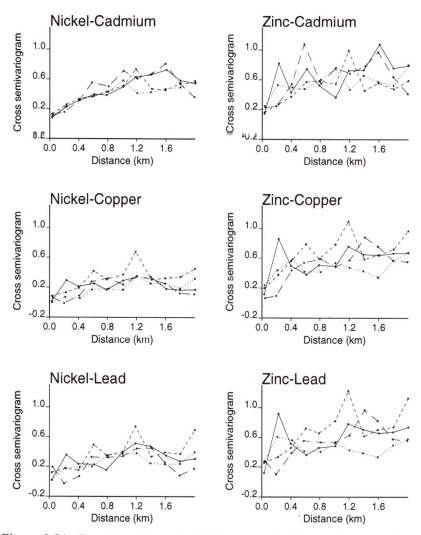

Figure 2.24: Experimental standardized cross semivariograms between the two exhaustively sampled metals (Ni, Zn) and the three metals (Cd, Cu, Pb) with widespread contamination in four directions (— : 22.5°, − − : 67.5°, - - - : 112.5°, . . . : 157.5° ; $\Delta\theta = 22.5°$).

were standardized to zero mean and unit variance so that the sill of each cross semivariogram $\gamma_{ij}(\mathbf{h})$ reflects the magnitude of the correlation between variables. The patterns of joint variation appear constant with direction (isotropy). The ranges of the cross semivariograms are larger when the secondary variable is Ni instead of Zn.

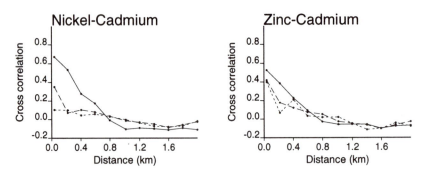

Figure 2.25: Experimental omnidirectional indicator cross correlograms between the two exhaustively sampled metals (Ni, Zn) and Cd: second- (—), fifth- (– –), and eighth-decile (- - -) thresholds.

Indicator cross correlograms

The concentrations of Ni, Zn, Cd, Cu, and Pb are transformed into indicator data using as thresholds the second, fifth, and eighth deciles of their respective cumulative distributions. Omnidirectional cross correlograms are then computed between indicator data related to the same decile. For example, Figure 2.25 (left graph) shows the cross correlogram between Cd and Ni indicator data for the second decile (solid line), fifth decile (large dashed line), and eighth decile (small dashed line). Small Cd and Ni concentrations are better connected in space than large concentrations. The same effect, though less pronounced, is observed for Zn-Cd. The cross dependence between Ni, Zn, and the two other metals Cu and Pb is weak and fairly similar across thresholds, hence their cross correlograms are not displayed here.

As expected, the difference between thresholds is most important for the two metals (Cd, Ni) that show important differences in the spatial continuity of small and large values (Figure 2.19).

2.5   Main Features of the Jura Data

The salient features of this exploratory data analysis are summarized as follows:

1. Many individual samples are contaminated with cadmium or lead, whereas a smaller proportion of samples exceeds the tolerable maximum for copper.

2. The distributions of Cd, Cu, and Pb concentrations are positively skewed.

3. The smallest metal concentrations are measured in forest soil or on Argovian rocks, whereas soil under pasture has the largest concentrations for all metals.

4. The metals with widespread contamination are positively related to the better sampled zinc. There is a positive relation between nickel and cadmium concentrations.

5. A small nugget effect, a short scale (range ≈ 200 m), and a regional scale (range ≈ 1 km) of spatial variability are observed on the semivariograms of metal concentrations. The short-range structure is the major component for the three metals with widespread contamination (Cd, Cu, and Pb) and Cr. The long-range structure dominates the semivariograms of Ni and Co concentrations. The Zn semivariogram combines the two structures in approximately equal proportions.

6. The short-range structure relates to the spatial distribution of rock types and land uses in the study area. The long-range structure reflects the influence of Argovian and Kimmeridgian rock types on metal concentrations.

7. Nickel concentrations vary more continuously in the SW-NE direction, which corresponds to the preferential orientation of the underlying geologic formations. The patterns of variation of other metals are fairly similar in all directions (isotropy).

8. Small concentrations in Cd, Ni, and Zn are better connected in space than larger concentrations. This suggests the existence of homogeneous areas of small concentrations and larger zones where high and median concentrations are intermingled.

9. The metals with widespread contamination (Cd, Cu, and Pb) show a short-range cross dependence with the better sampled Zn. There is a long-range cross dependence between Cd and Ni concentrations.

Chapter 3

The Random Function Model

Data description is rarely, if ever, the ultimate goal of a statistical study. Typically, one wants to go beyond the data to characterize the population from which the sample has been drawn. One essential step in this process is the quantitative modeling of the spatial statistics of that population from the data available over the study area. All subsequent applications, such as prediction or risk analysis, rely on the model (representation) chosen.

The deterministic and probabilistic approaches to modeling are compared in section 3.1. Section 3.2 introduces the random function model from which most geostatistical algorithms are built.

3.1 Deterministic and Probabilistic Models

Let $S_n = \{z(\mathbf{u}_\alpha), \ \alpha = 1, \ldots, n\}$ be the set of n measurements of attribute z over the study area $\mathcal{A}$. In Chapter 2 the population was identified as the sample set S_n. The population is now defined as the set of all measurements that could be made over $\mathcal{A}$, $\{z(\mathbf{u}), \ \forall \ \mathbf{u} \in \mathcal{A}\}$. The need for modeling the spatial distribution of z over $\mathcal{A}$ comes from the fact that the information available, S_n, is no longer exhaustive.

A model is but a representation of the (unknown) reality. Although that reality is unique, it has many possible representations, depending on the information available and the goal of the study. When building a model, one should use the following guidelines:

1. The model must incorporate all the relevant information. The pattern of spatial continuity of large values (e.g., pollutant concentrations, permeability, metal grades) is critical information for many applications in environmental studies, reservoir characterization, and selective mining, whereas the behavior of medium values may be more relevant to

sociological and demographic studies.

2. The model must be tractable. One should strike a balance between a congenial but possibly unrealistic model and a more representative model with too many parameters that are difficult to infer from sparse data.

3. The model must be tuned to the goal at hand. There is no need to model small-scale structures if the objective is to delineate target areas on a large scale or if the model is to be used, e.g., for flow simulation, a process insensitive to small-scale features. Conversely, a low-pass filter-type model would be inappropriate when the process under study depends on high-frequency (small-scale) variations.

Models can be classified as deterministic or probabilistic, depending on whether the representation is unique and deemed exact or whether the model consists of a set of alternative representations imaging the uncertainty about the unknown values.

Deterministic model

A deterministic model associates to any unsampled location $\mathbf{u}$ a single estimated value, say, $z^*(\mathbf{u})$ for the unknown value $z(\mathbf{u})$, without documenting the potential error $z^*(\mathbf{u}) - z(\mathbf{u})$. For all subsequent utilizations, that unique estimated value is taken as the true value; that is, the error is assumed to be nil or negligible. Such implicit disregard for the potential error is justifiable if the estimate $z^*(\mathbf{u})$ is based on either many data or some knowledge of the physics governing the spatial distribution of the attribute z. Unfortunately, in earth sciences such knowledge is limited, and the usually sparse information available does not allow one to ignore the error associated to any estimate $z^*(\mathbf{u})$, no matter how the variable is estimated.

Consider, for example, the problem of modeling the spatial distribution of Cd concentrations along the NE-SW transect shown in Figure 3.1 (top graph). A deterministic representation would attribute to the unsampled location $\mathbf{u}$ a single estimated value, say, $z^*(\mathbf{u}) = 0.7$ ppm. Such an estimate is likely to be in error because few data are available and our knowledge of the physical processes that control the spatial distribution of Cd values over the area is imperfect. In decision making, it is critical to assess the potential error $z^*(\mathbf{u}) - z(\mathbf{u})$ or, better, to know the range of values that the unknown $z(\mathbf{u})$ may take and the probability of occurrence of each outcome. In the example considered, it is critical to evaluate the probability that the Cd concentration at location $\mathbf{u}$ exceeds the tolerable maximum 0.8 ppm.

Probabilistic model

Instead of a single estimated value for the unknown $z(\mathbf{u})$, the probabilistic approach provides a set of possible values with the corresponding probabilities of occurrence. Such representation reflects our imperfect knowledge of

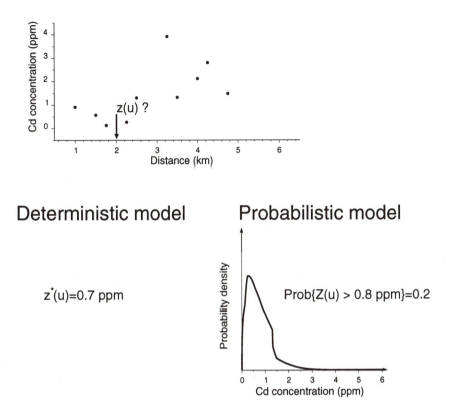

Figure 3.1: The problem of modeling the spatial distribution of a continuous attribute (Cd concentration) along a NE-SW transect. The deterministic model associates to the unsampled location u a single estimated value, say, a concentration of 0.7 ppm, whereas the probabilistic approach provides a probability distribution for the possible values at that location.

the unsampled value $z(\mathbf{u})$ and, more generally, of the distribution of z within the area. For example, Figure 3.1 (bottom graph) shows the distribution of probability for Cd concentrations at location $\mathbf{u}$. The unknown concentration has a 0.2 probability of exceeding the critical threshold 0.8 ppm. The location $\mathbf{u}$ would be classified as safe on the basis of the single estimated value $z^*(\mathbf{u}) =0.7$ ppm, yet the model of uncertainty indicates that there is a significant risk for the Cd concentration to exceed the tolerable maximum at $\mathbf{u}$. The decision of declaring the location $\mathbf{u}$ safe or contaminated would then depend on whether one chooses to ignore that risk.

Whereas deterministic models usually rely on the physics of the phenomenon, most of the information used in a probabilistic model comes from the data. The spatial distribution of Cd values may be modeled without any knowledge of the underlying physical processes; one would then capitalize on the spatial dependence between any two Cd values as inferred through the

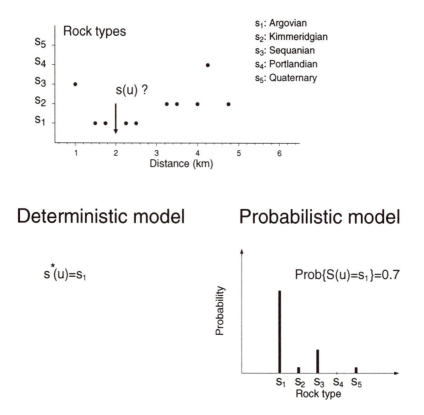

Figure 3.2: The problem of modeling the spatial distribution of a categorical attribute (rock type) along a NE-SW transect. The deterministic model associates to the unsampled location **u** a single value, say, a rock type s_1, whereas the probabilistic approach provides the probability of occurrence for all possible rock types.

sample variogram in section 2.3. Of course, considering a probabilistic model does not prevent using ancillary physical information, for example, the fact that particular rock types contain more cadmium than others. Also, geologic information, such as the orientation of rock types over the study area, may help in modeling the semivariogram of Cd concentrations.

Like continuous attributes, the spatial distribution of any categorical attribute s, say, rock type, can be modeled using either deterministic or probabilistic approaches. Consider, for example, the NE-SW transect in Figure 3.2 (top graph). Whereas the deterministic representation associates to the unsampled location **u** a single category, say, s_1, the probabilistic approach gives the probability for any of the five rock types to prevail at that location (Figure 3.2, bottom graph). The category s_1 has a large probability of prevailing at **u**, yet the probability that **u** actually belongs to any one of the four other categories is not zero.

Remarks

Good discussions about the philosophy and practice of modeling spatial data can be found in Journel (1986a, 1994a), Isaaks and Srivastava (1989, p. 196–236), and Matheron (1989).

1. Deterministic and probabilistic models may be used together. For example, a deterministic representation of better known large-scale structures can be complemented with a probabilistic modeling of small-scale variability.

2. A model is not cast forever. It should be updated whenever the goal of the study changes, additional data become available, or the physics of the phenomenon becomes better known.

3. When drawing a conclusion from a model, one should always question how much of that conclusion actually originates from the data and how much comes from the model itself. A model based on assumptions that are not supported by the data may generate artificial features; for example, see the discussion on the maximum entropy property of the multiGaussian RF model in section 8.4.

3.2 The Random Function Model

Geostatistics is largely based on the concept of random function, whereby the set of unknown values is regarded as a set of spatially dependent random variables. The *local* uncertainty about the attribute value at any particular location $\mathbf{u}$ is modeled through the set of possible realizations of the random variable at that location. The random function concept allows us to account for structures in the spatial variation of the attribute. The set of realizations of the random function models the uncertainty about the *spatial* distribution of the attribute over the entire study area.

3.2.1 Random variable

A random variable (RV) is a variable that can take a series of outcome values according to some probability distribution. Two types are usually distinguished: a discrete, or categorical, variable and a continuous variable.

Discrete random variable

If the number of possible outcomes of the RV is finite without any ordering, e.g., K mutually exclusive soil types or geologic facies, the RV is said to be discrete or categorical. Let $S(\mathbf{u})$ denote such a discrete random variable[1] at

---

[1] By convention, the random variable is denoted by a capital letter, whereas its outcomes or realizations are represented by the corresponding lowercase letter.

u. The quantity $p(\mathbf{u}; s_k)$ represents the probability for the category s_k to prevail at location **u**:

$$p(\mathbf{u}; s_k) = \text{Prob}\{S(\mathbf{u}) = s_k\} \tag{3.1}$$

The K probabilities $p(\mathbf{u}; s_k)$ must lie in the range $[0, 1]$ and sum to one:

$$p(\mathbf{u}; s_k) \in [0, 1] \qquad k = 1, \ldots, K \tag{3.2}$$

$$\sum_{k=1}^{K} p(\mathbf{u}; s_k) = 1 \tag{3.3}$$

For any specific ordering of the K outcomes s_k, the cumulative probability distribution $F(\mathbf{u}; s_k)$ is defined as

$$
\begin{aligned}
F(\mathbf{u}; s_k) &= \text{Prob}\{(S(\mathbf{u}) = s_1) \cup (S(\mathbf{u}) = s_2) \cup \ldots \cup (S(\mathbf{u}) = s_k)\} \\
&= \sum_{k'=1}^{k} p(\mathbf{u}; s_{k'})
\end{aligned}
\tag{3.4}
$$

The quantity $F(\mathbf{u}; s_k)$ is the probability for any one of the categories $s_{k'}$ ordered lesser or equal to s_k to prevail at **u**. Accounting for conditions (3.2) and (3.3), any cumulative probability $F(\mathbf{u}; s_k)$ must lie within $[0, 1]$, and the cumulative probability distribution of type (3.4) must be a non-decreasing function of the threshold s_k:

$$F(\mathbf{u}; s_k) \in [0, 1] \qquad k = 1, \ldots, K \tag{3.5}$$

$$F(\mathbf{u}; s_k) \leq F(\mathbf{u}; s_{k'}) \qquad \forall\, k' > k \tag{3.6}$$

Continuous random variable

If the attribute has a continuous range of possible outcomes with a natural ordering, it is modeled by a continuous RV, say, $Z(\mathbf{u})$. Examples are metal concentrations or porosity values. The random variable $Z(\mathbf{u})$ is fully characterized by its cumulative distribution function (cdf), which gives the probability that the variable Z at location **u** is no greater than any given threshold z:

$$F(\mathbf{u}; z) = \text{Prob}\{Z(\mathbf{u}) \leq z\} \qquad \forall\, z \tag{3.7}$$

The derivative of the cdf, when it exists, is the probability density function (pdf) $f(\mathbf{u}; z) = F'(\mathbf{u}; z)$. By analogy with the case of discrete random variables, the following order relations are satisfied:

$$F(\mathbf{u}; z) \in [0, 1] \qquad \forall\, z \tag{3.8}$$

$$F(\mathbf{u}; z) \leq F(\mathbf{u}; z') \qquad \forall\, z' > z \tag{3.9}$$

Any cumulative probability $F(\mathbf{u}; z)$ must lie within $[0, 1]$, and the cdf must be a non-decreasing function of the threshold z.

Figure 3.3 shows cdf and pdf representations for both categorical (top graphs) and continuous RVs (bottom graphs) at location $\mathbf{u}$. The cdf representation is typically used for continuous variables, whereas the pdf representation is more appropriate for categorical attributes with non-ordered states.

Indicator random variable

An indicator RV is a discrete binary RV with only two possible outcomes: 0 and 1. The probability (3.1) that a category s_k prevails at $\mathbf{u}$ can be expressed as the expected value of the indicator RV $I(\mathbf{u}; s_k)$ at that location:

$$
\begin{aligned}
E\{I(\mathbf{u}; s_k)\} &= 1 \cdot \text{Prob}\{I(\mathbf{u}; s_k) = 1\} + 0 \cdot \text{Prob}\{I(\mathbf{u}; s_k) = 0\} \\
&= \text{Prob}\{S(\mathbf{u}) = s_k\} = p(\mathbf{u}; s_k)
\end{aligned}
\tag{3.10}
$$

where the RV $I(\mathbf{u}; s_k)$ is defined as

$$
I(\mathbf{u}; s_k) = \begin{cases} 1 & \text{if } S(\mathbf{u}) = s_k \\ 0 & \text{otherwise} \end{cases} \qquad k = 1, \ldots, K
\tag{3.11}
$$

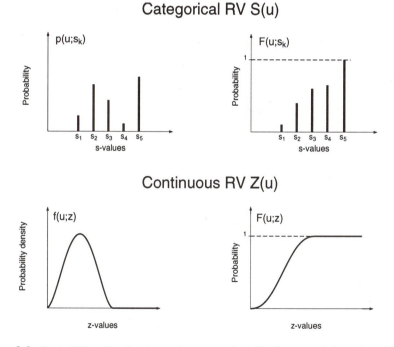

Figure 3.3: Probability distributions of a categorical RV (top graphs), and probability distribution functions of a continuous RV (bottom graphs).

Similarly, the probability (3.7) that the variable $Z(\mathbf{u})$ is no greater than any given threshold z can be expressed as

$$F(\mathbf{u}; z) = \text{Prob}\,\{Z(\mathbf{u}) \le z\} = \text{E}\,\{I(\mathbf{u}; z)\} \qquad (3.12)$$

where the RV $I(\mathbf{u}; z)$ is defined as

$$I(\mathbf{u}; z) = \left\{ \begin{array}{ll} 1 & \text{if } Z(\mathbf{u}) \le z \\ 0 & \text{otherwise} \end{array} \right. \quad \forall\, z \qquad (3.13)$$

Indicator geostatistics capitalizes on relations (3.10) and (3.12) to evaluate the probability for any category s_k to prevail at unsampled locations $\mathbf{u}$, or the probability for the variable Z to be no greater than any threshold z.

Two extreme probability distributions

Each datum $s(\mathbf{u}_\alpha)$ or $z(\mathbf{u}_\alpha)$ is viewed as a particular realization of the random variable $S(\mathbf{u}_\alpha)$ or $Z(\mathbf{u}_\alpha)$. Provided measurements are precise, there is *no uncertainty* about the sample value $s(\mathbf{u}_\alpha)$ or $z(\mathbf{u}_\alpha)$. Thus, the probability for the outcome value $s_k = s(\mathbf{u}_\alpha)$ to prevail at $\mathbf{u}_\alpha$ is one; the probability is zero for the remaining $(K - 1)$ categories:

$$p(\mathbf{u}_\alpha; s_k) = \left\{ \begin{array}{ll} 1 & \text{if } s(\mathbf{u}_\alpha) = s_k \\ 0 & \text{otherwise} \end{array} \right. \quad k = 1, \dots, K \qquad (3.14)$$

Similarly, the probability for the random variable Z at $\mathbf{u}_\alpha$ to be no greater than a threshold z is one for any threshold greater than or equal to the datum $z(\mathbf{u}_\alpha)$; the probability is zero for other thresholds $z < z(\mathbf{u}_\alpha)$:

$$F(\mathbf{u}_\alpha; z) = \left\{ \begin{array}{ll} 1 & \text{if } z(\mathbf{u}_\alpha) \le z \\ 0 & \text{otherwise} \end{array} \right. \quad \forall\, z \qquad (3.15)$$

These probability distributions with zero variance (no uncertainty) are depicted at the top of Figure 3.4.

Where no information whatsoever is available, all outcomes have the same probability of occurrence (*maximum uncertainty*). The K categories s_k then have the same probability $1/K$ to prevail at location $\mathbf{u}$:

$$p(\mathbf{u}; s_k) = \frac{1}{K} \quad k = 1, \dots, K \qquad (3.16)$$

Similarly, the probability for the random variable Z at $\mathbf{u}$ to be no greater than the threshold z increases linearly with that threshold:

$$F(\mathbf{u}; z) = \left\{ \begin{array}{ll} 0 & \text{if } z \le z_{min} \\[2mm] \dfrac{z - z_{min}}{(z_{max} - z_{min})} & \text{if } z \in (z_{min}, z_{max}] \\[2mm] 1 & \text{if } z > z_{max} \end{array} \right. \quad \forall\, z \qquad (3.17)$$

where z_{min} and z_{max} are the minimum and maximum values of the z-distribu-

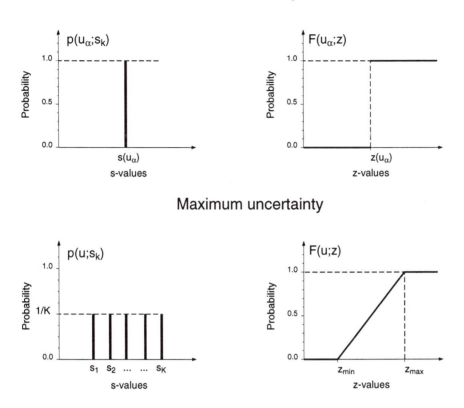

Figure 3.4: Probability distribution of categorical (left graphs) and continuous RVs (right graphs) when there is no uncertainty (top graphs) or maximum uncertainty (bottom graphs).

tion, and the notation $z \in (z_{min}, z_{max}]$ is equivalent to $z_{min} < z \leq z_{max}$. The corresponding uniform probability distributions are represented at the bottom of Figure 3.4.

The basic paradigm of the probabilistic approach is to model any unknown value $s(\mathbf{u})$ or $z(\mathbf{u})$ as a random variable $S(\mathbf{u})$ or $Z(\mathbf{u})$. The problem of assessing the uncertainty about an attribute value at $\mathbf{u}$ thus reduces to that of modeling the probability distribution of the random variable S or Z at that location. When no information is available, the uncertainty is maximum, with the same probability of occurrence for all outcome values. The idea is to use neighboring data values $s(\mathbf{u}_\alpha)$ or $z(\mathbf{u}_\alpha)$ to reduce the uncertainty at location $\mathbf{u}$, which amounts to updating the model of type (3.16) or (3.17) into a model that is conditional on (accounts for) the information available around $\mathbf{u}$. By accounting for the dependence between RVs at different locations, the random function model allows such updating. In the next section, the focus is on continuous random functions. Similar developments can be made for discrete random functions (e.g., see Isaaks and Srivastava, 1989, p. 218–236).

3.2.2   Random function

A random function (RF) is defined as a set of usually dependent random variables $Z(\mathbf{u})$, one for each location $\mathbf{u}$ in the study area $\mathcal{A}$, $\{Z(\mathbf{u}), \; \forall \; \mathbf{u} \in \mathcal{A}\}$. To any set of N locations $\mathbf{u}_k$, $k = 1, \ldots, N$ corresponds a vector of N random variables $\{Z(\mathbf{u}_1), \ldots, Z(\mathbf{u}_N)\}$ that is characterized by the N-variate or N-point cdf:

$$F(\mathbf{u}_1, \ldots, \mathbf{u}_N; z_1, \ldots, z_N) = \text{Prob}\,\{Z(\mathbf{u}_1) \leq z_1, \ldots, Z(\mathbf{u}_N) \leq z_N\} \qquad (3.18)$$

The multivariate cdf (3.18) characterizes the joint uncertainty about the N actual values $z(\mathbf{u}_1), \ldots, z(\mathbf{u}_N)$. The set of all such N-variate cdfs, for any positive integer N and for any choice of the locations $\mathbf{u}_k$, constitutes the spatial law of the RF $Z(\mathbf{u})$.

In practice, the analysis is limited to cdfs involving no more than two locations at a time and their corresponding moments:

- The one-point cdf:

$$F(\mathbf{u}; z) \; = \; \text{Prob}\,\{Z(\mathbf{u}) \leq z\} \; = \; \text{E}\,\{I(\mathbf{u}; z)\} \qquad (3.19)$$

  with the random variable $I(\mathbf{u}; z)$ defined in (3.13)

- The two-point cdf:

$$\begin{aligned} F(\mathbf{u}, \mathbf{u}'; z, z') \; &= \; \text{Prob}\,\{Z(\mathbf{u}) \leq z, Z(\mathbf{u}') \leq z'\} \\ &= \; \text{E}\,\{I(\mathbf{u}; z) \cdot I(\mathbf{u}'; z')\} \end{aligned} \qquad (3.20)$$

- The Z-expected value:

$$m(\mathbf{u}) \; = \; \text{E}\,\{Z(\mathbf{u})\} \qquad (3.21)$$

- The (two-point) Z-covariance:

$$C(\mathbf{u}, \mathbf{u}') \; = \; \text{E}\,\{Z(\mathbf{u}) \cdot Z(\mathbf{u}')\} - \text{E}\,\{Z(\mathbf{u})\} \cdot \text{E}\,\{Z(\mathbf{u}')\} \qquad (3.22)$$

- The (two-point) Z-correlogram:

$$\rho(\mathbf{u}, \mathbf{u}') \; = \; \frac{C(\mathbf{u}, \mathbf{u}')}{\sqrt{C(\mathbf{u}, \mathbf{u}) \cdot C(\mathbf{u}', \mathbf{u}')}} \qquad (3.23)$$

- The Z-variogram:

$$2\gamma(\mathbf{u}, \mathbf{u}') \; = \; \text{Var}\,\{Z(\mathbf{u}) - Z(\mathbf{u}')\} \qquad (3.24)$$

The *one-point* and *two-point* terminology reminds us that the two random variables relate to the same attribute z at two different points or locations rather than to two different attributes.

Dependence and linear correlation

The conditional cumulative distribution function (ccdf) $F(\mathbf{u}; z | Z(\mathbf{u}') \leq z')$ is the cdf of $Z(\mathbf{u})$ given knowledge about $Z(\mathbf{u}')$, specifically that $Z(\mathbf{u}') \leq z'$. According to the conditional probability definition, the ccdf is deduced from the previous one- and two-point cdfs by the relation

$$
\begin{aligned}
F(\mathbf{u}; z | Z(\mathbf{u}') \leq z') &= \text{Prob}\{Z(\mathbf{u}) \leq z | Z(\mathbf{u}') \leq z'\} \\
&= \frac{\text{Prob}\{Z(\mathbf{u}) \leq z, Z(\mathbf{u}') \leq z'\}}{\text{Prob}\{Z(\mathbf{u}') \leq z'\}} \\
&= \frac{F(\mathbf{u}, \mathbf{u}'; z, z')}{F(\mathbf{u}'; z')}
\end{aligned} \tag{3.25}
$$

Two RVs $Z(\mathbf{u})$ and $Z(\mathbf{u}')$ are said to be independent if the probability distribution of either one is not affected by any knowledge about the other one, that is, if and only if

$$
\begin{aligned}
F(\mathbf{u}; z | Z(\mathbf{u}') \leq z') &= F(\mathbf{u}; z) \quad \forall\, z' \\
F(\mathbf{u}'; z' | Z(\mathbf{u}) \leq z) &= F(\mathbf{u}'; z') \quad \forall\, z
\end{aligned}
$$

By substituting the first expression into relation (3.25), the independence condition becomes

$$
F(\mathbf{u}, \mathbf{u}'; z, z') = F(\mathbf{u}; z) \cdot F(\mathbf{u}'; z') \quad \forall\, z, z' \tag{3.26}
$$

A measure of dependence between $Z(\mathbf{u})$ and $Z(\mathbf{u}')$ is thus the centered two-point cdf, defined as

$$
F(\mathbf{u}, \mathbf{u}'; z, z') - F(\mathbf{u}; z) \cdot F(\mathbf{u}'; z') \quad \forall\, z, z' \tag{3.27}
$$

The two RVs $Z(\mathbf{u})$ and $Z(\mathbf{u}')$ are independent if expression (3.27) equals zero.

Using the indicator RV defined in (3.13), the measure of dependence (3.27) appears as the cross covariance between the two indicator RVs $I(\mathbf{u}; z)$ and $I(\mathbf{u}'; z')$:

$$
\mathrm{E}\{I(\mathbf{u}; z) \cdot I(\mathbf{u}'; z')\} - \mathrm{E}\{I(\mathbf{u}; z)\} \cdot \mathrm{E}\{I(\mathbf{u}'; z')\} = C_I(\mathbf{u}, \mathbf{u}'; z, z') \tag{3.28}
$$

Thus, the set of all indicator cross covariances $C_I(\mathbf{u}, \mathbf{u}'; z, z')$ for all thresholds z, z', provides a measure of *dependence* between the two RVs $Z(\mathbf{u})$ and $Z(\mathbf{u}')$. In contrast, the Z-covariance $C(\mathbf{u}, \mathbf{u}')$ is a measure only of *linear correlation* between the two RVs, i.e., a measure of the ability of a straight line to describe the relation between these two variables. The Z-covariance $C(\mathbf{u}, \mathbf{u}')$ and all indicator cross covariances $C_I(\mathbf{u}, \mathbf{u}'; z, z')$ vanish when the two RVs are independent. However, the condition $C(\mathbf{u}, \mathbf{u}') = 0$ does not necessarily imply relation (3.26): two linearly uncorrelated RVs may still be dependent, in which case the dependence relation is non-linear.

The decision of stationarity

The one- and two-point cdfs of the RF and their moments as defined by relations (3.19)–(3.24) are location-dependent. Their inference thus requires repetitive realizations at each location $\mathbf{u}$. For example, the inference of the Z-covariance $C(\mathbf{u}, \mathbf{u}')$ between the two RVs $Z(\mathbf{u})$ and $Z(\mathbf{u}')$ separated by a vector $\mathbf{h} = \mathbf{u}' - \mathbf{u}$ calls for a set of repetitive measurements $\{z^{(l)}(\mathbf{u}), z^{(l)}(\mathbf{u}');\ l = 1, \ldots, L\}$, which is never[2] available in practice. The idea is to use all pairs of measurements a vector $\mathbf{h}$ apart within the study area $\mathcal{A}$, $\{z(\mathbf{u}_\alpha), z(\mathbf{u}_\alpha + \mathbf{h});\ \alpha = 1, \ldots, n\}$, as a set of repetitions. The implicit assumption is that the corresponding pairs of RVs $\{Z(\mathbf{u}_\alpha), Z(\mathbf{u}_\alpha + \mathbf{h});\ \alpha = 1, \ldots, n\}$ originate from the same two-point distribution. Such pooling of data pairs regardless of their locations calls for the phenomenon under study to be spatially "homogeneous" within $\mathcal{A}$. In probabilistic terms, the RF model $Z(\mathbf{u})$ must be chosen to be stationary within $\mathcal{A}$.

The RF $Z(\mathbf{u})$ is said to be stationary within $\mathcal{A}$ if the multivariate cdf (3.18) is invariant under translation. This means that any two vectors of RVs $\{Z(\mathbf{u}_1), \ldots, Z(\mathbf{u}_N)\}$ and $\{Z(\mathbf{u}_1 + \mathbf{h}), \ldots, Z(\mathbf{u}_N + \mathbf{h})\}$ have the same multivariate cdf whatever the translation vector $\mathbf{h}$:

$$F(\mathbf{u}_1, \ldots, \mathbf{u}_N; z_1, \ldots, z_N) \;=\; F(\mathbf{u}_1 + \mathbf{h}, \ldots, \mathbf{u}_N + \mathbf{h}; z_1, \ldots, z_N)$$
$$\forall\ \mathbf{u}_1, \ldots, \mathbf{u}_N \text{ and } \mathbf{h} \quad (3.29)$$

The decision of stationarity is usually limited to the one-point and two-point cdfs, and the first two moments of the RF. The reference to a particular location $\mathbf{u}$ can then be dropped from expressions (3.19)–(3.24), and the two-point statistics now depend only on the separation vector $\mathbf{h}$:

- $F(z) \;=\; \text{Prob}\{Z(\mathbf{u}) \leq z\}$

- $F(\mathbf{h}; z, z') \;=\; \text{Prob}\{Z(\mathbf{u}) \leq z, Z(\mathbf{u} + \mathbf{h}) \leq z'\}$

- $m \;=\; \text{E}\{Z(\mathbf{u})\}$

- $C(\mathbf{h}) \;=\; \text{E}\{Z(\mathbf{u}) \cdot Z(\mathbf{u} + \mathbf{h})\} - \text{E}\{Z(\mathbf{u})\} \cdot \text{E}\{Z(\mathbf{u} + \mathbf{h})\}$

- $\rho(\mathbf{h}) \;=\; \dfrac{C(\mathbf{h})}{C(0)}$

- $2\gamma(\mathbf{h}) = \text{Var}\{Z(\mathbf{u}) - Z(\mathbf{u} + \mathbf{h})\} = \text{E}\left\{[Z(\mathbf{u}) - Z(\mathbf{u} + \mathbf{h})]^2\right\}$

Recall that the vector notation $\mathbf{h}$ accounts for both distance $|\mathbf{h}|$ and direction. The functions $C(\mathbf{h})$ and $2\gamma(\mathbf{h})$ are said to be *anisotropic* if they depend on both distance and direction. They are said to be *isotropic* if they depend only on the modulus of $\mathbf{h}$.

---

[2] When the set of repetitions consists of measurements recorded at the same location but at different times, the RF is actually defined in space-time and should be denoted $Z(\mathbf{u}, t)$.

An RF model is said to be stationary of order two when (1) the expected value $E\{Z(\mathbf{u})\}$ exists and is invariant within $\mathcal{A}$, and (2) the two-point covariance $C(\mathbf{h})$ exists and depends only on the separation vector $\mathbf{h}$. The covariance function, correlogram, and semivariogram of a stationary RF are related by

$$\gamma(\mathbf{h}) \;=\; C(0) - C(\mathbf{h}) \tag{3.30}$$

$$\rho(\mathbf{h}) \;=\; 1 - \frac{\gamma(\mathbf{h})}{C(0)} \tag{3.31}$$

As the separation distance $|\mathbf{h}|$ increases, the correlation between any two RVs $Z(\mathbf{u})$ and $Z(\mathbf{u} + \mathbf{h})$ generally tends to zero:[3]

$$C(\mathbf{h}) \;\longrightarrow\; 0 \ \text{ for } \ |\mathbf{h}| \;\longrightarrow\; \infty$$

Accounting for relation (3.30), the sill value of a bounded semivariogram tends toward the a priori variance $C(0)$:

$$\gamma(\mathbf{h}) \;\longrightarrow\; C(0) \ \text{ for } \ |\mathbf{h}| \;\longrightarrow\; \infty \tag{3.32}$$

Remarks

1. The definition of the semivariogram $\gamma(\mathbf{h})$ does not require the existence of a constant mean and finite variance for the RF $Z(\mathbf{u})$; a sufficient condition is that the RF increments $[Z(\mathbf{u}) - Z(\mathbf{u} + \mathbf{h})]$ are stationary of order two, a condition referred to as the *intrinsic hypothesis* (Journel and Huijbregts, 1978, p. 33). Second-order stationarity implies the intrinsic hypothesis but the reverse is not true: an intrinsic RF need not be stationary of order two (see discussion on unbounded semivariogram models in section 4.2.1).

2. Stationarity is a property of the RF model—a property needed for inference. It is not a characteristic of the phenomenon under study. Stationarity is a decision made by the user, not a hypothesis that can be proven or refuted from data.

3. The stationarity decision allows pooling data over areas that are deemed homogeneous. Exploratory data analysis may indicate the existence of several populations with significantly different statistics. One should then consider the possibility of subdividing the area $\mathcal{A}$ into more homogeneous subzones, each being modeled with a different RF. Such subdivision is conditioned by:

   - The availability of enough data to infer the parameters (mean, covariance function) of each separate RF

---

[3] One exception is a periodic phenomenon in one dimension (1-D) with cosine-type covariance (see Journel and Huijbregts, 1978, p. 169). Such periodic covariances will not be considered in this book.

- The ability to delineate the different populations both on the data and at unsampled locations; one must have a way to decide to which RF any location $\mathbf{u} \in \mathcal{A}$ pertains.

3.2.3 Multivariate random function

The notation and concepts introduced in the previous section are readily extended to the case where N_v continuous attributes are considered. The set of N_v interdependent RFs $\{Z_i(\mathbf{u}), \; i = 1, \ldots, N_v \; ; \; \forall \, \mathbf{u} \in \mathcal{A}\}$, denoted by the vector $\mathbf{Z}(\mathbf{u})$, is called a multivariate RF or a vector RF (of dimension N_v). The inference of the one-point and two-point cdfs and corresponding moments calls for a prior decision of joint stationarity. One then defines:

- The marginal cdfs:

$$F_i(z_i) \;\; = \;\; \mathrm{Prob}\,\{Z_i(\mathbf{u}) \leq z_i\} \qquad i = 1, \ldots, N_v \qquad (3.33)$$

- The two-point (joint) cdfs:

$$F_{ij}(\mathbf{h}; z_i, z_j) \;\; = \;\; \mathrm{Prob}\,\{Z_i(\mathbf{u}) \leq z_i, Z_j(\mathbf{u}+\mathbf{h}) \leq z_j\} \qquad \forall \, i,j \tag{3.34}$$

  where $F_{ij}(\mathbf{h}; z_i, z_j)$ is not necessarily equal to $F_{ji}(\mathbf{h}; z_j, z_i)$.

- The expected values:

$$m_i \;\; = \;\; \mathrm{E}\,\{Z_i(\mathbf{u})\} \qquad i = 1, \ldots, N_v \tag{3.35}$$

- The (cross) covariance functions:

$$C_{ij}(\mathbf{h}) \;\; = \;\; \mathrm{E}\,\{[Z_i(\mathbf{u}) - m_i] \cdot [Z_j(\mathbf{u}+\mathbf{h}) - m_j]\} \qquad \forall \, i,j \tag{3.36}$$

  where $C_{ij}(\mathbf{h})$ is not necessarily equal to $C_{ji}(\mathbf{h})$.

- The (cross) variograms:

$$
\begin{aligned}
2\gamma_{ij}(\mathbf{h}) \;\; &= \;\; \mathrm{Cov}\,\{[Z_i(\mathbf{u}) - Z_i(\mathbf{u}+\mathbf{h})], [Z_j(\mathbf{u}) - Z_j(\mathbf{u}+\mathbf{h})]\} \\
&= \;\; \mathrm{E}\,\{[Z_i(\mathbf{u}) - Z_i(\mathbf{u}+\mathbf{h})] \cdot [Z_j(\mathbf{u}) - Z_j(\mathbf{u}+\mathbf{h})]\} \;\; \forall \, i,j
\end{aligned}
\tag{3.37}
$$

The terms *joint* or *cross* apply when $i \neq j$, whereas the terms *direct* or *auto* apply when $i = j$.

Lag effect

The cross covariance function and the cross semivariogram are related through the expression

$$\gamma_{ij}(\mathbf{h}) \;\; = \;\; C_{ij}(0) - \frac{1}{2}\,[C_{ij}(\mathbf{h}) + C_{ij}(-\mathbf{h})] \tag{3.38}$$

The cross covariance function $C_{ij}(\mathbf{h})$ can be written as the sum of an even function of $\mathbf{h}$, $[C_{ij}(\mathbf{h})+C_{ij}(-\mathbf{h})]$ and an odd function of $\mathbf{h}$, $[C_{ij}(\mathbf{h})-C_{ij}(-\mathbf{h})]$:

$$C_{ij}(\mathbf{h}) = \frac{1}{2}[C_{ij}(\mathbf{h}) + C_{ij}(-\mathbf{h})] + \frac{1}{2}[C_{ij}(\mathbf{h}) - C_{ij}(-\mathbf{h})]$$

The cross semivariogram $\gamma_{ij}(\mathbf{h})$ incorporates only the even term of the cross covariance function, see relation (3.38), hence it is symmetric in $(\mathbf{h}, -\mathbf{h})$.

In practice, asymmetry of the cross covariance function (recall section 2.4.2) is most often ignored for these reasons:

1. The usual descriptive tools are the direct and cross semivariograms which are symmetric.

2. Lack of data typically prevents asserting the physical reality of a lag effect.

3. Modeling asymmetric cross covariances, although possible, is difficult; for example, see Journel and Huijbregts (1978, p. 173), Grzebyk (1993).

Therefore, all the models developed in this book have the following characteristics:

$$F_{ij}(\mathbf{h}; z_i, z_j) = F_{ji}(\mathbf{h}; z_j, z_i) \quad \forall\, i, j$$
$$C_{ij}(\mathbf{h}) = C_{ji}(\mathbf{h}) \quad \forall\, i, j$$

Note that an isotropic cross covariance function is always symmetric. Indeed, isotropy entails that

$$C_{ij}(\mathbf{h}) = C_{ij}(-\mathbf{h}) = C_{ji}(|\mathbf{h}|) \quad \text{since} \quad |\mathbf{h}| = |-\mathbf{h}|$$

As in the single-variable case, the correlation between any two RVs $Z_i(\mathbf{u})$ and $Z_j(\mathbf{u} + \mathbf{h})$ is assumed to tend to zero as the separation distance $|\mathbf{h}|$ increases:

$$C_{ij}(\mathbf{h}) \longrightarrow 0 \quad \text{as} \quad |\mathbf{h}| \longrightarrow \infty$$

Accounting for relation (3.38), the sill value of a bounded cross semivariogram tends toward the covariance value $C_{ij}(0)$:

$$\gamma_{ij}(\mathbf{h}) \longrightarrow C_{ij}(0) \quad \text{as} \quad |\mathbf{h}| \longrightarrow \infty \qquad (3.39)$$

The cross correlogram

A unit-free measure of linear correlation between any two RVs $Z_i(\mathbf{u})$ and $Z_j(\mathbf{u} + \mathbf{h})$ is the cross correlogram, defined as

$$\rho_{ij}(\mathbf{h}) = \frac{C_{ij}(\mathbf{h})}{\sqrt{C_{ii}(0) \cdot C_{jj}(0)}} \in [-1, +1] \qquad (3.40)$$

At $|\mathbf{h}| = 0$, expression (3.40) is the linear correlation coefficient between the two variables Z_i and Z_j. Note that $\rho_{ij}(0) = 0$ does not necessarily imply that $\rho_{ij}(\mathbf{h}) = 0$.

Semivariogram and correlation function matrix

Under the decision of joint second-order stationarity for all N_v RFs, one defines:

- The mean-value vector:

$$\mathbf{m} \ = \ \mathrm{E}\left\{\mathbf{Z}(\mathbf{u})\right\}$$

- The covariance function matrix:

$$\mathbf{C}(\mathbf{h}) \ = \ \mathrm{E}\left\{[\mathbf{Z}(\mathbf{u}) - \mathbf{m}] \cdot [\mathbf{Z}(\mathbf{u} + \mathbf{h}) \ - \mathbf{m}]^T\right\}$$

- The semivariogram matrix:

$$\boldsymbol{\Gamma}(\mathbf{h}) \ = \ \frac{1}{2}\,\mathrm{E}\left\{[\mathbf{Z}(\mathbf{u}) - \mathbf{Z}(\mathbf{u} + \mathbf{h})] \cdot [\mathbf{Z}(\mathbf{u}) - \mathbf{Z}(\mathbf{u} + \mathbf{h})]^T\right\}$$

with $\mathbf{Z}(\mathbf{u}) = [Z_1(\mathbf{u}), Z_2(\mathbf{u}), \dots, Z_{N_v}(\mathbf{u})]^T$, $\mathbf{m} = [m_1, m_2, \dots, m_{N_v}]^T$; the superscript T denotes matrix transposition.

The covariance function matrix is an $N_v \times N_v$ matrix that contains the autocovariance functions along its major diagonal and the cross covariance functions off that diagonal:

$$\mathbf{C}(\mathbf{h}) = \begin{bmatrix} C_{11}(\mathbf{h}) & \cdots & C_{1N_v}(\mathbf{h}) \\ \vdots & & \vdots \\ C_{N_v 1}(\mathbf{h}) & \cdots & C_{N_v N_v}(\mathbf{h}) \end{bmatrix} \tag{3.41}$$

At $|\mathbf{h}| = 0$, the matrix $\mathbf{C}(0)$ is the traditional variance–covariance matrix.

The semivariogram matrix $\boldsymbol{\Gamma}(\mathbf{h})$ is an $N_v \times N_v$ symmetric matrix that contains the direct semivariograms along its major diagonal and the cross semivariograms off that diagonal:

$$\boldsymbol{\Gamma}(\mathbf{h}) = \begin{bmatrix} \gamma_{11}(\mathbf{h}) & \cdots & \gamma_{1N_v}(\mathbf{h}) \\ \vdots & & \vdots \\ \gamma_{N_v 1}(\mathbf{h}) & \cdots & \gamma_{N_v N_v}(\mathbf{h}) \end{bmatrix}. \tag{3.42}$$

The covariance function matrix and semivariogram matrix are related by

$$\begin{aligned} \boldsymbol{\Gamma}(\mathbf{h}) \ &= \ \mathbf{C}(0) - \frac{1}{2}\,[\mathbf{C}(\mathbf{h}) + \mathbf{C}(-\mathbf{h})] \\ &= \ \mathbf{C}(0) - \mathbf{C}(\mathbf{h})\,, \ \text{if lag effect is ignored.} \end{aligned} \tag{3.43}$$

Chapter 4

Inference and Modeling

Once a random function model has been chosen, the next step consists of inferring its parameters from the available information. The focus of this chapter is on inference of the two first moments (mean, covariance) of the multivariate RF $Z(\mathbf{u})$, which are required by the interpolation (kriging) algorithms introduced in Chapters 5 and 6.

Section 4.1 addresses the problem of determining statistics representative of the study area and not only of the sample available. Theoretical and practical issues of modeling the spatial variability of continuous attributes are discussed in section 4.2. The discussion is extended to the modeling of cross correlations in section 4.3.

4.1 Statistical Inference

The inference process aims at estimating the parameters of the RF model from the sample information available over the study area. In contrast to Chapter 2, here the sample statistics are no longer population parameters since the sample is not considered exhaustive any more. The distinction between population and sample statistics is made clear by adding the ^ superscript to the latter.

The use of sample statistics as estimates of population parameters requires that the sample be representative of the underlying area or population. Such representativity can be achieved by carefully designing the sampling scheme; the interested reader should refer to Ripley (1981, p. 19–27) or Webster and Oliver (1990, p. 272–290) for a presentation of main sampling schemes. In this book, one considers the situation where data have already been collected, possibly with no statistical treatment in mind. The representativity of the sample should be questioned whenever the data are not spread evenly over the area, which is unfortunately often the case in earth sciences applications.

4.1.1   Preferential sampling

The sampling is said to be *preferential* whenever data locations are neither regularly nor randomly distributed over the study area. Several factors may cause specific subareas to be preferentially sampled:

1. Conditions of accessibility; fields bordering roads or farmlands are easier to sample than rugged terrain or dense forests.

2. Expected attribute values; sampling is often denser in areas that are deemed critical, for example, where high grades or large metal concentrations are expected to occur.

3. Sampling strategy; clustered locations may have been sampled to characterize short-range variability.

Preferential sampling should always be clearly documented by surveyors, since it may skew the results of any exploratory data analysis. For the Jura data set, 38 clusters[1] of 5 locations each were sampled to characterize the small-scale variability (see section 2.3.1). Furthermore, the sampling design was such that the clusters are random stratified, which explains their somewhat uneven distribution over the area (Figure 4.1).

Even if high- or low-valued areas were not purposely targeted, any preferential sampling is likely to impact sample statistics. For example, lead

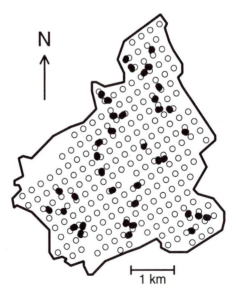

Figure 4.1: Distribution of clustered data (black dots) over the study area.

---

[1] Although the actual sampling was nested, non-gridded locations are hereafter considered as clustered for illustration.

concentrations in soils bordering roads may be larger than in open fields because of road traffic pollution. Similarly, oversampling of accessible farmlands relative to forest soils, which have low levels of contamination, would lead to overestimation of the average concentrations of most metals.

Preferential sampling of specific classes of values can be detected by comparing the quantiles of the distributions of clustered versus non-clustered (gridded) data. Recall that the p-quantile value of a distribution, q_p, is the value below which a proportion p of the data falls; that is, q_p is such that $F(q_p) = p$. Let $q_p^{(g)}$ and $q_p^{(c)}$ denote the p-quantiles of the distributions of gridded and clustered data, respectively. Similar distributions should have similar quantiles; thus the graph of $q_p^{(g)}$ versus $q_p^{(c)}$, called a Q-Q plot, should appear as the straight line $q_p^{(g)} = q_p^{(c)} \ \forall \ p$.

The Jura sample set was split into $n_c = 152$ clustered data (38 clusters of 4 locations each) and $n_g = 107$ gridded data ($n_g = 207$ for nickel and zinc concentrations). Figure 4.2 (left graphs) shows the Q-Q plots for the three metals with widespread contamination (Cd, Cu, and Pb) and for chromium. All metals show discrepancies of the upper tails of the two distributions: the large quantile values of the clustered data are larger than the corresponding quantile values of the gridded data (dots are above the 45° line). Such Q-Q plots indicate that the larger metal concentrations are measured preferentially at clustered locations. This difference is caused by preferential location of the clusters in farmland or grassland with higher metal concentrations (see conditional means in Table 2.4, page 18). Indeed, only 8.3% of the clustered data have been collected under forest, whereas that proportion is 18.4% for gridded data (global proportion of forest is 12.8%). This difference, however, is not pronounced enough to significantly affect statistics such as mean concentration or percentage of contaminated locations (Table 4.1). Note that the standard deviation of the clustered data is also larger than that of the gridded data.

4.1.2 Histogram declustering

Assume that the continuous attribute z has been preferentially sampled in high- or low-valued areas. Thus, the equal-weighted linear average $\hat{m}$ of the n data $z(\mathbf{u}_\alpha)$ is a biased estimate of the average value of z over the area $\mathcal{A}$, with

$$\hat{m} = \frac{1}{n} \sum_{\alpha=1}^{n} z(\mathbf{u}_\alpha) \tag{4.1}$$

More generally, the sample marginal distribution $\hat{F}(z)$ is not representative of the distribution of z-values over $\mathcal{A}$, with

$$\hat{F}(z) = \frac{1}{n} \sum_{\alpha=1}^{n} i(\mathbf{u}_\alpha; z) \tag{4.2}$$

where the indicator datum $i(\mathbf{u}_\alpha; z)$ is defined as in equation (2.19).

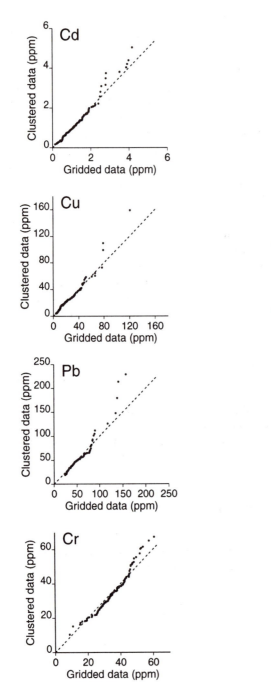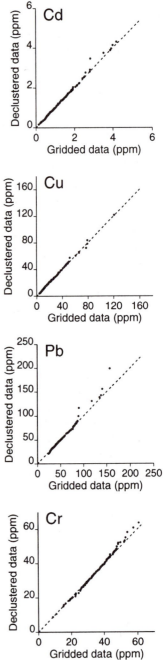

Figure 4.2: Q-Q plots of the distribution of gridded data versus that of clustered data (left graphs) and versus the declustered sample distribution (cell-declustering technique; right graphs). Similar distributions should plot on the 45° line.

Table 4.1: Statistics of clustered and gridded data for the seven heavy metals (units=ppm).

	Cd	Cu	Pb	Co	Cr	Ni	Zn
Mean							
Clustered data	1.30	24.4	54.2	9.3	35.0	19.1	75.3
Gridded data	1.31	22.9	53.5	9.3	35.2	20.7	76.3
Std deviation							
Clustered data	0.96	21.9	33.1	3.7	11.6	8.6	31.2
Gridded data	0.85	18.9	24.8	3.3	10.0	7.6	30.4
% contam.							
Clustered data	64.1	9.0	41.4	0.0	0.0	0.7	0.7
Gridded data	67.5	7.9	43.0	0.0	0.0	0.0	0.5

One procedure to correct for preferential sampling consists of retaining only the regularly spaced data. This approach is appropriate for data sets that include enough gridded data for reliable inference, for example, the Jura data set with more than 100 gridded data.

When data sparsity does not allow one to ignore the clustered values, the equal weights $1/n$ in expressions (4.1) and (4.2) should be replaced by weights that account for data clustering. Intuitively, data in densely sampled areas should receive less weight than those in sparsely sampled areas. Such weighting amounts to "declustering" the data. Two commonly used declustering techniques are the polygonal method (Isaaks and Srivastava, 1989, p. 238–239) and the cell-declustering method (Journel, 1983; Deutsch, 1989).

The polygonal method

The polygonal method first delineates the polygon of influence of each datum location $\mathbf{u}_\alpha$, that is, the area constituted by all locations $\mathbf{u} \in \mathcal{A}$ closer to $\mathbf{u}_\alpha$ than to any other datum location. The area of the polygon centered at location $\mathbf{u}_\alpha$ is then used as a declustering weight for datum value $z(\mathbf{u}_\alpha)$:

$$\widehat{m} = \frac{1}{|\mathcal{A}|} \sum_{\alpha=1}^{n} \omega_\alpha \cdot z(\mathbf{u}_\alpha) \qquad (4.3)$$

$$\widehat{F}(z) = \frac{1}{|\mathcal{A}|} \sum_{\alpha=1}^{n} \omega_\alpha \cdot i(\mathbf{u}_\alpha; z) \qquad (4.4)$$

where ω_α is the measure (area) of the polygon centered at $\mathbf{u}_\alpha$, and $|\mathcal{A}| = \int_{\mathcal{A}} du = \sum_\alpha \omega_\alpha$ is the measure of $\mathcal{A}$. For example, Figure 4.3 (top graph) shows the polygons of influence for seven data locations. Clustered data with

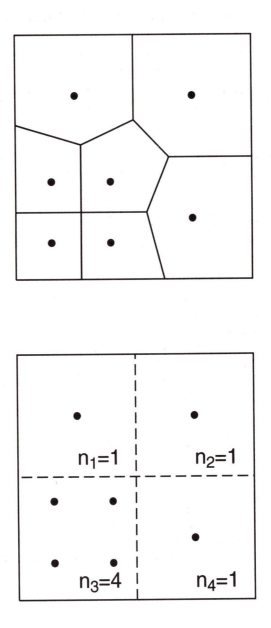

Figure 4.3: Two techniques that correct for the clustering of data locations (black dots): the polygonal method (top graph) and the cell-declustering technique (bottom graph). The declustering weights are proportional to the area of the polygon of influence of each datum in the first case; they are inversely proportional to the number n_b, $b = 1, \ldots, 4$, of data that fall within the same cell in the second case.

small polygons of influence receive less weight than isolated locations with large polygons of influence.

The cell-declustering technique

The cell-declustering approach calls for dividing the study area $\mathcal{A}$ into rectangular cells, and counting the number B of cells that contains at least one datum and the number n_b of data falling within each cell b. Each datum location $\mathbf{u}_\alpha$ then receives a weight $\lambda_\alpha = 1/(B \cdot n_b)$, which gives more importance to isolated locations:

$$\widehat{m} = \sum_{\alpha=1}^{n} \lambda_\alpha \cdot z(\mathbf{u}_\alpha) = \frac{1}{B} \sum_{b=1}^{B} \widehat{m}_b \qquad (4.5)$$

$$\widehat{F}(z) = \sum_{\alpha=1}^{n} \lambda_\alpha \cdot i(\mathbf{u}_\alpha; z) = \frac{1}{B} \sum_{b=1}^{B} \widehat{F}_b(z) \qquad (4.6)$$

where $\widehat{m}_b$ and $\widehat{F}_b(z)$ are the equal-weighted mean (4.1) and equal-weighted cumulative distribution (4.2) of z-values within cell b. Figure 4.3 (bottom graph) shows the seven data locations and four declustering cells ($B=4$). Each of the four clustered data values receives a weight $\lambda_\alpha = 1/16$ since they share the same cell ($n_b = 4$). The three other cells contain only one datum ($n_b = 1$), hence that isolated datum receives a weight of $1/4$. The total weight sums to 1 as it should.

Two key parameters of the cell-declustering technique are the cell size and the location (origin, orientation) of the grid. Clusters are frequently added on an underlying pseudo-regular grid. A natural cell size would then be the spacing of the grid: $\sqrt{|\mathcal{A}|/n_g}$, where n_g is the number of gridded data. Provided data are regularly spaced, the center of the cells should correspond to grid nodes.

When the sampling pattern does not suggest a natural cell size, several cell sizes and origins must be tried. The combination that yields the smallest or largest declustered mean is retained according to whether the high- or low-valued areas were preferentially sampled. To avoid erratic results caused by extreme values falling into specific cells, it is useful to average results for several different grid origins for each cell size.

The Jura data were declustered using square cells of 250 m. The sample grid was rotated to be parallel to the N-S direction, then each cell was centered on a grid node. The declustered distribution is compared with the distribution of the 107 gridded data using a Q-Q plot (Figure 4.2, right graphs). Using only the gridded data or applying the cell declustering technique provide similar results: both distributions plot close to the 45° line with similar statistics.

Remarks

1. The cell declustering technique reverts to equal-weighted estimators of

type (4.1) or (4.2) if the cell size is very small (each cell contains at most one datum) or very large (all data fall into the same cell).

2. By analogy with the mean, the declustered variance is computed as

$$\widehat{\sigma}^2 = \sum_{\alpha=1}^{n} \lambda_\alpha \cdot [z(\mathbf{u}_\alpha) - \widehat{m}]^2$$

where $\widehat{m}$ is the declustered mean, and λ_α are the corresponding declustering weights.

4.1.3   Semivariogram inference

The sample semivariogram $\widehat{\gamma}(\mathbf{h})$ is computed as

$$\widehat{\gamma}(\mathbf{h}) = \frac{1}{2N(\mathbf{h})} \sum_{\alpha=1}^{N(\mathbf{h})} [z(\mathbf{u}_\alpha) - z(\mathbf{u}_\alpha + \mathbf{h})]^2 \qquad (4.7)$$

where $N(\mathbf{h})$ is the number of pairs of data locations a vector $\mathbf{h}$ apart. In section 2.3.3, several techniques (data transformation, h-scattergram "cleaning") for reducing the influence of extreme data on sample semivariogram values were discussed. Like other summary statistics, $\widehat{\gamma}(\mathbf{h})$ values are also sensitive to clustering of data values, particularly when such clustering is combined with a proportional effect.

Proportional effect

Most often, the local variability of data changes across the study area, a feature known as *heteroscedasticity*. The *proportional effect* (Journel and Huijbregts, 1978, p. 186–189) is a particular form of heteroscedasticity where the local variance of data is related to their local mean. For positively skewed distributions, the local variance increases with the local mean (direct proportional effect). Conversely, if the distribution is negatively skewed, larger variances generally correspond to smaller means (inverse proportional effect).

A proportional effect can be detected from a scatterplot of local means versus local variances as calculated from moving window statistics. The area is divided into windows of equal size, and the mean and variance are computed within each window. Each window should include enough data for reliable inference, yet there should be enough windows to detect any spatial trend. For small data sets, the windows may have to overlap.

Fourteen non-overlapping 1 km × 1 km windows, each including between 10 and 28 data values, were defined over the study area. The positively skewed variables (Cd, Cu, Pb, Zn) show a clear direct relation between local average concentrations in metal and local variance (Figure 4.4). There is no significant proportional effect for the other metals.

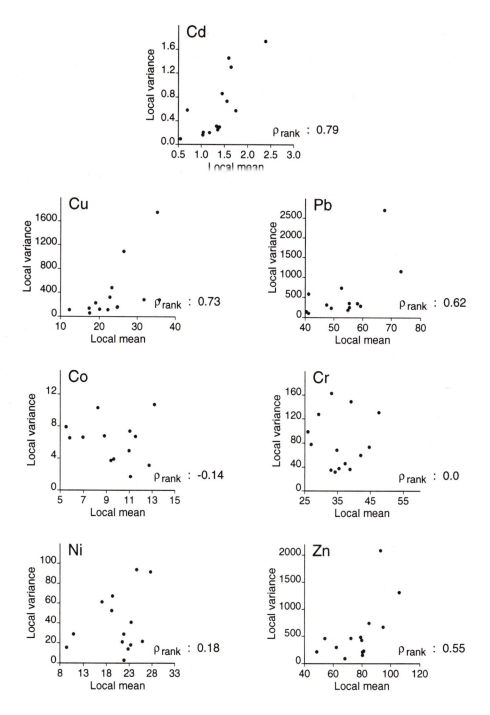

Figure 4.4: Plots of local variances versus local means computed from 1 km × 1 km moving windows overlaid across the study area.

Proportional effect and clustering

When combined with preferential sampling of high-valued areas, a proportional effect may render the sample semivariogram uninterpretable. Indeed, the clustering of high values entails that most data pairs that contribute to small lags come from high-valued areas. The corresponding lag mean is large and, because of the proportional effect, the lag variance is also large. As the distance $|\mathbf{h}|$ increases, the data that contribute to the lag become more representative of the entire area, and both lag mean and lag variance decrease. Such a trend in the lag variance results in overestimating the semivariogram value at short lags hence also its relative nugget effect. In the worst case, $\widehat{\gamma}(\mathbf{h})$ values at small lags may even appear higher than those at larger lags, giving the impression that the phenomenon is spatially unstructured or presents a hole effect (periodicity) (e.g., see Journel and Huijbregts, 1978, p. 168).

For the four attributes (Cd, Cu, Pb, and Zn) with proportional effect, the lag mean and lag variance are plotted as a function of $|\mathbf{h}|$ in Figure 4.5. Both statistics are standardized to their global values $\widehat{m}$ and $\widehat{\sigma}^2$, and they

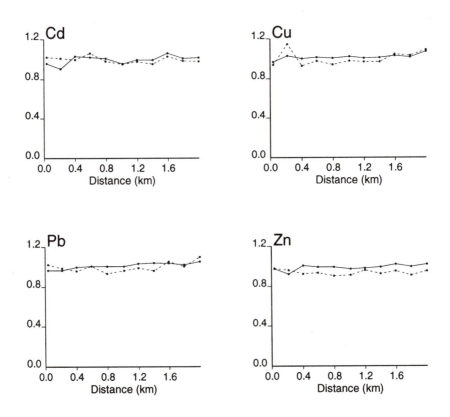

Figure 4.5: Lag mean (solid line) and lag variance (dashed line) as a function of the lag h. Both statistics are divided by their global values $\widehat{m}$ and $\widehat{\sigma}^2$.

fluctuate around the unit value with no obvious change with $|\mathbf{h}|$. Thus, for the Jura data, although there is a proportional effect (see Figure 4.4), it does not significantly affect the lag means and variances. If the impact on lag means were more significant, it could be corrected using only the regularly spaced data. However, this would amount to ignoring most of the critical short-range information. A better alternative consists of correcting the lag mean variation by computing relative semivariograms instead of the traditional semivariogram (4.7); see Isaaks and Srivastava (1989) p. 164–166; Srivastava and Parker (1989).

Relative semivariograms

The general relative semivariogram scales each semivariogram value by a function of the lag mean $\widehat{m}(\mathbf{h})$:

$$\widehat{\gamma}_{GR}(\mathbf{h}) = \frac{\widehat{\gamma}(\mathbf{h})}{f(\widehat{m}(\mathbf{h}))} \tag{4.8}$$

where $\widehat{m}(\mathbf{h})$ is the mean of all data values that contribute to the lag $\mathbf{h}$, that is, the average of the tail and head means as defined in expression (2.14):

$$\widehat{m}(\mathbf{h}) = \frac{1}{2N(\mathbf{h})} \sum_{\alpha=1}^{N(\mathbf{h})} [z(\mathbf{u}_\alpha) + z(\mathbf{u}_\alpha + \mathbf{h})] = \frac{m_{-\mathbf{h}} + m_{+\mathbf{h}}}{2}$$

The function f may be determined from the scattergram of local means versus local variances (Figure 4.4). For positively skewed distributions, that function is usually taken as the squared lag mean $[\widehat{m}(\mathbf{h})]^2$.

The pairwise relative semivariogram scales each $\mathbf{h}$-increment by the squared average of the head and tail values:

$$\widehat{\gamma}_{PR}(\mathbf{h}) = \frac{1}{2N(\mathbf{h})} \sum_{\alpha=1}^{N(\mathbf{h})} \frac{[z(\mathbf{u}_\alpha) - z(\mathbf{u}_\alpha + \mathbf{h})]^2}{\left[\dfrac{z(\mathbf{u}_\alpha) + z(\mathbf{u}_\alpha + \mathbf{h})}{2}\right]^2} \tag{4.9}$$

The measure $\widehat{\gamma}_{PR}(\mathbf{h})$ reduces directly the influence of each large value in the computation of the semivariogram.

Because of the denominators in expressions (4.8) and (4.9) utilization of relative semivariograms should be limited to strictly positive variables.

Remarks

1. The semivariogram behavior near the origin is important for later interpolation algorithms. As was done for the Jura data, it is good practice to sample a few randomly located clustered locations to allow inference of semivariogram values at short distances. Data clusters must,

however, be located with care since their unfortunate location in preferentially low- or high-valued areas may affect the representativity of
the sample statistics.

2. Relative semivariograms are not substitutes for the traditional semivariogram $\widehat{\gamma}(\mathbf{h})$ in the sense that kriging requires a model for the traditional
   semivariogram. However, these robust measures may provide a clearer
   description of the spatial continuity, revealing ranges and anisotropy
   whenever overestimation of the relative nugget effect renders the traditional semivariogram erratic. Such information supplements that provided by the rodogram or madogram on large-scale features (range,
   anisotropy).

3. All robust measures, such as relative semivariograms, madogram, or
   rodogram, consider only one attribute at a time. The sensitivity of cross
   semivariograms to proportional effects and clustering of high values has
   rarely been investigated. Intuitively, this sensitivity would depend on
   the sign of the correlation between attributes. When two variables are
   positively correlated, they are likely to show similar proportional effects and impacts of data clustering. Their cross semivariogram would
   then combine the adverse effects shown on the two direct semivariograms. Conversely, cross semivariograms of negatively correlated variables would be less sensitive to preferential clustering since small lag
   means of one variable might be balanced by large lag means of the other.

4.1.4 Covariance inference

The sample covariance $\widehat{C}(\mathbf{h})$ is computed as

$$\widehat{C}(\mathbf{h}) = \frac{1}{N(\mathbf{h})} \sum_{\alpha=1}^{N(\mathbf{h})} z(\mathbf{u}_\alpha) \cdot z(\mathbf{u}_\alpha + \mathbf{h}) - \widehat{m}_{-\mathbf{h}} \cdot \widehat{m}_{+\mathbf{h}} \qquad (4.10)$$

This statistic is called the *non-ergodic* covariance (Srivastava, 1987b) because
it does not assume that the means of head and tail values are identical to
the global mean. Because it accounts for changes in lag means, the non-
ergodic covariance estimate tends to be less sensitive than the traditional
semivariogram estimate (4.7) to both the proportional effect and clustering
of high values.

The non-ergodic correlogram $\widehat{\rho}(\mathbf{h})$ accounts for both lag means and lag
variances:

$$\widehat{\rho}(\mathbf{h}) = \frac{\widehat{C}(\mathbf{h})}{\widehat{\sigma}_{-\mathbf{h}} \cdot \widehat{\sigma}_{+\mathbf{h}}} \in [-1, +1] \qquad (4.11)$$

Srivastava and Parker (1989) have shown that the non-ergodic correlogram and general relative semivariogram are very resistant to a combination
of proportional effect and preferential sampling of high values. Similarly, because they account for lag means and lag variances, the cross covariance (2.25)

and cross correlogram (2.26) provide more robust cross correlation measures than the traditional cross semivariogram (2.28).

4.2 Modeling a Regionalization

Semivariogram or covariance inference provides a set of experimental values $\widehat{\gamma}(\mathbf{h}_k)$ or $\widehat{C}(\mathbf{h}_k)$ for a finite number of lags, $\mathbf{h}_k, k = 1, \ldots, K$, and directions. Continuous functions must be fitted to these experimental values so as to deduce semivariogram or covariance values for any possible lag $\mathbf{h}$ required by interpolation algorithms, and also to smooth out sample fluctuations.

In this section, the conditions that any semivariogram or covariance model must satisfy are first established, and the linear model of regionalization is introduced. Practical issues of modeling are addressed.

4.2.1 Permissible models

The positive definite condition

Let $\{Z(\mathbf{u}), \mathbf{u} \in \mathcal{A}\}$ be a stationary RF with a covariance function $C(\mathbf{h})$. The variance of any finite linear combination Y of random variables $Z(\mathbf{u}_\alpha)$, $\mathbf{u}_\alpha \in \mathcal{A}$, is expressed as a linear combination of the covariance values and must be non-negative:

$$
\begin{aligned}
\operatorname{Var}\{Y\} &= \operatorname{Var}\left\{\sum_{\alpha=1}^{n} \lambda_\alpha Z(\mathbf{u}_\alpha)\right\} \\
&= \sum_{\alpha=1}^{n} \sum_{\beta=1}^{n} \lambda_\alpha \lambda_\beta C(\mathbf{u}_\alpha - \mathbf{u}_\beta) \geq 0 \quad (4.12)
\end{aligned}
$$

for any choice of n locations $\mathbf{u}_\alpha \in \mathcal{A}$ and any weights λ_α. To ensure that this variance is non-negative, the covariance model $C(\mathbf{h})$ must be positive definite.

Accounting for the relation $\gamma(\mathbf{h}) = C(0) - C(\mathbf{h})$, the variance (4.12) is rewritten in terms of the semivariogram model $\gamma(\mathbf{h})$:

$$
\operatorname{Var}\{Y\} = C(0) \sum_{\alpha=1}^{n} \lambda_\alpha \sum_{\beta=1}^{n} \lambda_\beta - \sum_{\alpha=1}^{n} \sum_{\beta=1}^{n} \lambda_\alpha \lambda_\beta \gamma(\mathbf{u}_\alpha - \mathbf{u}_\beta) \geq 0 \quad (4.13)
$$

Some semivariogram models, such as the power model (Figure 4.6, bottom graph), have no sill, hence no covariance counterpart. For such semivariogram models, the variance (4.13) can still be expressed in terms of the semivariogram model under the condition that the weights λ_α sum to zero:

$$
\sum_{\alpha=1}^{n} \lambda_\alpha = 0 \quad (4.14)
$$

This condition on the weights filters the variance term $C(0)$ from expression (4.13) which then becomes

$$\text{Var}\{Y\} = -\sum_{\alpha=1}^{n} \sum_{\beta=1}^{n} \lambda_\alpha \lambda_\beta \, \gamma(\mathbf{u}_\alpha - \mathbf{u}_\beta) \geq 0 \qquad (4.15)$$

Relations (4.14) and (4.15) show that, to ensure the non-negativity of the variance of Y, the semivariogram model $\gamma(\mathbf{h})$ must be conditionally negative definite, the condition being that the sum of the weights λ_α is zero.

Basic semivariogram models

To avoid having to test a posteriori the permissibility of a semivariogram model (Christakos, 1984), a common practice consists of using only linear combinations of basic models that are known to be permissible. The following are the five most frequently used basic models:

- Nugget effect model

$$g(h) = \begin{cases} 0 & \text{if } h = 0 \\ 1 & \text{otherwise} \end{cases} \qquad (4.16)$$

- Spherical model with range a

$$g(h) = \text{Sph}\left(\frac{h}{a}\right) = \begin{cases} 1.5 \cdot \frac{h}{a} - 0.5 \cdot \left(\frac{h}{a}\right)^3 & \text{if } h \leq a \\ 1 & \text{otherwise} \end{cases} \qquad (4.17)$$

- Exponential model with practical range a

$$g(h) = 1 - \exp\left(\frac{-3h}{a}\right) \qquad (4.18)$$

- Gaussian model with practical range a

$$g(h) = 1 - \exp\left(\frac{-3h^2}{a^2}\right) \qquad (4.19)$$

- Power model

$$g(h) = h^\omega \quad \text{with } 0 < \omega < 2 \qquad (4.20)$$

All these models are permissible in three dimensions and are for now expressed in their isotropic form, that is, as a function of the scalar $h = |\mathbf{h}|$. These five models can be classified according to their behavior at infinity (bounded, unbounded) and at the origin (linear, quadratic, discontinuous); see Figure 4.6.

The first four models are bounded, which means that a sill, here set to 1 (standardized models), is actually or practically reached at a distance a called the range:

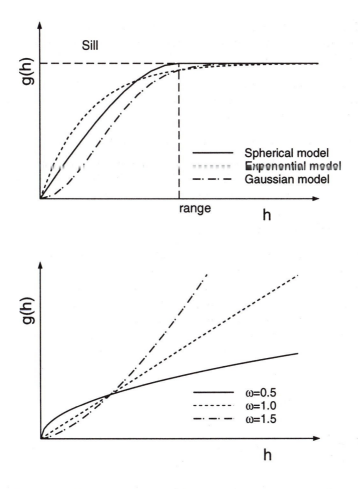

Figure 4.6: Bounded semivariogram models with the same practical range (top graph), and power models for different values of the parameter ω (bottom graph).

- For the nugget effect model, the sill is reached as soon as $h > 0$.

- The spherical model reaches its sill at distance a (*actual range*).

- The exponential and Gaussian models reach their sill asymptotically. A *practical range a* is defined as the distance at which the model value is at 95% of the sill.

Bounded models are also referred to as *transition* models, and their covariance counterpart is $c(h) = 1 - g(h)$. In contrast, the power model has no sill, hence no covariance counterpart.

Three types of behavior near the origin are distinguished:

1. Parabolic behavior, e.g., a Gaussian model. Such behavior is characteristic of highly regular phenomena such as topographic elevation of
   gently undulating hills.

2. Linear behavior, e.g., spherical or exponential model. Note that for
   the same practical range, the exponential model starts increasing faster
   than the spherical model (Figure 4.6, top graph).

3. Discontinuous behavior, e.g., nugget effect model.

The behavior near the origin of the power model changes with the value of
the parameter ω. It is linear for $\omega=1$ (linear model) and approaches parabolic
behavior as ω increases toward 2 (Figure 4.6, bottom graph).

4.2.2   Anisotropic models

A phenomenon is said to be anisotropic when its pattern of spatial variability
changes with direction. For example, nickel concentrations were found to vary
more continuously in the SW-NE direction corresponding to the elongation
of the geologic outcrops. Modeling anisotropy calls for functions that depend
on the vector $\mathbf{h}$ rather than on the distance $h = |\mathbf{h}|$ only. The following
presentation is limited to two-dimensional anisotropic models with vector
$\mathbf{h} = (h_x, h_y)^T$. Three-dimensional anisotropic models are discussed in Isaaks
and Srivastava (1989), Chapter 16.

Geometric anisotropy

An anisotropy is said to be geometric when:

1. the directional semivariograms (covariances) have the same shape and
   sill but different range values, and

2. the rose diagram of ranges[2], that is, the plot of range values versus the
   azimuth θ of the direction, is an ellipse.

Consider, for example, the elliptical rose diagram of ranges in Figure 4.7 (top
right graph). The major axis of the ellipse corresponding to the direction
of maximum continuity forms an angle θ with the coordinate axis y (north
direction). By convention, the azimuth angle θ is measured in degrees clockwise from the y-axis. The minor direction of anisotropy is perpendicular to
the major axis of the ellipse and has an azimuth $\phi = \theta + 90°$.
   The two semivariograms computed in the directions of azimuth θ and ϕ are
fitted by spherical models with the same unit sill but different range values;
see Figure 4.7 (top left graph). The major and minor ranges of anisotropy,
a_θ and a_ϕ, are plotted as the major and minor radii of the ellipse. The
anisotropy factor λ is defined as the ratio of the minor range to the major
range, $\lambda = a_\phi/a_\theta < 1$.

---

[2]For linear semivariogram models $g(\mathbf{h}) = b \cdot \mathbf{h}$, it is the slope b that changes with
direction, and the rose diagram of the inverse of slope $1/b$ is considered.

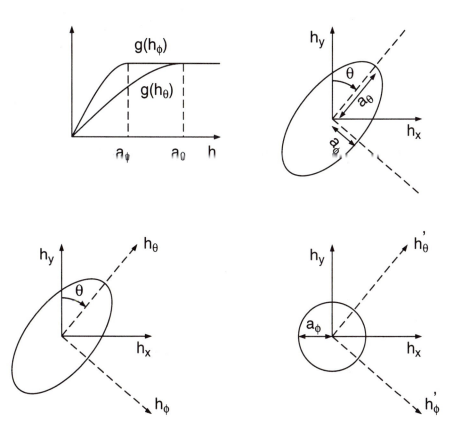

Figure 4.7: Example of geometric anisotropy. The major direction of anisotropy is at azimuth θ of axis y (top graphs). The elliptical rose plot of ranges is corrected by linear transform (rotation followed by rescaling) into the isotropic circle of radius a_ϕ (bottom graphs).

The anisotropy correction consists of transforming the vector of original coordinates $\mathbf{h} = (h_x, h_y)^T$ into a new vector $\mathbf{h}' = (h_\phi', h_\theta')^T$, so that the value of the anisotropic semivariogram model $g(\mathbf{h})$ identifies that of an isotropic model $g^t(|\mathbf{h}'|)$ in the new system of coordinates:

$$g(\mathbf{h}) \; = \; g^t(|\mathbf{h}'|) \quad \text{with} \quad |\mathbf{h}'| = \sqrt{h_\phi'^2 + h_\theta'^2} \tag{4.21}$$

where $g^t(.)$ is an isotropic model with a range equal to the minor range a_ϕ of anisotropy.

The coordinate transformation calls for two key parameters: the azimuth angle θ of the direction of maximum continuity and the anisotropy factor λ. The transformation proceeds in two steps:

1. The coordinate axes are rotated clockwise so as to identify the main axes of the ellipse (Figure 4.7, left bottom graph). The rotation angle corresponds to the azimuth angle θ. The new vector of coordinates, noted $(h_\phi, h_\theta)^T$, is deduced as

$$\mathbf{h}_\theta = \begin{bmatrix} h_\phi \\ h_\theta \end{bmatrix} = \begin{bmatrix} cos\theta & -sin\theta \\ sin\theta & cos\theta \end{bmatrix} \cdot \begin{bmatrix} h_x \\ h_y \end{bmatrix} \quad \text{or} \quad \mathbf{h}_\theta = \mathbf{Rot}_\theta \cdot \mathbf{h},$$

where $\mathbf{Rot}_\theta$ is the rotation matrix of angle θ.

2. The ellipse is then rescaled to a circle of radius equal to the minor range a_ϕ (Figure 4.7, right bottom graph). The rescaling of the new coordinates $(h_\phi, h_\theta)^T$ is written as

$$\mathbf{h}' = \begin{bmatrix} h'_\phi \\ h'_\theta \end{bmatrix} = \begin{bmatrix} 1 & 0 \\ 0 & \lambda \end{bmatrix} \cdot \begin{bmatrix} h_\phi \\ h_\theta \end{bmatrix} \quad \text{or} \quad \mathbf{h}' = \mathbf{D}_\lambda \cdot \mathbf{h}_\theta$$

where $\mathbf{D}_\lambda$ is the diagonal matrix of affinity.

The total transform is linear and is expressed as

$$\mathbf{h}' = \mathbf{D}_\lambda \cdot \mathbf{Rot}_\theta \cdot \mathbf{h} = \mathbf{A}_{\lambda,\theta} \cdot \mathbf{h} \tag{4.22}$$

where $\mathbf{A}_{\lambda,\theta} = \mathbf{D}_\lambda \cdot \mathbf{Rot}_\theta$.

Let $g(\mathbf{h})$ denote the anisotropic model shown in Figure 4.7; $g(\mathbf{h})$ is a spherical model of range a_θ (a_ϕ) in the major (minor) direction of anisotropy. Subject to the change of coordinates (4.22), the value of the anisotropic model $g(\mathbf{h})$ in various directions is equal to the value of the isotropic spherical model of range a_ϕ:

$$g(\mathbf{h}) = g^t(|\mathbf{h}'|) = \text{Sph}\left(\frac{|\mathbf{h}'|}{a_\phi}\right) \tag{4.23}$$

with

- $\text{Sph}\left(\frac{|\mathbf{h}'|}{a_\phi}\right) = \begin{cases} 1.5 \cdot \dfrac{|\mathbf{h}'|}{a_\phi} - 0.5 \cdot \left(\dfrac{|\mathbf{h}'|}{a_\phi}\right)^3 & \text{if } |\mathbf{h}'| \leq a_\phi \\ 1 & \text{otherwise} \end{cases}$

- $|\mathbf{h}'| = \sqrt{{h'_\phi}^2 + {h'_\theta}^2} = \sqrt{h_\phi^2 + (\lambda\, h_\theta)^2}$ and $\lambda = \dfrac{a_\phi}{a_\theta} < 1$.

The equality (4.23) is easily checked for the major and minor directions of anisotropy.

1. In the major direction of azimuth θ, the distance h_ϕ is equal to zero, hence $|\mathbf{h}'| = \frac{a_\phi}{a_\theta} h_\theta$. The model $g^t(|\mathbf{h}'|)$ resumes to a spherical model of range a_θ:

$$g(\mathbf{h}) = \text{Sph}\left(\frac{a_\phi}{a_\theta} \frac{h_\theta}{a_\phi}\right) = \text{Sph}\left(\frac{h_\theta}{a_\theta}\right)$$

as it should.

2. In the minor direction of azimuth ϕ, the distance h_θ is equal to zero, hence $|\mathbf{h}'| = h_\phi$. The model $g^i(|\mathbf{h}'|)$ resumes to a spherical model of range a_ϕ:

$$g(\mathbf{h}) \quad = \quad \text{Sph}\left(\frac{h_\phi}{a_\phi}\right)$$

as it should.

3. For any other directions such that $h_\phi \cdot h_\theta \neq 0$, say, the direction of azimuth ζ, the model $g^i(|\mathbf{h}'|)$ yields a spherical model with a range value a_ζ that would plot on the rose diagram of Figure 4.7 (top right graph), with $a_\phi < a_\zeta < a_\theta$.

Any isotropic model can be interpreted as a particular case of the geometric anisotropic model (4.21) where the anisotropy factor λ is equal to one ($a_\phi = a_\theta$). If $\lambda = 1$, it does not matter what θ is.

Zonal anisotropy

An anisotropy that involves sill values varying with direction is said to be zonal (Figure 4.8, right graph). The semivariogram in the direction of azimuth ϕ has a longer range a_ϕ and also a larger sill than in other directions. Such anisotropy can be modeled as the sum of an isotropic transition model $g_1(|\mathbf{h}|)$ and a "zonal" model $g_2(h_\phi)$, which depends only on the distance h_ϕ in the direction of greater variance:

$$g(\mathbf{h}) \quad = \quad g_1(|\mathbf{h}|) + g_2(h_\phi) \qquad (4.24)$$

where the model $g_2(.)$ has a range a_ϕ.

The component $g_2(h_\phi)$ can be seen as an extreme case of the geometric anisotropic model (4.21); its modeling proceeds in two steps:

1. The first step consists of rotating clockwise the coordinate axes so that the y-axis (north) identifies the direction of maximum continuity defined as the direction perpendicular to that of greater variance (highest sill). For the example in Figure 4.8 (left graph), the rotation angle is the azimuth angle $\theta = \phi - 90°$. As for the geometric anisotropy, the new vector of coordinates $(h_\phi, h_\theta)^T$ is computed as

$$\mathbf{h}_\theta \quad = \quad \begin{bmatrix} h_\phi \\ h_\theta \end{bmatrix} = \begin{bmatrix} cos\theta & -sin\theta \\ sin\theta & cos\theta \end{bmatrix} \cdot \begin{bmatrix} h_x \\ h_y \end{bmatrix}$$

2. The new axes are then rescaled so that the zonal model does not contribute to the direction of maximum continuity (azimuth θ). Such rescaling amounts to setting the range a_θ in that direction to infinity, hence the anisotropy factor $\lambda = a_\phi/a_\theta$ to zero:

$$\mathbf{h}' \quad = \quad \begin{bmatrix} h'_\phi \\ h'_\theta \end{bmatrix} = \begin{bmatrix} 1 & 0 \\ 0 & 0 \end{bmatrix} \cdot \begin{bmatrix} h_\phi \\ h_\theta \end{bmatrix}$$

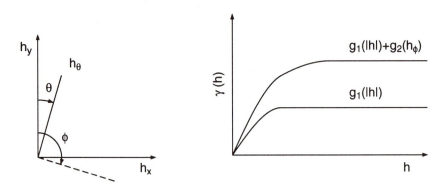

Figure 4.8: Example of zonal anisotropy in the direction of azimuth ϕ (greater variance). The anisotropic model consists of an isotropic model $g_1(|\mathbf{h}|)$ and a zonal component $g_2(h_\phi)$, which depends only on the distance h_ϕ.

Consider, for example, the zonal component $g_2(h_\phi)$ in Figure 4.8 (right graph). This model can be seen as an anisotropic model $g_2(\mathbf{h})$ that identifies a spherical model of range a_ϕ in the direction of greater variance but does not contribute in the perpendicular direction of azimuth θ. With the new system of coordinates $(h'_\phi, h'_\theta)^T$, the anisotropic model $g_2(\mathbf{h})$ can be expressed as an isotropic spherical model $g_2^t(|\mathbf{h}'|)$ with a range a_ϕ:

$$g_2(\mathbf{h}) \; = \; g_2^t(|\mathbf{h}'|) \; = \; \text{Sph}\left(\frac{|\mathbf{h}'|}{a_\phi}\right) \tag{4.25}$$

with

- $\text{Sph}\left(\dfrac{|\mathbf{h}'|}{a_\phi}\right) = \begin{cases} 1.5 \cdot \dfrac{|\mathbf{h}'|}{a_\phi} - 0.5 \cdot \left(\dfrac{|\mathbf{h}'|}{a_\phi}\right)^3 & \text{if } |\mathbf{h}'| \leq a_\phi \\ 1 & \text{otherwise} \end{cases}$

- $|\mathbf{h}'| = \sqrt{{h'_\phi}^2 + {h'_\theta}^2} = h_\phi$

The model has the following characteristics:

1. The model $g_2^t(|\mathbf{h}'|)$ is a spherical model with range a_ϕ in the direction of azimuth ϕ (greater variance):

$$g_2(\mathbf{h}) \; = \; \text{Sph}\left(\frac{h_\phi}{a_\phi}\right)$$

2. The model $g_2^t(|\mathbf{h}'|)$ does not contribute to the direction of azimuth θ (maximum continuity). Indeed, the distance h_ϕ is equal to zero in that direction and expression (4.25) becomes

$$g_2(\mathbf{h}) \; = \; \text{Sph}\left(\frac{0}{a_\phi}\right) \; = \; 0$$

3. In other directions, where both distances h_ϕ and h_θ are different from zero, the model $g_2^t(|\mathbf{h'}|)$ yields values that are intermediate between those obtained along the minor and major directions of anisotropy; for example, in the direction of azimuth ζ:

$$g_2(h_\theta) = 0 \leq g_2(h_\zeta) \leq g_2(h_\phi) \qquad \forall\, h_\phi = h_\zeta = h_\theta$$

The directional semivariogram $g(h_\zeta)$ would then plot between the two models of Figure 4.8.

4.2.3 The linear model of regionalization

In most situations, two or more basic models $g(\mathbf{h})$ or $c(\mathbf{h})$ must be combined (nested) to fit the shape of the experimental semivariogram or covariance function. However, not all combinations of permissible semivariogram or covariance models result in a permissible semivariogram or covariance function. The easiest way to build a permissible model consists in first building a random function. The covariance function or semivariogram of that random function is then, by definition, permissible.

The linear model of regionalization builds the random function $Z(\mathbf{u})$ as a linear combination of $(L + 1)$ independent random functions $Y^l(\mathbf{u})$, each with zero mean and basic covariance function $c_l(\mathbf{h})$:

$$Z(\mathbf{u}) = \sum_{l=0}^{L} a^l Y^l(\mathbf{u}) + m \tag{4.26}$$

with

- $\mathrm{E}\{Z(\mathbf{u})\} = m$

- $\mathrm{E}\{Y^l(\mathbf{u})\} = 0 \qquad \forall\, l$

- $\begin{aligned} \mathrm{Cov}\{Y^l(\mathbf{u}), Y^{l'}(\mathbf{u}+\mathbf{h})\} &= c_l(\mathbf{h}) \quad \text{if } l = l' \\ &= 0 \qquad \text{otherwise} \end{aligned}$

The latter condition expresses the mutual (two by two) independence of the $(L + 1)$ random functions $Y^l(\mathbf{u})$. The decomposition (4.26) entails that the covariance function of the RF $Z(\mathbf{u})$ is a linear combination of the $(L + 1)$ basic covariance functions $c_l(\mathbf{h})$:

$$\begin{aligned} C(\mathbf{h}) &= \mathrm{Cov}\{Z(\mathbf{u}), Z(\mathbf{u}+\mathbf{h})\} \\ &= \sum_{l=0}^{L} \sum_{l'=0}^{L} a^l\, a^{l'}\, \mathrm{Cov}\{Y^l(\mathbf{u}), Y^{l'}(\mathbf{u}+\mathbf{h})\} \\ &= \sum_{l=0}^{L} a^l\, a^l\, c_l(\mathbf{h}) \end{aligned}$$

since the $(L + 1)$ RFs $Y^l(\mathbf{u})$ are mutually independent.

The covariance model $C(\mathbf{h})$ is then written

$$C(\mathbf{h}) = \sum_{l=0}^{L} b^l\, c_l(\mathbf{h}) \quad \text{with} \quad b^l = (a^l)^2 \geq 0 \qquad (4.27)$$

where b^l is the positive sill of the basic covariance model $c_l(\mathbf{h})$. By convention, the superscript $l = 0$ denotes the nugget effect model. From expression (4.27), sufficient conditions for a linear covariance model $C(\mathbf{h})$ to be a permissible model of regionalization are:

1. the basic functions $c_l(\mathbf{h})$ are permissible covariance models, and

2. the sill b^l of each basic covariance model $c_l(\mathbf{h})$ is positive.

A similar development can be done in terms of semivariograms. Let $g_l(\mathbf{h})$ denote the semivariogram of the RF $Y^l(\mathbf{u})$, with the cross semivariogram between any two different RFs $Y^l(\mathbf{u})$ and $Y^{l'}(\mathbf{u})$ equal to zero:

$$E\left\{[Y^l(\mathbf{u}) - Y^l(\mathbf{u}+\mathbf{h})].[Y^{l'}(\mathbf{u}) - Y^{l'}(\mathbf{u}+\mathbf{h})]\right\} = g_l(\mathbf{h}) \quad \text{if} \quad l = l'$$
$$= 0 \qquad \text{otherwise}$$

The semivariogram model $\gamma(\mathbf{h})$ is then expressed as a positive linear combination of the basic semivariogram models $g_l(\mathbf{h})$:

$$\gamma(\mathbf{h}) = \frac{1}{2}E\{[Z(\mathbf{u}) - Z(\mathbf{u}+\mathbf{h})]^2\}$$
$$= \sum_{l=0}^{L} b^l\, g_l(\mathbf{h}) \quad \text{with} \quad b^l = (a^l)^2 \geq 0 \qquad (4.28)$$

where the positive coefficient b^l is the variance contribution of the corresponding basic semivariogram model $g_l(\mathbf{h})$.

Example

Consider the random function $Z(\mathbf{u})$ built as a linear combination of three independent RFs:

$$Z(\mathbf{u}) = 3 \cdot Y^0(\mathbf{u}) - 4 \cdot Y^1(\mathbf{u}) + 2 \cdot Y^2(\mathbf{u}) + 0,$$

with the three corresponding semivariogram models: a nugget effect, a spherical model of range $a = 1$, and a power model with $\omega = 1$. The three basic semivariogram models are isotropic and are shown in Figure 4.9 (left graph). According to relation (4.28) the semivariogram model is

$$\gamma(\mathbf{h}) = 9\, g_0(\mathbf{h}) + 16\, \text{Sph}(\frac{\mathbf{h}}{1}) + 4\,\mathbf{h}$$

where $g_0(\mathbf{h})$ is the nugget effect model (4.16). The model $\gamma(\mathbf{h})$ is, by construction, permissible and is depicted in Figure 4.9 (right graph).

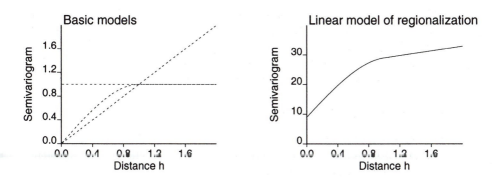

Figure 4.9: Three basic semivariogram models and their positive linear combination, which is a permissible semivariogram model.

Remarks

1. Each basic model of a linear model can be isotropic or can display either type of anisotropy. For example, the model in Figure 4.8 is a combination of an isotropic model and a zonal model. Accounting for the concept of linear rescaling of coordinates introduced in section 4.2.2, the general linear model is defined as

$$\gamma(\mathbf{h}) = \sum_{l=0}^{L} b^l \, g_l^t(|\mathbf{h}_l'|) \quad \text{with } b^l \geq 0, \qquad g_l^t(|\mathbf{h}_l'|) = g_l(\mathbf{h})$$

2. Under second-order stationarity, the coefficients b^l in both expressions (4.27) and (4.28) are identical, and their sum equals the a priori variance $C(0)$. Indeed, at $|\mathbf{h}| = 0$, the value of each basic model $c_l(\mathbf{h})$ is 1; then

$$C(0) = \sum_{l=0}^{L} b^l$$

4.2.4 The practice of modeling

Three key ingredients of the modeling process are:

1. experimental semivariogram or covariance values,

2. permissible semivariogram or covariance models, see section 4.2.1, and

3. utilization of ancillary information, such as provided by

   - physical knowledge of the area and phenomenon under study (e.g., directions of anisotropy, importance of measurement errors)

   - robust measures, such as the madogram (for evaluation of anisotropy directions, ratio, and range) or relative semivariograms

The art of modeling consists of capitalizing on these different sources of information to build a permissible model that captures the major spatial features of the attribute under study. The following discussion focuses on the semivariogram, which is the most frequently used structural tool. A similar approach can be adopted for covariance functions.

Automatic or visual modeling?

Too often, the modeling process is viewed as a mere exercise in fitting a curve to experimental values. The semivariogram is then modeled using statistical fitting procedures, which can be roughly classified into two categories:

- *Full black-box* procedures which involve an automatic choice and fitting of the model

- *Semi-automatic* procedures limited to the estimation of the parameters (sill, range) of models chosen by the user

Black-box procedures must be avoided because they cannot take into account ancillary information that is critical when sparse or preferential sampling makes the experimental semivariogram unreliable. Semi-automatic procedures may facilitate the determination of model parameters when the form of the experimental semivariogram is clear. However, such procedures contribute little to the actual modeling, since the most important decisions regarding number, type, and anisotropy of basic structures must be taken by the user. With the help of a good interactive graphical program (e.g., Chu, 1993), the user would do better than sophisticated fully automatic fitting procedures.

The modeling process relies critically on a series of user's *decisions*, which must be backed by experimental data or ancillary information, as follows:

1. whether to fit an isotropic or anisotropic model,

2. which number $(L + 1)$ and type (e.g., spherical, exponential) of basic semivariogram models $g_l(\mathbf{h})$ to use,

3. which parameters (sill, anisotropy, range or slope) for each basic semivariogram model to use.

Isotropic or anisotropic models?

The conventional approach for detecting anisotropy is to compare experimental semivariograms computed in several directions (Figure 4.10, top graphs). To decide whether geometric anisotropy is present, at least three directions must be considered. Directions of major and minor spatial continuity are often suggested by banding of large or small values on a location map, or by ancillary information, such as orientation of lithologic formations or prevailing

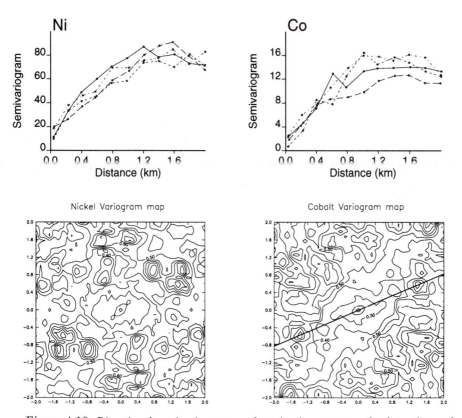

Figure 4.10: Directional semivariograms and semivariogram maps for detecting and identifying anisotropy directions. Both metals (Ni, Co) display an isotropic short-range structure, whereas there is a better long-range spatial continuity (smaller semivariogram values) for Co concentrations in the SW-NE direction (– –). This better continuity in the SW-NE direction (bold line) is apparent on the semivariogram map whose values are standardized to the sample variance.

wind direction for airborne pollution. In other cases, semivariograms should be computed for several sets of directions, such as $0°, 45°, 90°$, and $135°$ with, for example, an angle tolerance $\Delta\theta = \pm 22.5°$.

Semivariogram maps (Isaaks and Srivastava, 1989, p. 149–151; Chu, 1993) may greatly facilitate the detection of anisotropy directions. A semivariogram map is a plot of the experimental semivariogram values in the system of coordinates (h_x, h_y); see Figure 4.10 (bottom graphs). The center of the map corresponds to the origin of the semivariogram: $\gamma(0) = 0$. Any cross section appears as a traditional one-dimensional semivariogram. Semivariogram values are small near the origin $(0, 0)$ and increase with the distance from the origin. When the variation is isotropic, the increase is fairly similar in every direction; hence the map shows concentric contour lines (Figure 4.10, bottom left graph). Conversely, geometric anisotropy appears as elliptical contour lines whose major axis indicates the direction of maximum continuity (Figure 4.10, bottom right graph). Note that cobalt concentrations show a combination of short-range isotropy (small circles) and long-range geometric anisotropy (large ellipses) (Figure 4.10, right graphs).

The computation of a semivariogram map requires considering many directions and lags. Thus, such a representation is best suited for large gridded data sets. When data are sparse and irregularly spaced, large angular and lag tolerances are needed to fill in the semivariogram map, hence the spatial resolution of the map may be drastically reduced. Beware that the semivariogram map provides information only about the directions of major and minor spatial continuity. Anisotropic models still must be fitted to directional semivariograms.

An anisotropy that is not clearly apparent on experimental semivariograms nor backed by any qualitative information is better ignored. Conversely, strong prior qualitative information may lead one to adopt an anisotropic model even if data sparsity prevents seeing anisotropy from the experimental semivariograms.

How many and which basic structures?

One should avoid overfitting experimental semivariograms. The objective is to capture the major spatial features of the attribute, not to model any (possibly spurious) details of the sample semivariograms. When different nested models provide similar fits, one should select the simplest one. Where a spherical model suffices, there is no need to sum two exponential models (Figure 4.11). The more complicated model usually does not lead to more accurate estimates.

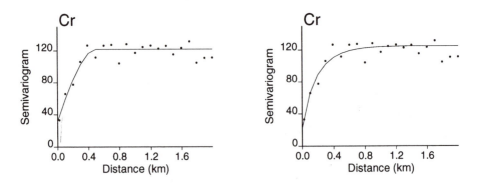

Figure 4.11: Two models that fit equally well the experimental semivariogram for Cr concentrations. The first model (left graph) is more parsimonious in the sense that only two basic structures are involved: a nugget effect and a spherical model. For interpolation, such a model would be preferred to the second model (right graph), which includes three basic structures: a nugget effect and two exponential models.

In earth sciences, there is rarely physical ground for choosing any particular basic model $g_l(\mathbf{h})$. For interpolation, a critical feature is the behavior of the semivariogram model at the origin. Models with a parabolic behavior at the origin should be used only for phenomena that are known to be highly continuous, for example, the surface of a water table. In most other cases, any model with a linear behavior at the origin would be appropriate.

Which parameter values?

The last step in the modeling process consists of determining the parameters (sill, range, anisotropy ratio) of the selected models. It bears repeating that a good interactive graphical program is more useful than any sophisticated statistical fitting procedure. The following tips should guide the user in the inference of the parameters of the semivariogram model.

1. Nugget effect
The nugget effect is usually determined by extrapolating the behavior of the first few semivariogram values to the vertical axis. The following should be taken into consideration:

1. The relative nugget effect on experimental semivariograms tends to increase with the lag tolerance (Figure 4.12) and with data sparsity. Typically, the relative nugget effect decreases as more and better data become available.

2. Where data are clustered and there is a proportional effect, the relative nugget effect is better inferred from relative semivariograms (see discussion in section 4.1.3).

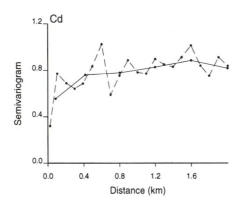

Figure 4.12: Effect of lag tolerance on the inference of the nugget effect. A large lag tolerance ($\Delta h = 200$ m, solid line) leads to overestimation of the relative nugget component, compare with dashed line ($\Delta h = 50$ m).

3. A Gaussian semivariogram model must **always** be combined with a small nugget effect to avoid numerical instabilities in interpolation algorithms (Posa, 1989) and generation of artifacts in interpolated maps (Wackernagel, 1995, p. 110–112). An alternative consists of using a power model of type (4.20).

4. Data sparsity or measurement errors or both may lead to noisy semivariograms that appear as pure nugget effect. Choosing a pure nugget effect model is an extreme modeling decision that precludes usage of kriging. Most natural phenomena are spatially structured. Noisy semivariograms typically reflect a lack of spatial resolution (the range of correlation is smaller than the shortest sampling interval) or the inadequacy of measurement devices. The objective is not to model the spatial continuity of data, but that of the attribute. A model of spatial continuity should be adopted whenever reliable qualitative information indicates that the phenomenon is indeed spatially structured.

5. The nugget effect is usually modeled as an isotropic component (4.16). The nugget effect may, however, relate to sources of variation that operate over distances smaller than the shortest sampling interval, say, h_{min}, and need not be isotropic. Based on ancillary information, the nugget effect could be modeled as an anisotropic spatial structure with ranges smaller than h_{min}.

Besides its contribution to the total variance, the nature of the nugget effect is of interest for later interpolation (see section 5.6). The nugget component of a semivariogram arises from measurement errors and spatial variation at distances smaller than the shortest sampling interval. Let us represent these two sources of variability as independent stationary random processes $\epsilon_i(\mathbf{u})$ and $Y_i^0(\mathbf{u})$. The RF $Z_i(\mathbf{u})$ is expressed as the sum of three independent processes:

$$Z_i(\mathbf{u}) \; = \; \epsilon_i(\mathbf{u}) \; + \; Y_i^0(\mathbf{u}) \; + \; Y_i^1(\mathbf{u})$$

where $Y_i^1(\mathbf{u})$ includes all sources of variability at large distances (greater than the shortest sampling interval h_{min}). The semivariogram of $Z_i(\mathbf{u})$, $\gamma_i(\mathbf{h})$, is then

$$\gamma_i(\mathbf{h}) \; = \; \sigma_{\epsilon_i}^2 + \; \gamma_i^0(\mathbf{h}) \; + \; \gamma_i^1(\mathbf{h})$$

where $\sigma_{\epsilon_i}^2$ is the measurement error variance, $\gamma_i^0(\mathbf{h}) = b_i^0 g_0(\mathbf{h})$ and $\gamma_i^1(\mathbf{h}) = b_i^1 g_1(\mathbf{h})$ are the non-standardized semivariograms of RFs $Y_i^0(\mathbf{u})$ and $Y_i^1(\mathbf{u})$. The range of the semivariogram $\gamma_i^0(\mathbf{h})$ is smaller than the shortest sampling interval, hence

$$\gamma_i(\mathbf{h}) \; = \; \underbrace{\sigma_{\epsilon_i}^2 \; + \; C_i^0(0)}_{D_i} + \gamma_i^1(\mathbf{h}) \qquad \forall \; |\mathbf{h}| \geq h_{min}$$

The micro-scale variance, $C_i^0(0)$, and the measurement error variance, $\sigma_{\epsilon_i}^2$, form a discontinuity D_i at the origin of the semivariogram $\gamma_i(\mathbf{h})$. Thus, micro-scale and measurement error components cannot be distinguished in practice unless the measurement error is evaluated separately, e.g., by replicating the measurement process.

If a second attribute z_j has been measured at the same locations as z_i, the relative contribution of measurement errors to the nugget effect on $\gamma_i(\mathbf{h})$ may be deduced from the relative nugget effect on the cross semivariogram $\gamma_{ij}(\mathbf{h})$. Indeed, the cross semivariogram $\gamma_{ij}(\mathbf{h})$ is expressed as

$$\gamma_{ij}(\mathbf{h}) \;=\; \mathrm{Cov}\{\epsilon_i(\mathbf{u}), \epsilon_j(\mathbf{u})\} + \mathrm{Cov}\{Y_i^0(\mathbf{u}), Y_j^0(\mathbf{u})\} + \gamma_{ij}^1(\mathbf{h}) \quad \forall \, |\mathbf{h}| \geq h_{min}$$

Under the assumption that errors associated with the measurement of different attribute values are independent, i.e., $\mathrm{Cov}\{\epsilon_i(\mathbf{u}), \epsilon_j(\mathbf{u})\} = 0$, the nugget effect on the cross semivariogram is due only to micro-scale variation common to both variables:

$$\gamma_{ij}(\mathbf{h}) \;=\; \mathrm{Cov}\{Y_i^0(\mathbf{u}), Y_j^0(\mathbf{u})\} + \gamma_{ij}^1(\mathbf{h}) \quad \forall \, |\mathbf{h}| \geq h_{min}$$

A large nugget effect on the cross semivariogram would suggest that the nugget effect on direct semivariograms is due mainly to micro-scale variation common to the two variables (Goovaerts and Chiang, 1993). Conversely, a small nugget effect would indicate that measurement errors contribute much to the nugget effect on the direct semivariograms or that micro-scale variations of both attributes are independent.

Whenever measurement error and micro-scale components can be distinguished, the latter should be modeled using a transition model with a finite range smaller than the shortest sampling interval.

2. Sill values
The contribution of non-nugget structures to experimental semivariograms is generally much easier to infer. An unreliable practice consists of forcing the sum of the $(L + 1)$ sills b^l to equal the sample variance $\hat{\sigma}^2$. The sill of the semivariogram, if it exists, is usually not equal to the sample variance; see Journel and Huijbregts (1978, p. 67), Barnes (1991).

3. Ranges
The range of a transition model is easy to infer from well-behaved experimental semivariograms; see, for example, the Cr semivariogram of Figure 4.11 (page 100). When data sparsity or outliers make the traditional semivariogram erratic, large-scale features (range and anisotropy) may be better inferred from measures such as the madogram or relative semivariograms (see section 2.3.3). The range of the experimental semivariogram typically tends to increase as more and better data become available.

4. Anisotropy parameters

Modeling a geometric anisotropy proceeds in two steps:

1. The minor and major ranges (or slopes) are first inferred from a series of experimental directional semivariograms (Figure 4.13, top graphs).

2. Model values are calculated in other directions and are checked against the experimental semivariograms computed in the same directions (Figure 4.13, bottom graphs).

Anisotropy ratios tend to be underestimated whenever directional semivariograms with large angular tolerances are computed. Indeed, pooling data pairs of different directions reduces the discrepancy between the most continuous and less continuous directions. In Figure 4.13, the angular tolerance was reduced to $\pm 10°$ to better capture the anisotropy. If data sparsity prohibits using small angular tolerances, one may decide arbitrarily to increase the anisotropy ratio inferred from the experimental directional semivariograms.

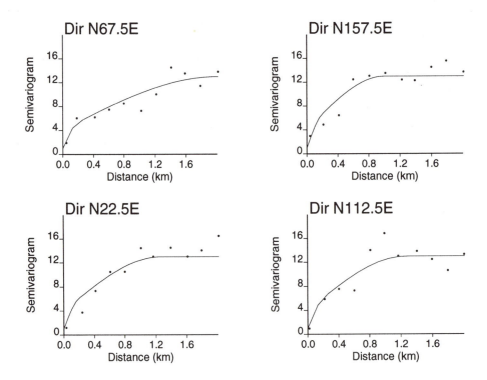

Figure 4.13: Experimental directional Co semivariograms and the geometric anisotropy model fitted. The major and minor directions of anisotropy are N67.5E and N157.5E.

How good is the model?

There is always uncertainty attached to the parameters of the semivariogram model: many models can match equally well the sample information; see, for example, Figure 4.11 (page 100). There is a tendency to justify the choice of a particular model using statistical criteria.

Weighted least-squares criteria
A criterion commonly used in conjunction with automatic fitting is the weighted sum of squares (WSS) of differences between experimental $\hat{\gamma}(\mathbf{h}_k)$ and model $\gamma(\mathbf{h}_k)$ semivariogram values:

$$\text{WSS} = \sum_{k=1}^{K} \omega(\mathbf{h}_k) \cdot [\hat{\gamma}(\mathbf{h}_k) - \gamma(\mathbf{h}_k)]^2 \qquad (4.29)$$

The weight $\omega(\mathbf{h}_k)$ given to each lag h_k is usually taken proportional to the number $N(\mathbf{h}_k)$ of data pairs that contribute to the estimate $\hat{\gamma}(\mathbf{h}_k)$. The implicit assumption is that the reliability of an experimental semivariogram value increases with statistical mass. An alternative that gives more weight to the first lags consists of dividing the number of data pairs by the squared model value: $N(\mathbf{h}_k)/[\gamma(\mathbf{h}_k)]^2$ (Cressie, 1985).

The WSS criterion is but a measure of the goodness of the fit; other measures can also be used. Recall that the objective is to capture the major spatial features of the attribute, not to build a semivariogram model that is the closest possible to experimental values. For example, a model of spatial continuity that accounts for reliable ancillary information should be preferred to a nugget-like model that closely fits a noisy experimental semivariogram.

The value of the WSS criterion depends on the number of lags considered and on the weights selected by the user. The model that yields the smallest WSS value need not be the same for various choices of parameters K or $\omega(\mathbf{h}_k)$. Therefore, the ranking of alternative models, though based on statistical criteria, still depends on prior user's decisions that are necessarily subjective.

Cross validation
Semivariogram modeling is rarely a goal per se. The ultimate objective is, for example, to estimate metal concentrations at unsampled locations. Cross validation allows one to compare the impact of different models on interpolation results; see Davis (1987), Journel (1987), Isaaks and Srivastava (1989, p. 351–368). The idea consists of removing one datum at a time from the data set and re-estimating this value from remaining data using the different semivariogram models. Interpolated and actual values are compared, and the model that yields the most accurate predictions is retained.

The use of cross validation to select semivariogram models suffers, however, from these severe restrictions:

- A rescaling of the semivariogram model does not influence kriging weights (see section 5.8.1). Thus, total sill values cannot be cross validated from re-estimation scores.

- Values of the semivariogram model for lags smaller than the shortest sampling interval do not intervene in interpolation algorithms. Hence, critical model parameters, such as relative nugget effect and the semivariogram behavior at the origin, cannot be cross validated.

- Sample data, particularly when they are scarce and preferentially located, may not be representative of the study area. Therefore, the model that produces the best cross-validated results may not yield the best predictions at unsampled locations.

- The re-estimation scores depend on the semivariogram model but also critically on implementation parameters related to the search strategy and the specific interpolation algorithm used. If the re-estimation scores of all models are deemed unsatisfactory, it is not clear what is faulty: the decision of stationarity, the semivariogram models, or the implementation of the algorithm. If the model is inadequate, then it is unclear which parameter should be changed.

Remarks

The "goodness" character of a model is elusive and cannot be measured by rigorous tests; there is no best semivariogram model. Rather than relying on an elusive objectivity provided by questionable statistical criteria, the user should decide on a model according to the following:

1. The user's experience and the information available; particular features of the experimental semivariogram may be deemed spurious and not worth modeling, whereas ancillary information may lead to model features that are not apparent on the experimental curves.

2. The objective of the study; modeling short-range features may not be relevant if one seeks a smooth representation of the long-range features.

A subjective decision that is clearly documented is preferable to a blind decision based on elusive tests. One must keep in mind that the choice of a semivariogram model is posterior and secondary to the fundamental choice of the RF model for modeling uncertainty and to the critical decision of stationarity. Such prior decisions are far more consequential for interpolation results than the use of an exponential model instead of a spherical model, or than setting the range value to 10 rather than 12.

Semivariogram models for metals

Figure 4.14 shows the experimental omnidirectional semivariograms and the models fitted. Two structures suffice to capture the major features of the Cr

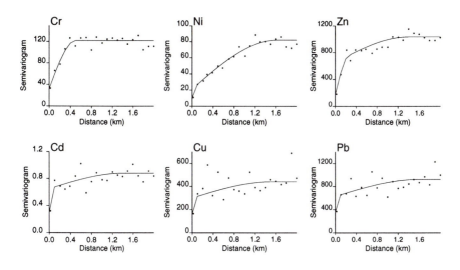

Figure 4.14: Experimental omnidirectional semivariograms for metal concentrations and the isotropic models fitted.

semivariogram. A third structure is needed to reproduce both short-range and long-range components of Ni and Zn semivariograms.

The semivariograms of the three other metals (Cd, Cu, and Pb) are more erratic and could be modeled either as a combination of a nugget effect and a long-range exponential model or as a combination of a nugget effect and two spherical models with short and long ranges. The latter model was chosen because the short-range structure could be related to the short-range distribution of human contaminant sources as featured by indicator semivariograms of land uses in Figure 2.17 (page 43).

4.3   Modeling a Coregionalization

Multivariate covariance or semivariogram inference provides a set of matrices $\widehat{\mathbf{C}}(\mathbf{h}_k) = [\widehat{C}_{ij}(\mathbf{h}_k)]$ and $\widehat{\mathbf{\Gamma}}(\mathbf{h}_k) = [\widehat{\gamma}_{ij}(\mathbf{h}_k)]$ for a finite number of lags, $\mathbf{h}_k, k = 1, \ldots, K$, and directions. As in the univariate case, modeling provides covariance or semivariogram values for any lag $\mathbf{h}$ as required by interpolation algorithms.

Modeling a coregionalization calls for inferring $N_v(N_v + 1)/2$ direct and cross semivariogram (covariance) models ignoring any lag effect. The difficulty does not lie in the number of models to infer, but in the fact that these models cannot be built independently from one another. In this section, the conditions that any matrix of semivariogram or covariance models must satisfy are first established. The linear model of coregionalization is then introduced, and practical issues of determining its parameters are discussed.

4.3.1   Permissible models

Let $\{Z_i(\mathbf{u}),\ i = 1, \ldots, N_v\}$ be a set of N_v intercorrelated random functions and $\{\mathbf{u}_\alpha,\ \alpha = 1, \ldots, n\}$ be a set of n data locations. The variance of any finite linear combination Y of the RVs $Z_i(\mathbf{u}_\alpha)$, $\mathbf{u}_\alpha \in \mathcal{A}$, $i = 1, \ldots, N_v$, can be expressed as a linear combination of cross covariance values:

$$
\begin{aligned}
\mathrm{Var}\,\{Y\} \;=\; & \mathrm{Var}\left\{ \sum_{i=1}^{N_v} \sum_{\alpha=1}^{n} \lambda_{\alpha i}\, Z_i(\mathbf{u}_\alpha) \right\} \\
=\; & \sum_{i=1}^{N_v} \sum_{j=1}^{N_v} \sum_{\alpha=1}^{n} \sum_{\beta=1}^{n} \lambda_{\alpha i}\, \lambda_{\beta j}\, C_{ij}(\mathbf{u}_\alpha - \mathbf{u}_\beta) \;\geq\; 0
\end{aligned}
$$

for any choice of n locations $\mathbf{u}_\alpha \in \mathcal{A}$ and any weights $\lambda_{\alpha i}$. Using matrix notation, the variance of Y is written as

$$
\mathrm{Var}\,\{Y\} \;=\; \sum_{\alpha=1}^{n} \sum_{\beta=1}^{n} \boldsymbol{\lambda}_\alpha^T\, \mathbf{C}(\mathbf{u}_\alpha - \mathbf{u}_\beta)\, \boldsymbol{\lambda}_\beta \;\geq\; 0 \tag{4.30}
$$

where $\boldsymbol{\lambda}_\alpha = [\lambda_{\alpha 1}, \ldots, \lambda_{\alpha N_v}]^T$ is an $N_v \times 1$ vector of weights $\lambda_{\alpha i}$, and $\mathbf{C}(\mathbf{u}_\alpha - \mathbf{u}_\beta) = [C_{ij}(\mathbf{u}_\alpha - \mathbf{u}_\beta)]$ is the $N_v \times N_v$ matrix of stationary cross covariances between any two RVs $Z_i(\mathbf{u}_\alpha)$ and $Z_j(\mathbf{u}_\beta)$. To ensure that the variance of Y is non-negative, the matrix of auto and cross covariance models $\mathbf{C}(\mathbf{h})$ must be positive semi-definite.

Accounting for the relation $\mathbf{C}(\mathbf{h}) = \mathbf{C}(0) - \boldsymbol{\Gamma}(\mathbf{h})$, the variance (4.30) is rewritten in terms of the matrix of semivariogram models $\boldsymbol{\Gamma}(\mathbf{h})$:

$$
\mathrm{Var}\,\{Y\} = \mathbf{C}(0) \sum_{\alpha=1}^{n} \boldsymbol{\lambda}_\alpha^T \sum_{\beta=1}^{n} \boldsymbol{\lambda}_\beta - \sum_{\alpha=1}^{n} \sum_{\beta=1}^{n} \boldsymbol{\lambda}_\alpha^T\, \boldsymbol{\Gamma}(\mathbf{h})\, \boldsymbol{\lambda}_\beta \;\geq\; 0 \tag{4.31}
$$

where $\mathbf{C}(0)$ is the variance–covariance matrix. As in the univariate case, there are semivariogram models that have no covariance counterpart. For such models, the variance of Y is defined on the condition that the vectors of weights $\boldsymbol{\lambda}_\alpha$ sum to the null vector, which allows the filtering of the term $\mathbf{C}(0)$ from expression (4.31):

$$
\mathrm{Var}\,\{Y\} = -\sum_{\alpha=1}^{n} \sum_{\beta=1}^{n} \boldsymbol{\lambda}_\alpha^T\, \boldsymbol{\Gamma}(\mathbf{h})\, \boldsymbol{\lambda}_\beta \quad \text{with} \quad \sum_{\alpha=1}^{n} \boldsymbol{\lambda}_\alpha = 0 \tag{4.32}
$$

Expression (4.32) shows that to ensure non-negativity of the variance of Y, the matrix of semivariogram models must be conditionally negative semi-definite, the condition being that the sum of the vectors $\boldsymbol{\lambda}_\alpha$ is the null vector.

4.3.2   The linear model of coregionalization

By analogy with the univariate case, the easiest way to build a permissible model of coregionalization consists in building a set of intercorrelated random functions $Z_i(\mathbf{u})$. The corresponding covariance function matrix $\mathbf{C}(\mathbf{h})$ or semivariogram matrix $\boldsymbol{\Gamma}(\mathbf{h})$ is then, by definition, permissible.

The linear model of coregionalization hereafter developed derives from one particular set of intercorrelated RFs $Z_i(\mathbf{u})$. Other sets have been built, custom-made for specific needs, resulting in different models of coregionalization; for example, see Zhu and Journel (1993) or the Markov model introduced subsequently in section 6.2.6.

The linear model of coregionalization builds each random function $Z_i(\mathbf{u})$ as a linear combination of independent random functions $Y_k^l(\mathbf{u})$, each with zero mean and basic covariance function $c_l(\mathbf{h})$ (Journel and Huijbregts, 1978, p. 172):

$$Z_i(\mathbf{u}) = \sum_{l=0}^{L} \sum_{k=1}^{n_l} a_{ik}^l Y_k^l(\mathbf{u}) + m_i \qquad \forall\, i \qquad (4.33)$$

with

- $E\{Z_i(\mathbf{u})\} = m_i$

- $E\{Y_k^l(\mathbf{u})\} = 0 \quad \forall\, k, l$

- $\mathrm{Cov}\{Y_k^l(\mathbf{u}), Y_{k'}^{l'}(\mathbf{u}+\mathbf{h})\} = c_l(\mathbf{h}) \quad \text{if } k = k' \text{ and } l = l'$
  $$= 0 \qquad \text{otherwise.}$$

The latter condition expresses the mutual (two by two) independence of the random functions $Y_k^l(\mathbf{u})$. Some of the RFs $Y_k^l(\mathbf{u})$, say, n_l $(n_l \leq N_v)$, may share the same covariance function $c_l(\mathbf{h})$, yet they remain independent of one another. The total number of independent RFs $Y_k^l(\mathbf{u})$ is thus $\sum_{l=0}^{L} n_l \leq (L+1) \cdot N_v$. The univariate decomposition (4.26) is but a particular case of the model (4.33) for $N_v = n_l = 1$.

The cross covariance between any two RVs $Z_i(\mathbf{u})$ and $Z_j(\mathbf{u}+\mathbf{h})$ can be expressed as a linear combination of cross covariances between any two RVs $Y_k^l(\mathbf{u})$ and $Y_{k'}^{l'}(\mathbf{u}+\mathbf{h})$:

$$\begin{aligned} C_{ij}(\mathbf{h}) &= \mathrm{Cov}\{Z_i(\mathbf{u}), Z_j(\mathbf{u}+\mathbf{h})\} \\ &= \sum_{l=0}^{L} \sum_{l'=0}^{L} \sum_{k=1}^{n_l} \sum_{k'=1}^{n_l} a_{ik}^l\, a_{jk'}^{l'}\, \mathrm{Cov}\{Y_k^l(\mathbf{u}), Y_{k'}^{l'}(\mathbf{u}+\mathbf{h})\} \quad (4.34) \end{aligned}$$

Because the RFs $Y_k^l(\mathbf{u})$ are mutually independent, expression (4.34) reduces to a linear combination of $(L+1)$ basic covariance models $c_l(\mathbf{h})$:

$$C_{ij}(\mathbf{h}) = \sum_{l=0}^{L} \sum_{k=1}^{n_l} a_{ik}^l\, a_{jk}^l\, c_l(\mathbf{h})$$

The linear model of coregionalization is the set of $N_v \times N_v$ auto and cross covariance models $C_{ij}(\mathbf{h})$ defined as

$$C_{ij}(\mathbf{h}) = \sum_{l=0}^{L} b_{ij}^l\, c_l(\mathbf{h}) \qquad \forall\, i, j \qquad (4.35)$$

where the sill b_{ij}^l of the basic covariance model $c_l(\mathbf{h})$ is

$$b_{ij}^l = \sum_{k=1}^{n_l} a_{ik}^l \, a_{jk}^l \qquad \forall \, l, i, j \tag{4.36}$$

By construction, the coefficients b_{ij}^l and b_{ji}^l are identical, hence the two cross covariance models $C_{ij}(\mathbf{h})$ and $C_{ji}(\mathbf{h})$ are the same. Furthermore, relation (4.36) is the general definition of a positive semi-definite $N_v \times N_v$ matrix $\mathbf{B}_l = \left[b_{ij}^l\right]$, called a coregionalization matrix.

The conditions sufficient for the matrix of functions $C_{ij}(\mathbf{h})$ defined in (4.35) to be a permissible model of coregionalization are:

1. the functions $c_l(\mathbf{h})$ are permissible covariance models, and

2. the $(L+1)$ coregionalization matrices $\mathbf{B}_l$ are all positive semi-definite.

The linear model of regionalization (4.27) is but a particular case of the linear model of coregionalization (4.35) for $N_v = 1$. As for the univariate case, each basic covariance model $c_l(\mathbf{h})$ may have its own pattern of anisotropy.

A similar development can be done in terms of semivariograms. Let $g_l(\mathbf{h})$ denote the semivariogram of the n_l RFs $Y_k^l(\mathbf{u})$, with the cross semivariogram between any two RFs $Y_k^l(\mathbf{u})$ and $Y_{k'}^{l'}(\mathbf{u})$ equal to zero:

$$
\begin{aligned}
\frac{1}{2} \, & \mathrm{E}\{[Y_k^l(\mathbf{u}) - Y_k^l(\mathbf{u}+\mathbf{h})] \\
& \cdot [Y_{k'}^{l'}(\mathbf{u}) - Y_{k'}^{l'}(\mathbf{u}+\mathbf{h})]\} \; = \; g_l(\mathbf{h}) \quad \text{if } k = k' \text{ and } l = l' \\
& \phantom{\cdot [Y_{k'}^{l'}(\mathbf{u}) - Y_{k'}^{l'}(\mathbf{u}+\mathbf{h})]\} \;} = \; 0 \qquad \text{otherwise}
\end{aligned}
$$

The linear model of coregionalization is then defined as the set of $N_v \times N_v$ direct and cross semivariogram models $\gamma_{ij}(\mathbf{h})$, such that

$$\gamma_{ij}(\mathbf{h}) = \sum_{l=0}^{L} b_{ij}^l \, g_l(\mathbf{h}) \qquad \forall \, i, j \tag{4.37}$$

where each function $g_l(\mathbf{h})$ is a permissible semivariogram model, and the $(L+1)$ matrices of coefficients b_{ij}^l, corresponding to the sill or slope of the model $g_l(\mathbf{h})$, are all positive semi-definite.

Example

Consider the two RFs $Z_1(\mathbf{u})$ and $Z_2(\mathbf{u})$ that are built as linear combinations of the same set of four independent RFs $\{Y_k^l(\mathbf{u}), k = 1, 2 \, , \, l = 0, 1\}$ and the two stationary means m_1 and m_2:

$$
\begin{aligned}
Z_1(\mathbf{u}) &= a_{11}^0 Y_1^0(\mathbf{u}) \; + \; a_{12}^0 Y_2^0(\mathbf{u}) \; + \; a_{11}^1 Y_1^1(\mathbf{u}) \; + \; a_{12}^1 Y_2^1(\mathbf{u}) \; + \; m_1 \\
Z_2(\mathbf{u}) &= a_{21}^0 Y_1^0(\mathbf{u}) \; + \; a_{22}^0 Y_2^0(\mathbf{u}) \; + \; a_{21}^1 Y_1^1(\mathbf{u}) \; + \; a_{22}^1 Y_2^1(\mathbf{u}) \; + \; m_2
\end{aligned}
$$

where $Y_1^0(\mathbf{u})$ and $Y_2^0(\mathbf{u})$ have the same semivariogram $g_0(\mathbf{h})$, and the two other independent RFs $Y_1^1(\mathbf{u})$ and $Y_2^1(\mathbf{u})$ share the same semivariogram $g_1(\mathbf{h})$. According to expression (4.37), the two direct semivariograms $\gamma_{11}(\mathbf{h})$, $\gamma_{22}(\mathbf{h})$ and the cross semivariogram $\gamma_{12}(\mathbf{h})$ are defined as weighted sums of the two basic semivariogram models $g_0(\mathbf{h})$ and $g_1(\mathbf{h})$:

$$\begin{aligned} \gamma_{11}(\mathbf{h}) &= b_{11}^0 \, g_0(\mathbf{h}) + b_{11}^1 \, g_1(\mathbf{h}) \\ \gamma_{22}(\mathbf{h}) &= b_{22}^0 \, g_0(\mathbf{h}) + b_{22}^1 \, g_1(\mathbf{h}) \\ \gamma_{12}(\mathbf{h}) &= b_{12}^0 \, g_0(\mathbf{h}) + b_{12}^1 \, g_1(\mathbf{h}) \end{aligned}$$

Consider the following numerical example:

$$\begin{aligned} Z_1(\mathbf{u}) &= 5 \cdot Y_1^0(\mathbf{u}) + 0 \cdot Y_2^0(\mathbf{u}) + 2 \cdot Y_1^1(\mathbf{u}) + 4 \cdot Y_2^1(\mathbf{u}) + 0 \\ Z_2(\mathbf{u}) &= 0 \cdot Y_1^0(\mathbf{u}) + 3 \cdot Y_2^0(\mathbf{u}) + 5 \cdot Y_1^1(\mathbf{u}) + 2 \cdot Y_2^1(\mathbf{u}) + 0 \\ g_0(\mathbf{h}) &= \text{nugget effect model} \\ g_1(\mathbf{h}) &= \text{Sph}\left(\frac{|\mathbf{h}|}{5}\right) \end{aligned}$$

Substituting the numerical values of coefficients a_{ik}^l into equation (4.36), one deduces the sills (contributions) of the three basic semivariogram models:

$$\begin{aligned} \gamma_{11}(\mathbf{h}) &= 25 \, g_0(\mathbf{h}) + 20 \, g_1(\mathbf{h}) \\ \gamma_{22}(\mathbf{h}) &= 9 \, g_0(\mathbf{h}) + 29 \, g_1(\mathbf{h}) \\ \gamma_{12}(\mathbf{h}) &= 18 \, g_1(\mathbf{h}) \end{aligned}$$

This linear model of coregionalization is, by construction, permissible and is displayed in Figure 4.15 (top graphs). Unlike the two semivariogram models, the cross semivariogram model does not include a nugget effect. As discussed in section 4.2.4, the nugget effect on a cross semivariogram is due only to micro-scale variability common to the two variables. In this example, the nugget components of $Z_1(\mathbf{u})$ and $Z_2(\mathbf{u})$ are built as $5Y_1^0(\mathbf{u})$ and $3Y_2^0(\mathbf{u})$, respectively. Because the two RFs $Y_1^0(\mathbf{u})$ and $Y_2^0(\mathbf{u})$ are independent, the two nugget components of $Z_1(\mathbf{u})$ and $Z_2(\mathbf{u})$ are independent, hence their cross covariance does not contribute to the nugget effect of the cross semivariogram.

Matrix notation

Using matrix notation, the RF vector $[Z_1(\mathbf{u}), Z_2(\mathbf{u})]^T$ is expressed as

$$\begin{bmatrix} Z_1(\mathbf{u}) \\ Z_2(\mathbf{u}) \end{bmatrix} = \begin{bmatrix} a_{11}^0 & a_{12}^0 \\ a_{21}^0 & a_{22}^0 \end{bmatrix} \cdot \begin{bmatrix} Y_1^0(\mathbf{u}) \\ Y_2^0(\mathbf{u}) \end{bmatrix} + \begin{bmatrix} a_{11}^1 & a_{12}^1 \\ a_{21}^1 & a_{22}^1 \end{bmatrix} \cdot \begin{bmatrix} Y_1^1(\mathbf{u}) \\ Y_2^1(\mathbf{u}) \end{bmatrix}$$
$$+ \begin{bmatrix} m_1 \\ m_2 \end{bmatrix}$$

$$\mathbf{Z}(\mathbf{u}) = \mathbf{A}_0 \, \mathbf{Y}_0(\mathbf{u}) + \mathbf{A}_1 \, \mathbf{Y}_1(\mathbf{u}) + \mathbf{m}$$

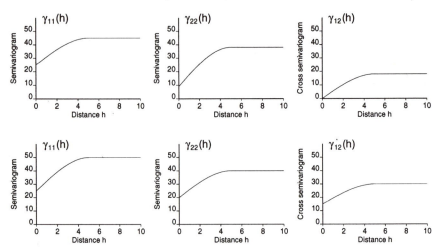

Figure 4.15: Examples of a linear model of coregionalization (top graphs) and an intrinsic coregionalization model (bottom graphs) in the bivariate case ($N_v = 2$). The intrinsic coregionalization model requires that each basic model (nugget effect, spherical model) contributes equally to the two direct semivariograms and to the cross semivariogram.

where $\mathbf{A}_0$ is the 2×2 matrix of coefficients a_{ik}^0, and $\mathbf{Y}_0(\mathbf{u})$ is the RF vector $[Y_1^0(\mathbf{u}), Y_2^0(\mathbf{u})]^T$. The corresponding linear model of coregionalization is

$$
\begin{bmatrix} \gamma_{11}(\mathbf{h}) & \gamma_{12}(\mathbf{h}) \\ \gamma_{21}(\mathbf{h}) & \gamma_{22}(\mathbf{h}) \end{bmatrix} = \begin{bmatrix} b_{11}^0 & b_{12}^0 \\ b_{21}^0 & b_{22}^0 \end{bmatrix} g_0(\mathbf{h}) + \begin{bmatrix} b_{11}^1 & b_{12}^1 \\ b_{21}^1 & b_{22}^1 \end{bmatrix} g_1(\mathbf{h})
$$

$$
\mathbf{\Gamma}(\mathbf{h}) = \mathbf{B}_0 \, g_0(\mathbf{h}) + \mathbf{B}_1 \, g_1(\mathbf{h})
$$

where $\mathbf{B}_0$ and $\mathbf{B}_1$ are the 2×2 coregionalization matrices that contain the coefficients of the basic semivariogram models $g_0(\mathbf{h})$ and $g_1(\mathbf{h})$.

The matrix formulation of the general decomposition (4.33) is written

$$
\mathbf{Z}(\mathbf{u}) = \sum_{l=0}^{L} \mathbf{A}_l \, \mathbf{Y}_l(\mathbf{u}) + \mathbf{m} \tag{4.38}
$$

with

- $E\{\mathbf{Z}(\mathbf{u})\} = \mathbf{m}$

- $E\{\mathbf{Y}_l(\mathbf{u})\} = 0 \quad \forall \, l$

- $E\{\mathbf{Y}_l(\mathbf{u}) \, [\mathbf{Y}_{l'}(\mathbf{u} + \mathbf{h})]^T\} = \mathbf{C}_Y^{ll'}(\mathbf{h}) = \delta_{ll'} \, \mathbf{I}_{n_l} \, c_l(\mathbf{h})$

where $\mathbf{Y}_l(\mathbf{u}) = [Y_1^l(\mathbf{u}), \ldots, Y_{n_l}^l(\mathbf{u})]^T$, $\delta_{ll'}$ is the Kronecker delta ($\delta_{ll'} = 1$ if $l = l'$ and $\delta_{ll'} = 0$ otherwise), and $\mathbf{I}_{n_l}$ is the $n_l \times n_l$ identity matrix. The latter condition expresses the mutual independence of the RFs $Y_k^l(\mathbf{u})$. The matrix

$\mathbf{C}_Y^{ll'}(\mathbf{h})$ is the $n_l \times n_l$ matrix of cross covariances $\mathrm{Cov}\{Y_k^l(\mathbf{u}), Y_{k'}^{l'}(\mathbf{u} + \mathbf{h})\}$. All the elements of $\mathbf{C}_Y^{ll'}(\mathbf{h})$ are zero whenever $l \neq l'$. For $l = l'$, the matrix $\mathbf{C}_Y^{ll}(\mathbf{h})$ is diagonal because the cross covariances between any two RFs $Y_k^l(\mathbf{u})$ and $Y_{k'}^l(\mathbf{u} + \mathbf{h})$ (off-diagonal elements) are zero.

The covariance function matrix of the multivariate random function $\mathbf{Z}(\mathbf{u})$ is then

$$
\begin{aligned}
\mathbf{C}(\mathbf{h}) &= \mathrm{E}\{[\sum_{l=0}^{L} \mathbf{A}_l \mathbf{Y}_l(\mathbf{u})] \cdot [\sum_{l'=0}^{L} \mathbf{A}_{l'} \mathbf{Y}_{l'}(\mathbf{u} + \mathbf{h})]^T\} \\
&= \sum_{l=0}^{L} \mathbf{A}_l \, \mathbf{C}_Y^{ll}(\mathbf{h}) \, \mathbf{A}_l^T \;=\; \sum_{l=0}^{L} \mathbf{B}_l \, c_l(\mathbf{h}) \quad \text{with} \quad \mathbf{B}_l \;=\; \mathbf{A}_l \, \mathbf{A}_l^T
\end{aligned}
$$

$$(4.39)$$

Similarly, the semivariogram matrix $\mathbf{\Gamma}(\mathbf{h})$ is expressed as

$$
\mathbf{\Gamma}(\mathbf{h}) \;=\; \sum_{l=0}^{L} \mathbf{B}_l \, g_l(\mathbf{h}) \tag{4.40}
$$

Under second-order stationarity, the sum of the $(L + 1)$ coregionalization matrices is equal to the variance–covariance matrix $\mathbf{C}(0)$. Indeed, at $|\mathbf{h}| = 0$, the value of each basic model $c_l(\mathbf{h})$ is 1, hence

$$
\mathbf{C}(0) \;=\; \sum_{l=0}^{L} \mathbf{B}_l
$$

The positive semi-definite condition

The linear model of coregionalization is very convenient because the conditions of permissibility are readily verified. Instead of checking the positive semi-definiteness of $\mathbf{C}(\mathbf{h})$ or the conditionally negative semi-definiteness of $\mathbf{\Gamma}(\mathbf{h})$, for all lags $\mathbf{h}$, we have only to verify that the $(L + 1)$ coregionalization matrices $\mathbf{B}_l$ are positive semi-definite.

A symmetric matrix is positive semi-definite if its determinant and all its principal minor determinants are non-negative. Consider, for example, the linear model of coregionalization for $N_v = 3$:

$$
\begin{bmatrix}
\gamma_{11}(\mathbf{h}) & \gamma_{12}(\mathbf{h}) & \gamma_{13}(\mathbf{h}) \\
\gamma_{21}(\mathbf{h}) & \gamma_{22}(\mathbf{h}) & \gamma_{23}(\mathbf{h}) \\
\gamma_{31}(\mathbf{h}) & \gamma_{32}(\mathbf{h}) & \gamma_{33}(\mathbf{h})
\end{bmatrix}
=
\sum_{l=0}^{L}
\begin{bmatrix}
b_{11}^l & b_{12}^l & b_{13}^l \\
b_{21}^l & b_{22}^l & b_{23}^l \\
b_{31}^l & b_{32}^l & b_{33}^l
\end{bmatrix}
g_l(\mathbf{h})
$$

Each coregionalization matrix $\mathbf{B}_l$ is positive semi-definite if the following seven inequalities are satisfied:

- All diagonal elements are non-negative:

$$
b_{11}^l \geq 0 \qquad b_{22}^l \geq 0 \qquad b_{33}^l \geq 0
$$

- All principal minor determinants of order 2 are non-negative:

$$\begin{vmatrix} b_{11}^l & b_{12}^l \\ b_{21}^l & b_{22}^l \end{vmatrix} = b_{11}^l\, b_{22}^l - [b_{21}^l]^2 \geq 0$$

$$\begin{vmatrix} b_{22}^l & b_{23}^l \\ b_{32}^l & b_{33}^l \end{vmatrix} = b_{22}^l\, b_{33}^l - [b_{32}^l]^2 \geq 0$$

$$\begin{vmatrix} b_{11}^l & b_{13}^l \\ b_{31}^l & b_{33}^l \end{vmatrix} = b_{11}^l\, b_{33}^l - [b_{31}^l]^2 \geq 0$$

- The determinant of order 3 is non-negative:

$$\begin{vmatrix} b_{11}^l & b_{12}^l & b_{13}^l \\ b_{21}^l & b_{22}^l & b_{23}^l \\ b_{31}^l & b_{32}^l & b_{33}^l \end{vmatrix} = b_{11}^l\,(b_{22}^l b_{33}^l - [b_{23}^l]^2) - b_{12}^l\,(b_{21}^l b_{33}^l - b_{31}^l b_{23}^l)$$

$$+ b_{13}^l\,(b_{21}^l b_{32}^l - b_{31}^l b_{22}^l) \geq 0$$

These inequalities provide four practical rules for choosing basic structures $g_l(\mathbf{h})$ in the linear model of coregionalization:

Rule 1: Every basic structure appearing on a cross semivariogram $\gamma_{ij}(\mathbf{h})$ must be present in both direct semivariogram models $\gamma_{ii}(\mathbf{h})$ and $\gamma_{jj}(\mathbf{h})$:

$$b_{ij}^l \neq 0 \implies b_{ii}^l \neq 0 \text{ and } b_{jj}^l \neq 0$$

Rule 2: If a basic structure $g_l(\mathbf{h})$ is absent on a direct semivariogram, it must be absent on all cross semivariograms involving this variable:

$$b_{ii}^l = 0 \implies b_{ij}^l = 0 \quad \forall\, j$$

Rule 3: Each direct or cross semivariogram model $\gamma_{ij}(\mathbf{h})$ need not include all $(L+1)$ basic structures:

$$b_{ij}^l \text{ may be equal to zero}, \qquad \forall\, i, j, l$$

Rule 4: There is no need for the structure $g_l(\mathbf{h})$ appearing on both direct semivariograms $\gamma_{ii}(\mathbf{h})$ and $\gamma_{jj}(\mathbf{h})$ to be present on the cross semivariogram $\gamma_{ij}(\mathbf{h})$:

$$b_{ii}^l \neq 0 \text{ and } b_{jj}^l \neq 0 \implies b_{ij}^l = 0 \text{ or } b_{ij}^l \neq 0$$

Similar rules apply for covariance models.

The intrinsic coregionalization model

The intrinsic coregionalization model is but a particular linear model (4.37) in which all the $N_v(N_v + 1)/2$ coefficients b_{ij}^l of any basic semivariogram model $g_l(\mathbf{h})$ are proportional to each other, that is, $b_{ij}^l = \varphi_{ij} \cdot b^l \ \forall\, i, j, l$. All direct and cross semivariogram models $\gamma_{ij}(\mathbf{h})$ are then obtained by simple rescaling of the same standardized linear model of regionalization $\gamma_0(\mathbf{h})$:

$$
\begin{aligned}
\gamma_{ij}(\mathbf{h}) &= \varphi_{ij} \cdot \gamma_0(\mathbf{h}) \qquad \forall\, i, j \\
&= \varphi_{ij} \sum_{l=0}^{L} b^l\, g_l(\mathbf{h}) \quad \text{with} \quad \sum_{l=0}^{L} b^l = 1 \qquad (4.41)
\end{aligned}
$$

For example, Figure 4.15 (page 112, bottom graphs) shows an intrinsic coregionalization model where all direct and cross semivariogram models are proportional to the same standardized model $\gamma_0(\mathbf{h}) = 0.5 g_0(\mathbf{h}) + 0.5 \mathrm{Sph}\,(|\mathbf{h}|/5)$. The nugget effect thus represents 50% of the sill value of all three models. Note that any linear model of coregionalization that consists of a single basic structure is an intrinsic model.

The intrinsic coregionalization model is much more restrictive than the linear model of coregionalization: the $N_v(N_v + 1)/2$ direct and cross semivariogram models must include *all* $(L + 1)$ structures in the *same proportions* b^l. This is not the case for the linear model (see Figure 4.15, top graphs). In particular, note the difference in the relative nugget effect between the three direct and cross semivariogram models.

The matrix formulation of the intrinsic coregionalization model (4.41) is

$$
\mathbf{\Gamma}(\mathbf{h}) = \mathbf{\Phi} \sum_{l=0}^{L} b^l\, g_l(\mathbf{h}) = \mathbf{\Phi}\, \gamma_0(\mathbf{h}) \qquad (4.42)
$$

where $\mathbf{\Phi} = [\varphi_{ij}]$ is the matrix of coefficients φ_{ij}. The intrinsic coregionalization model is similarly expressed in terms of basic covariance models $c_l(\mathbf{h})$ as

$$
\mathbf{C}(\mathbf{h}) = \mathbf{\Phi} \sum_{l=0}^{L} b^l\, c_l(\mathbf{h}) = \mathbf{\Phi}\, C_0(\mathbf{h}) \qquad (4.43)
$$

where $C_0(\mathbf{h})$ is a standardized covariance model. The matrix $\mathbf{\Phi}$ is expressed as the ratio $\mathbf{C}(\mathbf{h})/C_0(\mathbf{h})$, where $\mathbf{C}(\mathbf{h})$ is a positive semi-definite matrix and $C_0(\mathbf{h})$ is a positive definite function. Thus, the ratio that is the matrix of coefficients φ_{ij} is positive semi-definite.

Under second-order stationarity, the matrix $\mathbf{\Phi}$ is equal to the variance–covariance matrix $\mathbf{C}(0)$. Indeed, at $|\mathbf{h}| = 0$, the value of each basic model $c_l(\mathbf{h})$ is 1, hence

$$
\mathbf{C}(0) = \mathbf{\Phi} \sum_{l=0}^{L} b^l = \mathbf{\Phi}
$$

The proportionality coefficient φ_{ij} is equal to the variance $C_{ii}(0)$ of the RF $Z_i(\mathbf{u})$ for $i = j$ and is equal to the cross covariance $C_{ij}(0)$ for $i \neq j$.

4.3.3   The practice of modeling

Choosing the number and characteristics (type, range, anisotropy) of basic structures and determining their contribution (sill, slope) to each model is much more difficult than in the univariate case. Difficulties lie in the number of experimental covariance functions (semivariograms) and the constraints that the coefficients b_{ij}^l must satisfy. As in section 4.2.4, the following discussion focuses on semivariograms, which are most frequently used in practice. A similar approach could be adopted for covariance functions.

Selection of variables

Any multivariate modeling starts with a selection of the N_v variables to include in the coregionalization analysis. In most situations, there is no point in retaining all measured attributes for these reasons:

1. The number of variables that can be handled jointly by interpolation algorithms is limited.

2. Some variables are likely to be redundant and need not be considered, whereas independent variables are better handled separately.

3. Increasing the number of variables makes the fitting of all direct and cross semivariograms much more tedious.

It is good practice to define a hierarchy of variables according to the objective pursued. Consider, for example, the objective of estimating concentrations of Cd, which is called the *main* or *primary* variable. One should preferentially retain as secondary variables the one or two other metals better correlated with that primary variable and better sampled, say, Ni and Zn.

Different models of coregionalization could be built depending on which variable is selected as primary. A single model of coregionalization that poorly fits the direct and cross semivariograms of all attributes is better replaced by several models that provide better fits to fewer variables deemed important for the objective.

Selection of the model

Once the variables have been selected, the next decision concerns the type of model of coregionalization: intrinsic or linear.[3]

Intrinsic coregionalization model
The intrinsic coregionalization model is easier to fit than the linear model in the sense that only a single coregionalization matrix Φ need be inferred (see section 4.3.2). However, such a model is much more restrictive because it

---

[3] As mentioned previously, other models of coregionalization, possibly non-linear, might be considered. In the following, only the widely used linear model of coregionalization is retained.

requires that all experimental direct and cross semivariograms be proportional to each other. The proportionality condition can be visually checked from the shape of experimental curves. An alternative consists of verifying that the ratio between any two semivariograms is a constant independent of $\mathbf{h}$:

$$\frac{\gamma_{ij}(\mathbf{h})}{\gamma_{i'j'}(\mathbf{h})} = \text{Constant} \quad \forall\, i, i', j, j', \mathbf{h} \tag{4.44}$$

Similarly, all auto and cross correlograms are about equal.

Linear model of coregionalization
The intrinsic model rarely fits experimental coregionalizations. One reason is that the filtering of uncorrelated micro-structures by the cross semivariogram (see section 4.2.4) usually leads to a smaller relative nugget effect on the cross semivariogram than on direct semivariograms. The more flexible linear model of coregionalization would then be preferred. Note the following:

- Direct or cross semivariograms need not include all basic structures; see rule 3 (page 114).

- Variables that are well cross-correlated are more likely to show similar patterns of spatial variability.

- Slight differences in the shape of experimental direct and cross semi-variograms may be disregarded, in particular when data sparsity affects the reliability of the estimates.

For the three metals (Cd, Ni, and Zn), a quick inspection at the direct semivariograms suffices to invalidate the proportionality condition (4.44) of the intrinsic model (Figure 4.14, page 107). Indeed, the shape of the Ni semivariogram is dominated by the long-range structure, whereas the short-range structure is the major component for the other metals.

Fitting a linear model of coregionalization

To fit a linear model of coregionalization, proceed as follows:

1. Select the smallest set of basic structures $g_l(\mathbf{h})$ that captures the major features of all N_v omnidirectional direct semivariograms. There is no need to look at the experimental cross semivariograms in this step since their models can include only[4] the structures that are apparent on the direct semivariograms; see rule 2 (page 114). For Cd, Ni, and Zn, three structures are retained: the nugget effect, a spherical model with a range of 200 m, and a spherical model with a range of 1.3 km.

2. For each structure $g_l(\mathbf{h})$, consider anisotropy only if that anisotropy is clearly evident on all directional semivariograms. As for the univariate case, ancillary information may help to determine the directions of

---

[4] This represents a definite limitation of the linear model of coregionalization.

anisotropy. The pattern of variation of the three selected metals, Cd, Ni, and Zn, does not vary with the direction; see section 2.3.5.

3. Estimate the contributions (sill, slope) b^l_{ij} of the basic structures $g_l(\mathbf{h})$ building up each model $\gamma_{ij}(\mathbf{h})$ under the constraint of positive semi-definiteness of the coregionalization matrices $\mathbf{B}_l = [b^l_{ij}]$.

4. Appreciate visually the goodness-of-fit for all direct and cross semivariograms. One may then decide to modify a range or to change the type of basic semivariogram model, say, try an exponential rather than a spherical model, to improve the overall quality of the fit. The fit of any particular one of the $N_v(N_v + 1)/2$ experimental direct and cross semivariograms can always be improved, but it is generally done at the expense of poorer fits for other semivariograms. Whenever a compromise is needed, one should give priority to the direct semivariograms, particularly that of the primary variable.

Estimation of the coregionalization matrices

Bivariate case ($N_v = 2$)
In the bivariate case, the coregionalization matrix $\mathbf{B}_l$ is positive semi-definite if the following three inequalities are satisfied:

$$b^l_{11} \geq 0 \qquad b^l_{22} \geq 0 \tag{4.45}$$

$$b^l_{11}b^l_{22} - b^l_{12}b^l_{12} \geq 0 \Rightarrow b^l_{12} \leq \sqrt{b^l_{11}b^l_{22}} \tag{4.46}$$

Thus, the linear model of coregionalization is fitted in two steps:

1. Both direct semivariograms are first modeled as linear combinations of selected basic structures $g_l(\mathbf{h})$.

2. The same basic structures are then fitted to the cross semivariogram under the constraint (4.46).

A good interactive graphical program (e.g., Chu, 1993) and a pocket calculator suffice to model the coregionalization between two variables.

Figure 4.16 shows the linear model of coregionalization fitted to the pair Cd-Ni. The model is as follows:

$$\begin{bmatrix} \gamma_{\text{Cd-Cd}}(\mathbf{h}) & \gamma_{\text{Cd-Ni}}(\mathbf{h}) \\ \gamma_{\text{Ni-Cd}}(\mathbf{h}) & \gamma_{\text{Ni-Ni}}(\mathbf{h}) \end{bmatrix} = \begin{bmatrix} 0.3 & 0.6 \\ 0.6 & 11 \end{bmatrix} g_0(\mathbf{h})$$

$$+ \begin{bmatrix} 0.3 & 0.0 \\ 0.0 & 0.0 \end{bmatrix} \text{Sph}\left(\frac{h}{200\text{m}}\right) + \begin{bmatrix} 0.26 & 3.8 \\ 3.8 & 71 \end{bmatrix} \text{Sph}\left(\frac{h}{1.3\text{km}}\right) \tag{4.47}$$

where $g_0(\mathbf{h})$ is the nugget effect model (4.16). Only the semivariogram model for Cd includes the short-range structure (200 m). The diagonal elements

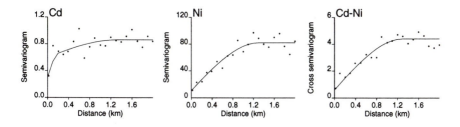

Figure 4.16: Experimental omnidirectional direct and cross semivariograms for Cd, and Ni, and the linear model of coregionalization fitted. The three experimental semivariograms were computed using the same set of 259 concentrations.

and the determinants of the three coregionalization matrices are all positive or zero:

$$\begin{vmatrix} 0.3 & 0.6 \\ 0.6 & 11 \end{vmatrix} = 2.94 \qquad \begin{vmatrix} 0.3 & 0.0 \\ 0.0 & 0.0 \end{vmatrix} = 0.0 \qquad \begin{vmatrix} 0.26 & 3.8 \\ 3.8 & 71 \end{vmatrix} = 4.02$$

hence, the linear model of coregionalization (4.47) is positive semi-definite.

Multivariate case ($N_v \geq 3$)
The bivariate approach is generalized to N_v variables as follows:

1. The N_v direct semivariograms $\gamma_{ii}(\mathbf{h})$ are first modeled as linear combinations of selected basic structures $g_l(\mathbf{h})$.

2. The same structures are then fitted to each of the $N_v(N_v - 1)/2$ cross semivariograms $\gamma_{ij}(\mathbf{h})$ under the constraint $b_{ij}^l \leq \sqrt{b_{ii}^l \, b_{jj}^l}$.

The latter condition is, however, insufficient to ensure positive semi-definiteness of matrices $\mathbf{B}_l = [b_{ij}^l]$ for $N_v > 2$. Indeed, all principal minor determinants, not only those of order 2, must be non-negative (recall page 114).
    Consider, for example, the linear model of coregionalization in Figure 4.17. The direct and cross semivariograms of Cd, Ni, and Zn were modeled using the two-step approach, yielding the following model:

$$\begin{bmatrix} \gamma_{Cd\text{-}Cd}(\mathbf{h}) & \gamma_{Cd\text{-}Ni}(\mathbf{h}) & \gamma_{Cd\text{-}Zn}(\mathbf{h}) \\ \gamma_{Co\text{-}Ni}(\mathbf{h}) & \gamma_{Ni\text{-}Ni}(\mathbf{h}) & \gamma_{Ni\text{-}Zn}(\mathbf{h}) \\ \gamma_{Zn\text{-}Cd}(\mathbf{h}) & \gamma_{Zn\text{-}Ni}(\mathbf{h}) & \gamma_{Zn\text{-}Zn}(\mathbf{h}) \end{bmatrix} = \begin{bmatrix} 0.3 & 0.6 & 4.4 \\ 0.6 & 11 & 20 \\ 4.4 & 20 & 180 \end{bmatrix} g_0(\mathbf{h})$$

$$+ \begin{bmatrix} 0.3 & 0.0 & 5.8 \\ 0.0 & 3.0 & 25 \\ 5.8 & 25 & 300 \end{bmatrix} \text{Sph}\left(\frac{h}{200\text{m}}\right)$$

$$+ \begin{bmatrix} 0.26 & 3.8 & 8.7 \\ 3.8 & 68 & 140 \\ 8.7 & 140 & 470 \end{bmatrix} \text{Sph}\left(\frac{h}{1.3\text{km}}\right) \qquad (4.48)$$

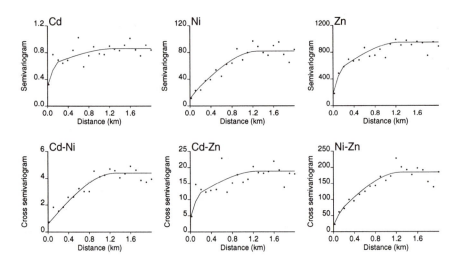

Figure 4.17: Experimental omnidirectional direct and cross semivariograms for Cd, Ni, and Zn ($n=259$). The linear model of coregionalization fitted using a two-step approach is not permissible.

The short-range structure now contributes to the Ni semivariogram model. The small contribution ($b^1_{Ni} = 3.0$) allows modeling the short-range structure seen on the cross semivariogram Ni-Zn.

By construction of the two-step approach, the three following conditions are satisfied for each basic semivariogram model $g_l(\mathbf{h})$:

$$b^l_{12} \leq \sqrt{b^l_{11} b^l_{22}} \qquad b^l_{13} \leq \sqrt{b^l_{11} b^l_{33}} \qquad b^l_{23} \leq \sqrt{b^l_{22} b^l_{33}}$$

The model (4.48) is, however, not permissible since the coregionalization matrix associated with the spherical model of range 200 m is not positive semi-definite; indeed, the determinant of the matrix is negative:

$$\begin{vmatrix} 0.30 & 0.0 & 5.8 \\ 0.0 & 3.0 & 25 \\ 5.8 & 25 & 300 \end{vmatrix} = -18.4 < 0$$

A current practice consists of modifying empirically the coefficients b^l_{ij} until the positive semi-definite condition is met. Such an approach has two drawbacks:

- It is tedious because it is unclear which coefficient should be modified first and which correction (increase, decrease) should be applied.

- It becomes unpractical as the number of variables, hence the number of coefficients b^l_{ij}, increases.

The burden of checking a posteriori each matrix $\mathbf{B}_l$ then applying empirical corrections is alleviated by an iterative procedure that fits the linear model of coregionalization *directly* under the constraint of positive semi-definiteness of all matrices $\mathbf{B}_l$ (Goulard, 1989; Goulard and Voltz, 1992). As with the univariate criterion (4.29), the algorithm attempts to minimize a weighted sum of differences between experimental $\hat{\gamma}_{ij}(\mathbf{h}_k)$ and model $\gamma_{ij}(\mathbf{h}_k)$ semivariogram values (see Appendix A).

This iterative procedure is used to model the coregionalization of the three metals Cd, Ni, and Zn (Figure 4.18). Although it fits the experimental curves as well as the model shown in Figure 4.17, the model of coregionalization is now positive semi-definite:

$$
\begin{bmatrix}
\gamma_{\text{Cd-Cd}}(\mathbf{h}) & \gamma_{\text{Cd-Ni}}(\mathbf{h}) & \gamma_{\text{Cd-Zn}}(\mathbf{h}) \\
\gamma_{\text{Co-Ni}}(\mathbf{h}) & \gamma_{\text{Ni-Ni}}(\mathbf{h}) & \gamma_{\text{Ni-Zn}}(\mathbf{h}) \\
\gamma_{\text{Zn-Cd}}(\mathbf{h}) & \gamma_{\text{Zn-Ni}}(\mathbf{h}) & \gamma_{\text{Zn-Zn}}(\mathbf{h})
\end{bmatrix}
=
\begin{bmatrix}
0.25 & 0.47 & 3.5 \\
0.47 & 7.9 & 15.4 \\
3.5 & 15.4 & 105
\end{bmatrix}
g_0(\mathbf{h})
$$

$$
+
\begin{bmatrix}
0.43 & 0.49 & 8.3 \\
0.49 & 2.0 & 24.5 \\
8.3 & 24.5 & 398
\end{bmatrix}
\text{Sph}\left(\frac{h}{200\text{m}}\right)
$$

$$
+
\begin{bmatrix}
0.18 & 3.3 & 6.2 \\
3.3 & 73.3 & 142.5 \\
6.2 & 142.5 & 398
\end{bmatrix}
\text{Sph}\left(\frac{h}{1.3\text{km}}\right)
\qquad (4.49)
$$

The determinants of order 3 of the coregionalization matrices are 78.8, 50.1, and 275.3, respectively.

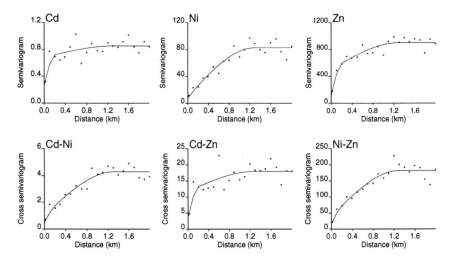

Figure 4.18: Experimental omnidirectional direct and cross semivariograms for Cd, Ni, and Zn ($n=259$). The linear model of coregionalization is fitted using an iterative procedure that ensures that the model is permissible.

Coregionalization of other metals

A similar approach is applied to the two other metals with widespread contamination (Cu, Pb). The linear model of coregionalization now includes
four variables: the two better sampled metals (Ni, Zn), and copper and lead,
which are closely related ($\rho=0.78$). Figure 4.19 shows the ten experimental
direct and cross semivariograms. The coregionalization model is fitted using
the iterative procedure. Despite the number of semivariograms, the overall
fit is satisfactory.

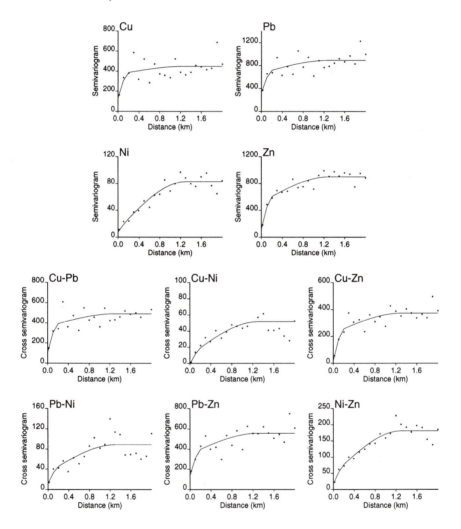

Figure 4.19: Experimental omnidirectional direct and cross semivariograms for Cu,
Pb, Ni, and Zn ($n=259$). The linear model of coregionalization is fitted using the
iterative procedure, hence is permissible.

Warning

The reason for the wide utilization of the linear model of coregionalization is that its permissibility can be easily checked by verifying the positive semi-definiteness of coregionalization matrices. The price to pay is the requirement that all direct and cross semivariograms share the same set of basic structures $g_l(\mathbf{h})$, which may represent a limitation.

Alternative linear models that do not require using a common set of basic structures have been proposed (e.g., Myers, 1982). Their greater flexibility is balanced by the need for checking the positive semi-definiteness condition of $\mathbf{C}(\mathbf{h})$ for **all** lags to be used (Goovaerts, 1994c). Beware that, even in the bivariate case, there is no easy way to perform such a check, especially in the presence of anisotropy. Therefore, it is better to stay with the well proven and yet reasonably flexible linear model of coregionalization.

Chapter 5

Local Estimation: Accounting for a Single Attribute

The existence of a model of spatial dependence allows one to tackle the problem of estimating attribute values at unsampled locations. This chapter presents least-squares linear regression (kriging) algorithms that account for data related solely to the continuous attribute being estimated. Algorithms for incorporating secondary information are introduced in Chapter 6.

Sections 5.1–5.4 introduce the linear regression paradigm and three of its most important variants: simple kriging, ordinary kriging, and kriging with a trend model. Estimation of average values over larger areas (block kriging) is addressed in section 5.5.

Section 5.6 presents factorial kriging, a method for filtering low- or high-frequency components from the spatial variation of the attribute. Filtering properties of kriging algorithms are further discussed in section 5.7 in the framework of the dual formalism of kriging.

Section 5.8 recalls some important properties of the kriging weights and kriging variance. The performance of ordinary kriging in estimating metal concentrations at test locations is discussed.

5.1   The Kriging Paradigm

Consider the problem of estimating the value of a continuous attribute z at any unsampled location $\mathbf{u}$ using only z-data available over the study area $\mathcal{A}$, say, the n data $\{z(\mathbf{u}_\alpha), \ \alpha = 1, \ldots, n\}$. Kriging is a generic name adopted by geostatisticians for a family of generalized least-squares regression algorithms in recognition of the pioneering work of Danie Krige (1951). All kriging estimators are but variants of the basic linear regression estimator $Z^*(\mathbf{u})$

defined as

$$Z^*(\mathbf{u}) - m(\mathbf{u}) \quad = \quad \sum_{\alpha=1}^{n(\mathbf{u})} \lambda_\alpha(\mathbf{u}) \; [Z(\mathbf{u}_\alpha) - m(\mathbf{u}_\alpha)] \tag{5.1}$$

where $\lambda_\alpha(\mathbf{u})$ is the weight assigned to datum $z(\mathbf{u}_\alpha)$ interpreted as a realization of the RV $Z(\mathbf{u}_\alpha)$. The quantities $m(\mathbf{u})$ and $m(\mathbf{u}_\alpha)$ are the expected values of the RVs $Z(\mathbf{u})$ and $Z(\mathbf{u}_\alpha)$. The number of data involved in the estimation as well as their weights may change from one location to another. In practice, only the $n(\mathbf{u})$ data closest to the location $\mathbf{u}$ being estimated are retained, i.e., the data within a given neighborhood or window $W(\mathbf{u})$ centered on $\mathbf{u}$.

The interpretation of the unknown value $z(\mathbf{u})$ and data values $z(\mathbf{u}_\alpha)$ as realizations of RVs $Z(\mathbf{u})$ and $Z(\mathbf{u}_\alpha)$ allows one to define the estimation error as a random variable $Z^*(\mathbf{u}) - Z(\mathbf{u})$. All flavors of kriging share the same objective of minimizing the estimation or error variance $\sigma_E^2(\mathbf{u})$ under the constraint of unbiasedness of the estimator; that is,

$$\sigma_E^2(\mathbf{u}) = \text{Var}\{Z^*(\mathbf{u}) - Z(\mathbf{u})\}$$

is minimized under the constraint that

$$E\{Z^*(\mathbf{u}) - Z(\mathbf{u})\} \; = \; 0 \tag{5.2}$$

The kriging estimator varies depending on the model adopted for the random function $Z(\mathbf{u})$ itself. The RF $Z(\mathbf{u})$ is usually decomposed into a residual component $R(\mathbf{u})$ and a trend component $m(\mathbf{u})$:

$$Z(\mathbf{u}) \; = \; R(\mathbf{u}) \; + \; m(\mathbf{u}) \tag{5.3}$$

The residual component is modeled as a stationary RF with zero mean and covariance $C_R(\mathbf{h})$:

$$E\{R(\mathbf{u})\} \quad = \quad 0$$
$$\text{Cov}\{R(\mathbf{u}), R(\mathbf{u}+\mathbf{h})\} \quad = \quad E\{R(\mathbf{u}) \cdot R(\mathbf{u}+\mathbf{h})\} \; = \; C_R(\mathbf{h})$$

The expected value of the RV Z at location $\mathbf{u}$ is thus the value of the trend component at that location:

$$E\{Z(\mathbf{u})\} \quad = \quad m(\mathbf{u})$$

Three kriging variants can be distinguished according to the model considered for the trend $m(\mathbf{u})$:

1. Simple kriging (SK) considers the mean $m(\mathbf{u})$ to be known and constant throughout the study area $\mathcal{A}$:

$$m(\mathbf{u}) \; = \; m, \; \text{known} \qquad \forall \, \mathbf{u} \in \mathcal{A} \tag{5.4}$$

2. Ordinary kriging (OK) accounts for local fluctuations of the mean by limiting the domain of stationarity of the mean to the local neighborhood $W(\mathbf{u})$:

$$m(\mathbf{u}') = \text{constant but unknown} \quad \forall \, \mathbf{u}' \in W(\mathbf{u}) \qquad (5.5)$$

Unlike simple kriging, here the mean is deemed unknown.

3. Kriging with a trend model[1] (KT) considers that the unknown local mean $m(\mathbf{u}')$ smoothly varies within each local neighborhood $W(\mathbf{u})$, hence over the entire study area $\mathcal{A}$. The trend component is modeled as a linear combination of functions $f_k(\mathbf{u})$ of the coordinates:

$$m(\mathbf{u}') = \sum_{k=0}^{K} a_k(\mathbf{u}') \, f_k(\mathbf{u}') \qquad (5.6)$$

$$\text{with } a_k(\mathbf{u}') \approx a_k \text{ constant but unknown} \quad \forall \, \mathbf{u}' \in W(\mathbf{u})$$

The coefficients $a_k(\mathbf{u}')$ are unknown and deemed constant within each local neighborhood $W(\mathbf{u})$. By convention, $f_0(\mathbf{u}') = 1$, hence the case $K = 0$ is equivalent to ordinary kriging (constant but unknown mean a_0).

5.2 Simple Kriging

The modeling of the trend component $m(\mathbf{u})$ as a known stationary mean m allows one to write the linear estimator (5.1) as a linear combination of $(n(\mathbf{u})+1)$ pieces of information: the $n(\mathbf{u})$ RVs $Z(\mathbf{u}_\alpha)$ and the mean value m:

$$Z_{SK}^*(\mathbf{u}) = \sum_{\alpha=1}^{n(\mathbf{u})} \lambda_\alpha^{SK}(\mathbf{u}) \, [Z(\mathbf{u}_\alpha) - m] + m$$

$$= \sum_{\alpha=1}^{n(\mathbf{u})} \lambda_\alpha^{SK}(\mathbf{u}) \, Z(\mathbf{u}_\alpha) + \left[1 - \sum_{\alpha=1}^{n(\mathbf{u})} \lambda_\alpha^{SK}(\mathbf{u})\right] m \qquad (5.7)$$

The $n(\mathbf{u})$ weights $\lambda_\alpha^{SK}(\mathbf{u})$ are then determined such as to minimize the error variance $\sigma_E^2(\mathbf{u}) = \text{Var}\{Z_{SK}^*(\mathbf{u}) - Z(\mathbf{u})\}$ under the unbiasedness constraint (5.2).

The simple kriging (SK) estimator (5.7) is already unbiased since the error mean is equal to zero. Indeed, utilizing the first expression (5.7), it comes:

$$E\{Z_{SK}^*(\mathbf{u}) - Z(\mathbf{u})\} = m - m = 0$$

---

[1] Though the algorithm is also known as *universal kriging* (Journel and Huijbregts, 1978, p. 313), the more appropriate terminology *kriging with a trend model*, proposed by Journel and Rossi (1989), will be used throughout this book.

The estimation error $Z_{SK}^*(\mathbf{u}) - Z(\mathbf{u})$ can be viewed as a linear combination of $(n(\mathbf{u}) + 1)$ residual RVs $R(\mathbf{u})$ and $R(\mathbf{u}_\alpha)$:

$$
\begin{aligned}
Z_{SK}^*(\mathbf{u}) - Z(\mathbf{u}) &= [Z_{SK}^*(\mathbf{u}) - m] - [Z(\mathbf{u}) - m] \\
&= \sum_{\alpha=1}^{n(\mathbf{u})} \lambda_\alpha^{SK}(\mathbf{u}) R(\mathbf{u}_\alpha) - R(\mathbf{u}) = R_{SK}^*(\mathbf{u}) - R(\mathbf{u})
\end{aligned}
$$

where $R(\mathbf{u}_\alpha) = Z(\mathbf{u}_\alpha) - m$ and $R(\mathbf{u}) = Z(\mathbf{u}) - m$. The error variance can thus be expressed as a double linear combination of residual covariance values:

$$
\begin{aligned}
\sigma_E^2(\mathbf{u}) &= \text{Var}\{R_{SK}^*(\mathbf{u})\} + \text{Var}\{R(\mathbf{u})\} - 2\,\text{Cov}\{R_{SK}^*(\mathbf{u}), R(\mathbf{u})\} \\
&= \sum_{\alpha=1}^{n(\mathbf{u})} \sum_{\beta=1}^{n(\mathbf{u})} \lambda_\alpha^{SK}(\mathbf{u})\, \lambda_\beta^{SK}(\mathbf{u})\, C_R(\mathbf{u}_\alpha - \mathbf{u}_\beta) + C_R(0) \\
&\quad - 2 \sum_{\alpha=1}^{n(\mathbf{u})} \lambda_\alpha^{SK}(\mathbf{u})\, C_R(\mathbf{u}_\alpha - \mathbf{u}) \\
&= Q(\lambda_\alpha^{SK}(\mathbf{u}), \alpha = 1, \ldots, n(\mathbf{u})) \quad\quad (5.8)
\end{aligned}
$$

The error variance (5.8) appears as a quadratic form $Q(.)$ in the $n(\mathbf{u})$ weights $\lambda_\alpha^{SK}(\mathbf{u})$. The optimal weights, i.e., those that minimize the error variance, are obtained by setting to zero each of the $n(\mathbf{u})$ partial first derivatives:

$$
\frac{1}{2} \frac{\partial Q(\mathbf{u})}{\partial \lambda_\alpha^{SK}(\mathbf{u})} = \sum_{\beta=1}^{n(\mathbf{u})} \lambda_\beta^{SK}(\mathbf{u})\, C_R(\mathbf{u}_\alpha - \mathbf{u}_\beta) - C_R(\mathbf{u}_\alpha - \mathbf{u}) = 0
$$

$$
\alpha = 1, \ldots, n(\mathbf{u}) \quad\quad (5.9)
$$

The system (5.9) of $n(\mathbf{u})$ linear equations is known as the system of normal equations (Luenberger, 1969, p. 56) or the simple kriging system.

Stationarity of the mean entails that the residual covariance function $C_R(\mathbf{h})$ is equal to the stationary covariance function $C(\mathbf{h})$ of the RF $Z(\mathbf{u})$:

$$
\begin{aligned}
C_R(\mathbf{h}) &= E\{R(\mathbf{u}) \cdot R(\mathbf{u} + \mathbf{h})\} \\
&= E\{[Z(\mathbf{u}) - m] \cdot [Z(\mathbf{u} + \mathbf{h}) - m]\} = C(\mathbf{h})
\end{aligned}
$$

Thus, the simple kriging system (5.9) can be written in terms of Z-covariances as

$$
\sum_{\beta=1}^{n(\mathbf{u})} \lambda_\beta^{SK}(\mathbf{u})\, C(\mathbf{u}_\alpha - \mathbf{u}_\beta) = C(\mathbf{u}_\alpha - \mathbf{u}) \quad \alpha = 1, \ldots, n(\mathbf{u}) \quad\quad (5.10)
$$

The minimum error variance, also called the SK variance, is deduced by substituting expression (5.10) into the definition (5.8) of the error variance:

$$
\sigma_{SK}^2(\mathbf{u}) = C(0) - \sum_{\alpha=1}^{n(\mathbf{u})} \lambda_\alpha^{SK}(\mathbf{u})\, C(\mathbf{u}_\alpha - \mathbf{u}) \quad\quad (5.11)
$$

Correlogram notation

By dividing all the equations of system (5.10) by the variance $C(0)$, the simple kriging system is rewritten in terms of the correlogram $\rho(\mathbf{h})$:

$$\sum_{\beta=1}^{n(\mathbf{u})} \lambda_\beta^{SK}(\mathbf{u})\, \rho(\mathbf{u}_\alpha - \mathbf{u}_\beta) = \rho(\mathbf{u}_\alpha - \mathbf{u}) \qquad \alpha = 1,\ldots,n(\mathbf{u}) \qquad (5.12)$$

The SK variance is then given by

$$\sigma_{SK}^2(\mathbf{u}) = C(0)\left[1 - \sum_{\alpha=1}^{n(\mathbf{u})} \lambda_\alpha^{SK}(\mathbf{u})\, \rho(\mathbf{u}_\alpha - \mathbf{u})\right]$$

Systems (5.10) and (5.12) yield the same kriging weights.

Matrix notation

Using matrix notation, the simple kriging system (5.10) is written as

$$\mathbf{K}_{SK}\, \boldsymbol{\lambda}_{SK}(\mathbf{u}) = \mathbf{k}_{SK} \qquad (5.13)$$

where $\mathbf{K}_{SK}$ is the $n(\mathbf{u}) \times n(\mathbf{u})$ matrix of data covariances:

$$\mathbf{K}_{SK} = \begin{bmatrix} C(\mathbf{u}_1 - \mathbf{u}_1) & \cdots & C(\mathbf{u}_1 - \mathbf{u}_{n(\mathbf{u})}) \\ \vdots & \vdots & \vdots \\ C(\mathbf{u}_{n(\mathbf{u})} - \mathbf{u}_1) & \cdots & C(\mathbf{u}_{n(\mathbf{u})} - \mathbf{u}_{n(\mathbf{u})}) \end{bmatrix}$$

$\boldsymbol{\lambda}_{SK}(\mathbf{u})$ is the vector of SK weights, and $\mathbf{k}_{SK}$ is the vector of data-to-unknown covariances:

$$\boldsymbol{\lambda}_{SK}(\mathbf{u}) = \begin{bmatrix} \lambda_1^{SK}(\mathbf{u}) \\ \vdots \\ \lambda_{n(\mathbf{u})}^{SK}(\mathbf{u}) \end{bmatrix} \qquad \mathbf{k}_{SK} = \begin{bmatrix} C(\mathbf{u}_1 - \mathbf{u}) \\ \vdots \\ C(\mathbf{u}_{n(\mathbf{u})} - \mathbf{u}) \end{bmatrix}$$

The kriging weights required by the SK estimator (5.7) are obtained by multiplying the inverse of the data covariance matrix by the vector of data-to-unknown covariances:

$$\boldsymbol{\lambda}_{SK}(\mathbf{u}) = \mathbf{K}_{SK}^{-1}\, \mathbf{k}_{SK}$$

The matrix formulation of the simple kriging variance (5.11) is correspondingly

$$\sigma_{SK}^2(\mathbf{u}) = C(0) - \boldsymbol{\lambda}_{SK}^T(\mathbf{u})\, \mathbf{k}_{SK} = C(0) - \mathbf{k}_{SK}^T\, \mathbf{K}_{SK}^{-1}\, \mathbf{k}_{SK}$$

The simple kriging system (5.10) has a unique solution and the resulting kriging variance is positive if the covariance matrix $\mathbf{K}_{SK} = [C(\mathbf{u}_\alpha - \mathbf{u}_\beta)]$ is positive definite, in practice if:

- no two data are colocated: $\mathbf{u}_\alpha \neq \mathbf{u}_\beta$ for $\alpha \neq \beta$.
- the covariance model $C(\mathbf{h})$ is permissible (see section 4.2.1).

Exactitude property

The SK estimator is an exact interpolator in that it honors data values $z(\mathbf{u}_\alpha)$ at their locations:

$$z^*_{SK}(\mathbf{u}) = z(\mathbf{u}_\alpha) \qquad \forall\, \mathbf{u} = \mathbf{u}_\alpha, \quad \alpha = 1, \ldots, n$$

Indeed, when the location $\mathbf{u}$ being estimated coincides with a datum location, say, $\mathbf{u}_{\alpha'}$, the SK system (5.10) becomes

$$\sum_{\substack{\beta=1 \\ \beta \neq \alpha'}}^{n(\mathbf{u})} \lambda^{SK}_\beta(\mathbf{u})\, C(\mathbf{u}_\alpha - \mathbf{u}_\beta) + \lambda^{SK}_{\alpha'}(\mathbf{u})\, C(\mathbf{u}_\alpha - \mathbf{u}_{\alpha'}) = C(\mathbf{u}_\alpha - \mathbf{u}_{\alpha'})$$

$$\alpha = 1, \ldots, n(\mathbf{u})$$

The unique solution is then

$$\lambda^{SK}_{\alpha'}(\mathbf{u}) = 1 \;\text{ and }\; \lambda^{SK}_\beta(\mathbf{u}) = 0 \qquad \forall\, \mathbf{u}_\beta \neq \mathbf{u}_{\alpha'}$$

Weight of the mean

The simple kriging estimator (5.7) can be rewritten as

$$Z^*_{SK}(\mathbf{u}) = \sum_{\alpha=1}^{n(\mathbf{u})} \lambda^{SK}_\alpha(\mathbf{u})\, Z(\mathbf{u}_\alpha) + \lambda^{SK}_m(\mathbf{u})\, m \qquad (5.14)$$

where the weight $\lambda^{SK}_m(\mathbf{u})$ assigned to the stationary mean m for estimation at location $\mathbf{u}$ is equal to 1 minus the sum of the $n(\mathbf{u})$ data weights $\lambda^{SK}_\alpha(\mathbf{u})$:

$$\lambda^{SK}_m(\mathbf{u}) = 1 - \sum_{\alpha=1}^{n(\mathbf{u})} \lambda^{SK}_\alpha(\mathbf{u})$$

As the location $\mathbf{u}$ being estimated gets farther away from data locations,

- the data-to-unknown covariances $C(\mathbf{u}_\alpha - \mathbf{u})$, which are elements of vector $\mathbf{k}$ in system (5.13), decrease.

- the data-to-data covariances $C(\mathbf{u}_\alpha - \mathbf{u}_\beta)$ remain unchanged.

Consequently, the SK weights $[\lambda^{SK}_\alpha(\mathbf{u})] = \boldsymbol{\lambda}_{SK}(\mathbf{u}) = \mathbf{K}^{-1}_{SK}\, \mathbf{k}_{SK}$ tend to decrease; hence the weight of the mean increases, and the estimate $z^*_{SK}(\mathbf{u})$ gets closer to the stationary mean m. The global information carried by the stationary mean becomes preponderant as remote neighboring data bring less information about the unknown value at $\mathbf{u}$.

The limiting case corresponds to a location $\mathbf{u}$ beyond the correlation range of any data location $\mathbf{u}_\alpha$, i.e., beyond the distance at which the covariance $C(\mathbf{u}_\alpha - \mathbf{u})$ vanishes. In such a case, the vector $\mathbf{k}_{SK}$ of data-to-unknown covariances $C(\mathbf{u}_\alpha - \mathbf{u})$ is a null vector and so is the vector $\boldsymbol{\lambda}_{SK}(\mathbf{u})$ of kriging weights: $\boldsymbol{\lambda}_{SK}(\mathbf{u}) = \mathbf{K}^{-1}_{SK}\, \mathbf{0} = \mathbf{0}$. The unique solution $\lambda^{SK}_\alpha(\mathbf{u}) = 0 \;\forall\, \alpha = 1, \ldots, n(\mathbf{u})$ entails that the weight of the mean is equal to 1, hence the SK estimator (5.14) is the stationary mean m.

Example

Throughout Chapter 5, kriging algorithms are illustrated from the one-dimensional data set shown in Figure 5.1 (left graph). The information available for the task of estimating Cd concentrations along the transect consists of:

- 10 Cd values (black dots), which show a general increase along the transect, and

- the Cd semivariogram model inferred in section 4.2.4 from all Cd data available over the study area (Figure 5.1, right graph):

$$\gamma_{\mathrm{Cd}}(\mathbf{h}) = 0.3\,g_0(\mathbf{h}) + 0.30\,\mathrm{Sph}\left(\frac{|\mathbf{h}|}{200\mathrm{m}}\right) + 0.26\,\mathrm{Sph}\left(\frac{|\mathbf{h}|}{1.3\mathrm{km}}\right) \quad (5.15)$$

where $g_0(\mathbf{h})$ is the nugget effect model (4.16).

The estimation is performed every 50 m using at each location $\mathbf{u}$ the five closest data, $n(\mathbf{u}) = 5\ \forall\ \mathbf{u}$.

Figure 5.2 shows the SK estimates of Cd concentrations (top graph) and the weight of the stationary mean (bottom graph) taken as the arithmetic mean of the 10 Cd values, $m = 1.49$ ppm. The SK estimates (solid line):

- identify the data (exactitude property),

- become closer to the mean (dashed line) away from the data because of the increase in the weight $\lambda_m^{SK}(\mathbf{u})$, and

- equal the mean beyond the correlation range, 1.3 km, of Cd data since the weight $\lambda_m^{SK}(\mathbf{u})$ is then equal to 1.

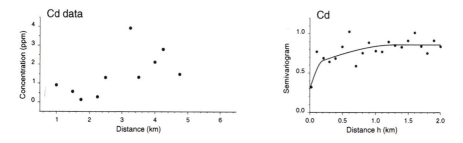

Figure 5.1: NE-SW transect of 10 Cd concentrations (left graph) and the Cd semivariogram inferred from 259 concentrations (right graph). Throughout Chapter 5, different kriging algorithms will utilize that information to estimate the concentration of cadmium every 50 m along that transect.

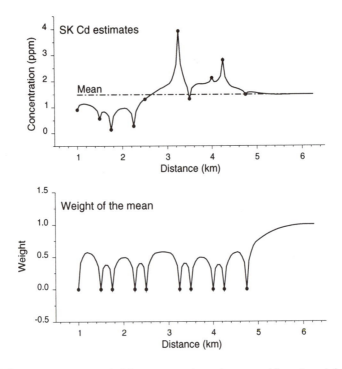

Figure 5.2: SK estimates of Cd concentrations (top graph) and weight of the stationary mean (bottom graph).

The exactitude property of the SK estimator[2] creates artifact discontinuities (peaks) at data locations (Figure 5.2, top graph). Discontinuities are important here because Cd concentrations vary in the semivariogram of Figure 5.1 (right graph). Indeed, such short-range variability entails that the data weights $\lambda_\alpha^{SK}(\mathbf{u})$ rapidly decrease as the location $\mathbf{u}$ being estimated gets farther away from data locations and the estimate gets closer to the stationary mean $m = 1.49$ ppm. Large discontinuities occur next to extreme data values—those that depart most from the global mean, for example, near the concentration of 4 ppm measured at 3.25 km. Alternatives for removing such discontinuities are discussed in the next sections.

5.3 Ordinary Kriging

As is apparent on the graphs of Figure 4.4 (page 83), the local mean may vary significantly over the study area. For example, the Cd local mean computed from 1 km × 1 km moving windows varies from 0.5 to 2.4 ppm, depending on the window location; recall that the overall Cd mean is 1.3 ppm. Ordinary

---

[2] This also applies to the OK and KT estimators introduced in the sections 5.3 and 5.4.

kriging (OK) allows one to account for such local variation of the mean by limiting the domain of stationarity of the mean to the local neighborhood $W(\mathbf{u})$ centered on the location $\mathbf{u}$ being estimated. The linear estimator (5.1) is then a linear combination of the $n(\mathbf{u})$ RVs $Z(\mathbf{u}_\alpha)$ plus the constant local mean $m(\mathbf{u})$:

$$Z^*(\mathbf{u}) = \sum_{\alpha=1}^{n(\mathbf{u})} \lambda_\alpha(\mathbf{u}) Z(\mathbf{u}_\alpha) + \left[1 - \sum_{\alpha=1}^{n(\mathbf{u})} \lambda_\alpha(\mathbf{u})\right] m(\mathbf{u})$$

The unknown local mean $m(\mathbf{u})$ is filtered from the linear estimator by forcing the kriging weights to sum to 1. The ordinary kriging estimator $Z_{OK}^*(\mathbf{u})$ is thus written as a linear combination only of the $n(\mathbf{u})$ RVs $Z(\mathbf{u}_\alpha)$:

$$Z_{OK}^*(\mathbf{u}) = \sum_{\alpha=1}^{n(\mathbf{u})} \lambda_\alpha^{OK}(\mathbf{u}) Z(\mathbf{u}_\alpha) \quad \text{with} \quad \sum_{\alpha=1}^{n(\mathbf{u})} \lambda_\alpha^{OK}(\mathbf{u}) = 1 \qquad (5.16)$$

Again, the $n(\mathbf{u})$ weights $\lambda_\alpha^{OK}(\mathbf{u})$ are determined such as to minimize the error variance checking for the unbiasedness constraint (5.2).

The OK estimator (5.16) is unbiased since the error mean is equal to zero:

$$
\begin{aligned}
E\{Z_{OK}^*(\mathbf{u}) - Z(\mathbf{u})\} &= \sum_{\alpha=1}^{n(\mathbf{u})} \lambda_\alpha^{OK}(\mathbf{u}) m(\mathbf{u}) - m(\mathbf{u}) \\
&= m(\mathbf{u}) - m(\mathbf{u}) = 0
\end{aligned}
$$

The minimization of the error variance (5.8) under the non-bias condition $\sum_{\alpha=1}^{n(\mathbf{u})} \lambda_\alpha^{OK}(\mathbf{u}) = 1$ calls for the definition of a Lagrangian $L(\mathbf{u})$, which is a function of the data weights $\lambda_\alpha^{OK}(\mathbf{u})$, and a Lagrange parameter $2\mu_{OK}(\mathbf{u})$ (e.g., see Edwards and Penney, 1982):

$$L(\lambda_\alpha^{OK}(\mathbf{u}), \alpha = 1, \ldots, n(\mathbf{u}); 2\mu_{OK}(\mathbf{u})) = \sigma_E^2(\mathbf{u}) + 2\mu_{OK}(\mathbf{u}) \left[\sum_{\alpha=1}^{n(\mathbf{u})} \lambda_\alpha^{OK}(\mathbf{u}) - 1\right]$$

The optimal weights $\lambda_\alpha^{OK}(\mathbf{u})$ are obtained by setting to zero each of the $(n(\mathbf{u}) + 1)$ partial first derivatives:

$$\frac{1}{2} \frac{\partial L(\mathbf{u})}{\partial \lambda_\alpha^{OK}(\mathbf{u})} = \sum_{\beta=1}^{n(\mathbf{u})} \lambda_\beta^{OK}(\mathbf{u}) C_R(\mathbf{u}_\alpha - \mathbf{u}_\beta) - C_R(\mathbf{u}_\alpha - \mathbf{u}) + \mu_{OK}(\mathbf{u}) = 0$$

$$\alpha = 1, \ldots, n(\mathbf{u})$$

$$\frac{1}{2} \frac{\partial L(\mathbf{u})}{\partial \mu_{OK}(\mathbf{u})} = \sum_{\alpha=1}^{n(\mathbf{u})} \lambda_\alpha^{OK}(\mathbf{u}) - 1 = 0$$

The ordinary kriging system includes $(n(\mathbf{u}) + 1)$ linear equations with $(n(\mathbf{u}) + 1)$ unknowns: the $n(\mathbf{u})$ weights $\lambda_\alpha^{OK}(\mathbf{u})$ and the Lagrange parameter $\mu_{OK}(\mathbf{u})$ that accounts for the constraint on the weights:

$$
\begin{cases}
\displaystyle\sum_{\beta=1}^{n(\mathbf{u})} \lambda_\beta^{OK}(\mathbf{u})\, C_R(\mathbf{u}_\alpha - \mathbf{u}_\beta) + \mu_{OK}(\mathbf{u}) = C_R(\mathbf{u}_\alpha - \mathbf{u}) \\[4pt]
\hspace{5cm} \alpha = 1, \ldots, n(\mathbf{u}) \\[4pt]
\displaystyle\sum_{\beta=1}^{n(\mathbf{u})} \lambda_\beta^{OK}(\mathbf{u}) = 1
\end{cases}
\tag{5.17}
$$

Although the mean $m(\mathbf{u})$ is assumed stationary only within the local neighborhood $W(\mathbf{u})$, in the practice of ordinary kriging the residual covariance is assimilated to the global z-covariance inferred from all data available (Journel and Huijbregts, 1978, p. 33–34), leading to the following system:

$$
\begin{cases}
\displaystyle\sum_{\beta=1}^{n(\mathbf{u})} \lambda_\beta^{OK}(\mathbf{u})\, C(\mathbf{u}_\alpha - \mathbf{u}_\beta) + \mu_{OK}(\mathbf{u}) = C(\mathbf{u}_\alpha - \mathbf{u}) \\[4pt]
\hspace{5cm} \alpha = 1, \ldots, n(\mathbf{u}) \\[4pt]
\displaystyle\sum_{\beta=1}^{n(\mathbf{u})} \lambda_\beta^{OK}(\mathbf{u}) = 1
\end{cases}
\tag{5.18}
$$

The resulting minimum error variance, called OK variance, is obtained by substituting the first $n(\mathbf{u})$ equations of the ordinary kriging system (5.18) into the error variance of type (5.8):

$$
\sigma_{OK}^2(\mathbf{u}) = C(0) - \sum_{\alpha=1}^{n(\mathbf{u})} \lambda_\alpha^{OK}(\mathbf{u})\, C(\mathbf{u}_\alpha - \mathbf{u}) - \mu_{OK}(\mathbf{u})
\tag{5.19}
$$

Semivariogram notation

Accounting for the relation $C(\mathbf{h}) = C(0) - \gamma(\mathbf{h})$, the ordinary kriging system (5.18) is expressed in terms of semivariograms as

$$
\begin{cases}
\displaystyle\sum_{\beta=1}^{n(\mathbf{u})} \lambda_\beta^{OK}(\mathbf{u})\, [C(0) - \gamma(\mathbf{u}_\alpha - \mathbf{u}_\beta)] + \mu_{OK}(\mathbf{u}) = C(0) - \gamma(\mathbf{u}_\alpha - \mathbf{u}) \\[4pt]
\hspace{5cm} \alpha = 1, \ldots, n(\mathbf{u}) \\[4pt]
\displaystyle\sum_{\beta=1}^{n(\mathbf{u})} \lambda_\beta^{OK}(\mathbf{u}) = 1
\end{cases}
$$

Thanks to the non-bias condition $\sum_{\beta=1}^{n(\mathbf{u})} \lambda_\beta^{OK}(\mathbf{u}) = 1$, the variance term $C(0)$ cancels out from the first $n(\mathbf{u})$ equations, yielding this system:

$$
\begin{cases}
\displaystyle\sum_{\beta=1}^{n(\mathbf{u})} \lambda_\beta^{OK}(\mathbf{u})\,\gamma(\mathbf{u}_\alpha - \mathbf{u}_\beta) \;-\; \mu_{OK}(\mathbf{u}) \;=\; \gamma(\mathbf{u}_\alpha - \mathbf{u}) \\[4pt]
\hspace{5cm} \alpha = 1, \ldots, n(\mathbf{u}) \\[4pt]
\displaystyle\sum_{\beta=1}^{n(\mathbf{u})} \lambda_\beta^{OK}(\mathbf{u}) \;=\; 1
\end{cases}
$$

In contrast, the SK system (5.10) can be expressed in terms of only covariances since there is no similar constraint on the simple kriging weights.

The common practice consists of inferring and modeling the semivariogram rather than the covariance function. Indeed, the semivariogram $\gamma(\mathbf{h})$ allows one to filter the unknown local mean $m(\mathbf{u})$ that is deemed constant but unknown over the local neighborhood $W(\mathbf{u})$:

$$
\begin{aligned}
2\gamma(\mathbf{h}) &= \mathrm{E}\left\{[Z(\mathbf{u}') - Z(\mathbf{u}' + \mathbf{h})]^2\right\} \\
&= \mathrm{E}\left\{[R(\mathbf{u}') + m(\mathbf{u}') - R(\mathbf{u}' + \mathbf{h}) - m(\mathbf{u}' + \mathbf{h})]^2\right\} \\
&= \mathrm{E}\left\{[R(\mathbf{u}') - R(\mathbf{u}' + \mathbf{h})]^2\right\}
\end{aligned}
$$

since $m(\mathbf{u}') = m(\mathbf{u}' + \mathbf{h})$, $\forall\ \mathbf{u}', \mathbf{u}' + \mathbf{h} \in W(\mathbf{u})$.

However, for reasons of computational efficiency, kriging systems are usually solved in terms of covariances. As discussed in section 4.2.1, there are semivariogram models, such as the power model, that have no covariance counterpart. For unbounded semivariogram models, a "pseudo covariance" $C(\mathbf{h})$ is defined by subtracting the semivariogram model $\gamma(\mathbf{h})$ from any positive value A, such that $A - \gamma(\mathbf{h}) \geq 0$, $\forall\ \mathbf{h}$. Again, the non-bias condition allows the constant A to cancel out from the ordinary kriging system, which is then written in terms of pseudo covariances. In summary, the common practice consists of (1) inferring and modeling the semivariogram, and (2) solving all ordinary kriging systems in terms of (pseudo) covariances.

Kriging the local mean

Instead of an estimate of the attribute z, one may be interested in estimating and mapping the local mean of the attribute. Such a map of trend estimates allows one to evaluate local departures from the overall mean and provides a smooth picture of global trends. Like the OK estimator (5.16) of attribute values, the estimator $m_{OK}^*(\mathbf{u})$ of the local mean is written as a linear combination of $n(\mathbf{u})$ random variables:

$$m_{OK}^*(\mathbf{u}) = \sum_{\alpha=1}^{n(\mathbf{u})} \lambda_{\alpha m}^{OK}(\mathbf{u}) \, Z(\mathbf{u}_\alpha) \qquad (5.20)$$

where $\lambda_{\alpha m}^{OK}(\mathbf{u})$ is the weight associated with the datum $z(\mathbf{u}_\alpha)$ in the OK estimation of the local mean at location $\mathbf{u}$.

The unbiasedness of the estimator (5.20) is ensured by forcing the kriging weights $\lambda_{\alpha m}^{OK}(\mathbf{u})$ to sum to 1:

$$E\{m_{OK}^*(\mathbf{u}) - m(\mathbf{u})\} = \sum_{\alpha=1}^{n(\mathbf{u})} \lambda_{\alpha m}^{OK}(\mathbf{u}) \, m(\mathbf{u}) - m(\mathbf{u})$$

$$= \quad 0 \quad \text{if} \ \sum_{\alpha=1}^{n(\mathbf{u})} \lambda_{\alpha m}^{OK}(\mathbf{u}) = 1$$

The error variance $\sigma_E^2 = \mathrm{Var}\{m_{OK}^*(\mathbf{u}) - m(\mathbf{u})\}$ is expressed as a double linear combination of covariance values:

$$\sigma_E^2(\mathbf{u}) = \mathrm{Var}\{m_{OK}^*(\mathbf{u})\} + \mathrm{Var}\{m(\mathbf{u})\} - 2\,\mathrm{Cov}\{m_{OK}^*(\mathbf{u}), m(\mathbf{u})\}$$

$$= \sum_{\alpha=1}^{n(\mathbf{u})} \sum_{\beta=1}^{n(\mathbf{u})} \lambda_{\alpha m}^{OK}(\mathbf{u}) \, \lambda_{\beta m}^{OK}(\mathbf{u}) \, C(\mathbf{u}_\alpha - \mathbf{u}_\beta) \qquad (5.21)$$

The last two terms of the error variance are zero since the trend $m(\mathbf{u})$ is viewed as a deterministic component.

The minimization of the error variance (5.21) under the non-bias condition yields the following system of $(n(\mathbf{u}) + 1)$ linear equations:

$$\begin{cases} \displaystyle\sum_{\beta=1}^{n(\mathbf{u})} \lambda_{\beta m}^{OK}(\mathbf{u}) \, C(\mathbf{u}_\alpha - \mathbf{u}_\beta) + \mu_m^{OK}(\mathbf{u}) = 0 \\[2mm] \qquad\qquad\qquad\qquad\qquad\qquad\qquad \alpha = 1,\ldots,n(\mathbf{u}) \\[2mm] \displaystyle\sum_{\beta=1}^{n(\mathbf{u})} \lambda_{\beta m}^{OK}(\mathbf{u}) = 1 \end{cases} \qquad (5.22)$$

System (5.22) is identical to the OK system (5.18) except for the right-hand-side covariances $C(\mathbf{u}_\alpha - \mathbf{u})$ being set to zero.

Because all data-to-unknown covariance terms $C(\mathbf{u}_\alpha - \mathbf{u})$ are zero, the location $\mathbf{u}$ being estimated does not appear in the ordinary kriging system (5.22). Provided the same set of data is used to estimate the local mean at two different locations $\mathbf{u}$ and $\mathbf{u}'$, the system (5.22) remains unchanged. Thus, the two sets of kriging weights and the two trend estimates are identical: $\lambda_{\alpha m}^{OK}(\mathbf{u}) = \lambda_{\alpha m}^{OK}(\mathbf{u}')$, $\forall\ \alpha$ and $m_{OK}^*(\mathbf{u}) = m_{OK}^*(\mathbf{u}')$.

Simple kriging versus ordinary kriging

Ordinary kriging is usually preferred to simple kriging because it requires neither knowledge nor stationarity of the mean over the entire area $\mathcal{A}$. Several authors (Matheron, 1970, p. 129; Journel and Rossi, 1989) showed that ordinary kriging with local search neighborhoods amounts to:

1. estimating the local mean $m_{OK}^*(\mathbf{u})$ at each location $\mathbf{u}$ using ordinary kriging with data specific to the neighborhood of $\mathbf{u}$, then

2. applying the simple kriging estimator (5.7) using that estimate of the mean rather than the stationary mean m, that is,

$$
\begin{aligned}
Z_{OK}^*(\mathbf{u}) &= \sum_{\alpha=1}^{n(\mathbf{u})} \lambda_\alpha^{SK}(\mathbf{u}) \left[Z(\mathbf{u}_\alpha) - m_{OK}^*(\mathbf{u}) \right] + m_{OK}^*(\mathbf{u}) \\
&= \sum_{\alpha=1}^{n(\mathbf{u})} \lambda_\alpha^{SK}(\mathbf{u}) \, Z(\mathbf{u}_\alpha) + \lambda_m^{SK}(\mathbf{u}) \, m_{OK}^*(\mathbf{u}) \qquad (5.23)
\end{aligned}
$$

where $\lambda_m^{SK}(\mathbf{u}) = 1 - \sum_{\alpha=1}^{n(\mathbf{u})} \lambda_\alpha^{SK}(\mathbf{u})$. Accounting for the definition (5.14) of the simple kriging estimator, one deduces the following relation between the SK and OK estimators:

$$
\begin{aligned}
Z_{OK}^*(\mathbf{u}) &= Z_{SK}^*(\mathbf{u}) - \lambda_m^{SK}(\mathbf{u}) \, m + \lambda_m^{SK}(\mathbf{u}) \, m_{OK}^*(\mathbf{u}) \\
&= Z_{SK}^*(\mathbf{u}) + \lambda_m^{SK}(\mathbf{u}) \left[m_{OK}^*(\mathbf{u}) - m \right]
\end{aligned}
$$

The difference between the SK and OK estimates of z at $\mathbf{u}$ is caused by a departure of the local mean $m_{OK}^*(\mathbf{u})$ from the global mean m. More precisely, since $\lambda_m^{SK}(\mathbf{u})$ is usually positive, the OK estimate is smaller than the SK estimate in low-valued areas where the local data mean is smaller than the global mean. Conversely, the OK estimate is larger than the SK estimate in high-valued areas where the local mean is larger than the global mean. In this sense, the commonly used OK algorithm with local search neighborhoods already accounts for trends (varying mean) in the z-data values (see discussion in section 5.4).

The discrepancy between the two estimates $z_{SK}^*(\mathbf{u})$ and $z_{OK}^*(\mathbf{u})$ increases as the weight $\lambda_m^{SK}(\mathbf{u})$ of the mean increases, i.e., as the location $\mathbf{u}$ being estimated gets farther away from data locations (section 5.2).

Example

Figure 5.3 (top graph) shows ten Cd concentrations at locations $\mathbf{u}_1$ to $\mathbf{u}_{10}$ along the NE-SW transect. The local mean is estimated every 50 m along that transect using the five closest data values; the resulting estimate is depicted by a solid line on the middle graph of Figure 5.3. The vertical dashed lines delineate the OK trend estimates that are based on the same set of five

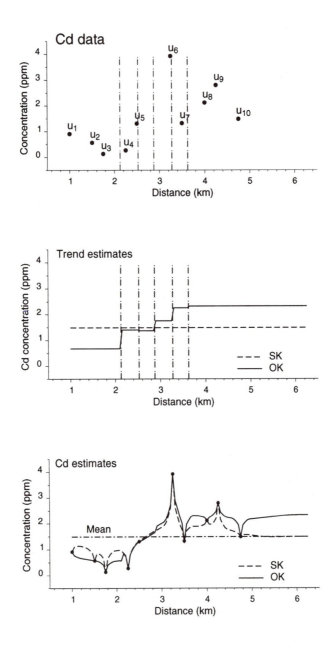

Figure 5.3: SK and OK estimates of the trend (middle graph) and of Cd concentrations (bottom graph). The vertical dashed lines delineate the segments that are estimated using the same five Cd concentrations. For example, the first segment, 1–2.1 km, includes all estimates that are based on the data at locations u_1 to u_5. The local mean is constant within each segment.

neighboring data. For example, Cd concentrations at locations $\mathbf{u}_1$ to $\mathbf{u}_5$ are used to estimate the local mean at all locations within the segment 1–2.1 km. Similarly, the next segment, 2.1–2.5 km, includes all trend values that are estimated from Cd concentrations at locations $\mathbf{u}_2$ to $\mathbf{u}_6$.

As mentioned previously, the ordinary kriging system (5.22) is identical at all locations where the same neighboring data are involved in the estimation. Consequently, the OK trend estimate $m_{OK}^*(\mathbf{u})$ is constant within each segment and changes from one segment to another depending on the neighboring data retained. This procedure yields a trend estimate that follows the general increase in Cd concentrations along the transect. In contrast, the mean of 10 data values, $m=1.49$ ppm (horizontal dashed line in the middle graph), overestimates the local mean in the low-valued (left) part of the transect and underestimates the local mean in the high-valued (right) part of the transect.

Figure 5.3 (bottom graph) shows both OK (solid line) and SK (dashed line) estimates of Cd concentrations along the NE-SW transect. Note that:

- Both estimators are exact.

- OK estimates are smaller than SK estimates in the left part of the transect where the local mean $m_{OK}^*(\mathbf{u})$ is smaller than the overall mean $m = 1.49$.

- OK estimates are larger than SK estimates in the right part of the transect where the local mean $m_{OK}^*(\mathbf{u})$ is larger than the overall mean $m = 1.49$.

- The departure between the two estimates is most important beyond the extreme right datum $\mathbf{u}_{10}$. Indeed, away from the data, the weight $\lambda_m^{SK}(\mathbf{u})$ of the mean increases (Figure 5.2, bottom graph); hence the SK estimate is closer to the overall mean $m = 1.49$. In contrast, the OK estimate is closer to the local mean $m_{OK}^*(\mathbf{u})$, which is estimated from the last and large Cd concentrations.

In summary, the use of a stationary mean yields SK estimates that are close to that mean value (1.49 ppm) away from the data. In contrast, local estimation of the mean within search neighborhoods yields OK estimates that better follow the data fluctuations: small values in the left part of the transect and large values in the right part.

5.4   Kriging with a Trend Model

The local estimation of the mean in ordinary kriging allows one to account for any "global" trend in the data over the study area $\mathcal{A}$. Thus, the OK algorithm implicitly considers a non-stationary random function model, where stationarity is limited within each search neighborhood $W(\mathbf{u})$. In some situations, it may be inappropriate to consider the local mean $m(\mathbf{u})$ as constant even within small search neighborhoods. Kriging with a trend (KT) consists of modeling the "local" trend within the neighborhood $W(\mathbf{u})$ as a smoothly varying function of the coordinates; recall expression (5.6):

$$m(\mathbf{u}') = \sum_{k=0}^{K} a_k(\mathbf{u}') f_k(\mathbf{u}') \quad \text{with } a_k(\mathbf{u}') \approx a_k \ \forall \, \mathbf{u}' \in W(\mathbf{u})$$

The functions $f_k(\mathbf{u}')$ are known, whereas the coefficients $a_k(\mathbf{u}')$ are unknown and deemed constant within each local neighborhood $W(\mathbf{u})$. The linear estimator (5.1) is thus written

$$Z^*(\mathbf{u}) = \sum_{k=0}^{K} a_k(\mathbf{u}) f_k(\mathbf{u}) + \sum_{\alpha=1}^{n(\mathbf{u})} \lambda_\alpha(\mathbf{u}) \left[Z(\mathbf{u}_\alpha) - \sum_{k=0}^{K} f_k(\mathbf{u}_\alpha) a_k(\mathbf{u}) \right]$$

$$= \sum_{\alpha=1}^{n(\mathbf{u})} \lambda_\alpha(\mathbf{u}) Z(\mathbf{u}_\alpha) + \sum_{k=0}^{K} a_k(\mathbf{u}) \left[f_k(\mathbf{u}) - \sum_{\alpha=1}^{n(\mathbf{u})} \lambda_\alpha(\mathbf{u}) f_k(\mathbf{u}_\alpha) \right]$$

The $(K+1)$ unknown coefficients $a_k(\mathbf{u})$ are filtered from the linear estimator by imposing the following $(K+1)$ constraints:

$$\sum_{\alpha=1}^{n(\mathbf{u})} \lambda_\alpha(\mathbf{u}) f_k(\mathbf{u}_\alpha) = f_k(\mathbf{u}) \qquad k = 0, \dots, K \tag{5.24}$$

By convention, the first trend function $f_0(\mathbf{u})$ is the unit constant, that is, $f_0(\mathbf{u}) = 1$. Hence the first condition is similar to the OK constraint on the weights: $\sum_{\alpha=1}^{n(\mathbf{u})} \lambda_\alpha(\mathbf{u}) = 1$.

The constraints (5.24) allow one to express the KT estimator as a linear combination of only the $n(\mathbf{u})$ RVs $Z(\mathbf{u}_\alpha)$:

$$Z^*_{KT}(\mathbf{u}) = \sum_{\alpha=1}^{n(\mathbf{u})} \lambda_\alpha^{KT}(\mathbf{u}) Z(\mathbf{u}_\alpha) \tag{5.25}$$

with

$$\sum_{\alpha=1}^{n(\mathbf{u})} \lambda_\alpha^{KT}(\mathbf{u}) f_k(\mathbf{u}_\alpha) = f_k(\mathbf{u}) \qquad k = 0, \dots, K$$

The kriging with trend estimator (5.25) is unbiased since the error mean is equal to zero:

$$E\{Z^*_{KT}(\mathbf{u}) - Z(\mathbf{u})\} = \sum_{\alpha=1}^{n(\mathbf{u})} \lambda_\alpha^{KT}(\mathbf{u}) m(\mathbf{u}_\alpha) - m(\mathbf{u})$$

$$= \sum_{\alpha=1}^{n(\mathbf{u})} \lambda_\alpha^{KT}(\mathbf{u}) \sum_{k=0}^{K} a_k(\mathbf{u}) f_k(\mathbf{u}_\alpha) - \sum_{k=0}^{K} a_k(\mathbf{u}) f_k(\mathbf{u})$$

$$= \sum_{k=0}^{K} a_k(\mathbf{u}) \left[\sum_{\alpha=1}^{n(\mathbf{u})} \lambda_\alpha^{KT}(\mathbf{u}) f_k(\mathbf{u}_\alpha) - f_k(\mathbf{u}) \right] = 0$$

because of the $(K+1)$ constraints (5.24).

The minimization of the corresponding error variance, of type (5.8), under the $(K+1)$ non-bias conditions (5.24) calls for the definition of a Lagrangian $L(\mathbf{u})$. The procedure is similar to that for ordinary kriging except that there are now $(K+1)$ Lagrange parameters $\mu_k^{KT}(\mathbf{u})$ accounting for the $(K+1)$ constraints on the weights. Setting the $(n(\mathbf{u})+K+1)$ partial first derivatives to zero yields the following system of $(n(\mathbf{u})+K+1)$ linear equations:

$$
\begin{cases}
\displaystyle\sum_{\beta=1}^{n(\mathbf{u})} \lambda_\beta^{KT}(\mathbf{u})\, C_R(\mathbf{u}_\alpha - \mathbf{u}_\beta) + \sum_{k=0}^{K} \mu_k^{KT}(\mathbf{u})\, f_k(\mathbf{u}_\alpha) = C_R(\mathbf{u}_\alpha - \mathbf{u}) \\
\qquad\qquad\qquad\qquad\qquad\qquad\qquad\qquad \alpha = 1,\ldots,n(\mathbf{u}) \\[2mm]
\displaystyle\sum_{\beta=1}^{n(\mathbf{u})} \lambda_\beta^{KT}(\mathbf{u}) = 1 \\[2mm]
\displaystyle\sum_{\beta=1}^{n(\mathbf{u})} \lambda_\beta^{KT}(\mathbf{u})\, f_k(\mathbf{u}_\beta) = f_k(\mathbf{u}) \qquad k = 1,\ldots,K
\end{cases}
$$

$$(5.26)$$

Accounting for the first $n(\mathbf{u})$ equations in system (5.26), the minimized error variance becomes

$$
\sigma_{KT}^2(\mathbf{u}) = C_R(0) - \sum_{\alpha=1}^{n(\mathbf{u})} \lambda_\alpha^{KT}(\mathbf{u})\, C_R(\mathbf{u}_\alpha - \mathbf{u}) - \sum_{k=0}^{K} \mu_k^{KT}(\mathbf{u})\, f_k(\mathbf{u}) \quad (5.27)
$$

Note that for $K = 0$, system (5.26) reverts to the ordinary kriging system (5.17). Thus, the KT estimator (5.25) and kriging variance (5.27) are equal to the OK estimator (5.16) and kriging variance (5.19).

Kriging with a trend model requires a prior determination of (1) the K trend functions $f_k(\mathbf{u})$, and (2) the covariance of the residual component $R(\mathbf{u})$, $C_R(\mathbf{h})$.

Modeling the trend

The type of functions $f_k(\mathbf{u})$ may be directly suggested by the physics of the problem. For example, a series of sine and cosine functions can be used to model a periodic trend of an attribute; for example, see Séguret and Huchon (1990). In most spatial situations, the earth scientist has no physical ground for choosing a particular type of analytical trend function. Because the concept of trend is usually associated with a smoothly varying component of the z-variability, low-order (≤ 2) polynomials are typically used to model the trend, for example,

- a linear trend in $\mathcal{R}^2$ ($K = 2$):

$$
m(\mathbf{u}) = m(x,y) = a_0 + a_1 x + a_2 y
$$

- a quadratic trend in $\mathcal{R}^2$ ($K = 5$):

$$m(\mathbf{u}) = a_0 + a_1 x + a_2 y + a_3 x^2 + a_4 y^2 + a_5 xy$$

where (x, y) are the coordinates of the location $\mathbf{u}$. The trend may be limited to a particular direction, say, the direction of prevailing wind for airbone pollution or of steepest slope for soil properties. For example, a linear trend in the $45°$ direction would be defined as

$$m(\mathbf{u}) = a_0 + a_1(x + y)$$

Inferring the residual covariance

In practice, the residual semivariogram $\gamma_R(\mathbf{h})$ is first inferred, then the residual (pseudo) covariance $C_R(\mathbf{h})$ is deduced as $A - \gamma_R(\mathbf{h})$. Like in ordinary kriging, the first unbiasedness condition $\sum_{\beta=1}^{n(\mathbf{u})} \lambda_\beta^{KT}(\mathbf{u}) = 1$ filters out the arbitrary constant A from the first $n(\mathbf{u})$ equations in the KT system (5.26).

The computation of the residual semivariogram $\gamma_R(\mathbf{h})$ is not straightforward because available data are z-values, not residual values. The experimentally available z-semivariogram $\gamma(\mathbf{h})$ is related to $\gamma_R(\mathbf{h})$:

$$
\begin{aligned}
2\gamma(\mathbf{h}) &= \mathrm{E}\left\{[Z(\mathbf{u}) - Z(\mathbf{u}+\mathbf{h})]^2\right\} \\
&= \mathrm{E}\left\{[R(\mathbf{u}) + m(\mathbf{u}) - R(\mathbf{u}+\mathbf{h}) - m(\mathbf{u}+\mathbf{h})]^2\right\} \\
&= 2\gamma_R(\mathbf{h}) + [m(\mathbf{u}) - m(\mathbf{u}+\mathbf{h})]^2
\end{aligned}
$$

where the trend values $m(\mathbf{u})$ and $m(\mathbf{u}+\mathbf{h})$ are unknown.

A first solution to the problem of inferring $\gamma_R(\mathbf{h})$ consists of selecting data pairs that are unaffected or slightly affected by the trend, i.e., data pairs such that

$$m(\mathbf{u}_\alpha) \approx m(\mathbf{u}_\alpha + \mathbf{h}), \text{ hence}$$
$$r(\mathbf{u}_\alpha) - r(\mathbf{u}_\alpha + \mathbf{h}) \approx z(\mathbf{u}_\alpha) - z(\mathbf{u}_\alpha + \mathbf{h}) \tag{5.28}$$

The residual semivariogram can then be inferred directly from the z-semivariogram computed from such pairs. The following guidelines are useful:

- Relation (5.28) is generally satisfied for small separation distances $|\mathbf{h}|$. Thus, the residual semivariogram for the first lags may be identified with the corresponding z-semivariogram $\gamma(\mathbf{h})$.

- For larger distances, the residual semivariogram would be inferred from pairs of z-values taken in subareas or along directions (e.g., perpendicular to the trend) where the influence of the trend can be ignored. In the latter case, the inaccessible residual semivariogram in the trend direction is deemed similar to that computed in the perpendicular direction; such a decision amounts to modeling the pattern of variation of the residuals as isotropic. Any anisotropy in the z-data is accounted for in the trend model, not in the residual covariance.

Another approach (Delfiner, 1976) consists of defining linear combinations of z-data that filter the trend $m(\mathbf{u})$. For example, a trend of order 1 such as $m(\mathbf{u}) = a_0 + a_1 \cdot \mathbf{u}$ would be filtered by "differences of order 2" such as $[z(\mathbf{u}) - 2z(\mathbf{u} + \mathbf{h}) + z(\mathbf{u} + 2\mathbf{h})]$. Indeed, each component of that difference of order 2 can be expressed as the sum of a residual and a trend component:

$$
\begin{aligned}
z(\mathbf{u}) &= r(\mathbf{u}) + a_0 + a_1 \cdot \mathbf{u} \\
z(\mathbf{u} + \mathbf{h}) &= r(\mathbf{u} + \mathbf{h}) + a_0 + a_1 \cdot \mathbf{u} + a_1 \cdot \mathbf{h} \\
z(\mathbf{u} + 2\mathbf{h}) &= r(\mathbf{u} + 2\mathbf{h}) + a_0 + a_1 \cdot \mathbf{u} + 2a_1 \cdot \mathbf{h}
\end{aligned}
$$

One verifies that the previous linear combination of z-values reverts to a linear combination of residual values, thereby filtering the trend component:

$$
z(\mathbf{u}) - 2z(\mathbf{u} + \mathbf{h}) + z(\mathbf{u} + 2\mathbf{h}) \;=\; r(\mathbf{u}) - 2r(\mathbf{u} + \mathbf{h}) + r(\mathbf{u} + 2\mathbf{h})
$$

More generally, a difference of order $(k + 1)$ filters any polynomial trend of order k. The variance of such differences is referred to as the generalized covariance of order k and is used as residual covariance in system (5.26). Beware that:

- Such high-order differences are not readily available when data are non gridded.

- The automatic modeling of generalized covariances from experimental high-order differences or related statistics typically results in very large (artifact) relative nugget effects.

Warning

An unreliable practice would consist of:

1. estimating the residual values as $\widehat{r}(\mathbf{u}_\alpha) = z(\mathbf{u}_\alpha) - \widehat{m}(\mathbf{u}_\alpha)$ from least-squares polynomial estimates $\widehat{m}(\mathbf{u}_\alpha)$ of the trend component, then

2. applying the KT algorithm using the residual semivariogram $\gamma_{\widehat{R}}(\mathbf{h})$ inferred from the estimated values $\widehat{r}(\mathbf{u}_\alpha)$.

Indeed, the semivariogram of estimated residuals, $\gamma_{\widehat{R}}(\mathbf{h})$, strongly depends on the algorithm used to estimate the trend component and may depart from the residual semivariogram $\gamma_R(\mathbf{h})$. A better alternative would be to interpolate the residual values $\widehat{r}(\mathbf{u}_\alpha)$ using SK and their semivariogram $\gamma_{\widehat{R}}(\mathbf{h})$, then add the trend estimates $\widehat{m}(\mathbf{u})$ back to the interpolated residuals $r^*_{SK}(\mathbf{u})$ to get the estimates $z^*(\mathbf{u})$. A similar approach is presented in Chapter 6, where the trend estimates $\widehat{m}(\mathbf{u})$ are deduced from exhaustively sampled secondary information rather than being modeled as a specific function of the spatial coordinates $\mathbf{u}$.

Matrix notation

As an example, consider the following linear trend model in two dimensions:

$$m(\mathbf{u}) = m(x, y) = a_0 + a_1 x + a_2 y$$

where (x, y) are the coordinates of the location $\mathbf{u}$ being estimated. The KT system (5.26) includes $(n(\mathbf{u}) + 3)$ linear equations:

$$
\left\{
\begin{array}{l}
\displaystyle\sum_{\beta=1}^{n(\mathbf{u})} \lambda_\beta^{KT}(\mathbf{u}) C_R(\mathbf{u}_\alpha - \mathbf{u}_\beta) + \mu_0^{KT}(\mathbf{u}) + \mu_1^{KT}(\mathbf{u}) x_\alpha + \mu_2^{KT}(\mathbf{u}) y_\alpha = C_R(\mathbf{u}_\alpha - \mathbf{u}) \\[2pt]
\qquad\qquad\qquad\qquad\qquad\qquad\qquad\qquad\qquad\qquad \alpha = 1, \ldots, n(\mathbf{u}) \\[4pt]
\displaystyle\sum_{\beta=1}^{n(\mathbf{u})} \lambda_\beta^{KT}(\mathbf{u}) = 1 \\[4pt]
\displaystyle\sum_{\beta=1}^{n(\mathbf{u})} \lambda_\beta^{KT}(\mathbf{u})\, x_\beta = x \\[4pt]
\displaystyle\sum_{\beta=1}^{n(\mathbf{u})} \lambda_\beta^{KT}(\mathbf{u})\, y_\beta = y
\end{array}
\right.
$$

$$(5.29)$$

where (x_α, y_α) are the coordinates of any datum location $\mathbf{u}_\alpha$. Using matrix notation, the KT system (5.29) is written as

$$\mathbf{K}_{KT}\, \boldsymbol{\lambda}_{KT}(\mathbf{u}) = \mathbf{k}_{KT} \qquad\qquad (5.30)$$

where $\mathbf{K}_{KT}$ is the $(n(\mathbf{u}) + 3) \times (n(\mathbf{u}) + 3)$ matrix:

$$
\mathbf{K}_{KT} =
\begin{bmatrix}
C_R(\mathbf{u}_1 - \mathbf{u}_1) & \cdots & C_R(\mathbf{u}_1 - \mathbf{u}_{n(\mathbf{u})}) & 1 & x_1 & y_1 \\
\vdots & \vdots & \vdots & \vdots & \vdots & \vdots \\
C_R(\mathbf{u}_{n(\mathbf{u})} - \mathbf{u}_1) & \cdots & C_R(\mathbf{u}_{n(\mathbf{u})} - \mathbf{u}_{n(\mathbf{u})}) & 1 & x_{n(\mathbf{u})} & y_{n(\mathbf{u})} \\
1 & \cdots & 1 & 0 & 0 & 0 \\
x_1 & \cdots & x_{n(\mathbf{u})} & 0 & 0 & 0 \\
y_1 & \cdots & y_{n(\mathbf{u})} & 0 & 0 & 0
\end{bmatrix}
$$

$\boldsymbol{\lambda}_{KT}(\mathbf{u})$ is the vector of KT weights and Lagrange parameters, and $\mathbf{k}_{KT}$ is the vector of data-to-unknown covariances and trend functions:

$$
\boldsymbol{\lambda}_{KT}(\mathbf{u}) =
\begin{bmatrix}
\lambda_1^{KT}(\mathbf{u}) \\
\vdots \\
\lambda_{n(\mathbf{u})}^{KT}(\mathbf{u}) \\
\mu_0^{KT}(\mathbf{u}) \\
\mu_1^{KT}(\mathbf{u}) \\
\mu_2^{KT}(\mathbf{u})
\end{bmatrix}
\qquad
\mathbf{k}_{KT} =
\begin{bmatrix}
C_R(\mathbf{u}_1 - \mathbf{u}) \\
\vdots \\
C_R(\mathbf{u}_{n(\mathbf{u})} - \mathbf{u}) \\
1 \\
x \\
y
\end{bmatrix}
$$

The matrix $\mathbf{K}_{KT}$ in system (5.30) is obtained by adding three rows and three columns to the covariance matrix $\mathbf{K}_{SK} = [C(\mathbf{u}_\alpha - \mathbf{u}_\beta)] = [C_R(\mathbf{u}_\alpha - \mathbf{u}_\beta)]$

of the simple kriging system (5.13). Adding only the first row and the first column of 1's to the data covariance matrix $\mathbf{K}_{SK}$ yields the OK system (5.18). Thus, a continuum exists between unconstrained simple kriging and kriging with a trend model. The number of constraints on the weights increases with the complexity of the trend model, i.e., as the number $(K+1)$ of trend functions $f_k(\mathbf{u})$ retained increases. Such constraints on the weights amount to selecting only those linear combinations of data that can filter out the unknown trend $m(\mathbf{u})$.

With more concise notation, the matrix formulation (5.30) of the KT system for any number $(K+1)$ of trend functions is written

$$
\begin{bmatrix}
[C_R(\mathbf{u}_\alpha - \mathbf{u}_\beta)] & [f_k(\mathbf{u}_\alpha)]^T \\
[f_k(\mathbf{u}_\beta)] & [0]
\end{bmatrix}
\begin{bmatrix}
[\lambda_\beta^{KT}(\mathbf{u})]^T \\
[\mu_k^{KT}(\mathbf{u})]^T
\end{bmatrix}
=
\begin{bmatrix}
[C_R(\mathbf{u}_\alpha - \mathbf{u})]^T \\
[f_k(\mathbf{u})]^T
\end{bmatrix}
\tag{5.31}
$$

The kriging weights are obtained as $\boldsymbol{\lambda}_{KT}(\mathbf{u}) = \mathbf{K}_{KT}^{-1}\,\mathbf{k}_{KT}$ and the kriging variance $\sigma_{KT}^2(\mathbf{u})$ is computed as

$$
\sigma_{KT}^2(\mathbf{u}) = C_R(0) - \boldsymbol{\lambda}_{KT}^T(\mathbf{u})\,\mathbf{k}_{KT} = C_R(0) - \mathbf{k}_{KT}^T\,\mathbf{K}_{KT}^{-1}\,\mathbf{k}_{KT}
$$

The kriging system (5.31) has a unique solution if these two conditions are met:

1. The covariance matrix $[C_R(\mathbf{u}_\alpha - \mathbf{u}_\beta)]$ is positive definite, in practice if:

   - no two data are colocated: $\mathbf{u}_\alpha \neq \mathbf{u}_\beta$ for $\alpha \neq \beta$.
   - the residual covariance model $C_R(\mathbf{h})$ is permissible.

2. The $(K+1)$ functions $f_k(\mathbf{u})$ are linearly independent on the set of $n(\mathbf{u})$ data; that is, the relations $\sum_{k=0}^K c_k f_k(\mathbf{u}_\alpha) = 0 \ \forall \alpha = 1, \ldots, n(\mathbf{u})$, would require that $c_k = 0 \ \forall k = 0, \ldots, K$. Such a condition means that a drift along a particular direction cannot be estimated if all $n(\mathbf{u})$ data are aligned perpendicular to that direction (Journel and Huijbregts, 1978, p. 319). Moreover, the number $(K+1)$ of trend functions cannot exceed the number $n(\mathbf{u})$ of data; for example, the modeling of a quadratic trend in two dimensions requires at least six data values.

Kriging the trend

The trend component $m(\mathbf{u})$ can be estimated explicitly using an approach that is similar to kriging the local mean in relation (5.20). The KT estimator of the trend is expressed as a linear combination of $n(\mathbf{u})$ random variables:

$$
m_{KT}^*(\mathbf{u}) = \sum_{\alpha=1}^{n(\mathbf{u})} \lambda_{\alpha m}^{KT}(\mathbf{u})\, Z(\mathbf{u}_\alpha)
\tag{5.32}
$$

where $\lambda_{\alpha m}^{KT}(\mathbf{u})$ is the weight associated with the datum $z(\mathbf{u}_\alpha)$. The kriging weights are obtained by solving a kriging system identical to the KT system (5.26) except that the right-hand-side covariances $C_R(\mathbf{u}_\alpha - \mathbf{u})$ are set to zero:

$$
\begin{cases}
\displaystyle\sum_{\beta=1}^{n(\mathbf{u})} \lambda_{\beta m}^{KT}(\mathbf{u})\, C_R(\mathbf{u}_\alpha - \mathbf{u}_\beta) + \sum_{k=0}^{K} \mu_{km}^{KT}(\mathbf{u})\, f_k(\mathbf{u}_\alpha) = 0 \\[4pt]
\hspace{6cm} \alpha = 1, \ldots, n(\mathbf{u}) \\[6pt]
\displaystyle\sum_{\beta=1}^{n(\mathbf{u})} \lambda_{\beta m}^{KT}(\mathbf{u}) = 1 \\[4pt]
\displaystyle\sum_{\beta=1}^{n(\mathbf{u})} \lambda_{\beta m}^{KT}(\mathbf{u})\, f_k(\mathbf{u}_\beta) = f_k(\mathbf{u}) \qquad k = 1, \ldots, K
\end{cases}
\tag{5.33}
$$

where $\mu_{km}^{KT}(\mathbf{u})$ is the Lagrange parameter that accounts for the $(k+1)$th KT constraint on the kriging weights. For $K = 0$, system (5.33) reverts to the ordinary kriging system (5.22) for estimating the local mean.

Estimation of the trend component $m(\mathbf{u})$ amounts to first estimating the $(K+1)$ trend coefficients $a_k(\mathbf{u})$, then computing the trend estimate as a linear combination of the known trend functions $f_k(\mathbf{u})$:

$$
m_{KT}^*(\mathbf{u}) = \sum_{k=0}^{K} a_k^*(\mathbf{u})\, f_k(\mathbf{u})
$$

Like the trend component, the unknown trend coefficients $a_k(\mathbf{u})$ can be estimated as linear combinations of z-values. For example, the linear estimator of the coefficient $a_{k'}$ at location $\mathbf{u}$ is

$$
a_{k'}^*(\mathbf{u}) = \sum_{\alpha=1}^{n(\mathbf{u})} \lambda_{\alpha k'}^{KT}(\mathbf{u})\, Z(\mathbf{u}_\alpha)
\tag{5.34}
$$

where $\lambda_{\alpha k'}^{KT}(\mathbf{u})$ is the weight associated with the z-datum at location $\mathbf{u}_\alpha$ for the KT estimation of the trend coefficient $a_{k'}$ at $\mathbf{u}$. The estimator $a_{k'}^*(\mathbf{u})$ is a RV, being a linear combination of the RVs $Z(\mathbf{u}_\alpha)$, whereas $a_k(\mathbf{u})$ is a deterministic, though unknown, value.

The estimator (5.34) is unbiased if the error mean is zero, that is, if

$$
\begin{aligned}
\mathrm{E}\{a_{k'}^*(\mathbf{u}) - a_{k'}(\mathbf{u})\} &= \sum_{\alpha=1}^{n(\mathbf{u})} \lambda_{\alpha k'}^{KT}(\mathbf{u}) \sum_{k=0}^{K} a_k(\mathbf{u})\, f_k(\mathbf{u}_\alpha) - a_{k'}(\mathbf{u}) \\[6pt]
&= a_{k'}(\mathbf{u}) \left[\sum_{\alpha=1}^{n(\mathbf{u})} \lambda_{\alpha k'}^{KT}(\mathbf{u})\, f_{k'}(\mathbf{u}_\alpha) - 1 \right] \\[6pt]
&\quad + \sum_{\substack{k=0 \\ k \neq k'}}^{K} a_k(\mathbf{u}) \sum_{\alpha=1}^{n(\mathbf{u})} \lambda_{\alpha k'}^{KT}(\mathbf{u}) f_k(\mathbf{u}_\alpha) \\[6pt]
&= 0
\end{aligned}
$$

which leads to the following $(K + 1)$ constraints on the kriging weights:

$$
\begin{cases}
\displaystyle\sum_{\alpha=1}^{n(\mathbf{u})} \lambda_{\alpha k'}^{KT}(\mathbf{u}) \ f_{k'}(\mathbf{u}_\alpha) = 1 \\
\displaystyle\sum_{\alpha=1}^{n(\mathbf{u})} \lambda_{\alpha k'}^{KT}(\mathbf{u}) \ f_k(\mathbf{u}_\alpha) = 0 \qquad \forall \ k \neq k'
\end{cases}
\tag{5.35}
$$

The minimization of the error variance $\mathrm{Var}\{a_{k'}^*(\mathbf{u}) - a_{k'}(\mathbf{u})\}$ under the $(K + 1)$ constraints (5.35) yields a kriging system identical to the KT system (5.33) except for the $(K + 1)$ non-bias conditions:

$$
\begin{cases}
\displaystyle\sum_{\beta=1}^{n(\mathbf{u})} \lambda_{\beta k'}^{KT}(\mathbf{u}) \ C_R(\mathbf{u}_\alpha - \mathbf{u}_\beta) + \sum_{k=0}^{K} \mu_{k k'}^{KT}(\mathbf{u}) \ f_k(\mathbf{u}_\alpha) = 0 \\
\hspace{6cm} \alpha = 1, \ldots, n(\mathbf{u}) \\
\displaystyle\sum_{\beta=1}^{n(\mathbf{u})} \lambda_{\beta k'}^{KT}(\mathbf{u}) \ f_{k'}(\mathbf{u}_\beta) = 1 \\
\displaystyle\sum_{\beta=1}^{n(\mathbf{u})} \lambda_{\beta k'}^{KT}(\mathbf{u}) \ f_k(\mathbf{u}_\beta) = 0 \qquad \forall \ k \neq k'
\end{cases}
\tag{5.36}
$$

Like the kriging system (5.33) for the trend component, the right-hand-side covariance terms $\mathrm{Cov}\{Z(\mathbf{u}_\alpha), a(\mathbf{u})\}$ in system (5.36) are zero because the actual trend coefficients are viewed as deterministic. Note the following:

- Both kriging systems (5.33) and (5.36) are identical for $K = 0$. Indeed, the constant trend component $m(\mathbf{u})$ is then the single trend coefficient $a_0(\mathbf{u})$.

- The location $\mathbf{u}$ being estimated does not appear in the kriging system (5.36). Provided the same set of data is used to estimate any trend coefficient a_k at two different locations $\mathbf{u}$ and $\mathbf{u}'$, their estimates are identical: $a_k^*(\mathbf{u}) = a_k^*(\mathbf{u}') \ \forall \ k = 0, \ldots, K$. For $K = 0$, the trend estimates at $\mathbf{u}$ and $\mathbf{u}'$ are then identical (section 5.3). For polynomials of higher order, say, a linear trend in $\mathcal{R}^1$, the trend estimates at locations $\mathbf{u} = (x, 0)$ and $\mathbf{u}' = (x', 0)$ appear as the same linear functions of coordinates x and x', $m_{KT}^*(\mathbf{u}) = a_0^* + a_1^* \cdot x$ and $m_{KT}^*(\mathbf{u}') = a_0^* + a_1^* \cdot x'$.

Ordinary kriging versus kriging with a trend

As mentioned in section 5.3, ordinary kriging amounts to estimating, within each search neighborhood $W(\mathbf{u})$, the local constant mean $m_{OK}^*(\mathbf{u})$, then performing SK on the corresponding residuals:

$$
Z_{OK}^*(\mathbf{u}) - m_{OK}^*(\mathbf{u}) = \sum_{\alpha=1}^{n(\mathbf{u})} \lambda_\alpha^{SK}(\mathbf{u}) \ [Z(\mathbf{u}_\alpha) - m_{OK}^*(\mathbf{u})]
$$

Similarly, kriging with a trend amounts to estimating, within each search neighborhood $W(\mathbf{u})$, the trend components $m_{KT}^*(\mathbf{u})$ and $m_{KT}^*(\mathbf{u}_\alpha)$, then performing SK on the corresponding residuals:

$$Z_{KT}^*(\mathbf{u}) - m_{KT}^*(\mathbf{u}) = \sum_{\alpha=1}^{n(\mathbf{u})} \lambda_\alpha^{SK}(\mathbf{u}) \, [Z(\mathbf{u}_\alpha) - m_{KT}^*(\mathbf{u}_\alpha)]$$

with

$$m_{KT}^*(\mathbf{u}_\alpha) = \sum_{k=0}^{K} a_k^*(\mathbf{u}) \, f_k(\mathbf{u}_\alpha)$$

Unlike in ordinary kriging, the trend component in KT is not constant within the search neighborhood. Rather, it depends on the coordinates of the location being estimated and of the data locations.

Both OK and KT estimates $z_{OK}^*(\mathbf{u})$ and $z_{KT}^*(\mathbf{u})$ can be expressed as the sum of two terms: the same linear combination of $n(\mathbf{u})$ data $z(\mathbf{u}_\alpha)$ and a function of the trend estimates:

$$z_{OK}^*(\mathbf{u}) = \sum_{\alpha=1}^{n(\mathbf{u})} \lambda_\alpha^{SK}(\mathbf{u}) \, z(\mathbf{u}_\alpha) + \underbrace{m_{OK}^*(\mathbf{u}) - \sum_{\alpha=1}^{n(\mathbf{u})} \lambda_\alpha^{SK}(\mathbf{u}) \, m_{OK}^*(\mathbf{u})}_{f\,(m_{OK}^*(\mathbf{u}))}$$

$$(5.37)$$

$$z_{KT}^*(\mathbf{u}) = \sum_{\alpha=1}^{n(\mathbf{u})} \lambda_\alpha^{SK}(\mathbf{u}) \, z(\mathbf{u}_\alpha) + \underbrace{m_{KT}^*(\mathbf{u}) - \sum_{\alpha=1}^{n(\mathbf{u})} \lambda_\alpha^{SK}(\mathbf{u}) \, m_{KT}^*(\mathbf{u}_\alpha)}_{f\,(m_{KT}^*(\mathbf{u}))}$$

$$(5.38)$$

The difference between the two estimates is thus

$$z_{OK}^*(\mathbf{u}) - z_{KT}^*(\mathbf{u}) = [m_{OK}^*(\mathbf{u}) - m_{KT}^*(\mathbf{u})]$$
$$- \sum_{\alpha=1}^{n(\mathbf{u})} \lambda_\alpha^{SK}(\mathbf{u}) \, [m_{OK}^*(\mathbf{u}) - m_{KT}^*(\mathbf{u}_\alpha)] \quad (5.39)$$

Any difference between the OK and KT estimates originates from a difference between the two trend estimates, hence from the usually arbitrary decision of modeling the local trend $m(\mathbf{u})$ within a neighborhood $W(\mathbf{u})$ as a constant or a particular polynomial of order K.

Numerical example
Consider, for example, the modeling of the trend component within two search neighborhoods centered at $\mathbf{u}_0 = (2.0, 0)$ and $\mathbf{u}_0' = (2.3, 0)$ along the NE-SW

transect of Figure 5.4. The trend is arbitrarily modeled as a linear function of the coordinate x along that transect: $m(\mathbf{u}) = m(x, 0) = a_0(\mathbf{u}) + a_1(\mathbf{u}) \cdot x$. The semivariogram model of Figure 5.1 is used for both OK and KT.

The first search neighborhood (Figure 5.4, left graph) includes Cd data at locations $\mathbf{u}_1$ to $\mathbf{u}_5$. The solid line depicts the local constant mean estimated by ordinary kriging: $m_{OK}^*(\mathbf{u}) = 0.68 \; \forall \; \mathbf{u} \in W(\mathbf{u}_0)$. For the same search neighborhood, kriging with the aforementioned trend model yields the following trend coefficients: $a_0^*(\mathbf{u}) = 0.49$, $a_1^*(\mathbf{u}) = 0.10$. The corresponding linear trend estimate, depicted by the dashed line, is then $m_{KT}^*(\mathbf{u}) = 0.49 + 0.1 \cdot x$. The small positive slope of that model reflects the slight increase in Cd concentrations from locations $\mathbf{u}_1$ to $\mathbf{u}_5$. At the central location $\mathbf{u}_0$, the KT estimate of the trend is $m_{KT}^*(2.0, 0) = 0.69$, which is close to the OK estimate of the local mean at that same location, $m_{OK}^*(2.0, 0) = 0.68$. The difference between the OK and KT estimates of the trend, i.e., between the solid and dashed lines, increases away from $\mathbf{u}_0$. That difference, however, is small because of the flatness (small slope) of the estimated linear trend model.

Conversely, the difference between OK and KT trend models is much larger for the second search neighborhood centered on $\mathbf{u}_0'$ and including locations $\mathbf{u}_2$ to $\mathbf{u}_6$ (Figure 5.4, right graph). The large Cd concentration at $\mathbf{u}_6$ greatly influences the trend fitted by both algorithms. The local constant mean estimated by ordinary kriging is twice the estimate for the first neighborhood: $m_{OK}^*(\mathbf{u}) = 1.40 \; \forall \; \mathbf{u} \in W(\mathbf{u}_0')$. The estimated linear trend model is also much steeper than for the first neighborhood: $m_{KT}^*(\mathbf{u}) = -3.22 + 2.01 \cdot x$. Again, at the central location $\mathbf{u}_0'$, both OK and KT estimates of the trend are similar: $m_{OK}^*(2.3, 0) = 1.40$ and $m_{KT}^*(2.3, 0) = 1.40$. However, the two trend models differ considerably away from $\mathbf{u}_0'$.

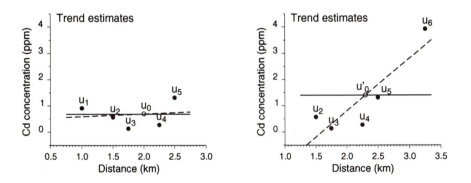

Figure 5.4: OK and KT modeling of the local trend within two successive search neighborhoods centered on locations $\mathbf{u}_0$ (left graph) and $\mathbf{u}_0'$ (right graph). The OK trend is a constant mean (solid line), whereas the KT trend is a linear function of the x-coordinate (dashed line).

Consider now the estimation of the trend component at all locations $\mathbf{u}$ along the NE-SW transect. Figure 5.5 (top graph) shows the staircase OK estimate of the local mean. The KT estimate of the trend is depicted by the solid line in Figure 5.5 (second graph). As in Figure 5.3, the vertical dashed lines delineate the trend estimates that are based on the same five closest Cd values. Within each segment, the estimates of the trend coefficients $a_0^*(\mathbf{u})$ and $a_1^*(\mathbf{u})$ are identical (see previous discussion and Figure 5.5, two bottom graphs). The KT trend estimate thus appears as a series of linear segments with different slopes. For the first segment, 1–2.1 km, the small positive slope, $a_1^*(\mathbf{u}) = 0.10$, reflects the slight increase in Cd concentrations from $\mathbf{u}_1$ to $\mathbf{u}_5$. This flat trend contrasts with the steep positive gradient of the next segment, 2.1–2.5 km, whose large slope, $a_1^*(\mathbf{u}) = 2.3$, reflects the large increase in Cd values from $\mathbf{u}_2$ to $\mathbf{u}_6$. The linear downward trend within the extreme right segment is fitted from data at locations $\mathbf{u}_6$ to $\mathbf{u}_{10}$. The large Cd concentration at $\mathbf{u}_6$ now yields a decreasing trend estimate with a substantial negative slope, $a_1^*(\mathbf{u}) = -1.0$. Together with the slope estimate $a_1^*(\mathbf{u})$, the trend coefficient estimate $a_0^*(\mathbf{u})$ also changes along the NE-SW transect (Figure 5.5, two bottom graphs).

Figure 5.6 shows both OK (dashed line) and KT (solid line) estimates of the trend (top graph) and of Cd concentrations (bottom graph). When comparing the performances of OK and KT estimators, it is important to distinguish between interpolation and extrapolation.

Interpolation
Interpolation corresponds to cases where the location $\mathbf{u}$ being estimated is surrounded by data and is within the correlation range of these data, e.g., the locations belonging to the segment 1–5 km of the NE-SW transect. Under such conditions, OK and KT yield similar estimates for both the trend component $m(\mathbf{u})$ and the attribute value $z(\mathbf{u})$. These results confirm that in interpolation kriging results are not influenced by the choice of a particular representation for the trend (Journel and Rossi, 1989).

Extrapolation
Extrapolation corresponds to situations where the location $\mathbf{u}$ being estimated is outside the geographic range of data, e.g., all locations that are beyond the extreme right datum in Figure 5.6 (top graph). In this case, the parameters $a_k(\mathbf{u})$ of the trend model ($a_0(\mathbf{u})$ for OK, $a_0(\mathbf{u})$ and $a_1(\mathbf{u})$ for KT) are estimated from the closest data and extrapolated toward the location $\mathbf{u}$ being estimated. For example, beyond the right extreme datum, the OK estimator extrapolates the constant local mean that is evaluated from the last five data values. In contrast, the KT estimator extrapolates the linear trend (decrease) fitted to the last five data values. Unlike in interpolation, the choice of a trend model is here critical.

In summary, no matter what trend is present in data, ordinary kriging with local search neighborhoods is preferred in interpolation situations be-

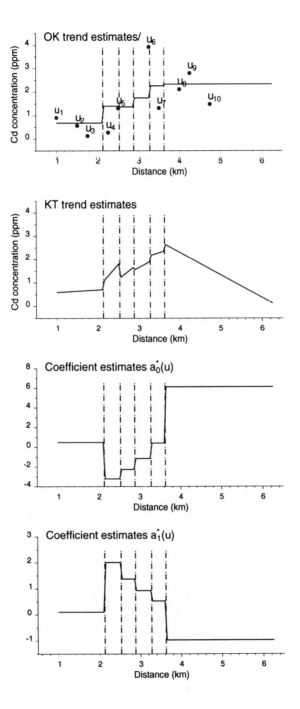

Figure 5.5: OK and KT estimates of the trend (two top graphs). The KT trend is a local linear function of the x-coordinate, $m(\mathbf{u}) = a_0(\mathbf{u}) + a_1(\mathbf{u}) \cdot x$. The two bottom graphs show the KT trend coefficients $a_0(\mathbf{u})$ and $a_1(\mathbf{u})$ estimated every 50 m using the five closest data.

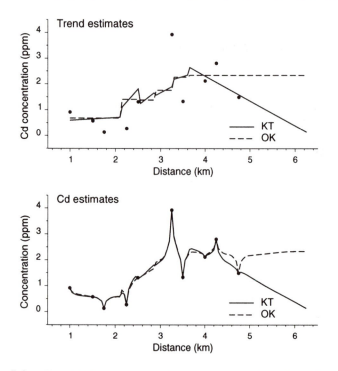

Figure 5.6: OK and KT estimates of the trend (top graph) and of Cd concentrations (bottom graph). Note the similarity between OK and KT estimates in interpolation situation 1–5 km, and how extrapolation results depend critically on the prior choice of a trend model: constant for OK or linear for KT.

cause it provides results similar to those of KT, but it is easier to implement. In extrapolation conditions, the KT estimator should be used whenever the physics of the phenomenon suggests a particular functional form for extrapolating a trend fitted from within the sampled area. In most earth science applications, however, there is usually no such physical basis for choosing a particular trend model, and the user should be aware that the estimated values depend heavily on the arbitrary trend being extrapolated.

Though the choice of a particular trend model cannot be validated when no data are available (extrapolation situation), mapping the trend estimate may draw attention to aberrant extrapolation results. For example, Figure 5.6 (bottom graph) indicates that the linear trend model is inappropriate much beyond the extreme right datum since its extrapolation would yield negative concentration estimates around 7 km.

5.5   Block Kriging

Any measurement $z(\mathbf{u}_\alpha)$ relates to a non-zero, finite sample volume, such as a piece of rock or a core of soil. Often the size or "support" of the datum may be assimilated to a point of coordinate $\mathbf{u}_\alpha$. Similarly, the support related to the attribute z to be estimated is usually assimilated to a point of coordinate $\mathbf{u}$. In several applications, however, the target quantity is the average value of attribute z over a block of specific dimensions, for example, the average Cd concentration over a 1-hectare field if remedial measures are applied to 1-hectare areas. Block kriging is a generic name for estimation of average z-values over a segment, a surface, or a volume of any size or shape. The term *point kriging* refers to estimation on point support.

Consider the problem of estimating the average value of attribute z over a block V centered at $\mathbf{u}$. Provided the averaging process is *linear*, the block value $z_V(\mathbf{u})$ is defined as

$$z_V(\mathbf{u}) = \frac{1}{|V|} \int_{V(\mathbf{u})} z(\mathbf{u}')d\mathbf{u}' \approx \frac{1}{N} \sum_{i=1}^{N} z(\mathbf{u}'_i) \qquad (5.40)$$

where $|V|$ is the measure (length, area, volume) of block V. The integral is, in practice, approximated by a discrete sum of z-values defined at N points $\mathbf{u}'_i$ discretizing the block $V(\mathbf{u})$. For example, the block $V(\mathbf{u})$ in Figure 5.7 is discretized by the four points $\mathbf{u}'_1$ to $\mathbf{u}'_4$.

The block value $z_V(\mathbf{u})$ could be estimated as the linear average of the N point estimates, say, OK estimates $z^*_{OK}(\mathbf{u}'_i)$:

$$z^*_V(\mathbf{u}) = \frac{1}{N} \sum_{i=1}^{N} z^*_{OK}(\mathbf{u}'_i) = \frac{1}{N} \sum_{i=1}^{N} \sum_{\alpha=1}^{n(\mathbf{u})} \lambda_\alpha(\mathbf{u}'_i)\, z(\mathbf{u}_\alpha) \qquad (5.41)$$

where the same $n(\mathbf{u})$ data are used for all N point estimates, and $\lambda_\alpha(\mathbf{u}'_i)$ is the weight[3] associated with the datum $z(\mathbf{u}_\alpha)$ for the OK estimation of

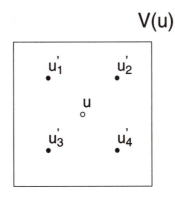

Figure 5.7: A two-dimensional block V centered on location $\mathbf{u}$ and its discretization by four points $\mathbf{u}'_1$ to $\mathbf{u}'_4$.

---

[3] To simplify notation, the upperscript OK is removed from all notations in section 5.5.

attribute z at location $\mathbf{u}_i'$. Such an approach requires solving, at each of the N locations $\mathbf{u}_i'$, an ordinary kriging system of dimension $(n(\mathbf{u}) + 1)$:

$$
\begin{cases}
\displaystyle\sum_{\beta=1}^{n(\mathbf{u})} \lambda_\beta(\mathbf{u}_i') \, C(\mathbf{u}_\alpha - \mathbf{u}_\beta) + \mu(\mathbf{u}_i') = C(\mathbf{u}_\alpha - \mathbf{u}_i') \\[2mm]
\hspace{4cm} \alpha = 1, \ldots, n(\mathbf{u}) \hspace{1cm} (5.42) \\[2mm]
\displaystyle\sum_{\beta=1}^{n(\mathbf{u})} \lambda_\beta(\mathbf{u}_i') = 1
\end{cases}
$$

The approach becomes computationally expensive as the number of blocks and discretizing points increase, hence it is preferable to estimate the block value directly from the data values $z(\mathbf{u}_\alpha)$ using an estimator of type

$$
Z_V^*(\mathbf{u}) = \sum_{\alpha=1}^{n(\mathbf{u})} \lambda_{\alpha V}(\mathbf{u}) \, Z(\mathbf{u}_\alpha) \tag{5.43}
$$

where $\lambda_{\alpha V}(\mathbf{u})$ is the block kriging weight assigned to the datum $z(\mathbf{u}_\alpha)$. Like the point estimator (5.16), the block estimator $Z_V^*(\mathbf{u})$ must be unbiased and such as to minimize the error variance $\sigma_E^2(\mathbf{u}) = \mathrm{Var}\{Z_V^*(\mathbf{u}) - Z_V(\mathbf{u})\}$.

The block ordinary kriging system is written as follows:

$$
\begin{cases}
\displaystyle\sum_{\beta=1}^{n(\mathbf{u})} \lambda_{\beta V}(\mathbf{u}) \, C(\mathbf{u}_\alpha - \mathbf{u}_\beta) + \mu_V(\mathbf{u}) = \overline{C}(\mathbf{u}_\alpha, V(\mathbf{u})) \\[2mm]
\hspace{4cm} \alpha = 1, \ldots, n(\mathbf{u}) \hspace{1cm} (5.44) \\[2mm]
\displaystyle\sum_{\beta=1}^{n(\mathbf{u})} \lambda_{\beta V}(\mathbf{u}) = 1
\end{cases}
$$

This "block" kriging system is identical to the "point" OK system (5.18) except for the right-hand-side term where the point-to-point covariance $C(\mathbf{u}_\alpha - \mathbf{u})$ is replaced by the point-to-block covariance $\overline{C}(\mathbf{u}_\alpha, V(\mathbf{u}))$, that is, the average covariance between the RV $Z(\mathbf{u}_\alpha)$ and the random variables $Z(\mathbf{u}')$ at all the points within the block V (Journel and Huijbregts, 1978, p. 54):

$$
\begin{aligned}
\overline{C}(\mathbf{u}_\alpha, V(\mathbf{u})) &= \mathrm{Cov}\{Z(\mathbf{u}_\alpha), Z_V(\mathbf{u})\} \\[2mm]
&= \frac{1}{|V|} \int_{V(\mathbf{u})} C(\mathbf{u}_\alpha - \mathbf{u}')d\mathbf{u}'
\end{aligned}
$$

In practice, this covariance $\overline{C}(\mathbf{u}_\alpha, V(\mathbf{u}))$ is approximated by the arithmetic average of the point support covariances $C(\mathbf{u}_\alpha - \mathbf{u}_i')$ defined between location $\mathbf{u}_\alpha$ and the N points $\mathbf{u}_i'$ discretizing the block $V(\mathbf{u})$:

$$
\overline{C}(\mathbf{u}_\alpha, V(\mathbf{u})) \simeq \frac{1}{N}\sum_{i=1}^{N} C(\mathbf{u}_\alpha - \mathbf{u}_i') \tag{5.45}
$$

For example, the covariance between location $\mathbf{u}_\alpha$ and the two-dimensional block in Figure 5.8 is approximated by the arithmetic average of point covariances between $\mathbf{u}_\alpha$ and the four discretizing points $\mathbf{u}_1'$ to $\mathbf{u}_4'$:

$$\overline{C}(\mathbf{u}_\alpha, V(\mathbf{u})) \simeq \frac{1}{4} \sum_{i=1}^{4} C(\mathbf{u}_\alpha - \mathbf{u}_i')$$

The block kriging variance is

$$\sigma_V^2(\mathbf{u}) = \overline{C}(V(\mathbf{u}), V(\mathbf{u})) - \sum_{\alpha=1}^{n(\mathbf{u})} \lambda_{\alpha V}(\mathbf{u}) \, \overline{C}(\mathbf{u}_\alpha, V(\mathbf{u})) - \mu_V(\mathbf{u})$$

where the block-to-block covariance $\overline{C}(V(\mathbf{u}), V(\mathbf{u}))$ is approximated by the arithmetic average of the covariances $C(\mathbf{u}_i' - \mathbf{u}_j')$ defined between any two discretizing points $\mathbf{u}_i'$ and $\mathbf{u}_j'$:

$$\overline{C}(V(\mathbf{u}), V(\mathbf{u})) = \frac{1}{N^2} \sum_{i=1}^{N} \sum_{j=1}^{N} C(\mathbf{u}_i' - \mathbf{u}_j') \tag{5.46}$$

Provided the same $n(\mathbf{u})$ data are used for all N point kriging systems (5.42) and for the block kriging system (5.44), each block kriging weight $\lambda_{\alpha V}(\mathbf{u})$ can be shown to be the average of the N point kriging weights $\lambda_\alpha(\mathbf{u}_i')$ (Journel and Huijbregts, 1978, p. 322):

$$\lambda_{\alpha V}(\mathbf{u}) = \frac{1}{N} \sum_{i=1}^{N} \lambda_\alpha(\mathbf{u}_i') \qquad \forall\, \alpha = 1, \ldots, n(\mathbf{u})$$

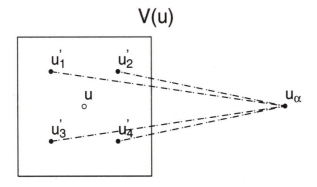

Figure 5.8: Approximation of the point-to-block covariance $\mathrm{Cov}\{Z(\mathbf{u}_\alpha), Z_V(\mathbf{u})\}$ by the average of point-to-point covariances $\mathrm{Cov}\{Z(\mathbf{u}_\alpha), Z(\mathbf{u}_i')\}$ between datum location $\mathbf{u}_\alpha$ and each of the four discretizing points $\mathbf{u}_1'$ to $\mathbf{u}_4'$.

Thus, the block kriging system yields an estimate identical to that obtained by averaging the N point estimates $z_{OK}^*(\mathbf{u}_i')$:

$$z_V^*(\mathbf{u}) \equiv \frac{1}{N} \sum_{i=1}^{N} z_{OK}^*(\mathbf{u}_i')$$

Remarks

1. Direct block kriging through system (5.44) yields an estimate of the linear average of z over the block V. When applied to an attribute such as pH that does not average linearly in space, the block kriging estimate is the average pH over the block V (*statistical* average), not the logarithm of the average concentration in H^+ over that block (*physical* average). Let $z(\mathbf{u})$ be the concentration in H^+ at location $\mathbf{u}$ and $y(\mathbf{u}) = -\log_{10}[z(\mathbf{u})]$ be the corresponding pH value. Block kriging performed on pH data $y(\mathbf{u}_\alpha)$ yields an estimate

$$y_V^*(\mathbf{u}) = \sum_{\alpha=1}^{n(\mathbf{u})} \lambda_{\alpha V}'(\mathbf{u}) \, y(\mathbf{u}_\alpha)$$

that is different from the pH of the block concentration estimate $z_V^*(\mathbf{u})$:

$$y_V^*(\mathbf{u}) \neq -\log_{10}[z_V^*(\mathbf{u})] = -\log_{10}\left[\sum_{\alpha=1}^{n(\mathbf{u})} \lambda_{\alpha V}(\mathbf{u}) \, z(\mathbf{u}_\alpha)\right]$$

The physical average of pH values over the block would be obtained by performing block kriging on the concentration $z(\mathbf{u})$ of H^+, not on the pH value $y(\mathbf{u})$.

2. One should strike a balance between too few discretizing points that provide a rough approximation of the point-to-block and block-to-block covariances and too many discretizing points that are computationally expensive. A rule of thumb is to select $(4)^n$ discretizing points, where n is the number of dimensions, 2 or 3, of the block, see Journel and Huijbregts (1978, p. 95–108) for further discussions. A good practice consists of computing covariance estimates for an increasing number of discretizing points: at some stage increasing the density of points will not significantly modify the results of the approximation.

3. The block kriging system (5.44) can be generalized to the case where the data themselves are defined on a support, say, $v(\mathbf{u}_\alpha)$, which cannot be considered as negligible with regard to the point support of the covariance model. The point-to-point $C(\mathbf{u}_\alpha - \mathbf{u}_\beta)$ and point-to-block $C(\mathbf{u}_\alpha, V(\mathbf{u}))$ covariance terms in system (5.44) are then replaced by the

averages

$$\overline{C}(v(\mathbf{u}_\alpha), v(\mathbf{u}_\beta)) = \frac{1}{|v_\alpha|\,|v_\beta|} \int_{v(\mathbf{u}_\alpha)} du \int_{v(\mathbf{u}_\beta)} C(\mathbf{u} - \mathbf{u}')\, du'$$

$$\overline{C}(v(\mathbf{u}_\alpha), V(\mathbf{u})) = \frac{1}{|v_\alpha|\,|V|} \int_{v(\mathbf{u}_\alpha)} du \int_{V(\mathbf{u})} C(\mathbf{u} - \mathbf{u}')\, du'$$

where $|v_\alpha|$ and $|v_\beta|$ are the measures of the data support at locations $\mathbf{u}_\alpha$ and $\mathbf{u}_\beta$.

Global estimation

The size of the block $V(\mathbf{u})$ could be increased until the block equals the entire study area $\mathcal{A}$. The block estimate $z_V^*(\mathbf{u})$ would then be an estimate of the linear average of z over $\mathcal{A}$. Theoretically, such a global mean z_A could be estimated directly from all data $z(\mathbf{u}_\alpha)$ in $\mathcal{A}$ by solving a block kriging system of the type in equation (5.44). In practice, several reasons prevent such direct kriging:

1. The covariance function needed can rarely be considered as stationary over the span of the entire study area.

2. Sample covariance values for large distances are usually unreliable because of the few data locations separated by such large distances.

3. Retaining all data makes the kriging system very large, lengthens the computation time, and often leads to instability of the kriging matrix.

For all these reasons, kriging is primarily used as a local estimation algorithm. A global kriging estimate $z_A^*(\mathbf{u})$ can, however, be obtained using a two-step procedure. First, the study area is discretized into small blocks and the average value of z is estimated within each such block. Second, the global estimate is computed as a linear combination of block estimates $z_V^*(\mathbf{u})$, with each estimate receiving a weight proportional to the area $|V|$ of that block. A more straightforward alternative consists of computing a declustered mean of the data using the polygonal method or cell-declustering technique introduced in section 4.1.2; see also Isaaks and Srivastava (1989), Chapter 10.

Example

Figure 5.9 shows the point (small dashed line) and block OK estimates of Cd concentrations along the NE-SW transect. The blocks are defined on segments of length 50 m (large dashed line) and 250 m (solid line). In both cases, four discretizing points were used to compute the point-to-block covariances. Note the following:

- The block estimates do not match point-data values (black dots) because the supports are different.

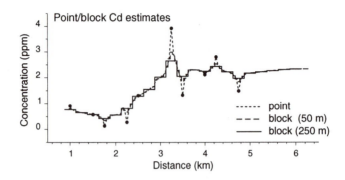

Figure 5.9: Point and block OK estimates of Cd concentrations. The blocks are defined as segments of length 50 m or 250 m.

- The block estimates vary more smoothly in space than the point estimates; that smoothing increases with increasing size of the block. The within-block averaging of block kriging smooths out the short-range variation of concentration, erasing the artifact discontinuities near data locations. If the objective is to map large-scale features of attribute z, block kriging is preferred to point kriging. An alternative consists of filtering the short-range variability component from the covariance model (see section 5.6).

5.6   Factorial Kriging

The kriging algorithms introduced so far are designed to estimate the unknown value of a continuous attribute z, say, metal concentration, on a point or a block support. In this section, the objective is no longer to estimate z (metal concentrations), but rather to understand the origins of that value. For example, trace metals in soils can originate naturally from rocks or they can result from human activities, such as mining, industrial waste, or farming. Whereas little can be done to correct for large natural metal concentrations, measures can be taken to prevent man-made pollutions from getting worse. Therefore, an early understanding of natural as distinct from human origins of contamination is critical.

Whereas large metal concentrations are often related to the nearby presence of a farm or a factory, it is generally much more difficult to discover the origin of medium concentrations, say, concentrations that just exceed the tolerable maximum. Such concentrations may result from rocks that are naturally rich in that metal. Medium concentrations may also originate from human pollution, the impact of which is temporarily balanced by small natural concentrations. For example, soil contamination by cobalt or nickel is more difficult to detect on Argovian rocks whose natural Co and Ni concentrations are half those measured on other rocks (Table 2.4, page 18).

If the scales at which the different factors (human, geologic) operate are very different from one another, then they should be apparent in the semivariograms of metal concentrations. The structural analysis performed in section 2.3.5 revealed the existence of three scales of spatial variation: micro-scale (range smaller than the first semivariogram lag of 50 m), local scale (short range ≈ 200 m), and regional scale (range ≈ 1 km). The short-range structure was related to the spatial distribution of rock types and land uses in the study area. The long-range structure was interpreted as the regional influence of geology, particularly Argovian and Kimmeridgian rocks, on metal concentrations. Such interpretation led us to model the experimental semivariograms of metal concentrations as linear combinations of three basic structures $g_l(\mathbf{h})$:

$$\gamma(\mathbf{h}) = \sum_{l=0}^{L=2} b^l g_l(\mathbf{h}) \tag{5.47}$$

where $g_0(\mathbf{h})$ is a nugget effect model, and $g_1(\mathbf{h})$ and $g_2(\mathbf{h})$ are spherical models with short and long ranges, a_1 and a_2. For example, Figure 5.1 (page 131, right graph) shows the experimental Cd semivariogram and the model fitted:

$$\gamma_{Cd}(\mathbf{h}) = 0.3 \, g_0(\mathbf{h}) + 0.30 \, \text{Sph}\left(\frac{|\mathbf{h}|}{200m}\right) + 0.26 \, \text{Sph}\left(\frac{|\mathbf{h}|}{1.3km}\right)$$

Under the linear model of regionalization (4.28), the RF $Z(\mathbf{u})$ with the nested semivariogram model (5.47) can be interpreted as a linear combination of three independent RFs $Y^l(\mathbf{u})$, each with zero mean and basic semivariogram $g_l(\mathbf{h})$:

$$Z(\mathbf{u}) = \sum_{l=0}^{L=2} a^l Y^l(\mathbf{u}) + m(\mathbf{u}) \tag{5.48}$$

where the trend component $m(\mathbf{u})$ is assumed locally constant as in the practice of ordinary kriging. Define the spatial component $Z^l(\mathbf{u})$ as the RF $a^l Y^l(\mathbf{u})$ with semivariogram $\gamma^l(\mathbf{h}) = b^l g_l(\mathbf{h})$. That new notation allows a more concise writing of the decompositions (5.47) and (5.48):

$$\gamma(\mathbf{h}) = \sum_{l=0}^{L=2} \gamma^l(\mathbf{h}) \tag{5.49}$$

$$Z(\mathbf{u}) = \sum_{l=0}^{L=2} Z^l(\mathbf{u}) + m(\mathbf{u}) \tag{5.50}$$

where $Z^0(\mathbf{u})$ is a micro-scale component, $Z^1(\mathbf{u})$ and $Z^2(\mathbf{u})$ are short-range and long-range spatial components associated with semivariograms $\gamma^1(\mathbf{h})$ and $\gamma^2(\mathbf{h})$, respectively.

In previous sections, the focus was either on the attribute estimate $z^*(\mathbf{u})$ or on the trend estimate $m^*(\mathbf{u})$. Factorial kriging amounts to splitting the residual component $R(\mathbf{u}) = Z(\mathbf{u}) - m(\mathbf{u})$ into several independent spatial components (or factors) on the basis of the semivariogram model $\gamma(\mathbf{h})$. Profiles or maps of estimates of these spatial components allow us to separate local and regional features of the phenomenon under study.

Estimating spatial components

Consider the problem of estimating the spatial component $Z^l(\mathbf{u})$ of decomposition (5.50). The OK estimator of that spatial component is

$$Z_{OK}^{l*}(\mathbf{u}) \;=\; \sum_{\alpha=1}^{n(\mathbf{u})} \lambda_{\alpha l}^{OK}(\mathbf{u}) \; Z(\mathbf{u}_\alpha) \tag{5.51}$$

where $\lambda_{\alpha l}^{OK}(\mathbf{u})$ is the weight assigned to datum $z(\mathbf{u}_\alpha)$ for the estimation of the lth component. The only data available are the z-values $z(\mathbf{u}_\alpha)$, which include the contributions of all $(L+1)$ components.

Each spatial component $Z^l(\mathbf{u})$ is defined as a RF with zero mean, hence its estimation error is

$$\mathrm{E}\{Z_{OK}^{l*}(\mathbf{u}) - Z^l(\mathbf{u})\} = \sum_{\alpha=1}^{n(\mathbf{u})} \lambda_{\alpha l}^{OK}(\mathbf{u}) \; m(\mathbf{u})$$

with $m(\mathbf{u})$ constant but unknown within the search neighborhood. The non-bias condition is then satisfied by forcing the kriging weights $\lambda_{\alpha l}^{OK}(\mathbf{u})$ to sum to zero.

The error variance $\sigma_E^2(\mathbf{u}) = \mathrm{Var}\{Z_{OK}^{l*}(\mathbf{u}) - Z^l(\mathbf{u})\}$ is typically expressed as a double linear combination of covariance values:

$$\begin{aligned}
\sigma_E^2(\mathbf{u}) \;=\;& \mathrm{Var}\{Z_{OK}^{l*}(\mathbf{u})\} \;+\; \mathrm{Var}\{Z^l(\mathbf{u})\} \;-\; 2\,\mathrm{Cov}\{Z_{OK}^{l*}(\mathbf{u}), Z^l(\mathbf{u})\} \\
=\;& \sum_{\alpha=1}^{n(\mathbf{u})} \sum_{\beta=1}^{n(\mathbf{u})} \lambda_{\alpha l}^{OK}(\mathbf{u}) \, \lambda_{\beta l}^{OK}(\mathbf{u}) \, C(\mathbf{u}_\alpha - \mathbf{u}_\beta) \;+\; C_l(0) \\
& - 2 \sum_{\alpha=1}^{n(\mathbf{u})} \lambda_{\alpha l}^{OK}(\mathbf{u}) \, \mathrm{Cov}\{Z(\mathbf{u}_\alpha), Z^l(\mathbf{u})\}
\end{aligned} \tag{5.52}$$

Because the $(L+1)$ spatial components $Z^l(\mathbf{u})$ are defined as mutually independent, the cross covariance between the RV $Z(\mathbf{u}_\alpha)$ and the spatial component $Z^l(\mathbf{u})$ reverts to the basic covariance $C_l(\mathbf{u}_\alpha - \mathbf{u})$:

$$\begin{aligned}
\mathrm{Cov}\{Z(\mathbf{u}_\alpha), Z^l(\mathbf{u})\} \;=\;& \sum_{l'=0}^{L} \mathrm{Cov}\{Z^{l'}(\mathbf{u}_\alpha), Z^l(\mathbf{u})\} \\
=\;& \mathrm{Cov}\{Z^l(\mathbf{u}_\alpha), Z^l(\mathbf{u})\} \;=\; C_l(\mathbf{u}_\alpha - \mathbf{u})
\end{aligned}$$

The linear model (5.50) is convenient because it allows us to infer the cross covariance between z-data and unsampled spatial components from the sole Z-covariance model.

Minimizing the estimation variance (5.52) under the constraint of unbiasedness leads to the following system of $(n(\mathbf{u}) + 1)$ equations:

$$\begin{cases} \sum_{\beta=1}^{n(\mathbf{u})} \lambda_{\beta l}^{OK}(\mathbf{u})\, C(\mathbf{u}_\alpha - \mathbf{u}_\beta) + \mu_l^{OK}(\mathbf{u}) = C_l(\mathbf{u}_\alpha - \mathbf{u}) \\ \qquad\qquad\qquad\qquad\qquad\qquad \alpha = 1, \ldots, n(\mathbf{u}) \qquad (5.53) \\ \sum_{\beta=1}^{n(\mathbf{u})} \lambda_{\beta l}^{OK}(\mathbf{u}) = 0 \end{cases}$$

where $\mu_l^{OK}(\mathbf{u})$ is the Lagrange parameter that accounts for the non-bias constraint.

Matrix notation

In matrix notation, the ordinary kriging system (5.53) is written as

$$\begin{bmatrix} [C(\mathbf{u}_\alpha - \mathbf{u}_\beta)] & [1]^T \\ [1] & 0 \end{bmatrix} \begin{bmatrix} \left[\lambda_{\beta l}^{OK}(\mathbf{u})\right]^T \\ \mu_l^{OK}(\mathbf{u}) \end{bmatrix} = \begin{bmatrix} [C_l(\mathbf{u}_\alpha - \mathbf{u})]^T \\ 0 \end{bmatrix}$$

or

$$\mathbf{K}_{OK}\, \boldsymbol{\lambda}_{OK}(\mathbf{u}) = \mathbf{k}_{OK} \qquad (5.54)$$

The kriging weights are obtained by multiplying the inverse of the covariance matrix $\mathbf{K}_{OK}$ by the vector $\mathbf{k}_{OK}$: $\boldsymbol{\lambda}_{OK}(\mathbf{u}) = \mathbf{K}_{OK}^{-1}\, \mathbf{k}_{OK}$.

When all separation distances $|\mathbf{u}_\alpha - \mathbf{u}|$ between the location $\mathbf{u}$ being estimated and data locations are larger than the correlation range a_Z^l of the component $Z^l(\mathbf{u})$, all right-hand-side covariance terms $C_l(\mathbf{u}_\alpha - \mathbf{u})$ in system (5.53) vanish; the vector $\mathbf{k}_{OK}$ is then a null vector, and so is the vector $\boldsymbol{\lambda}_{OK}(\mathbf{u})$ of kriging weights: $\boldsymbol{\lambda}_{OK}(\mathbf{u}) = \mathbf{K}_{OK}^{-1}\, \mathbf{0} = \mathbf{0}$. Thus, the kriging estimate of a spatial component $Z^l(\mathbf{u})$ at a remote location $\mathbf{u}$ is equal to its zero mean.

Using the matrix formulation (5.54), one can see that the decomposition (5.50) of the RV variable $Z(\mathbf{u})$ into spatial and trend components holds true in terms of kriging estimates, say, OK estimates:

$$\begin{aligned} z_{OK}^*(\mathbf{u}) &= \sum_{l=0}^{L=2} z_{OK}^{l*}(\mathbf{u}) + m_{OK}^*(\mathbf{u}) \\ &= z_{OK}^{0*}(\mathbf{u}) + z_{OK}^{1*}(\mathbf{u}) + z_{OK}^{2*}(\mathbf{u}) + m_{OK}^*(\mathbf{u}) \qquad (5.55) \end{aligned}$$

where all estimates are based on the same $n(\mathbf{u})$ data $z(\mathbf{u}_\alpha)$. Indeed, the right-hand-side vector of the OK system (5.18) is but the sum of the right-hand-side vectors of OK systems (5.22) and (5.53) for spatial and trend components:

$$\left[\begin{array}{c} [C(\mathbf{u}_\alpha - \mathbf{u})]^T \\ 1 \end{array} \right] = \sum_{l=0}^{L} \left[\begin{array}{c} [C_l(\mathbf{u}_\alpha - \mathbf{u})]^T \\ 0 \end{array} \right] + \left[\begin{array}{c} [0]^T \\ 1 \end{array} \right]$$

which entails the following equality between left-hand-side terms:

$$\mathbf{K}_{OK} \left[\begin{array}{c} [\lambda_\alpha^{OK}(\mathbf{u})]^T \\ \mu_{OK}(\mathbf{u}) \end{array} \right] = \sum_{l=0}^{L} \mathbf{K}_{OK} \left[\begin{array}{c} [\lambda_{\alpha l}^{OK}(\mathbf{u})]^T \\ \mu_l^{OK}(\mathbf{u}) \end{array} \right] + \mathbf{K}_{OK} \left[\begin{array}{c} [\lambda_{\alpha m}^{OK}(\mathbf{u})]^T \\ \mu_m^{OK}(\mathbf{u}) \end{array} \right]$$

Since the covariance matrix $\mathbf{K}_{OK}$ is common to all ordinary kriging systems, one deduces the following relation between kriging weights and Lagrange parameters:

$$\left[\begin{array}{c} [\lambda_\alpha^{OK}(\mathbf{u})]^T \\ \mu_{OK}(\mathbf{u}) \end{array} \right] = \sum_{l=0}^{L} \left[\begin{array}{c} [\lambda_{\alpha l}^{OK}(\mathbf{u})]^T \\ \mu_l^{OK}(\mathbf{u}) \end{array} \right] + \left[\begin{array}{c} [\lambda_{\alpha m}^{OK}(\mathbf{u})]^T \\ \mu_m^{OK}(\mathbf{u}) \end{array} \right] \qquad (5.56)$$

Multiplying both sides of equality (5.56) by the vector of data, one obtains relation (5.55) between kriging estimates.

Example

Figure 5.10 shows the decomposition of the OK estimates of Cd concentrations along the NE-SW transect (right top graph) into four components: a nugget component, short- and long-range components, and a trend component. Note the following:

- The short-range and long-range components are zero wherever the closest datum location $\mathbf{u}_\alpha$ is more than 200 m or 1.3 km away, respectively. The nugget component, which corresponds to a zero range, is zero at any unsampled location.

- The decomposition allows one to distinguish the trend increase of Cd concentrations along the transect from variations in either short- or long-range components.

Factorial kriging depends wholly on the somewhat arbitrary choice of the nested semivariogram model (5.47), hence remedial measures should not be decided solely on the basis of the maps of spatial components. Rather, these maps may draw attention to subareas that depart substantially from their regional background. For example, although similar Cd concentrations were measured at $\mathbf{u}_1$, $\mathbf{u}_7$, and $\mathbf{u}_{10}$ along the transect shown in Figure 5.10 (right top graph), the first measurement is in a low-valued subarea, whereas the two other concentrations belong to a high-valued subarea. This departure from the regional background is reflected in the profile of the short-range component (Figure 5.10, right middle graph).

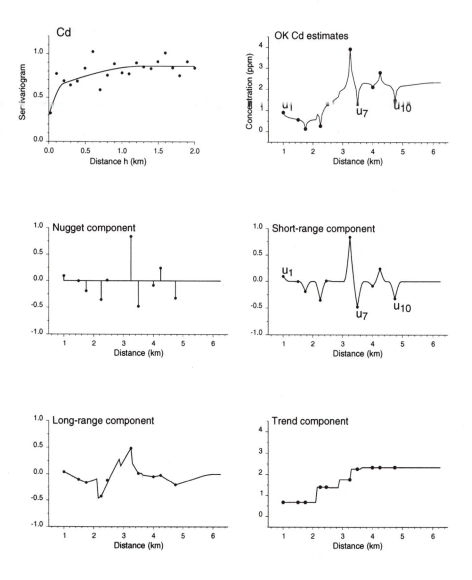

Figure 5.10: Decomposition of the OK estimates of Cd concentrations (top right graph) into three spatial components and a trend component on the basis of the Cd semivariogram model (top left graph).

Remarks

1. Again, beware that the spatial components $Z^l(\mathbf{u})$ are mathematical constructions with no a priori physical meaning. The decomposition (5.50) is but a model based on congenial hypotheses of independence and additivity of zero-mean spatial components. Physical processes are generally not independent, nor are they necessarily additive.

2. As mentioned in section 4.2.4, kriging estimates of an attribute depend essentially on the semivariogram behavior near the origin; the type and number $(L + 1)$ of basic semivariogram functions $g_l(\mathbf{h})$ are of less importance. In contrast, the decomposition (5.50), hence the subsequent interpretation of spatial components, relies fully on these basic semivariogram functions. Therefore, when adopting a decomposition of type (5.50), it is crucial to take into account any physical information related to the phenomenon and the study area.

The concept of local and regional scales (short versus long range) of spatial variation applies to many fields of earth sciences (Table 5.1). For example, geochemical exploration tries to distinguish large isolated values (pointwise anomalies) from groupwise anomalies that consist of two or more neighboring values just above the chemical detection limit. Factorial kriging can be used to map such groupwise anomalies that may point to the dispersion aureola of a mineral deposit (Wackernagel and Sanguinetti, 1993).

Table 5.1: Applications of kriging analysis to earth sciences.

Fields	Applications
Geochemistry	Distinguish between pointwise anomalies and groupwise anomalies; Sandjivy (1984), Wackernagel (1988), Wackernagel and Sanguinetti (1993).
Geophysics	Separate sharp anomalies (superficial sources) from broad anomalies (deeper sources); Chilès and Guillen (1984), Galli et al. (1984).
Hydrology	Distinguish between local sources of contamination linked to human activities and regional changes in geological properties of the aquifer; Goovaerts and Sonnet (1993), Goovaerts et al. (1993).
Soil science	Separate local variation in soil properties that result from field-to-field differences from regional variation related to different soil types; Goovaerts (1992, 1994d), Goovaerts and Webster (1994).

Filtering spatial components

As shown in Figure 5.2 (top graph) or Figure 5.6 (bottom graph), the exactitude property of the kriging estimator creates discontinuities (peaks) at data locations. One may want to remove such discontinuities by filtering the nugget (high-frequency) component at these locations. Such noise filtering is common in image processing to improve image clarity; for example, see Ma and Royer (1988). In other applications, one may want to filter the long-range (low-frequency) components to enhance the short-range variability of the phenomenon under study.

Consider the general problem of filtering the first l_0 spatial components from the OK estimate of z at location $\mathbf{u}$, $z^*_{OK}(\mathbf{u})$. Such filtering could proceed in two steps: first, the l_0 spatial components are estimated separately, then their sum is subtracted from the OK estimate at location $\mathbf{u}$:

$$z^*_{OK}(\mathbf{u}) - \sum_{l=0}^{l_0-1} z^{l*}_{OK}(\mathbf{u}) = w^{l_0*}_{OK}(\mathbf{u})$$

Such an approach requires solving $(l_0 + 1)$ ordinary kriging systems at each location, one of type (5.18) and l_0 of type (5.53). A more straightforward alternative consists of estimating directly the quantity $w^{l_0*}_{OK}(\mathbf{u})$ as a linear combination of the z-data. The estimator is then

$$W^{l_0*}_{OK}(\mathbf{u}) = \sum_{\alpha=1}^{n(\mathbf{u})} \nu^{OK}_{\alpha l_0}(\mathbf{u})\, Z(\mathbf{u}_\alpha) \qquad (5.57)$$

Because the spatial components are built as RFs with zero mean, the expected value of the estimator (5.57) is equal to $m(\mathbf{u}) = \mathrm{E}\{Z(\mathbf{u})\}$. Thus, the expected error is zero if the kriging weights $\nu^{OK}_{\alpha l_0}(\mathbf{u})$ sum to one:

$$\mathrm{E}\{W^{l_0*}_{OK}(\mathbf{u}) - W^{l_0}(\mathbf{u})\} = \sum_{\alpha=1}^{n(\mathbf{u})} \nu^{OK}_{\alpha l_0}(\mathbf{u})\, m(\mathbf{u}) - m(\mathbf{u})$$

$$= 0 \qquad \text{if } \sum_{\alpha=1}^{n(\mathbf{u})} \nu^{OK}_{\alpha l_0}(\mathbf{u}) = 1$$

The minimization of the error variance $\sigma^2_E(\mathbf{u}) = \mathrm{Var}\{W^{l_0*}_{OK}(\mathbf{u}) - W^{l_0}(\mathbf{u})\}$ under the unbiasedness constraint yields the following system:

$$\begin{cases} \displaystyle\sum_{\beta=1}^{n(\mathbf{u})} \nu^{OK}_{\beta l_0}(\mathbf{u})\, C(\mathbf{u}_\alpha - \mathbf{u}_\beta) + \mu^{OK}_{l_0}(\mathbf{u}) = C(\mathbf{u}_\alpha - \mathbf{u}) - \sum_{l=0}^{l_0-1} C_l(\mathbf{u}_\alpha - \mathbf{u}) \\ \qquad\qquad\qquad\qquad\qquad\qquad \alpha = 1, \ldots, n(\mathbf{u}) \\ \displaystyle\sum_{\beta=1}^{n(\mathbf{u})} \nu^{OK}_{\beta l_0}(\mathbf{u}) = 1 \end{cases}$$

$$(5.58)$$

System (5.58) is identical to the OK system (5.18) except for the right-hand-side covariance terms that are computed by subtracting from the Z-covariance $C(\mathbf{u}_\alpha - \mathbf{u})$ the covariances $C_l(\mathbf{u}_\alpha - \mathbf{u})$ of the l_0 spatial components being filtered. Two particular cases of the estimator (5.57) are:

1. No spatial component is filtered ($l_0 = 0$).
   The right-hand-side covariance terms are equal to the Z-covariances $C(\mathbf{u}_\alpha - \mathbf{u})$, hence the estimator $W_{OK}^{l_0*}(\mathbf{u})$ reverts to the OK estimator of the z-attribute value, $Z_{OK}^*(\mathbf{u})$.

2. All spatial components are filtered ($l_0 = L + 1$).
   The right-hand-side covariance terms are all equal to zero, the system (5.58) reverts to the OK system (5.22), and the estimator $W_{OK}^{l_0*}(\mathbf{u})$ is equal to OK estimator of the local mean, $m_{OK}^*(\mathbf{u})$.

Example

Figure 5.11 shows the OK estimates of Cd concentrations before (dashed line) and after (solid line) filtering the nugget component (case $l_0 = 1$). Note that both interpolation curves are identical except at data locations (black dots). Indeed, the OK estimate $z_{OK}^*(\mathbf{u})$ can be viewed as the sum of the estimate $w_{OK}^{1*}(\mathbf{u})$ plus the nugget component estimate $z_{OK}^{0*}(\mathbf{u})$:

$$z_{OK}^*(\mathbf{u}) \;=\; w_{OK}^{1*}(\mathbf{u}) + z_{OK}^{0*}(\mathbf{u})$$

As shown in Figure 5.10, the estimate of the nugget component is zero wherever the location $\mathbf{u}$ does not coincide with a datum location $\mathbf{u}_\alpha$. In such cases, both estimates $z_{OK}^*(\mathbf{u})$ and $w_{OK}^{1*}(\mathbf{u})$ are identical:

$$z_{OK}^*(\mathbf{u}) \;=\; w_{OK}^{1*}(\mathbf{u}) \qquad \forall\, \mathbf{u} \neq \mathbf{u}_\alpha$$

Filtering properties of the kriging estimator are further discussed in section 5.7, where the dual kriging formalism is introduced.

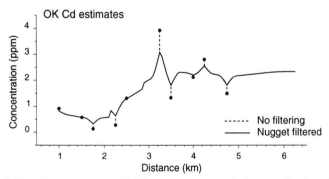

Figure 5.11: OK estimates of Cd concentrations before and after filtering the nugget component.

Because it relies on the linear decomposition (5.50), the noise filtering performed by factorial kriging regards that noise as an independent component added to the underlying signal. Another assumption is that the noise is "homoscedastic", that is, the noise variance is deemed constant over the whole range of variation of signal values. Whenever noise relates to measurement errors, as in contamination of sampling procedures, it may not be independent of the signal value (heteroscedastic noise). In environmental applications, small concentrations are typically overestimated, whereas large concentrations are underestimated. Furthermore, the error variance generally increases with the measured value.

When the sampling is exhaustive, for example, in image processing, Bourgault (1994) showed that factorial kriging succeeds in filtering the noise even if it is correlated with the signal and is heteroscedastic. Where data are sparse, a signal contaminated with heteroscedastic noise may be difficult to extract, especially in the presence of noise-to-signal correlation.

Kriging the regional components of metal concentrations

Metal concentrations are decomposed into spatial components on the basis of the linear models fitted to the experimental semivariograms in Figures 4.13 and 4.14. For each metal, the spatial component associated with the spherical model of long range (1.3 km) is added to the trend component that may also be viewed as a long-range (regional) component. The maps of the resulting regional components are shown in Figures 5.12 and Figure 5.13. Note the following:

- Regional components of copper and lead are fairly similar.

- Nickel and cobalt show common regional features linked to the spatial distribution of rock types (Figure 5.13, top graph). These two maps emphasize the impact of geology on the regional variation of metal concentrations, in particular the occurrence of the smallest concentrations on Argovian rocks.

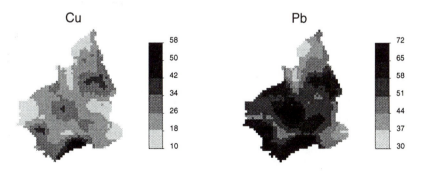

Figure 5.12: Maps of regional components for copper and lead.

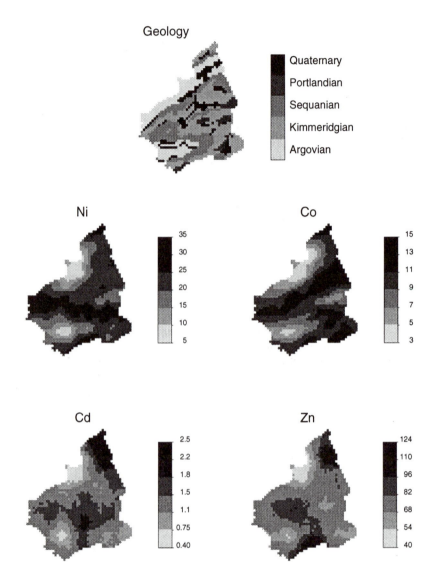

Figure 5.13: Spatial distribution of rock types over the study area and maps of regional components of four metals.

- Like the pair nickel-cobalt, small values of Cd and Zn regional components are confined to Argovian rocks. On the other hand, the spatial pattern of large Zn values is close to that displayed by the pair copper-lead. As already revealed by the structural analysis in section 2.3.5, the spatial distribution of zinc values shares the spatial characteristics of the pair copper-lead, as well as those of the group of other heavy metals.

5.7 Dual Kriging

Dual kriging is just another presentation of kriging whereby the estimates are expressed as linear combinations of covariance values instead of data values (Dubrule, 1983; Journel, 1989). The dual kriging formalism provides insight into the filtering properties of kriging and reduces the computational cost of kriging when used with a global search neighborhood.

Dual simple kriging

Consider the simple kriging estimate of z at location $\mathbf{u}$:

$$z^*_{SK}(\mathbf{u}) = \sum_{\alpha=1}^{n(\mathbf{u})} \lambda^{SK}_{\alpha}(\mathbf{u}) \left[z(\mathbf{u}_\alpha) - m \right] + m \qquad (5.59)$$

The SK weights $\lambda^{SK}_{\alpha}(\mathbf{u})$ are obtained by solving the system of $n(\mathbf{u})$ linear equations derived in section 5.2 and recalled here:

$$\sum_{\beta=1}^{n(\mathbf{u})} \lambda^{SK}_{\beta}(\mathbf{u}) \, C(\mathbf{u}_\alpha - \mathbf{u}_\beta) = C(\mathbf{u}_\alpha - \mathbf{u}) \qquad \alpha = 1, \ldots, n(\mathbf{u}) \qquad (5.60)$$

or in matrix form $\boldsymbol{\lambda}_{SK}(\mathbf{u}) = \mathbf{K}^{-1}_{SK} \, \mathbf{k}_{SK}$. The kriging weights appear as linear functions of the right-hand-side covariance values $C(\mathbf{u}_\alpha - \mathbf{u})$. Consequently, the SK estimate (5.59) can be expressed as a linear combination of these covariance values plus the stationary mean m:

$$z^*_{SK}(\mathbf{u}) = \sum_{\alpha=1}^{n(\mathbf{u})} d^{SK}_{\alpha}(\mathbf{u}) \, C(\mathbf{u}_\alpha - \mathbf{u}) + m \qquad (5.61)$$

where $d^{SK}_{\alpha}(\mathbf{u})$ is the dual weight associated with the covariance $C(\mathbf{u}_\alpha - \mathbf{u})$. Expression (5.61) is called the *dual* form of the traditional or *primal* simple kriging estimate (5.59). The dual weights $d^{SK}_{\alpha}(\mathbf{u})$ are obtained by identification of the data values by the kriging expression (5.61):

$$z^*_{SK}(\mathbf{u}_\alpha) = \sum_{\beta=1}^{n(\mathbf{u})} d^{SK}_{\beta}(\mathbf{u}) \, C(\mathbf{u}_\beta - \mathbf{u}_\alpha) + m = z(\mathbf{u}_\alpha) \quad \alpha = 1, \ldots, n(\mathbf{u}) \quad (5.62)$$

In contrast with the primal SK system (5.60), the dual system (5.62) does not result from minimizing an error variance; rather, it is established from the exactitude property of the kriging estimator.

Like the notation $\lambda^{SK}_{\beta}(\mathbf{u})$ used for the primal weights in system (5.60), the dependence on $\mathbf{u}$ of the dual weights $d^{SK}_{\beta}(\mathbf{u})$ refers to the fact that the $n(\mathbf{u})$ data retained may vary from one location $\mathbf{u}$ to another (see related discussion in section 5.1). For a given data configuration, the dual weights $d^{SK}_{\beta}(\mathbf{u})$ and $d^{SK}_{\beta}(\mathbf{u}')$ are the same for all β.

Dual ordinary kriging

By analogy with dual simple kriging, the ordinary kriging estimate is expressed in its dual form as a linear combination of covariances $C(\mathbf{u}_\alpha - \mathbf{u})$ plus the trend estimate $m^*_{OK}(\mathbf{u})$:

$$z^*_{OK}(\mathbf{u}) = \sum_{\alpha=1}^{n(\mathbf{u})} d^{OK}_\alpha(\mathbf{u}) \, C(\mathbf{u}_\alpha - \mathbf{u}) + m^*_{OK}(\mathbf{u}) \qquad (5.63)$$

The dual OK system includes $(n(\mathbf{u}) + 1)$ linear equations with $(n(\mathbf{u}) + 1)$ unknowns: the $n(\mathbf{u})$ dual weights $d^{OK}_\alpha(\mathbf{u})$ and the trend estimate $m^*_{OK}(\mathbf{u})$:

$$\begin{cases} \displaystyle\sum_{\beta=1}^{n(\mathbf{u})} d^{OK}_\beta(\mathbf{u}) \, C(\mathbf{u}_\beta - \mathbf{u}_\alpha) + m^*_{OK}(\mathbf{u}) = z(\mathbf{u}_\alpha) \\ \qquad\qquad\qquad\qquad\qquad\qquad \alpha = 1, \ldots, n(\mathbf{u}) \\ \displaystyle\sum_{\beta=1}^{n(\mathbf{u})} d^{OK}_\beta(\mathbf{u}) = 0 \end{cases} \qquad (5.64)$$

Like the dual SK system (5.62), the $n(\mathbf{u})$ first equations in the OK system (5.64) can be viewed as conditions for data identification. Recall that for a given set of $n(\mathbf{u})$ data, the weights $d^{OK}_\beta(\mathbf{u})$ and mean estimate $m^*_{OK}(\mathbf{u})$ do not depend on $\mathbf{u}$.

The last equation is deduced from the property that the ordinary kriging estimate $z^*_{OK}(\mathbf{u})$ identifies the trend estimate $m^*_{OK}(\mathbf{u})$ whenever all $n(\mathbf{u})$ data are equally correlated with the unknown value (Zhu, 1992). In such cases, all covariances $C(\mathbf{u}_\alpha - \mathbf{u})$ are equal to a constant q, and the dual estimate (5.63) becomes

$$z^*_{OK}(\mathbf{u}) = q \sum_{\alpha=1}^{n(\mathbf{u})} d^{OK}_\alpha(\mathbf{u}) + m^*_{OK}(\mathbf{u}) = m^*_{OK}(\mathbf{u})$$

which leads to the constraint that the dual kriging weights $d^{OK}_\alpha(\mathbf{u})$ must sum to zero. The dual formalism is readily extended to the case where the trend component is modeled as a linear combination of trend functions (dual kriging with trend); see Journel and Rossi (1989).

Provided the same data are used to estimate z at two different locations $\mathbf{u}$ and $\mathbf{u}'$, the dual system remains unchanged. The two sets of dual weights and the two trend estimates are identical: $d^{OK}_\alpha(\mathbf{u}) = d^{OK}_\alpha(\mathbf{u}')$, $\forall \, \alpha$ and $m^*_{OK}(\mathbf{u}) = m^*_{OK}(\mathbf{u}')$. More generally, if the same n data are used to estimate z at all locations, that is, $n(\mathbf{u}) = n$, $\forall \, \mathbf{u}$, the dual estimate (5.63) becomes

$$z^*_{OK}(\mathbf{u}) = \sum_{\alpha=1}^{n(\mathbf{u})} d^{OK}_\alpha \, C(\mathbf{u}_\alpha - \mathbf{u}) + m^*_{OK} \qquad (5.65)$$

where the n weights d^{OK}_α and the trend estimate m^*_{OK} are location-independent, hence must be computed only once. The estimator (5.65) appears as a deterministic function of $\mathbf{u}$. The estimated value $z^*_{OK}(\mathbf{u})$ at any location $\mathbf{u}$ is readily obtained by computing the n vectors $\mathbf{h} = \mathbf{u}_\alpha - \mathbf{u}$ and inserting these into expression (5.65). As in the case of global estimation, beware that the covariance model $C(\mathbf{h})$ is rarely known over large distances, and solution of a single but large dual system (5.64) would be unstable.

Dual factorial kriging

The dual representation is readily extended to estimators of spatial components introduced in section 5.6. For example, the dual form of the ordinary kriging estimate $z^{l*}_{OK}(\mathbf{u})$ of the lth component is a linear combination of the covariances $C_l(\mathbf{u}_\alpha - \mathbf{u})$ plus a trend component $m^*_{OK}(\mathbf{u})$ with a zero weight:

$$z^{l*}_{OK}(\mathbf{u}) = \sum_{\alpha=1}^{n(\mathbf{u})} d^{OK}_{\alpha l}(\mathbf{u})\, C_l(\mathbf{u}_\alpha - \mathbf{u}) + 0 \cdot m^*_{OK}(\mathbf{u})$$

$$= \sum_{\alpha=1}^{n(\mathbf{u})} d^{OK}_{\alpha l}(\mathbf{u})\, C_l(\mathbf{u}_\alpha - \mathbf{u}) \tag{5.66}$$

where $C_l(\mathbf{h})$ is the covariance function of the spatial component $Z^l(\mathbf{u})$. The dual weights $d^{OK}_{\alpha l}(\mathbf{u})$ are obtained by solving the dual OK system of type (5.64):

$$\begin{cases} \displaystyle\sum_{\beta=1}^{n(\mathbf{u})} d^{OK}_{\beta l}(\mathbf{u})\, C(\mathbf{u}_\beta - \mathbf{u}_\alpha) + m^*_{OK}(\mathbf{u}) = z(\mathbf{u}_\alpha) \\[4mm] \hspace{4cm} \alpha = 1,\ldots,n(\mathbf{u}) \tag{5.67} \\[4mm] \displaystyle\sum_{\beta=1}^{n(\mathbf{u})} d^{OK}_{\beta l}(\mathbf{u}) = 0 \end{cases}$$

Provided the same $n(\mathbf{u})$ data are used to estimate z and the $(L+1)$ spatial components $z^l(\mathbf{u})$ at $\mathbf{u}$, only one dual kriging system (5.64) need be solved. The dual weights do not depend on the spatial component that is estimated: $d^{OK}_{\alpha l}(\mathbf{u}) = d^{OK}_\alpha(\mathbf{u}),\ \forall\ l$. Thus, relation (5.66) can be rewritten:

$$z^{l*}_{OK}(\mathbf{u}) = \sum_{\alpha=1}^{n(\mathbf{u})} d^{OK}_\alpha(\mathbf{u})\, C_l(\mathbf{u}_\alpha - \mathbf{u}) \qquad l = 0,\ldots,L \tag{5.68}$$

In contrast, $(L+2)$ primal systems of type (5.18) and (5.53) would have to be solved at each location $\mathbf{u}$.

Whenever the distance between the location $\mathbf{u}$ being estimated and any data location $\mathbf{u}_\alpha$ is larger than the correlation range $a_{l'}$ of a specific covariance component $C_{l'}(\mathbf{h})$, all covariance terms $C_{l'}(\mathbf{u}_\alpha - \mathbf{u})$ in expression (5.68) vanish:

$$C_{l'}(\mathbf{u}_\alpha - \mathbf{u}) = 0 \quad \text{if} \quad |\mathbf{u}_\alpha - \mathbf{u}| \geq a_{l'} \quad \forall\ \alpha$$

Thus, the estimated value of the corresponding spatial component $Z^{l'}(\mathbf{u})$ is zero, as it should be; recall the discussion related to Figure 5.10.

Filtering properties of kriging

Accounting for the dual form (5.68), one can rewrite relation (5.55) between OK estimates as

$$z^*_{OK}(\mathbf{u}) = \sum_{l=0}^{L} z^{l*}_{OK}(\mathbf{u}) + m^*_{OK}(\mathbf{u})$$

$$= \sum_{\alpha=1}^{n(\mathbf{u})} d^{OK}_\alpha(\mathbf{u}) \sum_{l=0}^{L} C_l(\mathbf{u}_\alpha - \mathbf{u}) + m^*_{OK}(\mathbf{u}) \qquad (5.69)$$

Expression (5.69) allows a straightforward filtering of any spatial component $z^{l'*}_{OK}(\mathbf{u})$ from the kriging estimate $z^*_{OK}(\mathbf{u})$ by setting the corresponding covariance terms $C_{l'}(\mathbf{u}_\alpha - \mathbf{u})$ to zero. For example, the nugget component $z^{0*}_{OK}(\mathbf{u})$ is filtered at both data and unsampled locations by subtracting the nugget effect $C_0(\mathbf{u}_\alpha - \mathbf{u})$ from all covariance values $C(\mathbf{u}_\alpha - \mathbf{u})$. As shown for the nugget component in Figure 5.11 (page 166), spatial components are implicitly filtered in the usual ordinary kriging of z. Such filtering is, however, non-systematic: the component that is actually filtered depends on the local data configuration. For example, the nugget component $z^{0*}_{OK}(\mathbf{u})$ is filtered only at unsampled locations. The key factor is the separation distance, say, d_{min}, between the location $\mathbf{u}$ being estimated and the closest data location. Whenever the distance d_{min} is larger than the correlation range of a spatial component, say, $z^{l'}(\mathbf{u})$, the corresponding covariance terms $C_{l'}(\mathbf{u}_\alpha - \mathbf{u})$ vanish in the dual estimate (5.69); hence the spatial component $z^{l'*}_{OK}(\mathbf{u})$ is filtered.

Consider, for example, the ordinary kriging estimation of Cd concentrations along the transect in Figure 5.14. The dual OK estimate is written

$$z^*_{OK}(\mathbf{u}) = \sum_{\alpha=1}^{n(\mathbf{u})} d^{OK}_\alpha(\mathbf{u}) \left[C_0(\mathbf{u}_\alpha - \mathbf{u}) + C_1(\mathbf{u}_\alpha - \mathbf{u}) + C_2(\mathbf{u}_\alpha - \mathbf{u}) \right]$$

$$+ m^*_{OK}(\mathbf{u})$$

$$= z^{0*}_{OK}(\mathbf{u}) + z^{1*}_{OK}(\mathbf{u}) + z^{2*}_{OK}(\mathbf{u}) + m^*_{OK}(\mathbf{u}) \qquad (5.70)$$

where $C_0(\mathbf{h})$ is a nugget effect model, and $C_1(\mathbf{h})$ and $C_2(\mathbf{h})$ are spherical covariance models with ranges of 200 m and 1.3 km, respectively.

Four different situations can be distinguished in Figure 5.14, depending on the position of the location $\mathbf{u}$ being estimated relative to data locations:

1. Location $\mathbf{u}_1$: $d_{min} = 0$
   No component is filtered, and the datum is exactly honored:

   $$z^*_{OK}(\mathbf{u}_1) = z(\mathbf{u}_1)$$

2. Location $\mathbf{u}_2$: $0 < d_{min} \leq 200$ m
   All covariance terms $C_0(\mathbf{u}_\alpha - \mathbf{u})$ are zero, so only the nugget component is filtered from the estimate (5.70):

   $$z^*_{OK}(\mathbf{u}_2) = z^{1*}_{OK}(\mathbf{u}_2) + z^{2*}_{OK}(\mathbf{u}_2) + m^*_{OK}(\mathbf{u}_2)$$

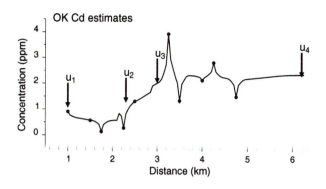

Figure 5.14: Impact of data configuration on filtering properties of kriging. As the location being estimated gets farther away from data locations depicted by black dots, high-frequency (short-range) components are progressively filtered, making the apparent spatial variability progressively smaller.

3. Location $\mathbf{u}_3$: 200 m $< d_{min} \leq 1.3$ km
   All covariance terms $C_0(\mathbf{u}_\alpha - \mathbf{u})$ and $C_1(\mathbf{u}_\alpha - \mathbf{u})$ are zero, hence both nugget and short-range components are filtered from the estimate (5.70):

$$z_{OK}^*(\mathbf{u}_3) = z_{OK}^{2*}(\mathbf{u}_3) + m_{OK}^*(\mathbf{u}_3)$$

4. Location $\mathbf{u}_4$: $d_{min} > 1.3$ km
   All covariance terms $C_l(\mathbf{u}_\alpha - \mathbf{u})$, $l = 0, 1, 2$, are zero, hence all spatial components are filtered and the estimate (5.70) includes only the trend component:

$$z_{OK}^*(\mathbf{u}_4) = m_{OK}^*(\mathbf{u}_4)$$

Thus the kriging estimator is a variable low-pass filter: the high-frequency (nugget, short-range) components are progressively filtered as the location $\mathbf{u}$ being estimated gets farther away from data locations. Such variable filtering creates artifact discontinuities near data locations. These discontinuities are particularly noticeable when most of the spatial variation occurs over short distances. For example, Figure 5.15 shows OK estimates of Cd concentrations using three arbitrary semivariogram models of type:

$$\gamma(\mathbf{h}) = b^0 \, g_0(\mathbf{h}) + (1 - b^0) \, \text{Sph}\left(\frac{|\mathbf{h}|}{2\text{km}}\right)$$

where b^0 is the proportion of the nugget effect model $g_0(\mathbf{h})$. For a zero relative nugget effect ($b^0 = 0$), the interpolation curve (solid line) varies smoothly without apparent discontinuity. Discontinuities appear as soon as a small nugget effect, say, $b^0 = 0.2$, is present (large dashed line), and they become increasingly important as the relative nugget effect increases.

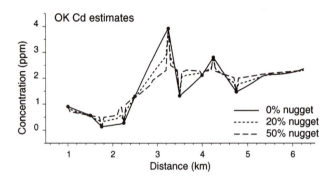

Figure 5.15: OK estimates of Cd concentrations using three spherical semivariogram models (range=2km) with increasing relative nugget effect.

Because of its variable (location-dependent) filtering property, the appearance (smoothness) of kriging profiles or maps depends on the local data configuration. For irregularly spaced data, the interpolated profile (map) is more variable where sampling is dense than where it is sparse. Such an effect may create structures that are pure artifacts of the data configuration. One solution consists of utilizing simulation algorithms, introduced in Chapter 8, which, as opposed to kriging algorithms, reproduce the full covariance everywhere.

5.8 Miscellaneous Aspects of Kriging

5.8.1 Kriging weights

Using the matrix formulation (5.31), the vector of ordinary kriging weights is computed as

$$
\begin{bmatrix} \left[\lambda_\beta^{OK}(\mathbf{u}) \right]^T \\ \mu_{OK}(\mathbf{u}) \end{bmatrix} = \begin{bmatrix} [C(\mathbf{u}_\alpha - \mathbf{u}_\beta)] & [1]^T \\ [1] & 0 \end{bmatrix}^{-1} \begin{bmatrix} [C(\mathbf{u}_\alpha - \mathbf{u})]^T \\ 1 \end{bmatrix}
$$

The kriging weighting system accounts for:

1. proximity of data to the location $\mathbf{u}$ being estimated through the covariance terms $C(\mathbf{u}_\alpha - \mathbf{u})$, and

2. data redundancy through the data covariance matrix $[C(\mathbf{u}_\alpha - \mathbf{u}_\beta)]$.

Instead of the Euclidean distance $|\mathbf{u}_\alpha - \mathbf{u}|$ common to all variables, the distance used in kriging is the semivariogram distance $\gamma(\mathbf{u}_\alpha - \mathbf{u})$, as modeled from the data and specific to the variable under study. The kriging weights depend only on the shape (relative nugget effect, anisotropy, correlation range) of the semivariogram, not on its global sill or any factor multiplying the semivariogram or covariance model.

Consider, for example, the ordinary kriging estimation of z at $\mathbf{u}_0$, using data at the five locations shown in Figure 5.16 (left graph). The ordinary kriging system (5.18) is

$$
\begin{bmatrix}
C_{11} & C_{12} & C_{13} & C_{14} & C_{15} & 1 \\
C_{21} & C_{22} & C_{23} & C_{24} & C_{25} & 1 \\
C_{31} & C_{32} & C_{33} & C_{34} & C_{35} & 1 \\
C_{41} & C_{42} & C_{43} & C_{44} & C_{45} & 1 \\
C_{51} & C_{52} & C_{53} & C_{54} & C_{55} & 1 \\
1 & 1 & 1 & 1 & 1 & 0
\end{bmatrix}
\begin{bmatrix}
\lambda_1^{OK}(\mathbf{u}_0) \\
\lambda_2^{OK}(\mathbf{u}_0) \\
\lambda_3^{OK}(\mathbf{u}_0) \\
\lambda_4^{OK}(\mathbf{u}_0) \\
\lambda_5^{OK}(\mathbf{u}_0) \\
\mu_{OK}(\mathbf{u}_0)
\end{bmatrix}
=
\begin{bmatrix}
C_{10} \\
C_{20} \\
C_{30} \\
C_{40} \\
C_{50} \\
1
\end{bmatrix}
\tag{5.71}
$$

where $C_{\alpha\beta}$ denotes the data-to-data covariance $C(\mathbf{u}_\alpha - \mathbf{u}_\beta)$, and $C_{\alpha 0}$ denotes the data-to-unknown covariance $C(\mathbf{u}_\alpha - \mathbf{u}_0)$. Let the semivariogram model be an isotropic spherical model with zero nugget effect, unit sill, and a 1 km range. The corresponding covariance model is

$$
C(h) = \begin{cases}
1 - \left[1.5 \cdot \dfrac{h}{a} - 0.5 \cdot \left(\dfrac{h}{a} \right)^3 \right] & \text{if } h < 1 \text{ km} \\
0 & \text{if } h \geq 1 \text{ km}
\end{cases}
$$

Given that covariance model and the data configuration shown in Figure 5.16 (left graph), the kriging system (5.71) is

$$
\begin{bmatrix}
1.00 & 0.06 & 0.05 & 0.00 & 0.00 & 1 \\
0.06 & 1.00 & 0.85 & 0.00 & 0.00 & 1 \\
0.05 & 0.85 & 1.00 & 0.00 & 0.00 & 1 \\
0.00 & 0.00 & 0.00 & 1.00 & 0.00 & 1 \\
0.00 & 0.00 & 0.00 & 0.00 & 1.00 & 1 \\
1 & 1 & 1 & 1 & 1 & 0
\end{bmatrix}
\begin{bmatrix}
\lambda_1^{OK}(\mathbf{u}_0) \\
\lambda_2^{OK}(\mathbf{u}_0) \\
\lambda_3^{OK}(\mathbf{u}_0) \\
\lambda_4^{OK}(\mathbf{u}_0) \\
\lambda_5^{OK}(\mathbf{u}_0) \\
\mu_{OK}(\mathbf{u}_0)
\end{bmatrix}
=
\begin{bmatrix}
0.43 \\
0.43 \\
0.42 \\
0.00 \\
0.00 \\
1
\end{bmatrix}
\tag{5.72}
$$

The large off-diagonal covariance value $C(\mathbf{u}_2 - \mathbf{u}_3) = 0.85$ informs the system on the redundancy of the two data at clustered locations $\mathbf{u}_2$ and $\mathbf{u}_3$. On the other hand, the zero covariance values in the right-hand-side vector inform the system that the point $\mathbf{u}$ to be estimated is beyond the correlation range of the two data at locations $\mathbf{u}_4$ and $\mathbf{u}_5$.

The following ordinary kriging weights are solutions of sytem (5.72):

$$
\lambda_1^{OK}(\mathbf{u}_0) = 0.446 \qquad \lambda_2^{OK}(\mathbf{u}_0) = 0.284 \qquad \lambda_3^{OK}(\mathbf{u}_0) = 0.190
$$
$$
\lambda_4^{OK}(\mathbf{u}_0) = 0.040 \qquad \lambda_5^{OK}(\mathbf{u}_0) = 0.040
$$

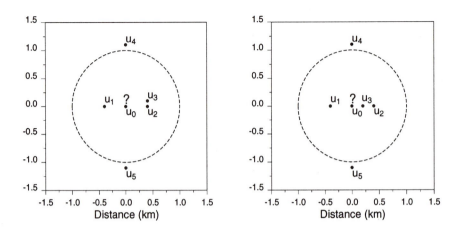

Figure 5.16: Two two-dimensional data configurations with different positions of u_3. In both cases, the attribute value at location u_0 is estimated using ordinary kriging and five data at locations u_1 to u_5. The radius of the dashed circle centered on u_0 corresponds to the correlation range (1 km).

As expected, the kriging weight decreases as the datum location gets farther from u_0. Note the following:

1. Both data at locations u_1 and u_2 are the same distance from u_0 (same right-hand-side covariance terms), yet the latter receives less weight because of its redundancy with the third datum at u_3.

2. Data at locations u_4 and u_5 get a non-zero weight, although they are beyond the correlation range of data. This non-zero weight is due to their contribution to the estimation of the trend component at location u_0. This trend estimation is implicit to ordinary kriging (see related discussion in section 5.3).

Consider a second data configuration where the location u_3 falls between the locations u_0 and u_2 (Figure 5.16, right graph). The new ordinary kriging weights are

$$\lambda_1^{OK}(u_0) = 0.311 \qquad \lambda_2^{OK}(u_0) = -0.057 \qquad \lambda_3^{OK}(u_0) = 0.702$$
$$\lambda_4^{OK}(u_0) = 0.022 \qquad \lambda_5^{OK}(u_0) = 0.022$$

Note the small negative weight given to the datum at u_2. Negative weights typically occur when the influence of a specific datum is screened by that of a closer one. In Figure 5.16 (right graph), the location u_3 screens the influence of u_2 for estimation at u_0. Negative weights allow the kriging estimate to take values outside the range of the data. Although this non-convexity of the estimator is generally a desirable property, it may yield unacceptable results,

such as negative concentrations or estimated proportions larger than 1. There
are three ways to deal with non-convexity problems:

1. Force all kriging weights to be positive; see Barnes and Johnson (1984),
   Xu (1994).

2. Add to all the weights a constant equal to the modulus of the largest
   negative weight, then reset the weights to sum to 1; see Journel and
   Rao (1996).

3. Reset any faulty estimate to the nearest bound, say, 0, if negative values
   are not acceptable, or 1 for excessive proportions; see Mallet (1980).

4. Impose constraints on the kriging estimates rather than on the kriging weights through the use of indicator constraint intervals; see section 7.4.2 and Journel (1986b).

For a particular data configuration, kriging weights may drastically change,
depending on the semivariogram model. For the two data configurations of
Figure 5.16, Figure 5.17 shows the evolution of the kriging weights as the
relative nugget effect increases. A larger nugget effect reduces the impact of
distance of data locations to $\mathbf{u}_0$ (left graph); it also reduces the screening
effect of location $\mathbf{u}_3$ on $\mathbf{u}_2$ (right graph). In the presence of a pure nugget
effect, all weights are equal to $1/n(\mathbf{u}) = 0.2$; the kriging estimate then reverts
to the arithmetic average of the data retained. The impact of semivariogram
parameters (shape, relative nugget effect, anisotropy) on kriging weights is
cogently discussed in Isaaks and Srivastava (1989, p. 296–313).

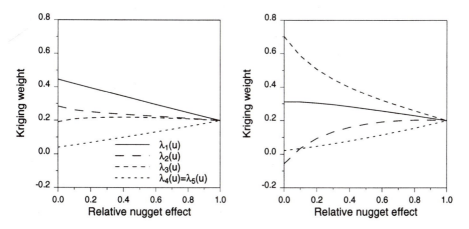

Figure 5.17: Impact of the relative nugget effect on the OK weights for the two
two-dimensional data configurations of Figure 5.16. All weights get close to each
other as the relative nugget effect increases.

5.8.2   Search neighborhood

As mentioned in the introduction to the linear estimator (5.1), common practice consists of using only the data closest to the location $\mathbf{u}$ being estimated, i.e., the data within a given neighborhood $W(\mathbf{u})$ centered on the location $\mathbf{u}$ to be estimated. There are several reasons for such a restriction:

- The covariance values for large distances are usually unreliable because of the few data pairs available for inference at such large distances.

- Using local search neighborhoods with ordinary kriging allows one to account for local departures from stationarity over the area $\mathcal{A}$.

- The closest data tend to screen the influence of those farther away; see previous discussion.

- The size of the kriging system, hence the computation time, drastically increases with the number of data retained, approximately in proportion to $[n(\mathbf{u})]^3$.

The search neighborhood is typically taken as a circle centered on the location being estimated. When the variation is anisotropic, it is better to consider an ellipse with its major axis oriented along the direction of maximum continuity: this provides more relevant data. To reduce the influence of data clusters (if any), it is good practice to split the search neighborhood into equal angle sectors, say, quadrants in two dimensions or octants in three dimensions, and retain within each sector a specified number of nearest data. When data are gridded and the estimation grid is aligned with the sampling grid, quadrant and octant searches may cause artifact discontinuities in the estimated maps because of sudden changes in the data retained for estimation from one location to the next.

One should avoid limiting a priori the maximum search distance to the correlation range of data. As featured in Figure 5.16, data beyond the correlation range contribute to the estimation of the local mean within the search neighborhood. As the relative nugget increases, the screening effect of closer data decreases, hence remote data get more weight (Figure 5.17). In subareas where sampling is sparse, the search distance should be increased to retain enough data. An alternative is to spiral away from the location $\mathbf{u}$ being estimated and to retain the $n(\mathbf{u})$ closest data whatever their distance to $\mathbf{u}$. For data selection, a useful practice consists of using the semivariogram distance $\gamma(\mathbf{u} - \mathbf{u}_\alpha)$ rather than the Euclidian distance $|\mathbf{u} - \mathbf{u}_\alpha|$, so that data are preferentially selected along the direction of maximum continuity.

In addition to size and orientation of the search neighborhood, one must specify the minimum and maximum number of data to use for estimation. The minimum must be equal at least to the number $(K + 1)$ of constraints on kriging weights. For example, the modeling of a quadratic trend in two dimensions requires at least six data values. In practice, 10 data values would be a reasonable minimum. That number should be larger where data are clustered so that one or two of the nearest data do not screen all others. The maximum number $n(\mathbf{u})$ of data values to retain depends on the objective pursued. When one aims at depicting local features of the attribute, that number should be limited; whereas more data and data farther away should be retained to depict long-range structures.

Before running any kriging over an entire area, it is good practice to try several search strategies on a test subarea. Cross validation techniques can then be used to evaluate the impact of different search parameters on the interpolation results. As mentioned in section 4.2.4, beware that the search strategy that produces the best cross-validated results may not yield the best predictions at unsampled locations.

Further discussion on the practice of kriging is available in Journel (1987), and Deutsch and Journel (1992a, section IV.6).

5.8.3 Kriging variance

Kriging provides not only a least-squares estimate of the attribute z but also the attached error variance; for example, for ordinary kriging:

$$\sigma_{OK}^2(\mathbf{u}) = C(0) - \sum_{\alpha=1}^{n(\mathbf{u})} \lambda_\alpha^{OK}(\mathbf{u}) \, C(\mathbf{u}_\alpha - \mathbf{u}) - \mu_{OK}(\mathbf{u}) \qquad (5.73)$$

That error variance is:

1. dependent on the covariance model.
   This is an excellent feature: the estimation precision should indeed depend on the complexity of the spatial variability of z as modeled by the covariance.

2. dependent on the data configuration.
   The terms $C(\mathbf{u}_\alpha - \mathbf{u})$ account for the relative geometry of data locations $\mathbf{u}_\alpha$ and their distances to the location $\mathbf{u}$ being estimated. This is also an excellent feature.

3. independent of data values.
   For a given covariance model, two identical data configurations would yield the same kriging variance no matter what the data were.

The latter two features are illustrated in Figure 5.18 (bottom graph), which shows the OK variance along the NE-SW transect. The pattern of variation of the kriging variance is fully controlled by the data configuration. The error variance is zero at data locations, increases away from the data, and reaches a maximum value beyond the extreme right datum (extrapolation situation). Indeed, as the location $\mathbf{u}$ being estimated gets farther away from data locations $\mathbf{u}_\alpha$, both the covariance term $C(\mathbf{u}_\alpha - \mathbf{u})$ and the kriging weight $\lambda_\alpha^{OK}(\mathbf{u})$ decrease, hence the kriging variance (5.73) increases.[4]

---

[4] Only the common case of positive weights and positive covariance values is considered here.

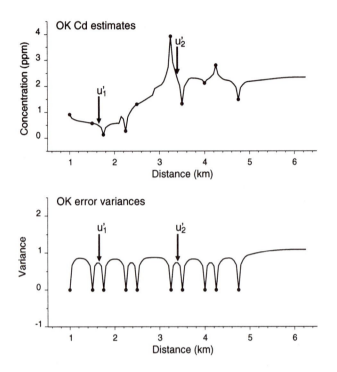

Figure 5.18: OK estimates of Cd concentrations (top graph) and the corresponding error variances (bottom graph). The kriging variance is independent of data values, hence it does not reflect the greater uncertainty that is expected at location $\mathbf{u}_2'$, which is surrounded by a very large and a small Cd values, compared to location $\mathbf{u}_1'$, which is surrounded by two consistently small Cd values.

Kriging yields similar error variances at any two or more locations with similar local data configurations. Intuitively, however, the potential for error is expected to be greater at a location $\mathbf{u}_2$ surrounded by data that are very different than at $\mathbf{u}_1$ surrounded by similarly valued data (Figure 5.18, top graph).

The problem with the kriging variance is that it is not conditioned to the data values used. The proper measure of local precision would be the conditional estimation variance specific to the data values and defined as

$$\text{Var} \left\{ Z^*(\mathbf{u}) - Z(\mathbf{u}) | Z(\mathbf{u}_\alpha) = z(\mathbf{u}_\alpha), \ \alpha = 1, \ldots, n(\mathbf{u}) \right\}$$

where the $z(\mathbf{u}_\alpha)$ are the data values. The kriging variance is but the average over all possible realizations of the $n(\mathbf{u})$ data RVs $Z(\mathbf{u}_\alpha)$. If the configurations of the $n(\mathbf{u}^{(l)})$ data retained at many different places $\mathbf{u}^{(l)}, l = 1, \ldots, L$ were the same, then the kriging variance (5.73) would be an estimate of the variance of the L errors $z^*(\mathbf{u}^{(l)}) - z(\mathbf{u}^{(l)})$. Because of the averaging over the L realizations of data values, the kriging variance retains, from the data used, only its geometry—not the data values. In this sense, the kriging variance is only a ranking index of data geometry (and size)—not a measure of the local spread of errors; see Journel (1986a), Deutsch and Journel (1992a, p. 15), and Chapter 7.

5.8.4 Re-estimation scores

Figure 5.19 shows the maps of OK estimates for the three metals that most exceed the Swiss guide values (Cd, Cu, and Pb). Note the following:

- Most kriging estimates exceed the critical threshold of 0.8 ppm for cadmium. The two low-valued zones correspond to clusters of small concentrations on Argovian rocks.

- Large estimates of copper concentrations are confined to the southern part of the region.

- Large estimated concentrations in Pb are in the central and south part of the region.

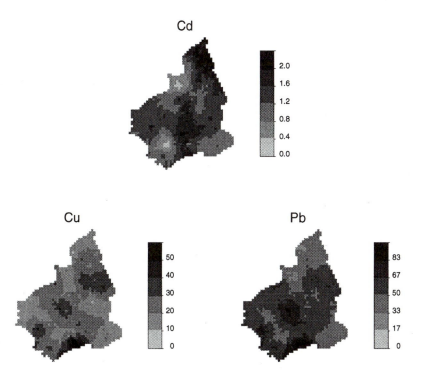

Figure 5.19: Maps of OK estimates. The tolerable maxima are 0.8 ppm for Cd and 50 ppm for Cu and Pb.

At the 100 test locations $\mathbf{u}_t$, both estimated values $z^*_{OK}(\mathbf{u}_t)$ and actual values $z(\mathbf{u}_t)$ are available. Statistics for both sets are given in Table 5.2. Besides the similarity of the means (global unbiasedness), note that the estimated values appear much less variable than actual values. This "smoothing effect" is due to the filtering of high-frequency (short-range) components; smoothing increases with increasing relative nugget effect and with decreasing sampling density. For example, smoothing is less pronounced for cobalt concentrations, the semivariogram of which has a larger range and a smaller relative nugget effect than those of the three other metals; compare the semivariograms of Figures 4.13 and 4.14.

The scattergrams of true $z(\mathbf{u}_t)$ versus estimated values $z^*_{OK}(\mathbf{u}_t)$ carry much more information than the summary statistics of Table 5.2 (Figure 5.20, left graphs). They all reveal an underestimation of large concentrations (values are below the 45° line) and an overestimation of small concentrations (values are above the 45° line). The balancing of these two effects results in the global unbiasedness seen in Table 5.2. Such a bias is called conditional because it depends on the class of values considered. The conditional bias drastically influences the evaluation of the extent of contamination (Table 5.2). The underestimation of large copper concentrations may lead one to declare safe all test locations, whereas 8% of them actually exceed the threshold value. In contrast, the overestimation of small cadmium concentrations may lead one to classify the majority (92%) of test locations as contaminated. The same effect, though less pronounced, is observed for lead.

Table 5.2: Statistics for true and estimated concentrations of heavy metals at 100 test locations.

	Cd	Cu	Pb	Co
Mean				
True values	1.23	23.2	56.5	9.8
OK estimates	1.36	24.0	55.4	9.4
Std deviation				
True values	0.69	25.8	40.3	3.5
OK estimates	0.41	7.5	11.7	2.4
% contamination				
True values	63.0	8.0	42.0	0.0
OK estimates	92.0	0.0	66.0	0.0
% misclassification				
OK estimates	35.0	8.0	36.0	0.0

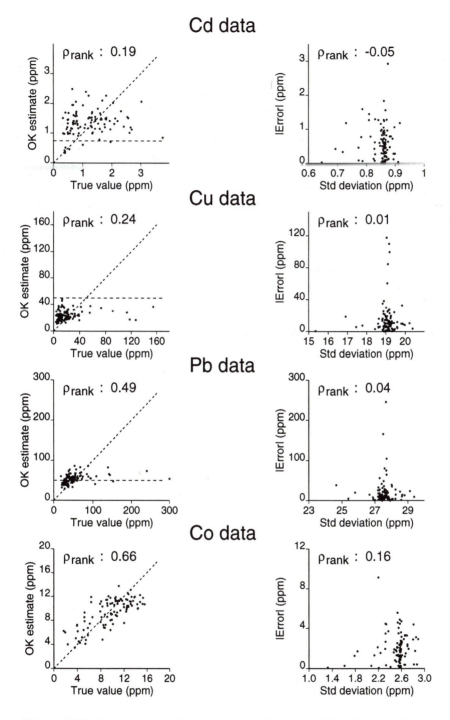

Figure 5.20: Scattergrams of true values $z(\mathbf{u}_t)$ versus OK estimates $z^*_{OK}(\mathbf{u}_t)$ at 100 test locations $\mathbf{u}_t$ (left graphs). For the same locations, note the lack of relation between the kriging error standard deviation $\sigma_{OK}(\mathbf{u}_t)$ and the absolute estimation error $|z^*_{OK}(\mathbf{u}_t) - z(\mathbf{u}_t)|$ (right graphs).

Conditional bias is a serious problem that relates to the smaller range of variation (smoothing effect) of estimated values relative to true values. As shown in section 5.7, interpolation algorithms are usually low-pass filters: they tend to smooth local spatial variation. The resulting overestimation of small values and underestimation of large ones is unfortunate since the focus of the study is typically on extreme values. Unlike kriging algorithms, simulation algorithms presented in later Chapter 8 allow one to reproduce the full covariance modeling the variability of data. Simulated maps would then correct for the conditional bias of estimated maps.

The rank correlation between actual and estimated values is weak for cadmium, copper, and lead. These poor re-estimation scores result from the short-range variation of the metals, which reduces the information carried by neighboring data. The rank correlation is stronger for cobalt, which varies more continuously in space (Figure 5.20, left bottom graph). In the presence of short-range variation, it becomes critical to account for any secondary information sampled with higher resolution and well correlated with the primary variable, for example, geology, land use, zinc or nickel concentrations. Kriging algorithms integrating such secondary information are presented in Chapter 6.

The error variance provided by kriging is unfortunately often misused as a measure of reliability of the kriging estimate. Figure 5.20 (right column) shows the scattergrams of the absolute error $|z^*_{OK}(\mathbf{u}_t) - z(\mathbf{u}_t)|$ versus the kriging standard deviation $\sigma_{OK}(\mathbf{u}_t)$, with the corresponding rank correlation coefficients. The weakness of the rank correlation confirms that the kriging standard deviation cannot be used as a direct measure of estimation precision.

Chapter 6

Local Estimation: Accounting for Secondary Information

Direct measurements of the primary attribute of interest are often supplemented by secondary information originating from other related categorical or continuous attributes. The estimation generally improves when this additional and usually denser information is taken into consideration, particularly when the primary data are sparse or poorly correlated in space.

Section 6.1 presents three kriging algorithms to incorporate exhaustively sampled secondary data: kriging within strata, simple kriging with varying local means, and kriging with an external drift. Section 6.2 introduces the cokriging algorithms that allow for non-exhaustive secondary information. All algorithms are used to interpolate metal concentrations at the test locations, and re-estimation scores are compared with ordinary kriging results.

6.1 Exhaustive Secondary Information

Consider the situation where primary data $\{z(\mathbf{u}_\alpha), \ \alpha = 1, \ldots, n\}$ are supplemented by secondary information exhaustively sampled. "Exhaustively sampled" refers to secondary information that is available at all primary data locations $\mathbf{u}_\alpha$ and at all locations $\mathbf{u}$ being estimated (Figure 6.1). Such information may relate to two types of attributes:

- A categorical attribute s with K mutually exclusives states s_k (e.g., rock types)

- A smoothly varying continuous attribute y (e.g., nickel concentration)

If the secondary information is densely available but not exhaustive, it is a reasonable approximation to complete it by interpolation or, better yet, by

185

simulation; see Almeida and Journel (1994). In the example of Figure 6.1, Ni concentrations were estimated in non-overlapping segments (blocks) of length 50 m so as to mimic a smoothly varying and exhaustively sampled variable.

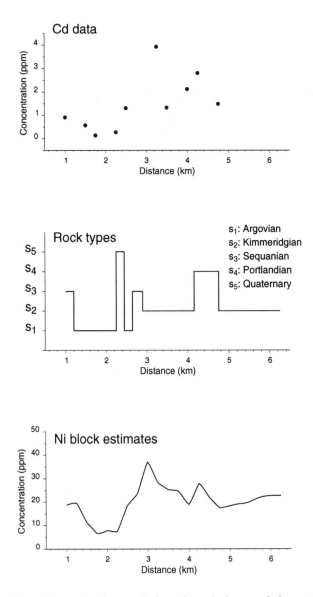

Figure 6.1: Ten primary Cd data and the exhaustively sampled secondary information: rock types and block OK estimates of Ni concentrations.

In this section, three variants of the kriging paradigm are developed to incorporate exhaustive secondary data in the estimation of the primary attribute z:

- *Kriging within strata* amounts to first stratifying the study area based on the secondary information then estimating the primary attribute within each specific stratum using the corresponding primary data and covariance model.

- *Simple kriging with varying local means* and *kriging with an external drift* use secondary data to characterize the spatial trend of the primary attribute.

6.1.1 Kriging within strata

Exploratory data analysis may have revealed significant differences in the average value and in the pattern of spatial continuity of the primary attribute z across the study area. In such a case, the original decision of stationarity should be reconsidered and the study area divided into more homogeneous regions (strata). The primary data should then be treated within each stratum as a separate population. That stratification is largely conditioned by the availability, within each stratum, of enough primary data to infer the necessary covariance. Moreover, one should be able to allocate each location $\mathbf{u}$ being estimated to a specific population. Where major scales of spatial variation are related to changes in land use, soil type, or lithology, secondary categorical information, such as land use, soil, or geologic maps, can be used to stratify the study area (e.g., see Stein et al., 1988; Voltz and Webster, 1990; Van Meirvenne et al., 1994).

Let the secondary information take the form of a categorical map of attribute s over the study area $\mathcal{A}$. Kriging within strata (KWS) proceeds in three steps:

1. The study area is first stratified according to the boundaries of the categorical map. The kth stratum $\mathcal{A}_k$ is defined as the set of locations $\mathbf{u}$ that belong to the category s_k, $\mathcal{A}_k = \{\mathbf{u} \in \mathcal{A}, \ s.t. \ s(\mathbf{u}) = s_k\}$.

2. The experimental semivariogram of z, $\widehat{\gamma}(\mathbf{h}; s_k)$, is computed within each stratum as

$$\widehat{\gamma}(\mathbf{h}; s_k) \ = \ \frac{1}{2N(\mathbf{h}; s_k)} \sum_{\alpha=1}^{N(\mathbf{h}; s_k)} [z(\mathbf{u}_\alpha) - z(\mathbf{u}_\alpha + \mathbf{h})]^2$$

where $N(\mathbf{h}; s_k)$ is the number of pairs of primary data locations $\mathbf{u}_\alpha$, a vector $\mathbf{h}$ apart, that jointly belong to the kth stratum.

3. The value of z at each location $\mathbf{u} \in k$th stratum is estimated using the semivariogram model $\gamma(\mathbf{h}; s_k)$ and the closest primary data $z(\mathbf{u}_\alpha)$ within that stratum.

Most often, data sparsity prevents computing reliable estimates $\hat{\gamma}(\mathbf{h}; s_k)$ within each geologic or soil category. One solution consists of combining the K experimental semivariograms $\hat{\gamma}(\mathbf{h}; s_k)$ into a single pooled within-stratum semivariogram $\hat{\gamma}_{ws}(\mathbf{h})$:

$$\hat{\gamma}_{ws}(\mathbf{h}) \quad = \quad \frac{\sum_{k=1}^{K} N(\mathbf{h}; s_k) \cdot \hat{\gamma}(\mathbf{h}; s_k)}{\sum_{k=1}^{K} N(\mathbf{h}; s_k)}$$

where each semivariogram value $\hat{\gamma}(\mathbf{h}; s_k)$ is weighted by the number of data pairs $N(\mathbf{h}; s_k)$ used for its estimation.

If the variance of z-values changes significantly from one stratum to another, each semivariogram $\hat{\gamma}(\mathbf{h}; s_k)$ could be scaled by the variance of the z-values that were used to build it:

$$\hat{\gamma}_{ws}(\mathbf{h}) \quad = \quad \frac{\sum_{k=1}^{K} N(\mathbf{h}; s_k) \cdot \hat{\gamma}(\mathbf{h}; s_k)/\hat{\sigma}_{|s_k}^2}{\sum_{k=1}^{K} N(\mathbf{h}; s_k)} \qquad (6.1)$$

where $\hat{\sigma}_{|s_k}^2$ is the variance of z-values within the kth stratum. This pooling of standardized semivariograms corresponds to a "proportional effect" correction (Journel and Huijbregts, 1978, p. 187).

As mentioned in section 5.8, the kriging estimate depends on the shape of the semivariogram model, not on its sill. Thus, the pooled semivariogram model $\gamma_{ws}(\mathbf{h})$ can be used for estimation within each stratum. However, the kriging variance specific to stratum k requires multiplication by the proper variance $\hat{\sigma}_{|s_k}^2$.

Example

Figure 6.2 (top graph) shows a stratification of the NE-SW transect based on geology. Data sparsity prevents considering more than two strata. The first stratum, denoted $\mathcal{A}_1$, includes all locations on Argovian and Quaternary rocks. The three other rock types with larger Cd concentrations and larger proportions of contaminated fields (Tables 2.4 and 2.5, pages 18–19), are regrouped into the second stratum $\mathcal{A}_2$, which consists of two disconnected segments.

Figure 6.2 (middle graph) shows the semivariogram of Cd concentrations of each stratum. Concentrations appear to vary more continuously within the stratum $\mathcal{A}_1$ (smaller relative nugget effect and larger range of the semivariogram). Recall that the smallest concentrations were measured on Argovian and Quaternary rocks, which form the first stratum. Thus, the better continuity of Cd concentrations within that stratum relates to the better connectivity of small Cd concentrations revealed by the analysis of indicator semivariograms (Figure 2.19, top graph, page 45)

Cadmium concentrations are estimated using ordinary kriging, stratum-specific primary data, and the semivariogram models shown in Figure 6.2 (middle graph). The resulting estimates are depicted by the solid line in the

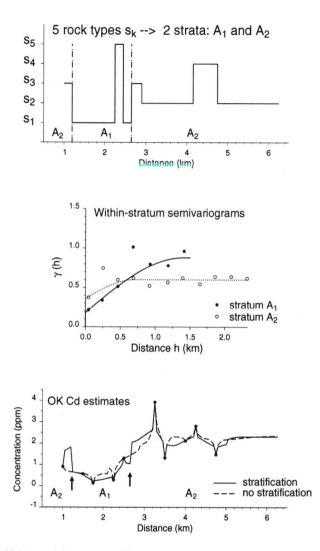

Figure 6.2: Kriging within strata. The transect is first split into two strata A_1 and A_2, according to geology (top graph). Within each stratum, the Cd semivariogram is inferred and modeled (middle graph), and Cd concentrations are estimated using ordinary kriging and stratum-specific data (bottom graph, solid line). Vertical arrows depict discontinuities at the strata boundaries. The dashed line represents the OK estimate without stratification.

bottom graph. The dashed line corresponds to the OK estimates computed without stratification in section 5.3. Note the following:

- Kriging within strata yields estimates that suddenly change at the strata boundaries depicted by vertical arrows. Indeed, estimates on different sides of a stratum boundary are based on two different sets of data usually with very different means.

- Away from strata boundaries, the stratification has little influence on the search strategy because primary data in the search neighborhood generally belong to the same stratum. Discrepancies between ordinary kriging estimates with and without stratification now result only from differences in the semivariogram models.

6.1.2   Simple kriging with varying local means

Recall the simple kriging (SK) estimator (5.7):

$$Z^*_{SK}(\mathbf{u}) - m = \sum_{\alpha=1}^{n(\mathbf{u})} \lambda^{SK}_\alpha(\mathbf{u}) \, [Z(\mathbf{u}_\alpha) - m]$$

Under the decision of stationarity, the mean m does not depend on location $\mathbf{u}$ but represents global information common to all unsampled locations. To account for the secondary information available at each location $\mathbf{u}$, the known stationary mean m may be replaced by known varying means $m^*_{SK}(\mathbf{u})$, leading to the simple kriging with varying local means (SKlm) estimator:

$$Z^*_{SKlm}(\mathbf{u}) - m^*_{SK}(\mathbf{u}) = \sum_{\alpha=1}^{n(\mathbf{u})} \lambda^{SK}_\alpha(\mathbf{u}) \, [Z(\mathbf{u}_\alpha) - m^*_{SK}(\mathbf{u}_\alpha)] \qquad (6.2)$$

Different estimates of the primary local mean $m(\mathbf{u})$ can be used, depending on the secondary information available:

1. If the secondary information relates to a categorical attribute s with K non-overlapping states s_k, the primary local mean can be identified with the mean of z-values within the category s_k prevailing at $\mathbf{u}$:

$$m^*_{SK}(\mathbf{u}) = m_{|s_k} \qquad \text{with } s(\mathbf{u}) = s_k$$

The conditional mean $m_{|s_k}$ is computed as

$$m_{|s_k} = \frac{1}{n_k} \sum_{\alpha=1}^{n} i(\mathbf{u}_\alpha; s_k) \cdot z(\mathbf{u}_\alpha)$$

where $n_k = \sum_{\alpha=1}^{n} i(\mathbf{u}_\alpha; s_k)$ is the number of primary data locations within category s_k, and $i(\mathbf{u}_\alpha; s_k)$ is the indicator (2.23) of category s_k.

2. In the case of a secondary continuous attribute y, the primary local mean can be a function (linear or not) of the secondary attribute value at $\mathbf{u}$:

$$m^*_{SK}(\mathbf{u}) = f(y(\mathbf{u}))$$

An alternative to using regression to determine the function $f(.)$ consists of discretizing the range of variation of the secondary attribute into K classes $(y_k, y_{k+1}]$. The primary local mean $m(\mathbf{u})$ is then identified with the mean of z-values with colocated y-values falling into class $(y_k, y_{k+1}]$:

$$m_{SK}^*(\mathbf{u}) \quad = \quad m_{|k} \qquad \text{with } y(\mathbf{u}) \in (y_k, y_{k+1}]$$

The conditional mean $m_{|k}$ is computed as

$$m_{|k} \quad = \quad \frac{1}{n_k} \sum_{\alpha=1}^{n} i(\mathbf{u}_\alpha, k) \cdot z(\mathbf{u}_\alpha)$$

The number of primary data $z(\mathbf{u}_\alpha)$, such as $y(\mathbf{u}_\alpha) \in (y_k, y_{k+1}]$, is n_k, and the y-indicator variable $i(\mathbf{u}_\alpha; k)$ is defined as

$$i(\mathbf{u}_\alpha; k) = \begin{cases} 1 & \text{if } y(\mathbf{u}_\alpha) \in (y_k, y_{k+1}] \\ 0 & \text{otherwise} \end{cases}$$

The kriging weights $\lambda_\alpha^{SK}(\mathbf{u})$ in expression (6.2) are obtained by solving a simple kriging system of type (5.9):

$$\sum_{\beta=1}^{n(\mathbf{u})} \lambda_\beta^{SK}(\mathbf{u}) \, C_R(\mathbf{u}_\alpha - \mathbf{u}_\beta) \quad = \quad C_R(\mathbf{u}_\alpha - \mathbf{u}) \qquad \alpha = 1, \ldots, n(\mathbf{u})$$

where $C_R(\mathbf{h})$ is the covariance function of the residual RF $R(\mathbf{u}) = Z(\mathbf{u}) - m(\mathbf{u})$, not that of $Z(\mathbf{u})$ itself.

Example

Consider the estimation of Cd concentrations along the NE-SW transect using as secondary information either geology (categorical attribute) or the block OK estimates of Ni concentration (continuous attribute), as defined in Figure 6.1. Figures 6.3 and 6.4 show the different steps of the corresponding kriging approaches:

1. The calibration between primary and secondary data amounts to determining either the average Cd concentration for each rock type (conditional means) or the regression function of Cd concentrations on Ni block estimates (top left graphs of Figures 6.3 and 6.4, respectively). This calibration results either in a "staircase" trend estimate (Figure 6.3, second row) or a more smoothly varying trend when derived from the continuous Ni attribute (Figure 6.4, second row).

2. At each primary datum location $\mathbf{u}_\alpha$, the residual value $r(\mathbf{u}_\alpha)$ is computed by subtracting the trend estimate $m_{SK}^*(\mathbf{u}_\alpha)$ from the primary datum $z(\mathbf{u}_\alpha)$, i.e., $r(\mathbf{u}_\alpha) = z(\mathbf{u}_\alpha) - m_{SK}^*(\mathbf{u}_\alpha)$. The semivariogram of residuals is then computed and modeled. Figures 6.3 and 6.4 (top right graph) show the semivariogram computed from 259 residual data and the model fitted.

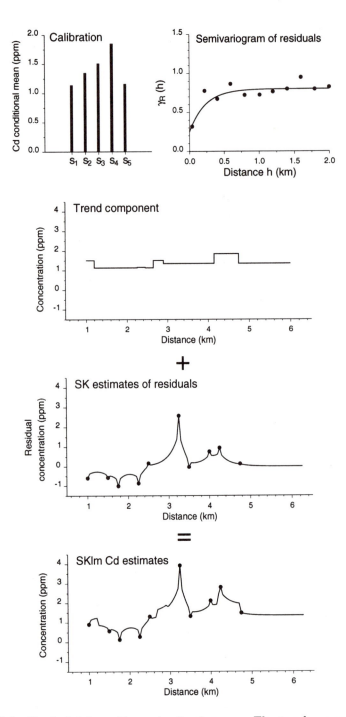

Figure 6.3: Simple kriging with varying local means. The trend component at location **u** is estimated by the mean of Cd concentrations for the rock type prevailing at that location.

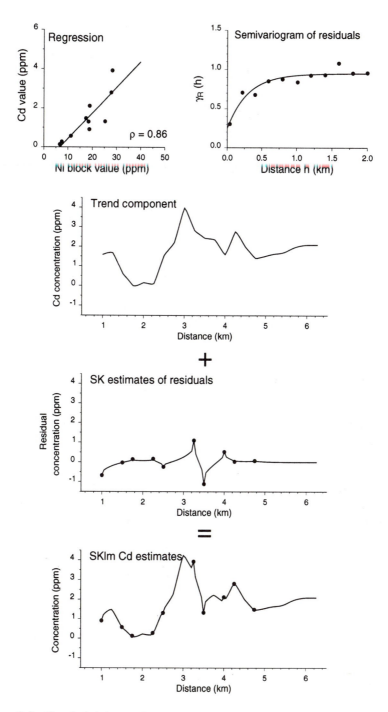

Figure 6.4: Simple kriging with varying local means. The trend component at location **u** is estimated by the regression of Cd concentration on the Ni block value at that location.

3. The residual values are then estimated along the transect using simple kriging and the five closest residual data $r(\mathbf{u}_\alpha)$ (third row). The final estimate of Cd concentration, $z^*_{SKlm}(\mathbf{u})$, is obtained by adding the trend estimate $m^*_{SK}(\mathbf{u})$ to the SK estimate of the residual $r^*_{SK}(\mathbf{u})$ (bottom graphs).

Figures 6.3 and 6.4 show the following:

- The estimated values in Figure 6.3 (bottom graph) show discontinuities at the geologic boundaries. For example, the sharp decrease around 4.8 km (extreme right datum) reflects the decrease in the mean of Cd concentrations from Portlandian to Kimmeridgian rocks. Such discontinuities are, however, less important than when kriging within strata (recall Figure 6.2, bottom graph) because data across geologic boundaries are now used.

- Because Cd data and Ni block estimates are strongly correlated across the transect ($\rho= 0.86$), the trend component in Figure 6.4 already accounts for most of the variance in the Cd data. Thus, most residual data are close to zero and the residual estimates contribute little to the estimates of z, which appear as a mere rescaling of the smoothly varying secondary variable shown in Figure 6.1 (bottom graph).

- As the location $\mathbf{u}$ being estimated gets farther away from data locations, the SK weights decrease and the kriging estimate gets closer to the trend estimate at that location; see right extreme part of the transect in Figures 6.3 and 6.4 (bottom graph).

6.1.3   Kriging with an external drift

Kriging with an external drift (KED) is but a variant of kriging with a trend model (KT), as presented in relation (5.25). The trend $m(\mathbf{u})$ is modeled as a linear function of a smoothly varying secondary (external) variable $y(\mathbf{u})$ instead of as a function of the spatial coordinates:

$$m(\mathbf{u}) \quad = \quad a_0(\mathbf{u}) + a_1(\mathbf{u})\, y(\mathbf{u}) \tag{6.3}$$

as opposed to $m(\mathbf{u}) = a_0(\mathbf{u}) + \sum_{k=1}^{K} a_k(\mathbf{u})\, f_k(\mathbf{u})$ for kriging with a trend model.

As with the KT approach, the two unknown trend coefficients $a_0(\mathbf{u})$ and $a_1(\mathbf{u})$ are deemed constant within the search neighborhood $W(\mathbf{u})$ and are implicitly estimated through the kriging system. Unlike the SK approach of the previous section, the mean $m(\mathbf{u})$ is not estimated through a calibration or regression process prior to the kriging of z.

The KED estimator is

$$Z^*_{KED}(\mathbf{u}) \quad = \quad \sum_{\alpha=1}^{n(\mathbf{u})} \lambda_\alpha^{KED}(\mathbf{u})\, Z(\mathbf{u}_\alpha) \tag{6.4}$$

The kriging weights $\lambda_\alpha^{KED}(\mathbf{u})$ are the solution of the following system of $(n(\mathbf{u}) + 2)$ linear equations (Maréchal, 1984; Deutsch and Journel, 1992a, p. 67):

$$
\begin{cases}
\displaystyle\sum_{\beta=1}^{n(\mathbf{u})} \lambda_\beta^{KED}(\mathbf{u})\, C_R(\mathbf{u}_\alpha - \mathbf{u}_\beta) \;+\; \mu_0^{KED}(\mathbf{u}) \;+ \mu_1^{KED}(\mathbf{u})\, y(\mathbf{u}_\alpha) \\[2mm]
\qquad\qquad\qquad\qquad = \; C_R(\mathbf{u}_\alpha - \mathbf{u}) \qquad \alpha = 1, \ldots, n(\mathbf{u}) \\[4mm]
\displaystyle\sum_{\beta=1}^{n(\mathbf{u})} \lambda_\beta^{KED}(\mathbf{u}) \;=\; 1 \\[4mm]
\displaystyle\sum_{\beta=1}^{n(\mathbf{u})} \lambda_\beta^{KED}(\mathbf{u})\, y(\mathbf{u}_\beta) \;=\; y(\mathbf{u})
\end{cases}
\tag{6.5}
$$

The KED system (6.5) is a particular case of the KT system (5.26) where $K = 1$ and the trend component $f_1(\mathbf{u})$ at any location $\mathbf{u}$ is identified with the value $y(\mathbf{u})$ of the secondary attribute there.

As with kriging with a trend (KT), two major issues are the choice of the trend function and the inference of the residual semivariogram $\gamma_R(\mathbf{h})$:

- The trend model (6.3) states that the local average of the primary variable z at location $\mathbf{u}$ is linearly related to the secondary datum $y(\mathbf{u})$. It is critical to validate that assumption from calibration data or, better, from physical rationale. For example, it makes sense to assume that the seismic travel time to a reflecting horizon is linearly related to the depth of that horizon. Seismic data can then be used as an external drift for mapping the top of that horizon from a few borehole data; see Galli and Meunier (1987), Chu et al. (1991). Another example relates to using elevation data to model trends in temperature (Hudson, 1993).

- The relation between primary trend and secondary variable must be linear. If this is not the case, an appropriate transformation of the secondary variable could make that relation linear.

- The value of the secondary variable y must be known at all primary data locations $\mathbf{u}_\alpha$ and at all locations $\mathbf{u}$ being estimated.

- The secondary variable should vary smoothly in space to avoid instability of the KED system. In an early application of kriging with an external drift, however, Maréchal (1984) considered the example of local trends of mineral grades controlled by faults blocks. The external variable was then an indicator variable that changed value only across the faults.

- As discussed in section 5.4, the residual semivariogram $\gamma_R(\mathbf{h})$ should be inferred from pairs of z-values that are unaffected or slightly affected by the trend, i.e., from data pairs such that $y(\mathbf{u}_\alpha) \approx y(\mathbf{u}_\alpha + \mathbf{h})$.

Kriging the trend

It is sometimes useful to compute and map the trend component $m(\mathbf{u})$ that is implicitly used in the expression of the KED estimator. The trend estimator is written

$$m_{KED}^*(\mathbf{u}) = \sum_{\alpha=1}^{n(\mathbf{u})} \lambda_{\alpha m}^{KED}(\mathbf{u}) \, Z(\mathbf{u}_\alpha) \tag{6.6}$$

where $\lambda_{\alpha m}^{KED}(\mathbf{u})$ is the weight assigned to datum $z(\mathbf{u}_\alpha)$. The kriging weights are obtained by solving a kriging system identical to the KT system (5.33) except that the trend function $f_1(\mathbf{u})$ is now set equal to the secondary variable $y(\mathbf{u})$:

$$
\begin{cases}
\displaystyle\sum_{\beta=1}^{n(\mathbf{u})} \lambda_{\beta m}^{KED}(\mathbf{u}) \, C_R(\mathbf{u}_\alpha - \mathbf{u}_\beta) \;+\; \mu_{0m}^{KED}(\mathbf{u}) \;+\; \mu_{1m}^{KED}(\mathbf{u}) \, y(\mathbf{u}_\alpha) \;=\; 0 \\[4pt]
\hspace{10em} \alpha = 1, \ldots, n(\mathbf{u}) \\[6pt]
\displaystyle\sum_{\beta=1}^{n(\mathbf{u})} \lambda_{\beta m}^{KED}(\mathbf{u}) \;=\; 1 \\[6pt]
\displaystyle\sum_{\beta=1}^{n(\mathbf{u})} \lambda_{\beta m}^{KED}(\mathbf{u}) \, y(\mathbf{u}_\beta) \;=\; y(\mathbf{u})
\end{cases}
\tag{6.7}
$$

Estimation of the trend $m(\mathbf{u})$ through system (6.7) amounts to estimating by least-squares regression the coefficients $a_0(\mathbf{u})$ and $a_1(\mathbf{u})$, defined in relation (6.3), within each search neighborhood $W(\mathbf{u})$:

$$m_{KED}^*(\mathbf{u}) = a_0^*(\mathbf{u}) + a_1^*(\mathbf{u}) \cdot y(\mathbf{u})$$

A zero slope $a_1^*(\mathbf{u})$ means that the secondary datum $y(\mathbf{u})$ does not influence the primary trend estimate at $\mathbf{u}$. As that slope increases, the influence of the secondary value becomes preponderant. A map of the scaling factor $a_1^*(\mathbf{u})$ allows one to depict the local influence of the secondary variable in the estimation of the primary trend component.

The estimator of the trend coefficient $a_1(\mathbf{u})$ is written

$$a_1^*(\mathbf{u}) = \sum_{\alpha=1}^{n(\mathbf{u})} \lambda_{\alpha 1}^{KED}(\mathbf{u}) \, Z(\mathbf{u}_\alpha)$$

The kriging weights are given by a system of $(n(\mathbf{u})+2)$ linear equations that is a particular case of the KT system (5.36) with $K = k' = 1$ and $f_1(\mathbf{u}) = y(\mathbf{u})$:

$$\begin{cases} \displaystyle\sum_{\beta=1}^{n(\mathbf{u})} \lambda_{\beta 1}^{KED}(\mathbf{u}) \, C_R(\mathbf{u}_\alpha - \mathbf{u}_\beta) \; + \; \mu_{01}^{KED}(\mathbf{u}) \; + \mu_{11}^{KED}(\mathbf{u}) \, y(\mathbf{u}_\alpha) \; = \; 0 \\ \qquad\qquad\qquad\qquad\qquad\qquad\qquad\qquad \alpha = 1,\ldots,n(\mathbf{u}) \\ \displaystyle\sum_{\beta=1}^{n(\mathbf{u})} \lambda_{\beta 1}^{KED}(\mathbf{u}) \; = \; 0 \\ \displaystyle\sum_{\beta=1}^{n(\mathbf{u})} \lambda_{\beta 1}^{KED}(\mathbf{u}) \, y(\mathbf{u}_\beta) \; = \; 1 \end{cases} \qquad (6.8)$$

Exchanging the 0 and 1 right-hand-side terms in the last two equations yields the kriging system for the estimation of the constant term $a_0(\mathbf{u})$. Note that the location $\mathbf{u}$ being estimated does not appear in the kriging system (6.8). Thus, the trend coefficients estimates $a_0^*(\mathbf{u})$ and $a_1^*(\mathbf{u})$ do not change as long as the same set of $n(\mathbf{u})$ primary data $z(\mathbf{u}_\alpha)$ is used.

SKlm versus KED estimator

Like the KT approach, kriging with an external drift amounts to:

1. evaluating, within each search neighborhood $W(\mathbf{u})$, the regression co-efficients $a_0(\mathbf{u})$ and $a_1(\mathbf{u})$ from the $n(\mathbf{u})$ data pairs $(z(\mathbf{u}_\alpha), y(\mathbf{u}_\alpha))$.

2. estimating the trend component at all $n(\mathbf{u})$ primary data locations $\mathbf{u}_\alpha$ and at the location $\mathbf{u}$ being estimated as

$$\begin{aligned} m_{KED}^*(\mathbf{u}_\alpha) &= a_0^*(\mathbf{u}) + a_1^*(\mathbf{u}) \cdot y(\mathbf{u}_\alpha) \qquad \forall\, \alpha = 1,\ldots,n(\mathbf{u}) \\ m_{KED}^*(\mathbf{u}) &= a_0^*(\mathbf{u}) + a_1^*(\mathbf{u}) \cdot y(\mathbf{u}) \end{aligned}$$

3. performing simple kriging on the corresponding residuals:

$$Z_{KED}^*(\mathbf{u}) - m_{KED}^*(\mathbf{u}) = \sum_{\alpha=1}^{n(\mathbf{u})} \lambda_\alpha^{SK}(\mathbf{u}) \, [Z(\mathbf{u}_\alpha) - m_{KED}^*(\mathbf{u}_\alpha)]$$

$$(6.9)$$

The KED estimator (6.9) is thus similar to an SK estimator with varying local means derived from a linear rescaling of the colocated y-datum:

$$Z_{SKlm}^*(\mathbf{u}) - m_{SK}^*(\mathbf{u}) = \sum_{\alpha=1}^{n(\mathbf{u})} \lambda_\alpha^{SK}(\mathbf{u}) \, [Z(\mathbf{u}_\alpha) - m_{SK}^*(\mathbf{u}_\alpha)]$$

where $m_{SK}^*(\mathbf{u}) = f(y(\mathbf{u})) = a_0^* + a_1^* \cdot y(\mathbf{u})$.

The two estimators differ by the definition of the trend component. The trend coefficients a_0^* and a_1^* are derived once and independently of the kriging system in the SKlm approach, whereas in the KED approach the regression coefficients $a_0^*(\mathbf{u})$ and $a_1^*(\mathbf{u})$ are implicitly estimated through the kriging system within each search neighborhood $W(\mathbf{u})$.

Example

Figure 6.5 shows both KED (solid line) and SKlm (dashed line) estimates of the trend (top graph) and of Cd concentrations (middle graph) along the

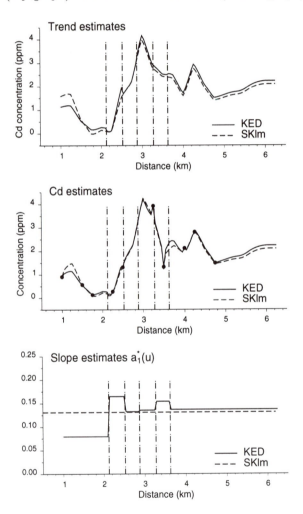

Figure 6.5: SKlm and KED estimates of the trend and of Cd concentrations. The vertical dashed lines delineate the segments that are estimated using the same five Cd concentrations. The estimated slope $a_1^*(\mathbf{u})$ is constant within each such segment.

NE-SW transect. As in Figure 5.3, the vertical dashed lines split the transect into six segments within which the same five primary data are used for estimation: for example, all trend and Cd estimates within the segment 1–2.1 km are based on Cd data at locations $\mathbf{u}_1$ to $\mathbf{u}_5$. As discussed previously, the kriging system (6.8) is identical at all locations where the same neighboring data are involved in the estimation. Therefore, the two KED-estimated trend coefficients $a_0^*(\mathbf{u})$ and $a_1^*(\mathbf{u})$ are constant within each segment and change from one segment to the next, depending on the five neighboring data retained; see the slope estimates (Figure 6.5, bottom graph, solid line). The slope of the trend model is multiplied by two from the first segment to the next because of the large Cd concentration at $\mathbf{u}_6$, then it gets closer to the constant slope of the global trend model that is used in simple kriging with varying local means (bottom graph, horizontal dashed line).

Both KED and SKlm trend and Cd estimates are fairly similar. The largest differences occur for the first segment, 1–2.1 km, where the two KED and SKlm slope estimates differ more. As with kriging with a trend model, beyond the extreme right datum at coordinate 4.8 km, kriging with an external drift extrapolates the linear trend model fitted to the last five Cd values. Such extrapolation would be dangerous if the last five values showed an atypical relation between Cd data and Ni block estimates, say, a negative correlation caused by an outlier. Simple kriging with varying local means would be more robust in that it extrapolates a relation that is fitted to all data along the transect.

6.1.4 Performance comparison

Figures 6.6 and 6.7 show different estimates of Cd concentrations accounting for the information provided by either geologic and land use maps or by the map of Zn block estimates. Three algorithms for integration of the secondary information are considered:

1. Kriging within strata. The study area is divided into two strata based on the geologic map in Figure 6.6 (left top graph). The first stratum includes Argovian rocks with the smallest proportion of contaminated locations; the four other rocks form the second stratum (Figure 6.6, left bottom graph). Ordinary kriging is performed separately within each stratum using a pooled within-stratum semivariogram; see right bottom graph of Figure 6.6.

2. Simple kriging with varying local means. The local means are determined as a linear function of the Zn block estimates shown in Figure 6.7 (left top graph).

3. Kriging with an external drift. The external drift consists of the map of Zn block estimates.

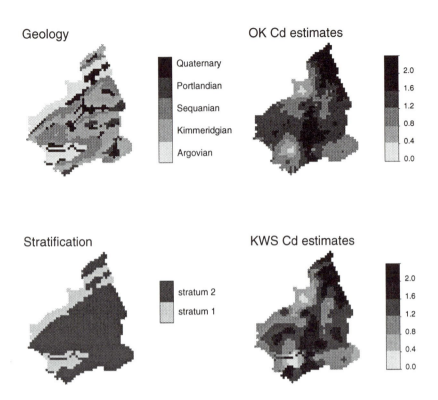

Figure 6.6: Maps of Cd estimates obtained using ordinary kriging (OK) across all strata and within two strata defined from geology (KWS).

Accounting for secondary information yields more detailed maps than the map of the reference OK estimates shown in Figures 6.6 and 6.7 (right top graphs). Kriging within strata (KWS) enhances the contrast between low-valued Argovian rocks and other rocks (Figure 6.6, right bottom graph). Wherever the two strata are intermingled, as in the SW part of the study area, discontinuities in Cd estimates occur.

Because they share the same secondary information, maps of SKlm and KED estimates in Figure 6.7 (middle graphs) show similar long-range features. However, kriging with an external drift yields more local detail. Such short-range variation results from the local re-evaluation of the linear regression of Cd concentrations on Zn block estimates. For example, the larger variation of KED estimates in the central part of the study area relates to larger slopes of the trend model (Figure 6.7, right bottom graph). The combination of a steep trend model and small Zn block estimates yields negative KED estimates of Cd concentrations; see white pixels in Figure 6.7 (right middle graph). Such unacceptable estimates are not produced by the other algorithms.

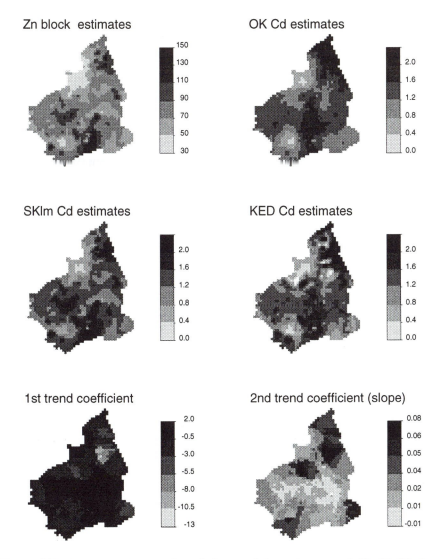

Figure 6.7: Accounting for an exhaustively sampled continuous variable (Zn block estimates) in the estimation of Cd concentrations: simple kriging with varying local means (SKlm) and kriging with an external drift (KED). The two bottom graphs depict the trend coefficients implicitly estimated locally in the KED approach.

Re-estimation scores

The same three algorithms for integrating secondary information are used to estimate metals with widespread contamination (Cd, Cu, Pb) and cobalt at the 100 test locations. Results in Table 6.1 show that accounting for secondary information

- corrects partly for the smoothing effect of ordinary kriging: this is most significant for the metals (Cd, Cu, Pb) with short-range variation,

- increases significantly the rank correlation between actual and estimated values, and

- decreases the percentage of test locations misclassified, that is, wrongly declared safe or contaminated.

Table 6.1: Statistics for true and estimated concentrations of heavy metals at 100 test locations; algorithms are the reference ordinary kriging (OK), kriging within strata (KWS), simple kriging with local means computed as a linear function of Zn block estimates (SKlm), and kriging with an external drift (KED).

	Cd	Cu	Pb	Co
Mean				
True values	**1.23**	**23.2**	**56.5**	**9.8**
OK estimates	1.36	24.0	55.4	9.4
KWS estimates	1.34	22.6	54.7	9.3
SKlm estimates	1.38	25.2	55.6	9.5
KED estimates	1.30	24.7	55.4	9.5
Std deviation				
True values	**0.69**	**25.8**	**40.3**	**3.5**
OK estimates	0.41	7.5	11.7	2.4
KWS estimates	0.41	10.9	13.3	2.6
SKlm estimates	0.59	12.6	17.6	2.7
KED estimates	0.65	18.4	25.0	2.8
ρ_{rank} *with true values*				
OK estimates	0.19	0.24	0.49	0.66
KWS estimates	0.20	0.47	0.35	0.69
SKlm estimates	0.54	0.66	0.70	0.82
KED estimates	0.53	0.66	0.69	0.80
% misclassification				
OK estimates	35	8	36	0
KWS estimates	30	9	39	0
SKlm estimates	28	7	28	0
KED estimates	24	5	25	0

6.2 The Cokriging Approach

The main limitation of the algorithms previously proposed for integration of secondary information is that secondary data must be available at all locations being estimated and in addition, for kriging with an external drift, at all primary data locations. Non-exhaustive secondary information can be incorporated using the cokriging approach that explicitly accounts for the spatial cross correlation between primary and secondary variables.

6.2.1 The cokriging paradigm

Consider the situation where primary data $\{z_1(\mathbf{u}_{\alpha_1}), \alpha_1 = 1, \ldots, n_1\}$ are supplemented by secondary data related to $(N_v - 1)$ continuous attributes z_i, $\{z_i(\mathbf{u}_{\alpha_i}), \alpha_i = 1, \ldots, n_i, i = 2 \ldots, N_v\}$, at any, possibly different, locations.[1] The linear estimator (5.1) is readily extended to incorporate that additional information:

$$Z_1^*(\mathbf{u}) - m_1(\mathbf{u}) = \sum_{\alpha_1=1}^{n_1(\mathbf{u})} \lambda_{\alpha_1}(\mathbf{u}) \left[Z_1(\mathbf{u}_{\alpha_1}) - m_1(\mathbf{u}_{\alpha_1})\right]$$

$$+ \sum_{i=2}^{N_v} \sum_{\alpha_i=1}^{n_i(\mathbf{u})} \lambda_{\alpha_i}(\mathbf{u}) \left[Z_i(\mathbf{u}_{\alpha_i}) - m_i(\mathbf{u}_{\alpha_i})\right] \quad (6.10)$$

where $\lambda_{\alpha_1}(\mathbf{u})$ is the weight assigned to the primary datum $z_1(\mathbf{u}_{\alpha_1})$ and $\lambda_{\alpha_i}(\mathbf{u})$, $i > 1$, is the weight assigned to the secondary datum $z_i(\mathbf{u}_{\alpha_i})$. The expected values of the RVs $Z_1(\mathbf{u})$ and $Z_i(\mathbf{u}_{\alpha_i})$ are denoted $m_1(\mathbf{u})$ and $m_i(\mathbf{u}_{\alpha_i})$, respectively. Typically, only the primary and secondary data closest to the location $\mathbf{u}$ being estimated are retained, i.e., $n_i(\mathbf{u})$ is usually smaller than n_i. The amount of data retained and the size of the search neighborhood need not be the same for all attributes.

All cokriging estimators are but variants of expression (6.10). They are all required to be unbiased and to minimize the error variance $\sigma_E^2(\mathbf{u})$, that is,

$$\sigma_E^2(\mathbf{u}) = \text{Var}\{Z_1^*(\mathbf{u}) - Z_1(\mathbf{u})\}$$

is minimized under the constraint that the expected error is zero:

$$\text{E}\{Z_1^*(\mathbf{u}) - Z_1(\mathbf{u})\} = 0$$

The various cokriging estimators differ in the random function model $Z_i(\mathbf{u})$ adopted for the various variables. Typically, each RF $Z_i(\mathbf{u})$ is decomposed into a residual component $R_i(\mathbf{u})$ and a trend component $m_i(\mathbf{u})$:

$$Z_i(\mathbf{u}) = R_i(\mathbf{u}) + m_i(\mathbf{u}) \qquad i = 1, \ldots, N_v$$

---

[1] Since data locations can be different from one variable to another, a more rigorous notation would be $\mathbf{u}_{\alpha_i}^{(i)}$. To simplify notation, the reference to the attribute z_i is hereafter limited to the subscript α_i.

The residual component $R_i(\mathbf{u})$ is then modeled as a stationary RF with zero mean and a covariance function $C_i^R(\mathbf{h})$:

$$
\begin{aligned}
\mathrm{E}\{R_i(\mathbf{u})\} &= 0 \\
\mathrm{Cov}\{R_i(\mathbf{u}), R_i(\mathbf{u}+\mathbf{h})\} &= \mathrm{E}\{R_i(\mathbf{u}) \cdot R_i(\mathbf{u}+\mathbf{h})\} = C_i^R(\mathbf{h})
\end{aligned}
$$

The cross covariance between any two residual RVs $R_i(\mathbf{u})$ and $R_j(\mathbf{u}+\mathbf{h})$ is

$$
C_{ij}^R(\mathbf{h}) = \mathrm{Cov}\{R_i(\mathbf{u}), R_j(\mathbf{u}+\mathbf{h})\}
$$

Three common cokriging variants can be distinguished according to the trend model $m_i(\mathbf{u})$:

1. Simple cokriging (SCK) considers each local mean $m_i(\mathbf{u})$ known and constant within the study area $\mathcal{A}$:

$$
m_i(\mathbf{u}) = m_i, \text{ known } \quad \forall\, \mathbf{u} \in \mathcal{A} \quad i = 1, \dots, N_v
$$

2. Ordinary cokriging (OCK) accounts for local variations of the means by limiting the domain of stationarity of both primary and secondary means (both unknown) to the local neighborhood $W(\mathbf{u})$:

$$
m_i(\mathbf{u}') = \text{ constant but unknown } \quad \forall\, \mathbf{u}' \in W(\mathbf{u}) \quad i = 1, \dots, N_v
$$

3. Cokriging with trend models[2] (CKT) consists of modeling the trend components as linear combinations of known functions $f_{ki}(\mathbf{u})$ of the spatial coordinates $\mathbf{u}$:

$$
m_i(\mathbf{u}') = \sum_{k=0}^{K} a_{ki}(\mathbf{u}')\, f_{ki}(\mathbf{u}'),
$$
$$
\text{with } a_{ki}(\mathbf{u}') \approx a_{ki} \quad \forall\, \mathbf{u}' \in W(\mathbf{u})
$$

The trend coefficients $a_{ki}(\mathbf{u}')$ are unknown but constant within each local neighborhood $W(\mathbf{u})$. By convention, the first trend function $f_{0i}(\mathbf{u}') = 1$, $\forall\, i$, hence the case $K = 0$ corresponds to ordinary cokriging (constant but unknown means a_{0i}). In section 5.4, the arbitrariness of the trend functions and the difficulties in inferring the corresponding residual covariance were discussed. Such problems become even more severe as two or more variables are considered. Ordinary cokriging with moving search neighborhoods allows one to account for spatial trends while being easier to implement, hence the cokriging with trend approach is not developped hereafter.

Note that no new concept is added to the kriging paradigm of section 5.1. The only additional complexity of cokriging is the heavier notation associated with having several variables.

---

[2] As for the single attribute case, that terminology is preferred to the term *universal cokriging*.

6.2.2 Simple cokriging

Consider first a single secondary attribute z_2. Modeling the primary and secondary trend components $m_1(\mathbf{u})$ and $m_2(\mathbf{u})$ as stationary means m_1 and m_2 allows one to write the linear estimator (6.10) as

$$Z_{SCK}^{(1)^*}(\mathbf{u}) - m_1 = \sum_{\alpha_1=1}^{n_1(\mathbf{u})} \lambda_{\alpha_1}^{SCK}(\mathbf{u}) [Z_1(\mathbf{u}_{\alpha_1}) - m_1]$$

$$+ \sum_{\alpha_2=1}^{n_2(\mathbf{u})} \lambda_{\alpha_2}^{SCK}(\mathbf{u}) [Z_2(\mathbf{u}_{\alpha_2}) - m_2] \qquad (6.11)$$

where $Z_{SCK}^{(1)^*}(\mathbf{u})$ is the simple cokriging estimator of the primary attribute z_1 at location $\mathbf{u}$. The $(n_1(\mathbf{u}) + n_2(\mathbf{u}))$ cokriging weights are determined such as to ensure unbiasedness and minimum error variance.

Unbiasedness is guaranteed by expression (6.11). Indeed

$$E\{Z_{SCK}^{(1)^*}(\mathbf{u}) - Z_1(\mathbf{u})\} = \sum_{\alpha_1=1}^{n_1(\mathbf{u})} \lambda_{\alpha_1}^{SCK}(\mathbf{u}) [E\{Z_1(\mathbf{u}_{\alpha_1})\} - m_1]$$

$$+ \sum_{\alpha_2=1}^{n_2(\mathbf{u})} \lambda_{\alpha_2}^{SCK}(\mathbf{u}) [E\{Z_2(\mathbf{u}_{\alpha_2})\} - m_2]$$

$$+ [m_1 - E\{Z_1(\mathbf{u})\}]$$

$$= 0$$

whatever the cokriging weights $\lambda_{\alpha_1}^{SCK}(\mathbf{u})$ and $\lambda_{\alpha_2}^{SCK}(\mathbf{u})$.

Let $R_i(\mathbf{u}_{\alpha_i}) = Z_i(\mathbf{u}_{\alpha_i}) - m_i$ be a primary $(i = 1)$ or secondary $(i = 2)$ residual RV at $\mathbf{u}_{\alpha_i}$. The estimation error $Z_{SCK}^{(1)^*}(\mathbf{u}) - Z_1(\mathbf{u})$ at location $\mathbf{u}$ can be written as a linear combination of the residual variables:

$$Z_{SCK}^{(1)^*}(\mathbf{u}) - Z_1(\mathbf{u}) = \sum_{\alpha_1=1}^{n_1(\mathbf{u})} \lambda_{\alpha_1}^{SCK}(\mathbf{u}) R_1(\mathbf{u}_{\alpha_1})$$

$$+ \sum_{\alpha_2=1}^{n_2(\mathbf{u})} \lambda_{\alpha_2}^{SCK}(\mathbf{u}) R_2(\mathbf{u}_{\alpha_2}) - [Z_1(\mathbf{u}) - m_1]$$

$$= \sum_{i=1}^{2} \sum_{\alpha_i=1}^{n_i(\mathbf{u})} \lambda_{\alpha_i}^{SCK}(\mathbf{u}) R_i(\mathbf{u}_{\alpha_i}) - R_1(\mathbf{u})$$

$$= R_{SCK}^{(1)^*}(\mathbf{u}) - R_1(\mathbf{u})$$

The error variance $\sigma_E^2(\mathbf{u})$ is then expressed as a linear combination of residual auto and cross covariance values:

$$\sigma_E^2(\mathbf{u}) = \text{Var}\{R_{SCK}^{(1)^*}(\mathbf{u})\} + \text{Var}\{R_1(\mathbf{u})\} - 2\,\text{Cov}\{R_{SCK}^{(1)^*}(\mathbf{u}), R_1(\mathbf{u})\}$$

$$= \sum_{i=1}^{2} \sum_{j=1}^{2} \sum_{\alpha_i=1}^{n_i(\mathbf{u})} \sum_{\alpha_j=1}^{n_j(\mathbf{u})} \lambda_{\alpha_i}^{SCK}(\mathbf{u}) \, \lambda_{\alpha_j}^{SCK}(\mathbf{u}) \, C_{ij}^{R}(\mathbf{u}_{\alpha_i} - \mathbf{u}_{\alpha_j})$$

$$+ \, C_{11}^{R}(0) - 2 \sum_{i=1}^{2} \sum_{\alpha_i=1}^{n_i(\mathbf{u})} \lambda_{\alpha_i}^{SCK}(\mathbf{u}) \, C_{i1}^{R}(\mathbf{u}_{\alpha_i} - \mathbf{u})$$

$$= \, Q\left(\lambda_{\alpha_i}^{SCK}(\mathbf{u}), \, \alpha_i = 1, \ldots, n_i(\mathbf{u}) \, ; \, i = 1, 2\right) \qquad (6.12)$$

The cokriging weights that minimize the error variance are obtained by setting to zero the $(n_1(\mathbf{u}) + n_2(\mathbf{u}))$ partial first derivatives of the quadratic form (6.12):

$$\frac{1}{2} \frac{\partial Q(\mathbf{u})}{\partial \lambda_{\alpha_1}^{SCK}(\mathbf{u})} = \sum_{\beta_1=1}^{n_1(\mathbf{u})} \lambda_{\beta_1}^{SCK}(\mathbf{u}) \, C_{11}^{R}(\mathbf{u}_{\alpha_1} - \mathbf{u}_{\beta_1})$$

$$+ \sum_{\beta_2=1}^{n_2(\mathbf{u})} \lambda_{\beta_2}^{SCK}(\mathbf{u}) \, C_{12}^{R}(\mathbf{u}_{\alpha_1} - \mathbf{u}_{\beta_2})$$

$$- \, C_{11}^{R}(\mathbf{u}_{\alpha_1} - \mathbf{u}) \qquad \alpha_1 = 1, \ldots, n_1(\mathbf{u})$$

$$\frac{1}{2} \frac{\partial Q(\mathbf{u})}{\partial \lambda_{\alpha_2}^{SCK}(\mathbf{u})} = \sum_{\beta_1=1}^{n_1(\mathbf{u})} \lambda_{\beta_1}^{SCK}(\mathbf{u}) \, C_{21}^{R}(\mathbf{u}_{\alpha_2} - \mathbf{u}_{\beta_1})$$

$$+ \sum_{\beta_2=1}^{n_2(\mathbf{u})} \lambda_{\beta_2}^{SCK}(\mathbf{u}) \, C_{22}^{R}(\mathbf{u}_{\alpha_2} - \mathbf{u}_{\beta_2})$$

$$- \, C_{21}^{R}(\mathbf{u}_{\alpha_2} - \mathbf{u}) \qquad \alpha_2 = 1, \ldots, n_2(\mathbf{u})$$

Because the primary and secondary trend components are assumed stationary (constant) and known, the two residual autocovariance functions, $C_{11}^{R}(\mathbf{h})$ and $C_{22}^{R}(\mathbf{h})$, and the residual cross covariance function $C_{12}^{R}(\mathbf{h})$ are equal to the auto and cross covariance functions of RFs $Z_1(\mathbf{u})$ and $Z_2(\mathbf{u})$:

$$C_{ij}^{R}(\mathbf{h}) = \, \mathrm{E}\left\{R_i(\mathbf{u}) \cdot R_j(\mathbf{u} + \mathbf{h})\right\}$$

$$= \, \mathrm{E}\left\{[Z_i(\mathbf{u}) - m_i] \cdot [Z_j(\mathbf{u} + \mathbf{h}) - m_j]\right\} = \, C_{ij}(\mathbf{h}) \quad i, j = 1, 2$$

Finally, the simple cokriging system is written as

$$\left\{ \begin{array}{l} \displaystyle\sum_{\beta_1=1}^{n_1(\mathbf{u})} \lambda_{\beta_1}^{SCK}(\mathbf{u}) \, C_{11}(\mathbf{u}_{\alpha_1} - \mathbf{u}_{\beta_1}) + \sum_{\beta_2=1}^{n_2(\mathbf{u})} \lambda_{\beta_2}^{SCK}(\mathbf{u}) \, C_{12}(\mathbf{u}_{\alpha_1} - \mathbf{u}_{\beta_2}) \\[2mm] \qquad = \, C_{11}(\mathbf{u}_{\alpha_1} - \mathbf{u}) \qquad \alpha_1 = 1, \ldots, n_1(\mathbf{u}) \\[2mm] \displaystyle\sum_{\beta_1=1}^{n_1(\mathbf{u})} \lambda_{\beta_1}^{SCK}(\mathbf{u}) \, C_{21}(\mathbf{u}_{\alpha_2} - \mathbf{u}_{\beta_1}) + \sum_{\beta_2=1}^{n_2(\mathbf{u})} \lambda_{\beta_2}^{SCK}(\mathbf{u}) \, C_{22}(\mathbf{u}_{\alpha_2} - \mathbf{u}_{\beta_2}) \\[2mm] \qquad = \, C_{21}(\mathbf{u}_{\alpha_2} - \mathbf{u}) \qquad \alpha_2 = 1, \ldots, n_2(\mathbf{u}) \end{array} \right.$$

$$(6.13)$$

The minimum error variance, called the simple cokriging (SCK) variance, is computed by substituting equations (6.13) into the expression of the error variance (6.12):

$$\sigma_{SCK}^2(\mathbf{u}) = C_{11}(0) - \sum_{\alpha_1=1}^{n_1(\mathbf{u})} \lambda_{\alpha_1}^{SCK}(\mathbf{u}) \, C_{11}(\mathbf{u}_{\alpha_1} - \mathbf{u})$$

$$- \sum_{\alpha_2=1}^{n_2(\mathbf{u})} \lambda_{\alpha_2}^{SCK}(\mathbf{u}) \, C_{21}(\mathbf{u}_{\alpha_2} - \mathbf{u}) \qquad (6.14)$$

That error variance generally decreases as the location $\mathbf{u}$ being estimated gets closer to the data (larger cokriging weights) and as the primary covariance $C_{11}(\mathbf{h})$ or the cross covariance $C_{12}(\mathbf{h})$ increases.

Several secondary variables

The simple cokriging estimator (6.11) is readily extended to more than one secondary variable; for example, for $(N_v - 1)$ variables,

$$Z_{SCK}^{(1)^*}(\mathbf{u}) - m_1 = \sum_{\alpha_1=1}^{n_1(\mathbf{u})} \lambda_{\alpha_1}^{SCK}(\mathbf{u}) \, [Z_1(\mathbf{u}_{\alpha_1}) - m_1]$$

$$+ \sum_{i=2}^{N_v} \sum_{\alpha_i=1}^{n_i(\mathbf{u})} \lambda_{\alpha_i}^{SCK}(\mathbf{u}) \, [Z_i(\mathbf{u}_{\alpha_i}) - m_i] \qquad (6.15)$$

Using an approach similar to that for two attributes, one derives the following cokriging system of $(\sum_{i=1}^{N_v} n_i(\mathbf{u}))$ linear equations:

$$\sum_{j=1}^{N_v} \sum_{\beta_j=1}^{n_j(\mathbf{u})} \lambda_{\beta_j}^{SCK}(\mathbf{u}) \, C_{ij}(\mathbf{u}_{\alpha_i} - \mathbf{u}_{\beta_j}) = C_{i1}(\mathbf{u}_{\alpha_i} - \mathbf{u})$$

$$\alpha_i = 1, \ldots, n_i(\mathbf{u}) \quad i = 1, \ldots, N_v \qquad (6.16)$$

The simple cokriging variance is

$$\sigma_{SCK}^2(\mathbf{u}) = C_{11}(0) - \sum_{i=1}^{N_v} \sum_{\alpha_i=1}^{n_i(\mathbf{u})} \lambda_{\alpha_i}^{SCK}(\mathbf{u}) \, C_{i1}(\mathbf{u}_{\alpha_i} - \mathbf{u})$$

Matrix notation

Using matrix notation, the simple cokriging system (6.13) is written

$$\mathbf{K}_{SCK} \, \boldsymbol{\lambda}_{SCK}(\mathbf{u}) = \mathbf{k}_{SCK} \qquad (6.17)$$

with

$$\mathbf{K}_{SCK} = \begin{bmatrix} [C_{11}(\mathbf{u}_{\alpha_1} - \mathbf{u}_{\beta_1})] & [C_{12}(\mathbf{u}_{\alpha_1} - \mathbf{u}_{\beta_2})] \\ [C_{21}(\mathbf{u}_{\alpha_2} - \mathbf{u}_{\beta_1})] & [C_{22}(\mathbf{u}_{\alpha_2} - \mathbf{u}_{\beta_2})] \end{bmatrix}$$

$$\boldsymbol{\lambda}_{SCK}(\mathbf{u}) \;=\; \begin{bmatrix} \left[\lambda_{\beta_1}^{SCK}(\mathbf{u})\right]^T \\ \left[\lambda_{\beta_2}^{SCK}(\mathbf{u})\right]^T \end{bmatrix} \qquad \mathbf{k}_{SCK} \;=\; \begin{bmatrix} [C_{11}(\mathbf{u}_{\alpha_1} - \mathbf{u})]^T \\ [C_{21}(\mathbf{u}_{\alpha_2} - \mathbf{u})]^T \end{bmatrix}$$

where $[C_{ij}(\mathbf{u}_{\alpha_i} - \mathbf{u}_{\beta_j})]$ is the $n_i(\mathbf{u}) \times n_j(\mathbf{u})$ matrix of data auto and cross covariances, $\left[\lambda_{\beta_i}^{SCK}(\mathbf{u})\right]^T$ is an $n_i(\mathbf{u}) \times 1$ vector of cokriging weights, and $[C_{i1}(\mathbf{u}_{\alpha_i} - \mathbf{u})]^T$ is an $n_i(\mathbf{u}) \times 1$ vector of data-to-unknown auto and cross covariances. The cokriging weights are obtained by multiplying the inverse of matrix $\mathbf{K}_{SCK}$ by the vector $\mathbf{k}_{SCK}$:

$$\boldsymbol{\lambda}_{SCK}(\mathbf{u}) \;=\; \mathbf{K}_{SCK}^{-1}\, \mathbf{k}_{SCK}$$

The simple cokriging variance (6.14) is then computed as

$$\begin{aligned} \sigma_{SCK}^2(\mathbf{u}) &= C_{11}(0) - \boldsymbol{\lambda}_{SCK}^T(\mathbf{u})\,\mathbf{k}_{SCK} \\ &= C_{11}(0) - \mathbf{k}_{SCK}^T\,\mathbf{K}_{SCK}^{-1}\,\mathbf{k}_{SCK} \end{aligned}$$

Similarly, the matrix formulation of the simple cokriging estimator (6.15) is

$$Z_{SCK}^{(1)*}(\mathbf{u}) \;=\; \sum_{i=1}^{N_v} [\mathbf{Z}_i(\mathbf{u}) - \mathbf{m}_i]^T \boldsymbol{\lambda}_i^{SCK}(\mathbf{u}) + m_1 \qquad (6.18)$$

where $\mathbf{Z}_i(\mathbf{u}) = [Z_i(\mathbf{u}_1), \dots, Z_i(\mathbf{u}_{n_i}(\mathbf{u}))]^T$, $\boldsymbol{\lambda}_i^{SCK}(\mathbf{u})$ is the $n_i(\mathbf{u}) \times 1$ vector of cokriging weights assigned to the z_i-data at locations $\mathbf{u}_{\alpha_i}$, $\alpha_i = 1 \dots, n_i(\mathbf{u})$, and $\mathbf{m}_i$ is the $n_i(\mathbf{u}) \times 1$ vector of stationary (constant) means m_i. The N_v vectors of cokriging weights are solutions of the system

$$\begin{bmatrix} \mathbf{K}_{11} & \cdots & \mathbf{K}_{1N_v} \\ \vdots & \vdots & \vdots \\ \mathbf{K}_{N_v 1} & \cdots & \mathbf{K}_{N_v N_v} \end{bmatrix} \begin{bmatrix} \boldsymbol{\lambda}_1^{SCK}(\mathbf{u}) \\ \vdots \\ \boldsymbol{\lambda}_{N_v}^{SCK}(\mathbf{u}) \end{bmatrix} = \begin{bmatrix} \mathbf{k}_{11} \\ \vdots \\ \mathbf{k}_{N_v 1} \end{bmatrix} \qquad (6.19)$$

with $\mathbf{K}_{ij} = [C_{ij}(\mathbf{u}_{\alpha_i} - \mathbf{u}_{\beta_j})]$, $\mathbf{k}_{i1} = [C_{i1}(\mathbf{u}_1 - \mathbf{u}), \dots, C_{i1}(\mathbf{u}_{n_i}(\mathbf{u}) - \mathbf{u})]^T$.

Remarks

1. The simple cokriging system (6.17) has a unique solution with positive cokriging variance if and only if the covariance matrix $\mathbf{K}_{SCK}$ is positive definite. That condition is satisfied by using permissible coregionalization models as introduced in section 4.3.2, provided that no two data values related to the same variable are colocated.

2. If primary and secondary variables are uncorrelated, the simple cokriging estimator reverts to the simple kriging estimator (5.7).

3. The simple cokriging estimator (6.15) is an exact interpolator in that it honors the primary data at their locations:

$$Z_{SCK}^{(1)*}(\mathbf{u}) = z_1(\mathbf{u}_{\alpha_1}) \qquad \forall\, \mathbf{u} = \mathbf{u}_{\alpha_1}, \quad \alpha_1 = 1, \dots, n_1$$

Correlogram notation

Because of differences in units of measurement, the variances of primary and secondary variables may differ by several orders of magnitude, leading to large differences between rows of the cokriging matrix $\mathbf{K}_{SCK}$. This may cause numerical instability when solving cokriging system (6.13) or (6.16). In such cases, it is good practice to rescale the auto and cross covariance values; for example, solve the cokriging system (6.13) in terms of correlograms:

$$
\begin{cases}
\displaystyle\sum_{\beta_1=1}^{n_1(\mathbf{u})} \nu_{\beta_1}^{SCK}(\mathbf{u})\, \rho_{11}(\mathbf{u}_{\alpha_1} - \mathbf{u}_{\beta_1}) + \sum_{\beta_2=1}^{n_2(\mathbf{u})} \nu_{\beta_2}^{SCK}(\mathbf{u})\, \rho_{12}(\mathbf{u}_{\alpha_1} - \mathbf{u}_{\beta_2}) \\
\qquad = \rho_{11}(\mathbf{u}_{\alpha_1} - \mathbf{u}) \qquad \alpha_1 = 1, \dots, n_1(\mathbf{u}) \\
\displaystyle\sum_{\beta_1=1}^{n_1(\mathbf{u})} \nu_{\beta_1}^{SCK}(\mathbf{u})\, \rho_{21}(\mathbf{u}_{\alpha_2} - \mathbf{u}_{\beta_1}) + \sum_{\beta_2=1}^{n_2(\mathbf{u})} \nu_{\beta_2}^{SCK}(\mathbf{u})\, \rho_{22}(\mathbf{u}_{\alpha_2} - \mathbf{u}_{\beta_2}) \\
\qquad = \rho_{21}(\mathbf{u}_{\alpha_2} - \mathbf{u}) \qquad \alpha_2 = 1, \dots, n_2(\mathbf{u})
\end{cases}
\tag{6.20}
$$

Recall that the cross correlogram $\rho_{ij}(\mathbf{h})$ is defined as the ratio $C_{ij}(\mathbf{h})/(\sigma_i \cdot \sigma_j)$, where σ_i^2 is the stationary variance of the RF $Z_i(\mathbf{u})$. Cokriging systems written in terms of correlograms or covariances yield different weights, yet the cokriging estimator remains the same.

Indeed, accounting for the definition of the cross correlogram, system (6.20) becomes

$$
\begin{cases}
\displaystyle\sum_{\beta_1=1}^{n_1(\mathbf{u})} \nu_{\beta_1}^{SCK}(\mathbf{u})\, C_{11}(\mathbf{u}_{\alpha_1} - \mathbf{u}_{\beta_1})/\sigma_1^2 + \sum_{\beta_2=1}^{n_2(\mathbf{u})} \nu_{\beta_2}^{SCK}(\mathbf{u})\, C_{12}(\mathbf{u}_{\alpha_1} - \mathbf{u}_{\beta_2})/\sigma_1\sigma_2 \\
\qquad = C_{11}(\mathbf{u}_{\alpha_1} - \mathbf{u})/\sigma_1^2 \qquad \alpha_1 = 1, \dots, n_1(\mathbf{u}) \\
\displaystyle\sum_{\beta_1=1}^{n_1(\mathbf{u})} \nu_{\beta_1}^{SCK}(\mathbf{u})\, C_{21}(\mathbf{u}_{\alpha_2} - \mathbf{u}_{\beta_1})/\sigma_1\sigma_2 + \sum_{\beta_2=1}^{n_2(\mathbf{u})} \nu_{\beta_2}^{SCK}(\mathbf{u})\, C_{22}(\mathbf{u}_{\alpha_2} - \mathbf{u}_{\beta_2})/\sigma_2^2 \\
\qquad = C_{21}(\mathbf{u}_{\alpha_2} - \mathbf{u})/\sigma_1\sigma_2 \qquad \alpha_2 = 1, \dots, n_2(\mathbf{u})
\end{cases}
$$

Multiplying the first $n_1(\mathbf{u})$ equations by σ_1^2 and the next $n_2(\mathbf{u})$ equations by $\sigma_1\sigma_2$, one deduces the following relation between the two sets of cokriging weights:

$$
\begin{aligned}
\lambda_{\alpha_1}^{SCK}(\mathbf{u}) &= \nu_{\alpha_1}^{SCK}(\mathbf{u}) & \alpha_1 &= 1, \dots, n_1(\mathbf{u}) \\
\lambda_{\alpha_2}^{SCK}(\mathbf{u}) &= \frac{\sigma_1}{\sigma_2}\, \nu_{\alpha_2}^{SCK}(\mathbf{u}) & \alpha_2 &= 1, \dots, n_2(\mathbf{u})
\end{aligned}
$$

Whereas the primary data weights are the same for both systems, the weights of the secondary data are rescaled by ratios of standard deviations. The cokriging estimator (6.11) is then written

$$Z_{SCK}^{(1)^*}(\mathbf{u}) - m_1 = \sum_{\alpha_1=1}^{n_1(\mathbf{u})} \nu_{\alpha_1}^{SCK}(\mathbf{u}) \left[Z_1(\mathbf{u}_{\alpha_1}) - m_1\right]$$

$$+ \sum_{\alpha_2=1}^{n_2(\mathbf{u})} \nu_{\alpha_2}^{SCK}(\mathbf{u}) \frac{\sigma_1}{\sigma_2} \left[Z_2(\mathbf{u}_{\alpha_2}) - m_2\right]$$

The standardized form of the SCK estimator is obtained by dividing both terms of the expression by σ_1:

$$\frac{Z_{SCK}^{(1)^*}(\mathbf{u}) - m_1}{\sigma_1} = \sum_{\alpha_1=1}^{n_1(\mathbf{u})} \nu_{\alpha_1}^{SCK}(\mathbf{u}) \left[\frac{Z_1(\mathbf{u}_{\alpha_1}) - m_1}{\sigma_1}\right]$$

$$+ \sum_{\alpha_2=1}^{n_2(\mathbf{u})} \nu_{\alpha_2}^{SCK}(\mathbf{u}) \left[\frac{Z_2(\mathbf{u}_{\alpha_2}) - m_2}{\sigma_2}\right] \quad (6.21)$$

Thus the estimator $Z_{SCK}^{(1)^*}(\mathbf{u})$ in the standardized form (6.21) with weights provided by cokriging system (6.20) identifies the estimator (6.11), as it should.

Weights of the means

The simple cokriging estimator (6.11) can be rewritten as

$$Z_{SCK}^{(1)^*}(\mathbf{u}) = \sum_{\alpha_1=1}^{n_1(\mathbf{u})} \lambda_{\alpha_1}^{SCK}(\mathbf{u}) \, Z_1(\mathbf{u}_{\alpha_1}) + \sum_{\alpha_2=1}^{n_2(\mathbf{u})} \lambda_{\alpha_2}^{SCK}(\mathbf{u}) \, Z_2(\mathbf{u}_{\alpha_2})$$

$$+ \lambda_{m_1}^{SCK}(\mathbf{u}) \, m_1 + \lambda_{m_2}^{SCK}(\mathbf{u}) \, m_2 \quad (6.22)$$

where the implicit weight $\lambda_{m_1}^{SCK}(\mathbf{u})$ given to the primary mean m_1 is

$$\lambda_{m_1}^{SCK}(\mathbf{u}) = 1 - \sum_{\alpha_1=1}^{n_1(\mathbf{u})} \lambda_{\alpha_1}^{SCK}(\mathbf{u})$$

Similarly, the weight $\lambda_{m_2}^{SCK}(\mathbf{u})$ given to the secondary mean m_2 is

$$\lambda_{m_2}^{SCK}(\mathbf{u}) = - \sum_{\alpha_2=1}^{n_2(\mathbf{u})} \lambda_{\alpha_2}^{SCK}(\mathbf{u})$$

The weights $\lambda_{\alpha_1}^{SCK}(\mathbf{u})$ and $\lambda_{\alpha_2}^{SCK}(\mathbf{u})$ generally decrease as the corresponding data locations get farther away from the location $\mathbf{u}$ being estimated. Therefore, the weight $\lambda_{m_1}^{SCK}(\mathbf{u})$ of the primary mean m_1 tends to increase where primary data are sparse. Conversely, the weight $\lambda_{m_2}^{SCK}(\mathbf{u})$ of the secondary mean m_2 is smaller where secondary data are sparse.

When the location $\mathbf{u}$ is beyond the correlation range of both primary and secondary data, all data-to-unknown covariance terms in the cokriging system (6.17) vanish. The right-hand-side vector $\mathbf{k}_{SCK}$ is a null vector, and so is the vector $\boldsymbol{\lambda}_{SCK}(\mathbf{u})$ of cokriging weights: $\boldsymbol{\lambda}_{SCK}(\mathbf{u}) = \mathbf{K}_{SCK}^{-1} \, \mathbf{0} = \mathbf{0}$. The SCK estimator then reverts to the primary mean m_1, as it should.

Example

The cokriging variants are now illustrated using the one-dimensional data set with a primary variable (cadmium concentration) and a single secondary variable (nickel concentration). The information available for estimating Cd concentrations along the NE-SW transect consists of the following:

1. Ten Cd concentrations and 16 Ni concentrations; see Figure 6.8 (top graphs). Closed squares depict locations where both primary and secondary variables are known. Open circles correspond to six locations where only the secondary variable is available.

2. The linear model of coregionalization (4.47) shown in Figure 4.16:

$$\begin{aligned}
\gamma_{\mathrm{Cd}}(\mathbf{h}) &= 0.3 \, g_0(\mathbf{h}) + 0.3 \, \mathrm{Sph}\,(h/200\mathrm{m}) + 0.26 \, \mathrm{Sph}\,(h/1.3\mathrm{km}) \\
\gamma_{\mathrm{Ni}}(\mathbf{h}) &= 11 \, g_0(\mathbf{h}) + 71 \, \mathrm{Sph}\,(h/1.3\mathrm{km}) \\
\gamma_{\mathrm{Cd\text{-}Ni}}(\mathbf{h}) &= 0.6 \, g_0(\mathbf{h}) + 3.8 \, \mathrm{Sph}\,(h/1.3\mathrm{km})
\end{aligned}$$

where $g_0(\mathbf{h})$ is a nugget effect model.

3. Primary and secondary stationary means identified with the sample means of Cd and Ni data along the transect: $m_1 = 1.49$ ppm, $m_2 = 19.6$ ppm.

The estimation is performed every 50 m using at each location the five closest primary data and five closest secondary data: $n_1(\mathbf{u}) = n_2(\mathbf{u}) = 5 \; \forall \, \mathbf{u}$.

Figure 6.8 (bottom graphs) shows the SCK estimates of Cd concentrations and the weights[3] given to the primary and secondary means. Note that the SCK estimates (solid line, third row):

- identify the primary data (exactitude property).

- identify the primary mean (horizontal dashed line) beyond the range of influence of the extreme right datum. The primary weight $\lambda_{m_1}^{SCK}(\mathbf{u})$ is then equal to one, whereas the weight of the secondary mean is zero.

---

[3] Secondary data weights are rescaled by the ratio of standard deviations of secondary to primary variables, $\lambda_{\alpha_2}^{SCK}(\mathbf{u}) \cdot \sigma_2/\sigma_1$, so their magnitude is comparable to that of primary data weights. Such rescaling amounts to solving the simple cokriging system in terms of correlograms; see relation (6.20).

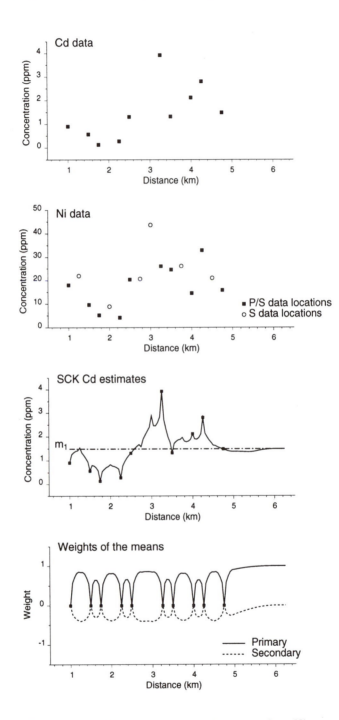

Figure 6.8: Ten primary Cd concentrations and 16 secondary Ni concentrations along the NE-SW transect. Open circles depict the six locations where only Ni concentrations are known. The two bottom graphs show the simple cokriging (SCK) estimates of Cd concentration and the weights of the primary and secondary means.

Estimating a vector of variables

Depending on the sampling density of primary and secondary variables, one can distinguish two situations in practice:

1. All N_v variables are recorded at every sampled location, a situation referred to as the *equally sampled* or *isotopic* case. The isotopic case is frequent in geochemical exploration, where the focus is on estimating a set of metal concentrations.

2. The primary variable is undersampled relative to secondary variables, a situation referred to as the *undersampled* or *heterotopic* case. Typically, a few expensive measurements of the attribute of interest (e.g., Cd concentrations or permeability values) are supplemented by more abundant data on correlated attributes that are cheaper to sample (e.g., pH or seismic data).

In the general form of cokriging, expression (6.15), primary and secondary data locations need not coincide. One particular case is when all variables are equally sampled, and the same $n(\mathbf{u})$ primary and secondary data locations are retained in the estimation:

$$n_i(\mathbf{u}) = n(\mathbf{u}) \qquad i = 1, \ldots, N_v$$
$$\mathbf{u}_{\alpha_i} = \mathbf{u}_\alpha \qquad \alpha_i = 1, \ldots, n_i(\mathbf{u}) \quad i = 1, \ldots, N_v$$

The simple cokriging estimator (6.15) can then be written

$$Z_{SCK}^{(1)^*}(\mathbf{u}) = \sum_{i=1}^{N_v} \sum_{\alpha=1}^{n(\mathbf{u})} \lambda_{\alpha_i}^{SCK}(\mathbf{u}) \, [Z_i(\mathbf{u}_\alpha) - m_i] + m_1$$

In such an isotopic case, one generally aims at estimating all N_v variables $Z_i(\mathbf{u})$ at each unsampled location $\mathbf{u}$. Such estimation calls for solving N_v cokriging systems of type (6.16), with each variable $Z_i(\mathbf{u})$ being considered as primary in turn. A more straightforward alternative consists of estimating directly the vector of N_v variables (Myers, 1982). In matrix notation,

$$\mathbf{Z}_{SCK}^*(\mathbf{u}) = \sum_{\alpha=1}^{n(\mathbf{u})} [\mathbf{L}_\alpha^{SCK}(\mathbf{u})]^T \, [\mathbf{Z}(\mathbf{u}_\alpha) - \mathbf{m}] + \mathbf{m} \qquad (6.23)$$

where $\mathbf{Z}_{SCK}^*(\mathbf{u}) = [Z_{SCK}^{(1)^*}(\mathbf{u}), \ldots, Z_{SCK}^{(N_v)^*}(\mathbf{u})]^T$ is the vector of simple cokriging estimators, $\mathbf{Z}(\mathbf{u}_\alpha) = [Z_1(\mathbf{u}_\alpha), \ldots, Z_{N_v}(\mathbf{u}_\alpha)]^T$ is the vector of data at $\mathbf{u}_\alpha$, and $\mathbf{m} = [m_1, \ldots, m_{N_v}]^T$ is the vector of means (known). Each $N_v \times N_v$ matrix $\mathbf{L}_\alpha^{SCK}(\mathbf{u})$ contains the cokriging weights, $[\lambda_{\alpha_{ij}}^{SCK}(\mathbf{u})]$, assigned to z_i-data at location $\mathbf{u}_\alpha$ for the estimation of attributes z_j at $\mathbf{u}$:

$$\mathbf{L}_\alpha^{SCK}(\mathbf{u}) = [\lambda_{\alpha_{ij}}^{SCK}(\mathbf{u})] = \begin{bmatrix} \lambda_{\alpha_{11}}^{SCK}(\mathbf{u}) & \cdots & \lambda_{\alpha_{1N_v}}^{SCK}(\mathbf{u}) \\ \vdots & \vdots & \vdots \\ \lambda_{\alpha_{N_v 1}}^{SCK}(\mathbf{u}) & \cdots & \lambda_{\alpha_{N_v N_v}}^{SCK}(\mathbf{u}) \end{bmatrix}$$

The matrices of cokriging weights are obtained by solving the single cokriging system:

$$\mathbf{C}_{SCK} \; \mathbf{L}_{SCK}(\mathbf{u}) \;\; = \;\; \mathbf{c}_{SCK} \tag{6.24}$$

where the matrix of data-to-data auto and cross covariances is

$$\mathbf{C}_{SCK} \;\; = \;\; \begin{bmatrix} \mathbf{C}(\mathbf{u}_1 - \mathbf{u}_1) & \cdots & \mathbf{C}(\mathbf{u}_1 - \mathbf{u}_{n(\mathbf{u})}) \\ \vdots & \vdots & \vdots \\ \mathbf{C}(\mathbf{u}_{n(\mathbf{u})} - \mathbf{u}_1) & \cdots & \mathbf{C}(\mathbf{u}_{n(\mathbf{u})} - \mathbf{u}_{n(\mathbf{u})}) \end{bmatrix}$$

The matrix of cokriging weights and the matrix of data-to-unknown auto and cross covariances are

$$\mathbf{L}_{SCK}(\mathbf{u}) \;\; = \;\; \begin{bmatrix} \mathbf{L}_1^{SCK}(\mathbf{u}) \\ \vdots \\ \mathbf{L}_{n(\mathbf{u})}^{SCK}(\mathbf{u}) \end{bmatrix} \qquad\qquad \mathbf{c}_{SCK} \;\; = \;\; \begin{bmatrix} \mathbf{C}(\mathbf{u}_1 - \mathbf{u}) \\ \vdots \\ \mathbf{C}(\mathbf{u}_{n(\mathbf{u})} - \mathbf{u}) \end{bmatrix}$$

where $\mathbf{C}(\mathbf{u}_\alpha - \mathbf{u}_\beta) = [C_{ij}(\mathbf{u}_\alpha - \mathbf{u}_\beta)]$ is the $N_v \times N_v$ matrix of auto and cross covariances between any two variables Z_i and Z_j at $\mathbf{u}_\alpha$ and $\mathbf{u}_\beta$, and $\mathbf{C}(\mathbf{u}_\alpha - \mathbf{u})$ is the $N_v \times N_v$ matrix of auto and cross covariances between any two variables Z_i and Z_j at $\mathbf{u}_\alpha$ and $\mathbf{u}$. The simple cokriging system (6.24) has the same form as the simple kriging system (5.10) except that the entries are matrices instead of scalars.

The error variances associated with the N_v simple cokriging estimators are the diagonal elements of the matrix:

$$\mathbf{S}_{SCK}^2 \;\; = \;\; \mathbf{C}(0) \; - \; \mathbf{L}_{SCK}^T(\mathbf{u}) \, \mathbf{c}_{SCK} \quad \text{with} \quad \mathbf{L}_{SCK}(\mathbf{u}) = \mathbf{C}_{SCK}^{-1} \, \mathbf{c}_{SCK}$$

Cokriging of linearly dependent variables

In some situations, the variables involved in cokriging are linearly dependent. For example, the z_i-values may sum to a constant D; e.g., relative concentrations of geochemical elements sum to one. Knowledge of any $(N_v - 1)$ attribute values at $\mathbf{u}$ allows one to deduce exactly the value of the remaining attribute as $D - \sum_{i=1}^{N_v-1} z_i(\mathbf{u})$. In such a case, beware that:

1. The constant sum constraint may induce a linear dependence in the rows and columns of the matrix of experimental covariance values for any lag $\mathbf{h}$, particularly when all N_v variables are equally sampled. The resulting matrix of covariance models may be close to singularity with risks of numerical instabilities for the cokriging system.

2. The cokriging estimates generally do not satisfy the constant sum constraint.

Singularity can be avoided by estimating one variable at a time, using a cokriging system of type (6.16) and discarding any variable that is weakly correlated with the variable being estimated. To satisfy the constant sum constraint, each estimate $z_{CK}^{(i)*}(\mathbf{u})$ is divided by the sum $\sum_{i=1}^{N_v} z_{CK}^{(i)*}(\mathbf{u})$ of estimated values at the same location $\mathbf{u}$, then it is multiplied by the constant D.

An alternative consists of estimating all variables but one, say, $Z_{i_0}(\mathbf{u})$, using a vector cokriging system of type (6.24). The z_{i_0}-value at $\mathbf{u}$ is then computed as the constraint value minus the sum of remaining attribute estimates at that location:

$$z_{CK}^{(i_0)*}(\mathbf{u}) \;=\; D - \sum_{\substack{i=1 \\ i \neq i_0}}^{N_v} z_{CK}^{(i)*}(\mathbf{u})$$

One drawback is that the estimate $z_{CK}^{(i_0)*}(\mathbf{u})$ tends to accumulate all estimation errors affecting the $(N_v - 1)$ estimates $z_{CK}^{(i)*}(\mathbf{u})$. Therefore, the variable Z_{i_0} should be the attribute of least interest or the one with the largest average value, or both.

Simple kriging versus simple cokriging

Cokriging is much more demanding than kriging for these reasons:

- $N_v(N_v + 1)/2$ direct and cross semivariograms must be inferred and jointly modeled.

- A large cokriging system must be solved.

It is then worth looking at the benefit one may gain from the additional modeling and computational effort necessitated by cokriging.

Theoretical advantage of cokriging
The cokriging estimator is theoretically better because its error variance is always smaller than or equal to the error variance of kriging which ignores all secondary information:

$$\sigma_{SCK}^2(\mathbf{u}) \;\leq\; \sigma_{SK}^2(\mathbf{u}) \qquad \forall\, \mathbf{u} \in \mathcal{A}$$

In the isotopic case, another advantage of cokriging is that the cokriging estimator of a sum of variables, say, $Y(\mathbf{u}) = \sum_{k=1}^{K} Z_k(\mathbf{u})$, is equal to the sum of the cokriging estimators of the K components $Z_k(\mathbf{u})$ (Matheron, 1979):

$$Y_{SCK}^*(\mathbf{u}) \;\equiv\; \sum_{k=1}^{K} Z_{SCK}^{(k)*}(\mathbf{u})$$

provided the $(K + 1)$ cokriging estimators $Y_{SCK}^*(\mathbf{u})$ and $Z_{SCK}^{(k)*}(\mathbf{u})$ are built from the same set of z_k-data. For example, the variable $Y(\mathbf{u})$ may be the thickness of a soil horizon defined as the difference between the depths of the bottom $Z_1(\mathbf{u})$ and top $Z_2(\mathbf{u})$ of that horizon:

$$Y(\mathbf{u}) \;=\; Z_1(\mathbf{u}) \;-\; Z_2(\mathbf{u})$$

Then, the cokriging estimator of the thickness using both depths as secondary variables would be consistent with the cokriging estimators of the two depths:

$$Y^*_{SCK}(\mathbf{u}) \;=\; Z^{(1)^*}_{SCK}(\mathbf{u}) \;-\; Z^{(2)^*}_{SCK}(\mathbf{u})$$

The consistency of cokriging estimates, although a desirable property, does not guarantee that the thickness estimate is positive.

Equivalence between kriging and cokriging estimates
Consider the SK and SCK estimates of attribute z_1 at location $\mathbf{u}$:

$$z^{(1)^*}_{SK}(\mathbf{u}) \;=\; \sum_{\alpha_1=1}^{n_1(\mathbf{u})} \lambda^{SK}_{\alpha_1}(\mathbf{u})\,[z_1(\mathbf{u}_{\alpha_1}) - m_1] \;+\; m_1$$

$$z^{(1)^*}_{SCK}(\mathbf{u}) \;=\; \sum_{\alpha_1=1}^{n_1(\mathbf{u})} \lambda^{SCK}_{\alpha_1}(\mathbf{u})\,[z_1(\mathbf{u}_{\alpha_1}) - m_1] \;+\; m_1$$

$$+ \sum_{i=2}^{N_v} \sum_{\alpha_i=1}^{n_i(\mathbf{u})} \lambda^{SCK}_{\alpha_i}(\mathbf{u})\,[z_i(\mathbf{u}_{\alpha_i}) - m_i]$$

where the same $n_1(\mathbf{u})$ primary data $z_1(\mathbf{u}_{\alpha_1})$ are used in both estimations. Simple kriging and cokriging estimates are identical in these two situations:

1. Primary and secondary variables are uncorrelated:

$$C_{1i}(\mathbf{h}) \;=\; C_{i1}(\mathbf{h}) \;=\; 0 \qquad \forall\,\mathbf{h} \quad i = 2, \ldots, N_v$$

2. Primary and secondary variables are measured at the same locations (isotopic case) and the cross covariance $C_{1i}(\mathbf{h})$ between the primary variable $Z_1(\mathbf{u})$ and each secondary variable $Z_i(\mathbf{u})$ is proportional to the primary autocovariance $C_{11}(\mathbf{h})$:

$$C_{1i}(\mathbf{h}) \;=\; C_{i1}(\mathbf{h}) \;=\; \varphi_{1i} \cdot C_{11}(\mathbf{h}) \qquad i = 2, \ldots, N_v \qquad (6.25)$$

This model is a particular case of the intrinsic coregionalization model (4.43) Cokriging system (6.19) then becomes

$$
\begin{bmatrix}
\mathbf{K}_{11} & \cdots & \varphi_{1N_v}\mathbf{K}_{11} \\
\vdots & \vdots & \vdots \\
\varphi_{1N_v}\mathbf{K}_{11} & \cdots & \mathbf{K}_{N_vN_v}
\end{bmatrix}
\begin{bmatrix}
\boldsymbol{\lambda}^{SCK}_1(\mathbf{u}) \\
\vdots \\
\boldsymbol{\lambda}^{SCK}_{N_v}(\mathbf{u})
\end{bmatrix}
=
\begin{bmatrix}
\mathbf{k}_{11} \\
\vdots \\
\varphi_{1N_v}\mathbf{k}_{11}
\end{bmatrix}
$$

The unique solution (Matheron, 1979; Wackernagel, 1994) of that system is

$$\lambda_1^{SCK}(\mathbf{u}) = \mathbf{K}_{11}^{-1}\,\mathbf{k}_{11} \equiv [\lambda_{\alpha_1}^{SK}(\mathbf{u})]$$
$$\lambda_i^{SCK}(\mathbf{u}) = [\lambda_{\alpha_i}^{SCK}(\mathbf{u})] = 0 \qquad i = 2, \ldots, N_v$$

The simple cokriging estimator (6.18) then reverts to the simple kriging estimator. If the N_v variables are intrinsically correlated, the vector of all N_v simple cokriging estimators (6.23) reverts to the vector of N_v simple kriging estimators $\left[Z_{SK}^{(1)}(\mathbf{u}), \ldots, Z_{SK}^{(N_v)}(\mathbf{u}) \right]^T$.

More generally, cokriging and kriging estimates are likely to be similar when variables are weakly correlated or in the isotopic case when direct and cross semivariograms have similar shape. Practice has shown that cokriging improves over kriging only when the primary variable is undersampled with regard to the secondary variables and those secondary data are well correlated with the primary value to be estimated (Journel and Huijbregts, 1978, p. 326).

Influence of secondary data
Intuitively, the contribution of a specific secondary attribute z_2 to the cokriging estimate of the primary attribute z_1 at location $\mathbf{u}$ should depend on the following characteristics:

1. The correlation coefficient $\rho_{12}(0)$ between primary and secondary variables

2. The pattern of spatial continuity of the attributes

3. The spatial configuration of primary and secondary data locations

4. The sampling density of each variable

The impact of these factors on cokriging weights will be illustrated using the examples given in Figures 6.9–6.11. The relative influence of secondary information is measured by the ratio of the sums of absolute values of the secondary and primary data weights:

$$\psi(\mathbf{u}) = \frac{\displaystyle\sum_{\alpha_2=1}^{n_2(\mathbf{u})} |\lambda_{\alpha_2}^{SCK}(\mathbf{u})|}{\displaystyle\sum_{\alpha_1=1}^{n_1(\mathbf{u})} |\lambda_{\alpha_1}^{SCK}(\mathbf{u})|} \cdot \frac{\sigma_2}{\sigma_1} = \frac{\displaystyle\sum_{\alpha_2=1}^{n_2(\mathbf{u})} |\nu_{\alpha_2}^{SCK}(\mathbf{u})|}{\displaystyle\sum_{\alpha_1=1}^{n_1(\mathbf{u})} |\nu_{\alpha_1}^{SCK}(\mathbf{u})|} \qquad (6.26)$$

where the rescaling factor σ_2/σ_1 corrects for differences in variances.

(1) *Correlation coefficient* $\rho_{12}(0)$
Consider the data configuration in Figure 6.9 (top graph). The primary attribute at location $\mathbf{u}$ is estimated using four primary data values (black

dots) and four secondary data values (open circles) all located on a circle of unit radius away from $\mathbf{u}$. Simple cokriging weights are computed using the following coregionalization model:

$$
\begin{aligned}
\gamma_{11}(\mathbf{h}) &= 0.2\, g_0(\mathbf{h}) + 0.8\, \mathrm{Sph}\,(h/2) \\
\gamma_{22}(\mathbf{h}) &= 1.0\, \mathrm{Sph}\,(h/2) \\
\gamma_{12}(\mathbf{h}) &= \rho_{12}(0)\, \mathrm{Sph}\,(h/2)
\end{aligned}
$$

Both direct semivariogram models are bounded and of unit sill, hence the sill of the cross semivariogram model is the correlation coefficient between primary and secondary variables; see relation (3.39).

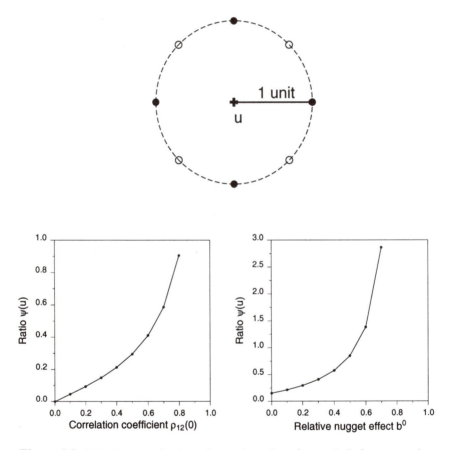

Figure 6.9: Relative contribution of secondary data (open circles) versus primary data (black dots) for the simple cokriging estimation of the primary attribute at location $\mathbf{u}$. That contribution, measured by the ratio $\psi(\mathbf{u})$ of secondary to primary cokriging weights (absolute values), increases as the two variables are better correlated and as the relative nugget effect on the primary semivariogram model increases.

Figure 6.9 (left bottom graph) shows the value of the ratio $\psi(\mathbf{u})$ as a function of the correlation coefficient $\rho_{12}(0)$. The weight of secondary data is zero when primary and secondary variables are uncorrelated. As the correlation coefficient increases, the secondary variable brings increasingly more information on the primary value, hence the secondary data have larger weight.

(2) *Pattern of spatial continuity*
Keep the same data configuration and vary the relative nugget effect b^0 of the primary semivariogram model $\gamma_{11}(\mathbf{h})$:

$$\gamma_{11}(\mathbf{h}) = b^0 \, g_0(\mathbf{h}) + (1 - b^0) \, \text{Sph}\,(h/2)$$
$$\gamma_{22}(\mathbf{h}) = 1.0 \, \text{Sph}\,(h/2)$$
$$\gamma_{12}(\mathbf{h}) = 0.5 \, \text{Sph}\,(h/2)$$

Note that the correlation coefficient $\rho_{12}(0)$ is fixed at 0.5.

For a zero nugget effect (smooth variation in space of the primary variable), the secondary data have little weight (Figure 6.9, right bottom graph). As the relative nugget effect b^0 increases, the increasingly noisy primary data carry less information and the secondary data have larger weight.

(3) *Data configuration*
The screening effect discussed in section 5.8 extends to the case of multiple attributes. The most common situation is the screening of secondary information by colocated primary data. Consider the data configuration shown in Figure 6.10 (top graph) with the azimuth angle θ varying between 0 and 45°. A zero azimuth angle means that the four primary and secondary data are colocated. As that angle increases, the secondary data get farther away from the primary data. All eight data locations are at the same one unit distance away from the location $\mathbf{u}$ being estimated, and the distances between primary (secondary) data locations remain unchanged for different values of θ. For each azimuth angle, the simple cokriging weights are computed using the coregionalization model:

$$\gamma_{11}(\mathbf{h}) = 0.1 \, g_0(\mathbf{h}) + 0.9 \, \text{Sph}\,(h/2)$$
$$\gamma_{22}(\mathbf{h}) = 0.4 \, g_0(\mathbf{h}) + 0.6 \, \text{Sph}\,(h/2)$$
$$\gamma_{12}(\mathbf{h}) = 0.1 \, g_0(\mathbf{h}) + 0.7 \, \text{Sph}\,(h/2)$$

The value of the ratio $\psi(\mathbf{u})$ as a function of the azimuth angle θ is depicted by the solid line in Figure 6.10 (left bottom graph). When the two sources of information are colocated ($\theta = 0$), the contribution of secondary data to the cokriging estimate is negligible. As the secondary data get farther away from the primary data (increasing θ value), the left-hand-side cross covariance terms $C_{12}(\mathbf{u}_{\alpha_1} - \mathbf{u}_{\beta_2})$ in system (6.13) decrease. Consequently, the secondary data are seen as less redundant with primary data and have larger weight.

In other situations, it is the secondary information that screens the influence of the colocated primary data. This is likely to happen when both primary and secondary variables are highly correlated and the secondary variable

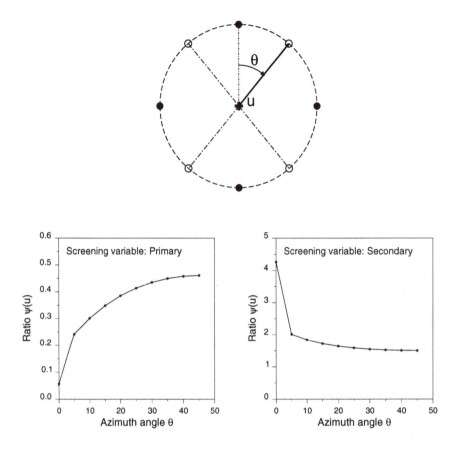

Figure 6.10: Screening effect of primary data (black dots) on secondary data (open circles) for the simple cokriging estimation of the primary attribute at location **u** using two different coregionalization models. When data types are colocated (θ=0), the most continuous variable screens the influence of the other. That screening effect decreases as secondary data locations are shifted (increasing angle θ).

varies more continuously in space than the primary variable. Consider the following coregionalization model where the previous direct semivariogram models were exchanged:

$$\gamma_{11}(\mathbf{h}) = 0.4\, g_0(\mathbf{h}) + 0.6\, \mathrm{Sph}\,(h/2)$$
$$\gamma_{22}(\mathbf{h}) = 0.1\, g_0(\mathbf{h}) + 0.9\, \mathrm{Sph}\,(h/2)$$
$$\gamma_{12}(\mathbf{h}) = 0.1\, g_0(\mathbf{h}) + 0.7\, \mathrm{Sph}\,(h/2)$$

For a zero azimuth angle, the ratio $\psi(\mathbf{u})$ now exceeds four (Figure 6.10, right bottom graph). This large ratio value indicates a screening of the primary data by the colocated secondary information. A small shift ($\theta = 5°$) of the secondary data drastically reduces that screening effect.

(4) *Sampling density*

Figure 6.11 (top graph) shows an isotopic data configuration where both primary and secondary variables are measured at four locations one unit distance away from the location **u** to be estimated. Two heterotopic data configurations are also considered:

- *Config 1*: Four additional secondary data are located on a circle of radius 0.5 (middle left graph) for a total of eight secondary data.

- *Config 2*: One single additional secondary datum is colocated with the location **u** being estimated (middle right graph) for a total of five secondary data.

For each data configuration, the impact of the correlation coefficient $\rho_{12}(0)$ and pattern of spatial continuity on cokriging weights are investigated using the ratio $\psi(\mathbf{u})$ defined in expression (6.26) and the two coregionalization models considered for Figure 6.9.

For all three data configurations, the ratio $\psi(\mathbf{u})$ increases with increasing correlation coefficient (left bottom graph) and increasing relative nugget effect of the semivariogram model $\gamma_{11}(\mathbf{h})$ (right bottom graph). The solid line depicts the isotopic case. The two undersampled cases are represented by small dashed line (Config 1) and large dashed line (Config 2). For both factors, the contribution of the secondary information increases when the secondary variable is better sampled than the primary variable. Indeed, closer secondary data screen the influence of primary and secondary data on the outer circle of radius 1.0. The colocated datum gets as much weight as the four remote secondary data.

Example

Figure 6.12 shows the simple kriging (dashed line) and cokriging (solid line) estimates of Cd concentrations along the NE-SW transect for three sampling densities:

- Ten pairs of colocated Cd and Ni data (equally sampled case)

- Ten pairs of colocated Cd and Ni data plus six additional Ni data (undersampled case 1)

- Six pairs of colocated Cd and Ni data plus ten additional Ni data (undersampled case 2)

The kriging and cokriging estimates are essentially the same in the isotopic case. Differences between estimates increase as secondary data become more numerous than primary data. When only six Cd values are available (undersampled case 2), the simple kriging estimates are close to the stationary mean. Accounting for the secondary nickel data allows a better resolution of the cokriging estimates.

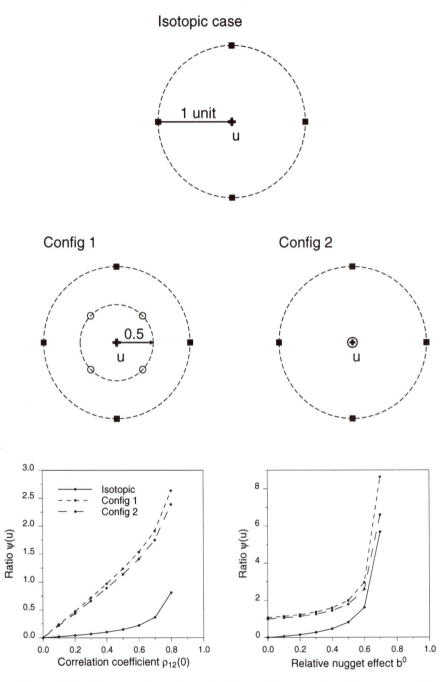

Figure 6.11: Impact of sampling density and data configuration on cokriging weights. The ratio of secondary to primary weights increases when the four pairs of colocated primary/secondary data (black squares) are supplemented by four closer secondary data (Config 1) or by a single secondary datum at the location being estimated (Config 2). Bottom graphs show the ratio of cokriging weights as a function of: the correlation coefficient between variables (left) and the relative nugget effect on the primary semivariogram model (right).

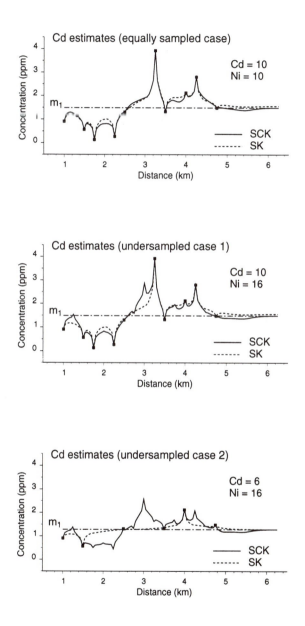

Figure 6.12: The difference between simple cokriging (solid line) and simple kriging (dashed line) estimates of Cd concentrations increases as the primary variable is more undersampled. Nickel data locations increase from 10 (upper graph) to 16 (cases 1 and 2). Cd data locations decrease from 10 (two upper graphs) to 6 (lower graph).

6.2.3   Ordinary cokriging

Ordinary cokriging (OCK) allows for local variability of the means by restricting the stationarity of both primary and secondary variables to a local neighborhood $W(\mathbf{u})$ centered on the location $\mathbf{u}$ being estimated. For the case of a single secondary attribute z_2, the cokriging estimator (6.10) is written

$$
\begin{aligned}
Z_1^*(\mathbf{u}) \;=\; & \sum_{\alpha_1=1}^{n_1(\mathbf{u})} \lambda_{\alpha_1}(\mathbf{u})\, Z_1(\mathbf{u}_{\alpha_1}) + \sum_{\alpha_2=1}^{n_2(\mathbf{u})} \lambda_{\alpha_2}(\mathbf{u})\, Z_2(\mathbf{u}_{\alpha_2}) \\
& + \lambda_{m_1}(\mathbf{u})\, m_1(\mathbf{u}) + \lambda_{m_2}(\mathbf{u})\, m_2(\mathbf{u})
\end{aligned}
$$

where primary and secondary means are deemed constant within each search neighborhood $W(\mathbf{u})$. The weights $\lambda_{m_1}(\mathbf{u})$ and $\lambda_{m_2}(\mathbf{u})$ are

$$
\lambda_{m_1}(\mathbf{u}) \;=\; 1 - \sum_{\alpha_1=1}^{n_1(\mathbf{u})} \lambda_{\alpha_1}(\mathbf{u}), \qquad \lambda_{m_2}(\mathbf{u}) \;=\; - \sum_{\alpha_2=1}^{n_2(\mathbf{u})} \lambda_{\alpha_2}(\mathbf{u})
$$

If the local means $m_1(\mathbf{u})$ and $m_2(\mathbf{u})$ are actually unknown, they can be filtered from the linear estimator by setting their respective weights to zero, i.e., by constraining the primary data weights $\lambda_{\alpha_1}(\mathbf{u})$ to sum to one and the secondary data weights $\lambda_{\alpha_2}(\mathbf{u})$ to sum to zero. The ordinary cokriging estimator is then written

$$
Z_{OCK}^{(1)*}(\mathbf{u}) \;=\; \sum_{\alpha_1=1}^{n_1(\mathbf{u})} \lambda_{\alpha_1}^{OCK}(\mathbf{u})\, Z_1(\mathbf{u}_{\alpha_1}) + \sum_{\alpha_2=1}^{n_2(\mathbf{u})} \lambda_{\alpha_2}^{OCK}(\mathbf{u})\, Z_2(\mathbf{u}_{\alpha_2}) \quad (6.27)
$$

with these two constraints:

$$
\sum_{\alpha_1=1}^{n_1(\mathbf{u})} \lambda_{\alpha_1}^{OCK}(\mathbf{u}) \;=\; 1 \qquad \sum_{\alpha_2=1}^{n_2(\mathbf{u})} \lambda_{\alpha_2}^{OCK}(\mathbf{u}) \;=\; 0 \qquad (6.28)
$$

The ordinary cokriging estimator (6.27) is unbiased:

$$
\begin{aligned}
\mathrm{E}\{Z_{OCK}^{(1)*}(\mathbf{u}) - Z_1(\mathbf{u})\} \;=\; & \Big[\sum_{\alpha_1=1}^{n_1(\mathbf{u})} \lambda_{\alpha_1}^{OCK}(\mathbf{u}) - 1\Big]\, m_1 + \sum_{\alpha_2=1}^{n_2(\mathbf{u})} \lambda_{\alpha_2}^{OCK}(\mathbf{u})\, m_2 \\
\;=\; & 0
\end{aligned}
$$

Minimization of the error variance of type (6.12) under the two constraints (6.28) calls for the definition of a Lagrangian $L(\mathbf{u})$ similar to that introduced for ordinary kriging in section 5.3:

$$
\begin{aligned}
L(\mathbf{u}) \;=\; & \sigma_E^2(\mathbf{u}) + 2\mu_1^{OCK}(\mathbf{u}) \left[\sum_{\alpha_1=1}^{n_1(\mathbf{u})} \lambda_{\alpha_1}^{OCK}(\mathbf{u}) - 1\right] \\
& + 2\mu_2^{OCK}(\mathbf{u}) \left[\sum_{\alpha_2=1}^{n_2(\mathbf{u})} \lambda_{\alpha_2}^{OCK}(\mathbf{u})\right]
\end{aligned}
$$

where the two Lagrange parameters $\mu_1^{OCK}(\mathbf{u})$ and $\mu_2^{OCK}(\mathbf{u})$ account for the two unbiasedness constraints. The Lagrangian $L(\mathbf{u})$ is minimized by setting to zero its partial first derivatives with respect to the $(n_1(\mathbf{u}) + n_2(\mathbf{u}) + 2)$ data weights and Lagrange parameters:

$$\frac{1}{2} \frac{\partial L(\mathbf{u})}{\partial \lambda_{\alpha_i}^{OCK}(\mathbf{u})} = 0 \qquad \alpha_i = 1, \ldots, n_i(\mathbf{u}) \quad i = 1, 2$$

$$\frac{1}{2} \frac{\partial L(\mathbf{u})}{\partial \mu_i^{OCK}(\mathbf{u})} = 0 \qquad i = 1, 2$$

The ordinary cokriging system is then written in terms of the stationary auto and cross covariances as

$$\left\{ \begin{array}{l} \displaystyle\sum_{\beta_1=1}^{n_1(\mathbf{u})} \lambda_{\beta_1}^{OCK}(\mathbf{u}) \, C_{11}(\mathbf{u}_{\alpha_1} - \mathbf{u}_{\beta_1}) + \sum_{\beta_2=1}^{n_2(\mathbf{u})} \lambda_{\beta_2}^{OCK}(\mathbf{u}) \, C_{12}(\mathbf{u}_{\alpha_1} - \mathbf{u}_{\beta_2}) \\ \qquad\qquad + \mu_1^{OCK}(\mathbf{u}) = C_{11}(\mathbf{u}_{\alpha_1} - \mathbf{u}) \qquad \alpha_1 = 1, \ldots, n_1(\mathbf{u}) \\[4pt] \displaystyle\sum_{\beta_1=1}^{n_1(\mathbf{u})} \lambda_{\beta_1}^{OCK}(\mathbf{u}) \, C_{21}(\mathbf{u}_{\alpha_2} - \mathbf{u}_{\beta_1}) + \sum_{\beta_2=1}^{n_2(\mathbf{u})} \lambda_{\beta_2}^{OCK}(\mathbf{u}) \, C_{22}(\mathbf{u}_{\alpha_2} - \mathbf{u}_{\beta_2}) \\ \qquad\qquad + \mu_2^{OCK}(\mathbf{u}) = C_{21}(\mathbf{u}_{\alpha_2} - \mathbf{u}) \qquad \alpha_2 = 1, \ldots, n_2(\mathbf{u}) \\[4pt] \displaystyle\sum_{\beta_1=1}^{n_1(\mathbf{u})} \lambda_{\beta_1}^{OCK}(\mathbf{u}) = 1 \\[4pt] \displaystyle\sum_{\beta_2=1}^{n_2(\mathbf{u})} \lambda_{\beta_2}^{OCK}(\mathbf{u}) = 0 \end{array} \right.$$

$$(6.29)$$

The minimum variance is computed as

$$\sigma_{OCK}^2(\mathbf{u}) = C_{11}(0) - \mu_1^{OCK}(\mathbf{u}) - \sum_{\alpha_1=1}^{n_1(\mathbf{u})} \lambda_{\alpha_1}^{OCK}(\mathbf{u}) \, C_{11}(\mathbf{u}_{\alpha_1} - \mathbf{u})$$

$$- \sum_{\alpha_2=1}^{n_2(\mathbf{u})} \lambda_{\alpha_2}^{OCK}(\mathbf{u}) \, C_{21}(\mathbf{u}_{\alpha_2} - \mathbf{u}) \qquad (6.30)$$

Several secondary variables

Similarly, for the case of two or more secondary variables, the ordinary cokriging estimator is written

$$Z_{OCK}^{(1)*}(\mathbf{u}) = \sum_{i=1}^{N_v} \sum_{\alpha_i=1}^{n_i(\mathbf{u})} \lambda_{\alpha_i}^{OCK}(\mathbf{u}) \, Z_i(\mathbf{u}_{\alpha_i}) \qquad (6.31)$$

with the N_v unbiasedness constraints:

$$\sum_{\alpha_1=1}^{n_1(\mathbf{u})} \lambda_{\alpha_1}^{OCK}(\mathbf{u}) \;=\; 1$$

$$\sum_{\alpha_i=1}^{n_i(\mathbf{u})} \lambda_{\alpha_i}^{OCK}(\mathbf{u}) \;=\; 0 \qquad i = 2, \ldots, N_v$$

Minimizing the estimation variance under these non-bias constraints leads to the following system of $(\sum_{i=1}^{N_v} n_i(\mathbf{u}) + N_v)$ linear equations:

$$\begin{cases} \displaystyle\sum_{j=1}^{N_v} \sum_{\beta_j=1}^{n_j(\mathbf{u})} \lambda_{\beta_j}^{OCK}(\mathbf{u})\, C_{ij}(\mathbf{u}_{\alpha_i} - \mathbf{u}_{\beta_j}) \;+\; \mu_i^{OCK}(\mathbf{u}) \;=\; C_{i1}(\mathbf{u}_{\alpha_i} - \mathbf{u}) \\[4pt] \qquad\qquad\qquad\qquad \alpha_i = 1, \ldots, n_i(\mathbf{u}) \quad i = 1, \ldots, N_v \\[6pt] \displaystyle\sum_{\beta_i=1}^{n_i(\mathbf{u})} \lambda_{\beta_i}^{OCK}(\mathbf{u}) \;=\; \delta_{i1} \qquad i = 1, \ldots, N_v \end{cases}$$

$$(6.32)$$

with $\delta_{i1} = 1$ for $i = 1$ and $\delta_{i1} = 0$ otherwise. The cokriging variance is

$$\sigma_{OCK}^2(\mathbf{u}) \;=\; C_{11}(0) \;-\; \mu_1^{OCK}(\mathbf{u}) \;-\; \sum_{i=1}^{N_v} \sum_{\alpha_i=1}^{n_i(\mathbf{u})} \lambda_{\alpha_i}^{OCK}(\mathbf{u})\, C_{i1}(\mathbf{u}_{\alpha_i} - \mathbf{u})$$

Matrix notation

Using the matrix formulation introduced in section 6.2.2, the ordinary cokriging system (6.29) is written as

$$\begin{bmatrix} \mathbf{K}_{11} & \mathbf{K}_{12} & \mathbf{1}_1 & \mathbf{0}_1 \\ \mathbf{K}_{21} & \mathbf{K}_{22} & \mathbf{0}_2 & \mathbf{1}_2 \\ \mathbf{1}_1^T & \mathbf{0}_2^T & 0 & 0 \\ \mathbf{0}_1^T & \mathbf{1}_2^T & 0 & 0 \end{bmatrix} \begin{bmatrix} \lambda_1^{OCK}(\mathbf{u}) \\ \lambda_2^{OCK}(\mathbf{u}) \\ \mu_1^{OCK}(\mathbf{u}) \\ \mu_2^{OCK}(\mathbf{u}) \end{bmatrix} = \begin{bmatrix} \mathbf{k}_{11} \\ \mathbf{k}_{21} \\ 1 \\ 0 \end{bmatrix} \qquad (6.33)$$

where $\mathbf{1}_i$ and $\mathbf{0}_i$ are $n_i(\mathbf{u}) \times 1$ unit and null vectors, respectively. The generalization to the case $N_v \geq 3$ is straightforward.

In the equally sampled case, the vector of ordinary cokriging estimators is written

$$\mathbf{Z}_{OCK}^*(\mathbf{u}) \;=\; \sum_{\alpha=1}^{n(\mathbf{u})} [\mathbf{L}_\alpha^{OCK}(\mathbf{u})]^T \, \mathbf{Z}(\mathbf{u}_\alpha)$$

where $\mathbf{Z}_{OCK}^*(\mathbf{u}) = \left[Z_{OCK}^{(1)*}(\mathbf{u}), \ldots, Z_{OCK}^{(N_v)*}(\mathbf{u}) \right]^T$ is the vector of cokriging estimators. As for simple cokriging, the $N_v \times N_v$ matrices of cokriging weights, $\mathbf{L}_\alpha^{OCK}(\mathbf{u}) = [\lambda_{\alpha_{ij}}^{OCK}(\mathbf{u})]$, are obtained by solving a single cokriging system:

$$\mathbf{K}_{OCK} \, \mathbf{L}_{OCK}(\mathbf{u}) \;=\; \mathbf{k}_{OCK}$$

with

$$\mathbf{K}_{OCK} = \begin{bmatrix} \mathbf{C}(\mathbf{u}_1 - \mathbf{u}_1) & \cdots & \mathbf{C}(\mathbf{u}_1 - \mathbf{u}_{n(\mathbf{u})}) & \mathbf{I} \\ \vdots & \vdots & \vdots & \vdots \\ \mathbf{C}(\mathbf{u}_{n(\mathbf{u})} - \mathbf{u}_1) & \cdots & \mathbf{C}(\mathbf{u}_{n(\mathbf{u})} - \mathbf{u}_{n(\mathbf{u})}) & \mathbf{I} \\ \mathbf{I} & \cdots & \mathbf{I} & \mathbf{0} \end{bmatrix}$$

$$\mathbf{L}_{OCK}(\mathbf{u}) = \begin{bmatrix} \mathbf{L}_1^{OCK}(\mathbf{u}) \\ \vdots \\ \mathbf{L}_{n(\mathbf{u})}^{OCK}(\mathbf{u}) \\ \mathbf{M}(\mathbf{u}) \end{bmatrix} \qquad \mathbf{k}_{OCK} = \begin{bmatrix} \mathbf{C}(\mathbf{u}_1 - \mathbf{u}) \\ \vdots \\ \mathbf{C}(\mathbf{u}_{n(\mathbf{u})} - \mathbf{u}) \\ \mathbf{I} \end{bmatrix}$$

where $\mathbf{I}$ is the $N_v \times N_v$ identity matrix, and $\mathbf{M}(\mathbf{u})$ is the $N_v \times N_v$ matrix of Lagrange parameters.

Semivariogram notation

The cokriging system (6.32) can be expressed in terms of direct and cross semivariograms provided the cross covariance models between primary and secondary variables are symmetric, i.e., $C_{ij}(\mathbf{h}) = C_{ji}(\mathbf{h}) \ \forall \ i \neq j$ (recall discussion about the lag effect in section 3.2.3). Accounting for the relations $C_{ij}(\mathbf{h}) = C_{ij}(0) - \gamma_{ij}(\mathbf{h})$, $i, j = 1, 2$, the ordinary cokriging system is written

$$\begin{cases} \sum_{j=1}^{N_v} \sum_{\beta_j=1}^{n_j(\mathbf{u})} \lambda_{\beta_j}^{OCK}(\mathbf{u}) \, \gamma_{ij}(\mathbf{u}_{\alpha_i} - \mathbf{u}_{\beta_j}) - \mu_i^{OCK}(\mathbf{u}) = \gamma_{i1}(\mathbf{u}_{\alpha_i} - \mathbf{u}) \\ \hspace{3cm} \alpha_i = 1, \ldots, n_i(\mathbf{u}) \quad i = 1, \ldots, N_v \\ \sum_{\beta_i=1}^{n_i(\mathbf{u})} \lambda_{\beta_i}^{OCK}(\mathbf{u}) = \delta_{i1} \quad i = 1, \ldots, N_v \end{cases}$$

Note that the non-bias conditions cancel out the covariance terms $C_{ij}(0)$ from all equations. These are similarities to the single attribute case:

- Unlike ordinary cokriging, a simple cokriging system can be expressed in terms of only auto and cross covariances.

- The structural analysis is traditionally conducted in terms of semivariograms, though the ordinary cokriging system is best solved in terms of covariances. For unbounded semivariogram models (e.g., the power model), "pseudo (cross) covariances" can be defined by subtracting the (cross) semivariogram model $\gamma_{ij}(\mathbf{h})$ from any positive value A_{ij}, such that $A_{ij} - \gamma_{ij}(\mathbf{h}) \geq 0$, $\forall \ \mathbf{h}$. Again, the non-bias conditions cancel out the constant terms A_{ij} from the ordinary cokriging system, which is then written in terms of pseudo auto and cross covariances.

Correlogram notation

It is good practice to rescale covariance values such that elements of the covariance matrix in system (6.33) are of the same order of magnitude. The standardized form of the ordinary cokriging estimator (6.27) is

$$
\frac{Z_{OCK}^{(1)*}(\mathbf{u}) - m_1}{\sigma_1} = \sum_{\alpha_1=1}^{n_1(\mathbf{u})} \nu_{\alpha_1}^{OCK}(\mathbf{u}) \left[\frac{Z_1(\mathbf{u}_{\alpha_1}) - m_1}{\sigma_1} \right]
$$

$$
+ \sum_{\alpha_2=1}^{n_2(\mathbf{u})} \nu_{\alpha_2}^{OCK}(\mathbf{u}) \left[\frac{Z_2(\mathbf{u}_{\alpha_2}) - m_2}{\sigma_2} \right] \quad (6.34)
$$

The corresponding cokriging weights $\nu_{\alpha_i}^{OCK}(\mathbf{u})$ are obtained by solving the ordinary cokriging system expressed in terms of correlograms.

Like simple cokriging weights, OCK weights of standardized and original variables are different, although the cokriging estimates are the same:

$$
\begin{aligned}
\nu_{\alpha_1}^{OCK}(\mathbf{u}) &= \lambda_{\alpha_1}^{OCK}(\mathbf{u}) & \alpha_1 &= 1, \ldots, n_1(\mathbf{u}) \\
\nu_{\alpha_2}^{OCK}(\mathbf{u}) &= \frac{\sigma_2}{\sigma_1} \lambda_{\alpha_2}^{OCK}(\mathbf{u}) & \alpha_2 &= 1, \ldots, n_2(\mathbf{u})
\end{aligned}
$$

A single unbiasedness constraint

In some cases, primary and secondary data relate to the same attribute. For example, a few precise laboratory measurements of pH or clay content are supplemented by more numerous field data collected using cheaper measurement devices. Measurement errors are likely to be larger for field data, and their semivariogram is likely to have a larger relative nugget effect. To account for such a difference in the patterns of spatial continuity, precise laboratory measurements and less precise field data are weighted differently through cokriging.

Provided both measurement processes are unbiased and the two primary and secondary means within the search neighborhood $W(\mathbf{u})$ are deemed equal, the linear estimator (6.10) is written

$$
Z^{(1)*}(\mathbf{u}) = \sum_{\alpha_1=1}^{n_1(\mathbf{u})} \lambda_{\alpha_1}(\mathbf{u}) \, Z_1(\mathbf{u}_{\alpha_1}) + \sum_{\alpha_2=1}^{n_2(\mathbf{u})} \lambda_{\alpha_2}(\mathbf{u}) \, Z_2(\mathbf{u}_{\alpha_2})
$$

$$
+ \left[1 - \sum_{\alpha_1=1}^{n_1(\mathbf{u})} \lambda_{\alpha_1}(\mathbf{u}) - \sum_{\alpha_2=1}^{n_2(\mathbf{u})} \lambda_{\alpha_2}(\mathbf{u}) \right] m(\mathbf{u})
$$

The unknown mean $m(\mathbf{u})$ is filtered from that estimator by forcing all primary and secondary data weights to sum to one (Journel and Huijbregts, 1978, p. 325). The corresponding expression of the ordinary cokriging estimator is similar to the usual expression (6.27):

$$Z_{OCK}^{(1)*}(\mathbf{u}) \;=\; \sum_{\alpha_1=1}^{n_1(\mathbf{u})} \zeta_{\alpha_1}^{OCK}(\mathbf{u})\, Z_1(\mathbf{u}_{\alpha_1}) \;+\; \sum_{\alpha_2=1}^{n_2(\mathbf{u})} \zeta_{\alpha_2}^{OCK}(\mathbf{u})\, Z_2(\mathbf{u}_{\alpha_2}) \quad (6.35)$$

except that the two non-bias conditions (6.28) are replaced by the single constraint:

$$\sum_{\alpha_1=1}^{n_1(\mathbf{u})} \zeta_{\alpha_1}^{OCK}(\mathbf{u}) \;+\; \sum_{\alpha_2=1}^{n_2(\mathbf{u})} \zeta_{\alpha_2}^{OCK}(\mathbf{u}) \;=\; 1 \qquad (6.36)$$

The ordinary cokriging system then includes only $(n_1(\mathbf{u}) + n_2(\mathbf{u}) + 1)$ linear equations:

$$\begin{cases}
\displaystyle\sum_{\beta_1=1}^{n_1(\mathbf{u})} \zeta_{\beta_1}^{OCK}(\mathbf{u})\, C_{11}(\mathbf{u}_{\alpha_1} - \mathbf{u}_{\beta_1}) \;+\; \sum_{\beta_2=1}^{n_2(\mathbf{u})} \zeta_{\beta_2}^{OCK}(\mathbf{u})\, C_{12}(\mathbf{u}_{\alpha_1} - \mathbf{u}_{\beta_2}) \\
\qquad\qquad\qquad +\; \mu^{OCK}(\mathbf{u}) \;=\; C_{11}(\mathbf{u}_{\alpha_1} - \mathbf{u}) \qquad \alpha_1 = 1, \dots, n_1(\mathbf{u}) \\
\displaystyle\sum_{\beta_1=1}^{n_1(\mathbf{u})} \zeta_{\beta_1}^{OCK}(\mathbf{u})\, C_{21}(\mathbf{u}_{\alpha_2} - \mathbf{u}_{\beta_1}) \;+\; \sum_{\beta_2=1}^{n_2(\mathbf{u})} \zeta_{\beta_2}^{OCK}(\mathbf{u})\, C_{22}(\mathbf{u}_{\alpha_2} - \mathbf{u}_{\beta_2}) \\
\qquad\qquad\qquad +\; \mu^{OCK}(\mathbf{u}) \;=\; C_{21}(\mathbf{u}_{\alpha_2} - \mathbf{u}) \qquad \alpha_2 = 1, \dots, n_2(\mathbf{u}) \\
\displaystyle\sum_{\beta_1=1}^{n_1(\mathbf{u})} \zeta_{\beta_1}^{OCK}(\mathbf{u}) \;+\; \sum_{\beta_2=1}^{n_2(\mathbf{u})} \zeta_{\beta_2}^{OCK}(\mathbf{u}) \;=\; 1
\end{cases}$$

$$(6.37)$$

Simple cokriging versus ordinary cokriging

Like ordinary kriging, ordinary cokriging amounts to:

1. estimating the local primary and secondary means, say, $m_{OCK}^{(1)*}(\mathbf{u})$ and $m_{OCK}^{(2)*}(\mathbf{u})$ for the case $N_v = 2$, at each location $\mathbf{u}$ using both primary and secondary data specific to that neighborhood, then

2. applying the simple cokriging estimator (6.11) using these estimates of the means rather than the stationary means m_1 and m_2:

$$Z_{OCK}^{(1)*}(\mathbf{u}) \;=\; \sum_{\alpha_1=1}^{n_1(\mathbf{u})} \lambda_{\alpha_1}^{SCK}(\mathbf{u})\, [Z_1(\mathbf{u}_{\alpha_1}) - m_{OCK}^{(1)*}(\mathbf{u})] \;+\; m_{OCK}^{(1)*}(\mathbf{u})$$

$$+\; \sum_{\alpha_2=1}^{n_2(\mathbf{u})} \lambda_{\alpha_2}^{SCK}(\mathbf{u})\, [Z_2(\mathbf{u}_{\alpha_2}) - m_{OCK}^{(2)*}(\mathbf{u})] \qquad (6.38)$$

Accounting for expression (6.22) of the simple cokriging estimator, one can deduce the following relation between the two estimators:

$$Z_{OCK}^{(1)*}(\mathbf{u}) \;=\; Z_{SCK}^{(1)*}(\mathbf{u}) \;+\; \lambda_{m_1}^{SCK}(\mathbf{u})\, \Delta_{m_1}^{OCK}(\mathbf{u}) \;+\; \lambda_{m_2}^{SCK}(\mathbf{u})\, \Delta_{m_2}^{OCK}(\mathbf{u})$$

with

$$\Delta_{m_1}^{OCK}(\mathbf{u}) \;=\; m_{OCK}^{(1)\bullet}(\mathbf{u}) \;-\; m_1 \qquad\qquad \Delta_{m_2}^{OCK}(\mathbf{u}) \;=\; m_{OCK}^{(2)\bullet}(\mathbf{u}) \;-\; m_2$$

As in the single attribute case, discrepancies between simple and ordinary cokriging estimators are related to the estimated primary and secondary local means being different from the stationary values m_1 and m_2.

The local means $m_{OCK}^{(1)\bullet}(\mathbf{u})$ and $m_{OCK}^{(2)\bullet}(\mathbf{u})$ that are implicitly used in estimator (6.38) can be explicitly estimated as linear combinations of primary and secondary data. For a single secondary attribute, the OCK estimator of the primary local mean $m_1(\mathbf{u})$ is

$$m_{OCK}^{(1)\bullet}(\mathbf{u}) \;=\; \sum_{\alpha_1=1}^{n_1(\mathbf{u})} \lambda_{\alpha_1 m_1}^{OCK}(\mathbf{u})\, Z_1(\mathbf{u}_{\alpha_1}) \;+\; \sum_{\alpha_2=1}^{n_2(\mathbf{u})} \lambda_{\alpha_2 m_1}^{OCK}(\mathbf{u})\, Z_2(\mathbf{u}_{\alpha_2})$$

(6.39)

where $\lambda_{\alpha_i m_1}^{OCK}(\mathbf{u})$ is the weight assigned to the datum $z_i(\mathbf{u}_{\alpha_i})$. To ensure unbiasedness, the primary and secondary data weights must meet the two constraints (6.28).

The cokriging weights for the primary mean are then obtained by solving a system similar to the ordinary cokriging system (6.29) for attribute values, except for the right-hand-side covariance terms $C_{i1}(\mathbf{u}_{\alpha_i} - \mathbf{u})$ being set to zero.

Example

Figure 6.13 (two top graphs) shows the ordinary cokriging estimates of Cd and Ni local means along the NE-SW transect. Like the single attribute case in Figure 5.3 (middle graph), both trend estimates have a staircase shape, each step corresponding to estimates based on the same neighboring primary and secondary data. The number of steps is, however, larger since 16 primary data locations are here considered instead of only 10 primary data locations for Figure 5.3. Both trend estimates follow the general increase in Cd and Ni concentrations along the transect. In contrast, the overall means (horizontal dashed line) overestimate the local mean in the low-valued (left) part of the transect and underestimate the local mean in the high-valued (right) part of the transect. This underestimation is less pronounced for Ni concentrations.

Figure 6.13 (bottom graph) shows the simple (dashed line) and ordinary (solid line) cokriging estimates of Cd concentrations. Note the following:

- Both interpolators are exact.

- OCK estimates are smaller than SCK estimates in the left part of the transect where Cd local means are smaller than the overall mean.

- OCK estimates are larger than SCK estimates in the right part of the transect where Cd local means are larger than the overall mean.

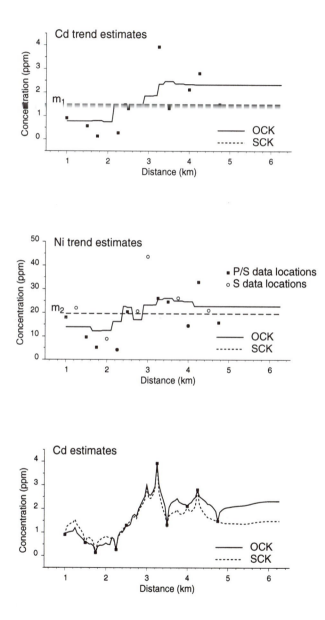

Figure 6.13: Estimates of Cd and Ni local means, and Cd concentrations using ordinary cokriging (solid line) and simple cokriging (dashed line).

6.2.4 Standardized ordinary cokriging

The ordinary cokriging constraint (6.28) that calls for the secondary data weights to sum to zero has two drawbacks:

1. Some of the secondary data weights are negative, thereby increasing the risk of getting unacceptable estimates; see related discussion in section 5.8.

2. Most of the weights $\lambda_{\alpha_2}^{OCK}(\mathbf{u})$ tend to be small, thus reducing the influence of the secondary data.

To reduce the occurrence of negative weights and avoid artificially limiting the impact of secondary data, Isaaks and Srivastava (1989, p. 416) proposed using a single constraint of type (6.36); for example, for a single secondary attribute:

$$\sum_{\alpha_1=1}^{n_1(\mathbf{u})} \lambda_{\alpha_1}(\mathbf{u}) + \sum_{\alpha_2=1}^{n_2(\mathbf{u})} \lambda_{\alpha_2}(\mathbf{u}) = 1 \qquad (6.40)$$

That single constraint suffices to ensure that the ordinary cokriging estimator (6.27) is unbiased if the secondary variable is rescaled so that its mean is equal to that of the primary variable. The OCK estimator is then written[4]

$$Z_{OCK}^{(1)*}(\mathbf{u}) = \sum_{\alpha_1=1}^{n_1(\mathbf{u})} \lambda_{\alpha_1}^{OCK}(\mathbf{u})\, Z_1(\mathbf{u}_{\alpha_1})$$

$$+ \sum_{\alpha_2=1}^{n_2(\mathbf{u})} \lambda_{\alpha_2}^{OCK}(\mathbf{u})\, [Z_2(\mathbf{u}_{\alpha_2}) - m_2 + m_1] \qquad (6.41)$$

Estimator (6.41) is called the "standardized" ordinary cokriging estimator. The cokriging weights are solutions of the ordinary cokriging system (6.37) with a single unbiasedness constraint. The estimator (6.41) is readily extended to two or more secondary variables, with the following single non-bias condition:

$$\sum_{i=1}^{N_v} \sum_{\alpha_i=1}^{n_i(\mathbf{u})} \lambda_{\alpha_i}^{OCK}(\mathbf{u}) = 1$$

Unlike traditional ordinary cokriging, estimator (6.41) calls for knowledge of the stationary means of primary and secondary variables. Provided the data are representative of the study area $\mathcal{A}$, these means can be estimated from the sample means. Although stationary means are used for the prior rescaling of secondary variables, the unbiasedness constraint (6.40) leads to

---

[4] To simplify notation, the corresponding cokriging weights $\lambda^{OCK}(\mathbf{u})$ are denoted with the same superscript OCK as the weight of estimator (6.27), although they are different.

estimation of the common local mean of primary and rescaled secondary variables within each search neighborhood $W(\mathbf{u})$. Thus, local departures from the overall means are still accounted for as they are in traditional ordinary cokriging.

Like simple or ordinary cokriging, it is good practice to solve the cokriging system (6.37) in terms of correlograms when the variances of primary and secondary variables differ by several orders of magnitude. The cokriging weights that are provided by the cokriging systems written in terms of covariances or correlograms are not linearly related, hence the resulting cokriging estimators are slightly different (Goovaerts, 1997).

6.2.5 Principal component kriging

Principal component analysis[5] aims at defining a few linear combinations of original variables $Z_i(\mathbf{u})$ that are uncorrelated at $|\mathbf{h}| = 0$ and account for most of the variance in the data. Principal component kriging (Davis and Greenes, 1983) capitalizes on the orthogonality of these components to reduce the cokriging of N_v variables to the kriging of N_v uncorrelated principal components, thereby alleviating the modeling and computational effort implied by cokriging.

Principal component kriging (PCK) proceeds in five steps:

1. The correlation matrix $\mathbf{R}$ of the N_v variables $Z_i(\mathbf{u})$ is computed from the set of n data locations where all N_v attributes z_i are jointly measured (isotopic data set).

2. At each datum location $\mathbf{u}_\alpha$, the set of N_v principal component scores is computed as linear combinations of the standardized attribute values at that location:

$$y_k(\mathbf{u}_\alpha) \;=\; \sum_{i=1}^{N_v} q_{ki} \left(\frac{z_i(\mathbf{u}_\alpha) - m_i}{\sigma_i} \right) \qquad k = 1, \ldots, N_v$$

where q_{ki} is the loading of variable Z_i on the kth principal component Y_k, and m_i and σ_i are the mean and standard deviation of the z_i-data. The matrix of loadings $\mathbf{Q} = [q_{ki}]$ originates from the spectral decomposition of the correlation matrix $\mathbf{R}$:

$$\mathbf{R} \;=\; \mathbf{Q}\,\boldsymbol{\Lambda}\,\mathbf{Q}^T \quad \text{with} \quad \mathbf{Q}^T\,\mathbf{Q} \;=\; \mathbf{I}$$

where $\mathbf{Q}$ is the orthogonal matrix of eigenvectors of the matrix $\mathbf{R}$, $\boldsymbol{\Lambda} = [\lambda_k]$ is the diagonal matrix of eigenvalues, and $\mathbf{I}$ is the $N_v \times N_v$ identity matrix.

3. The N_v semivariograms $\gamma_{kk}(\mathbf{h})$ of principal components are estimated and modeled from the scores $y_k(\mathbf{u}_\alpha)$.

---

[5] Presentations of principal component analysis for geologists and soil scientists are given in Davis (1986, p. 602) and Webster and Oliver (1990, p. 126).

4. Each of the N_v principal components is estimated separately at each location $\mathbf{u}$. For example, the ordinary kriging estimator of the kth component at $\mathbf{u}$ is written

$$Y_{OK}^{(k)^*}(\mathbf{u}) = \sum_{\alpha=1}^{n(\mathbf{u})} \lambda_{\alpha k}^{OK}(\mathbf{u}) \, Y_k(\mathbf{u}_\alpha)$$

where the kriging weights are solutions of an OK system of type (5.18).

5. The estimate of z_i at $\mathbf{u}$ is then reconstituted as a linear combination of the principal component estimates at that location plus the attribute mean m_i:

$$z_{PCK}^{(i)^*}(\mathbf{u}) = \sum_{k=1}^{K} a_{ki} \, y_{OK}^{(k)^*}(\mathbf{u}) \, \sigma_i + m_i$$

where the matrix of transformation coefficients $\mathbf{A} = [a_{ki}]$ is the inverse of the matrix of loadings, $\mathbf{A} = \mathbf{Q}^{-1}$.

Spatial orthogonality of principal components

As shown in section 6.2.2, kriging and cokriging estimators are identical if the N_v variables involved in the estimation are mutually (two by two) spatially independent. The N_v principal components $Y_k(\mathbf{u})$ are, by construction, orthogonal at lag $|\mathbf{h}| = 0$:

$$\text{Cov}\left\{Y_k(\mathbf{u}), Y_{k'}(\mathbf{u})\right\} = 0 \qquad \forall \, k \neq k'$$

The key assumption of principal component kriging is that this orthogonality extends to all other separation vectors $\mathbf{h}$, that is,

$$\text{Cov}\left\{Y_k(\mathbf{u}), Y_{k'}(\mathbf{u}+\mathbf{h})\right\} = C_{kk'}(\mathbf{h}) = 0 \qquad \forall \, k \neq k', \mathbf{h} \qquad (6.42)$$

In theory, condition (6.42) is satisfied if the N_v variables are intrinsically correlated (Goovaerts, 1993). In practice, condition (6.42) can be checked by looking at the cross correlogram between principal components. For example, Figure 6.14 (solid line) shows the experimental cross correlograms for three pairs of the first three principal components of the correlation matrix given in Table 2.6 (page 22). The principal components are, by construction, uncorrelated at $|\mathbf{h}| = 0$. The cross correlation between components increases at short lags reaching a maximum of 0.3 for the pair PC1-PC3.

Remarks

The recent development of user-friendly iterative fitting procedures (see Appendix A) lessens the practical advantage of modeling fewer semivariograms of principal components instead of the full semivariogram matrix of original

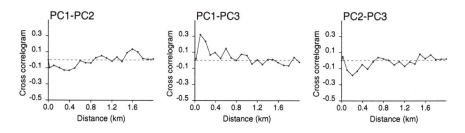

Figure 6.14: Spatial orthogonality of principal components.

variables. Although principal component kriging remains computationally less expensive than cokriging, it suffers from several limitations:

1. Only those data locations where all N_v attributes z_i are jointly measured can be considered (limitation to the isotopic case).

2. The cross correlation between principal components at $\mathbf{h} \neq 0$ may not be negligible.

3. Principal components are usually but mathematical constructions, hence the modeling of their semivariogram cannot capitalize on ancillary information available about the original variables.

6.2.6   Colocated cokriging

When secondary variables are much more densely sampled than the primary variable, the left-hand-side matrix of cokriging system (6.19) or (6.33) may be unstable because the correlation between close secondary data is much greater than the correlation between distant primary data. Moreover, secondary data that are close or even colocated with the unknown primary value $z_1(\mathbf{u})$ tend to screen the influence of secondary data that are farther away (recall discussion related to Figure 6.11).

One solution to the instability problems caused by highly redundant secondary information consists of retaining only the single secondary datum of any given type closest to the location $\mathbf{u}$ being estimated, say, the colocated data $z_i(\mathbf{u})$. The linear estimator (6.10) is then written

$$Z_1^*(\mathbf{u}) - m_1(\mathbf{u}) = \sum_{\alpha_1=1}^{n_1(\mathbf{u})} \lambda_{\alpha_1}(\mathbf{u}) \left[Z_1(\mathbf{u}_{\alpha_1}) - m_1(\mathbf{u}_{\alpha_1}) \right]$$

$$+ \sum_{i=2}^{N_v} \lambda_i(\mathbf{u}) \left[Z_i(\mathbf{u}) - m_i(\mathbf{u}) \right]$$

This variant of the cokriging estimator is called *colocated cokriging* (Xu et al., 1992; Almeida, 1993). As for previous developments, different colocated cokriging estimators can be built, depending on the trend model adopted. Hereafter, the case of a single exhaustively sampled secondary attribute z_2 is considered.

Colocated simple cokriging

The colocated simple cokriging estimator of the primary attribute z_1 at location $\mathbf{u}$ is

$$Z_{SCK}^{(1)^*}(\mathbf{u}) = \sum_{\alpha_1=1}^{n_1(\mathbf{u})} \lambda_{\alpha_1}^{SCK}(\mathbf{u}) \, [Z_1(\mathbf{u}_{\alpha_1}) - m_1]$$
$$+ \lambda_2^{SCK}(\mathbf{u}) \, [Z_2(\mathbf{u}) - m_2] + m_1 \qquad (6.43)$$

The cokriging weights are obtained by solving the following system of $(n_1(\mathbf{u})+1)$ linear equations:

$$\begin{cases} \displaystyle\sum_{\beta_1=1}^{n_1(\mathbf{u})} \lambda_{\beta_1}^{SCK}(\mathbf{u}) \, C_{11}(\mathbf{u}_{\alpha_1} - \mathbf{u}_{\beta_1}) + \lambda_2^{SCK}(\mathbf{u}) \, C_{12}(\mathbf{u}_{\alpha_1} - \mathbf{u}) \\ \qquad = C_{11}(\mathbf{u}_{\alpha_1} - \mathbf{u}) \qquad \alpha_1 = 1, \ldots, n_1(\mathbf{u}) \qquad (6.44) \\ \displaystyle\sum_{\beta_1=1}^{n_1(\mathbf{u})} \lambda_{\beta_1}^{SCK}(\mathbf{u}) \, C_{21}(\mathbf{u} - \mathbf{u}_{\beta_1}) + \lambda_2^{SCK}(\mathbf{u}) \, C_{22}(0) = C_{21}(0) \end{cases}$$

Unlike the simple cokriging system (6.13), the colocated simple cokriging system does not call for the covariance between secondary data for $|\mathbf{h}| > 0$, which alleviates the inference and modeling effort (see subsequent discussion related to the Markov model).

Colocated ordinary cokriging

The usual formalism of ordinary cokriging cannot be implemented in the colocated case since the constraint (6.28) that secondary data weights must sum to zero would impose a zero weight on the single colocated datum. Therefore, the standardized ordinary cokriging alternative (6.41) with one single unbiasedness constraint must be considered. The colocated OCK estimator of attribute z_1 at location $\mathbf{u}$ is

$$Z_{OCK}^{(1)^*}(\mathbf{u}) = \sum_{\alpha_1=1}^{n_1(\mathbf{u})} \lambda_{\alpha_1}^{OCK}(\mathbf{u}) \, Z_1(\mathbf{u}_{\alpha_1}) + \lambda_2^{OCK}(\mathbf{u}) \, [Z_2(\mathbf{u}) - m_2 + m_1]$$
$$(6.45)$$

with the single constraint that all weights must sum to one:

$$\sum_{\alpha_1=1}^{n_1(\mathbf{u})} \lambda_{\alpha_1}^{OCK}(\mathbf{u}) + \lambda_2^{OCK}(\mathbf{u}) = 1$$

Unless primary and secondary data means are equal, the secondary variable must be rescaled to the mean of the primary variable to ensure unbiased estimation. Such rescaling calls for a prior knowledge of the stationary means m_1 and m_2. As mentioned in section 6.2.4, the local primary mean is implicitly estimated within each search neighborhood, hence estimator (6.45) does account for local departures from the overall means m_1 and m_2.

The cokriging weights are obtained by solving the following system of $(n_1(\mathbf{u}) + 2)$ linear equations:

$$
\begin{cases}
\displaystyle\sum_{\beta_1=1}^{n_1(\mathbf{u})} \lambda_{\beta_1}^{OCK}(\mathbf{u})\, C_{11}(\mathbf{u}_{\alpha_1} - \mathbf{u}_{\beta_1}) + \lambda_2^{OCK}(\mathbf{u})\, C_{12}(\mathbf{u}_{\alpha_1} - \mathbf{u}) \\[4pt]
\qquad\qquad + \mu^{OCK}(\mathbf{u}) = C_{11}(\mathbf{u}_{\alpha_1} - \mathbf{u}) \qquad \alpha_1 = 1, \ldots, n_1(\mathbf{u}) \\[6pt]
\displaystyle\sum_{\beta_1=1}^{n_1(\mathbf{u})} \lambda_{\beta_1}^{OCK}(\mathbf{u})\, C_{21}(\mathbf{u} - \mathbf{u}_{\beta_1}) + \lambda_2^{OCK}(\mathbf{u})\, C_{22}(0) + \mu^{OCK}(\mathbf{u}) = C_{21}(0) \\[6pt]
\displaystyle\sum_{\beta_1=1}^{n_1(\mathbf{u})} \lambda_{\beta_1}^{OCK}(\mathbf{u}) + \lambda_2^{OCK}(\mathbf{u}) = 1
\end{cases}
$$

$$(6.46)$$

Both primary and secondary variables should be standardized to a zero mean and unit variance when their variances are significantly different. The cokriging weights would then be solutions of system (6.46) expressed in terms of correlograms.

Markov-type approximation

Unlike full cokriging, colocated cokriging calls for the inference and modeling only of the primary covariance function $C_{11}(\mathbf{h})$ and the cross covariance function $C_{12}(\mathbf{h})$. The modeling effort can be further alleviated by the following approximation:

$$
C_{12}(\mathbf{h}) \;\simeq\; \frac{C_{12}(0)}{C_{11}(0)}\, C_{11}(\mathbf{h})
$$

or, in terms of correlograms,

$$
\rho_{12}(\mathbf{h}) \;\simeq\; \rho_{12}(0)\, \rho_{11}(\mathbf{h}) \tag{6.47}
$$

This correlogram model corresponds to the linear regression model:

$$
Z_2(\mathbf{u}) = \rho_{12}(0)\, Z_1(\mathbf{u}) + R(\mathbf{u}) \tag{6.48}
$$

where the residual $R(\mathbf{u})$ is assumed orthogonal to $Z_1(\mathbf{u} + \mathbf{h})$, $\forall\, \mathbf{h}$. The linear regression (6.48) is not a requisite for model (6.47); rather equation (6.48) should be read as the definition of the residual $R(\mathbf{u})$.

A sufficient but not necessary condition for model (6.47) to hold is the independence relation:

$$E\{Z_2(\mathbf{u})|Z_1(\mathbf{u}) = z, \ Z_1(\mathbf{u}+\mathbf{h}) = z'\} \ = \ E\{Z_2(\mathbf{u})|Z_1(\mathbf{u}) = z\}$$
$$\forall \ \mathbf{h}, \ \forall \ z' \tag{6.49}$$

In words, dependence of the secondary variable on the primary is limited to the colocated primary datum, a Markov-type assumption.

Proof
For conciseness of notation, assume that the two RFs $Z_1(\mathbf{u})$ and $Z_2(\mathbf{u})$ are standardized and that there is no lag effect (see section 3.2.3). Their covariance is

$$
\begin{aligned}
\rho_{21}(\mathbf{h}) \ &= \ E\{Z_2(\mathbf{u}) \cdot Z_1(\mathbf{u}+\mathbf{h})\} \\
&= \ \int\int E\{Z_2(\mathbf{u}) \cdot Z_1(\mathbf{u}+\mathbf{h})|Z_1(\mathbf{u}) = z, Z_1(\mathbf{u}+\mathbf{h}) = z'\} \\
&\quad \cdot f_{11}(\mathbf{h}; z, z')dzdz'
\end{aligned}
$$

where $f_{11}(\mathbf{h}; z, z')$ is the bivariate pdf of $Z_1(\mathbf{u})$ and $Z_1(\mathbf{u}+\mathbf{h})$,

$$= \ \int\int z'E\{Z_2(\mathbf{u})|Z_1(\mathbf{u}) = z\} \cdot f_{11}(\mathbf{h}; z, z')dzdz'$$

if relation (6.49) holds true,

$$= \ \rho_{12}(0) \int\int zz'f_{11}(\mathbf{h}; z, z')dzdz'$$

if the regression of $Z_2(\mathbf{u})$ on $Z_1(\mathbf{u})$ is linear, as in relation (6.48),

$$= \ \rho_{12}(0) \cdot \rho_{11}(\mathbf{h})$$

by the definition of the covariance of $Z_1(\mathbf{u})$.

The model (6.48) is rewritten in terms of semivariograms:

$$\gamma_{12}(\mathbf{h}) \ = \ \frac{C_{12}(0)}{C_{11}(0)} \gamma_{11}(\mathbf{h}) \tag{6.50}$$

and is readily extended to several secondary variables. This model is very congenial in that only the semivariogram $\gamma_{11}(\mathbf{h})$ need be modeled.
The Markov-type model has the following characteristics:

- It entails the symmetry of the cross covariance model $C_{12}(\mathbf{h})$ since $C_{12}(0) = C_{21}(0)$. Such symmetry has little practical limitations in that lag effects are most often ignored, see discussion in section 3.2.3.

- The semivariogram of the primary variable is assumed to be proportional to the cross semivariogram. It is recommended to check whether such proportionality holds true; see example of Figure 6.15.

- It does not lead necessarily to a linear model of coregionalization if only because it does not specify the autocovariance function of the secondary variable.

Figure 6.15 shows the experimental Cd semivariogram and the experimental cross semivariograms for the pairs cadmium-zinc and cadmium-nickel. The two cross semivariogram models are computed from the Cd semivariogram model (5.15) using relation (6.50). Whereas the fit is satisfactory for the pair Cd-Zn, the Markov approximation underestimates the long-range structure for the pair Cd-Ni. In the latter case, a linear model of coregionalization is preferred (see Figure 4.18, page 121).

Colocated cokriging versus full cokriging

In the presence of densely sampled secondary information, colocated cokriging is a valuable alternative to full cokriging for these reasons:

1. Colocated cokriging avoids instability caused by highly redundant secondary data.

2. It is fast, since it calls for a smaller cokriging system.

3. Colocated cokriging does not call for the secondary covariance function $C_{22}(\mathbf{h})$ at lags $|\mathbf{h}| > 0$.

4. It does not require modeling the cross covariance function $C_{12}(\mathbf{h})$ as long as the Markov approximation is reasonable.

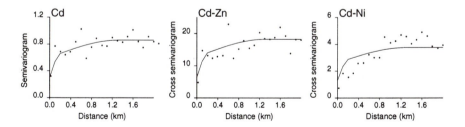

Figure 6.15: Markov model for the pairs cadmium-zinc and cadmium-nickel; the two cross semivariogram models (solid line) are rescalings of the Cd semivariogram model using the Markov model (6.50). The model fares poorly for the pair cadmium-nickel.

The trade-off costs are as follows:

1. The secondary variable must be known at all locations being estimated.

2. The rescaling of variables calls for knowledge of the stationary means of primary and secondary variables.

3. The information provided by secondary data beyond the colocated secondary datum $z_2(\mathbf{u})$ is ignored.

Figure 6.16 shows ordinary cokriging estimates of Cd concentrations using Ni block estimates for secondary information (as given in Figure 6.1, page 186, bottom graph). Whereas full cokriging (dashed line) uses the five closest block estimates of Ni concentrations, colocated cokriging (solid line) retains only the colocated secondary datum. In both cases, a single unbiasedness constraint and a linear model of coregionalization are considered. Full cokriging and the computationally less expensive colocated cokriging give similar results.

Colocated cokriging versus kriging with an external drift

Consider colocated ordinary cokriging and kriging with an external drift of attribute z_1 at location $\mathbf{u}$:

$$Z^{(1)*}_{OCK}(\mathbf{u}) = \sum_{\alpha_1=1}^{n_1(\mathbf{u})} \lambda^{OCK}_{\alpha_1}(\mathbf{u})\, Z_1(\mathbf{u}_{\alpha_1}) + \lambda^{OCK}_2(\mathbf{u})\, [Z_2(\mathbf{u}) - m_2 + m_1]$$

$$Z^{(1)*}_{KED}(\mathbf{u}) = \sum_{\alpha_1=1}^{n_1(\mathbf{u})} \lambda^{SK}_{\alpha_1}(\mathbf{u})\, [Z_1(\mathbf{u}_{\alpha_1}) - m_1(\mathbf{u}_{\alpha_1})] + m_1(\mathbf{u})$$

with $m_1(\mathbf{u}) = a_0(\mathbf{u}) + a_1(\mathbf{u})\, z_2(\mathbf{u})$.

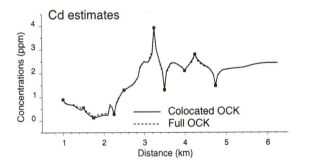

Figure 6.16: Ordinary cokriging estimates of Cd concentrations using the five closest Ni block estimates (full cokriging, dashed line) or the single colocated Ni block estimate (colocated cokriging, solid line).

Although both estimators are designed to incorporate exhaustively sampled secondary information, they differ in many aspects:

- The main difference lies in how the colocated z_2-datum is handled. Whereas that datum directly influences the primary cokriging estimate, in KED that secondary datum provides information only about the primary trend at location $\mathbf{u}$. As discussed in section 6.1.3, such interpretation should be validated by the physics of the phenomenon under study.

- The secondary information tends to influence strongly the KED estimate, especially when the estimated slope $a_1(\mathbf{u})$ of the local trend model is large. Colocated cokriging accounts for the global linear correlation between primary and secondary variables as captured by the cross semivariogram.

- The inference of the residual covariance required by kriging with an external drift is not straightforward. Modeling direct and cross semivariograms is straightforward though demanding. It is even less demanding if the Markov approximation (6.47) is appropriate.

6.2.7   Accounting for soft information

Information available about the primary attribute[6] z is usually not limited to precise (single-valued) measurements of that attribute. Related primary information can take two forms:

1. Constraint intervals $(z_k, z_{k+1}]$. These indicate that the primary attribute is valued between z_k and z_{k+1}. Interval-type data are typically provided by inexpensive measurement devices, such as colorimetric paper for measuring concentrations.

2. Indicators of occurrence of a particular facies or rock category. A calibration such as that in section 2.1.2 then yields, for each category, a specific conditional distribution of the primary attribute.

Both types of data are referred to as *soft* rather than secondary because they relate directly to the primary attribute value. Precise z-measurements are called *hard* data. Soft data, though imprecise, are usually more numerous than hard data and hence are worth considering. Both hard and soft data can be combined using the cokriging formalism.

Interval-type information

Consider the situation where interval-type data are collected at n' locations $\mathbf{u}'_\alpha$. More elaborate measurements are conducted at $n \ll n'$ locations,

---

[6]The primary attribute is denoted z in this section since no other continuous attribute is considered.

yielding a set of precise z-measurements $\{z(\mathbf{u}_\alpha), \ \alpha = 1,\ldots,n\}$. At each datum location $\mathbf{u}'_\alpha$, the soft information $z(\mathbf{u}'_\alpha) \in (z_k, z_{k+1}]$, equivalent to $z_k < z(\mathbf{u}'_\alpha) \le z_{k+1}$, can be coded as a vector of K indicator values $i(\mathbf{u}'_\alpha; z_k)$ defined as

$$i(\mathbf{u}'_\alpha; z_k) = \left\{ \begin{array}{ll} 1 & \text{if } z(\mathbf{u}'_\alpha) \le z_k \\ 0 & \text{otherwise} \end{array} \right. \quad k = 1, \ldots, K \qquad (6.51)$$

where the range of variation of the primary attribute is discretized into K intervals $(z_k, z_{k+1}]$, say, K reference colors on a colorimetric paper. Where a hard datum is available, the indicator coding (6.51) is based on the precise value $z(\mathbf{u}_\alpha)$ rather than on a constraint interval.

Zhu and Journel (1989) proposed to incorporate both hard data $z(\mathbf{u}_\alpha)$ and indicator data $i(\mathbf{u}'_\alpha; z_k)$ into an ordinary cokriging estimator:[7]

$$Z^*_{OCK}(\mathbf{u}) \ = \ \sum_{\alpha=1}^{n(\mathbf{u})} \lambda_\alpha(\mathbf{u}) \, Z(\mathbf{u}_\alpha) \ + \ \sum_{k=1}^{K} \sum_{\alpha'=1}^{n'(\mathbf{u})} \lambda_{\alpha'}(\mathbf{u}; z_k) \, I(\mathbf{u}'_\alpha; z_k)$$

where $\lambda^{OCK}_{\alpha'}(\mathbf{u}; z_k)$ is the cokriging weight assigned to the indicator datum $i(\mathbf{u}'_\alpha; z_k)$, interpreted as a realization of the indicator RV $I(\mathbf{u}'_\alpha; z_k)$.

Consider the case of a single indicator variable ($K = 1$), i.e., a single cutoff z_1; for example, the indicator data relate to exceedence of a single critical threshold value z_1. The cokriging estimator is written

$$Z^*_{OCK}(\mathbf{u}; z_1) \ = \ \sum_{\alpha=1}^{n(\mathbf{u})} \lambda_\alpha(\mathbf{u}; z_1) \, Z(\mathbf{u}_\alpha) \ + \ \sum_{\alpha'=1}^{n'(\mathbf{u}) \ \cdot} \lambda_{\alpha'}(\mathbf{u}; z_1) \, I(\mathbf{u}'_\alpha; z_1)$$

$$(6.52)$$

The data weights, hence the cokriging estimator, depend on the threshold z_1. Again, the $(n(\mathbf{u}) + n'(\mathbf{u}))$ cokriging weights are determined such that estimator (6.52) is unbiased and has minimum error variance.

As with the ordinary cokriging estimator (6.27), the non-bias condition is satisfied by forcing the primary data weights to sum to one and the secondary data weights to sum to zero:

$$\sum_{\alpha=1}^{n(\mathbf{u})} \lambda_\alpha(\mathbf{u}; z_1) \ = \ 1 \qquad \sum_{\alpha'=1}^{n'(\mathbf{u})} \lambda_{\alpha'}(\mathbf{u}; z_1) \ = \ 0$$

Minimizing the estimation variance under these two constraints yields the

---

[7]To simplify notation, the superscript OCK is removed from all notations in this section.

following ordinary cokriging system:

$$
\begin{cases}
\displaystyle\sum_{\beta=1}^{n(\mathbf{u})} \lambda_\beta(\mathbf{u}; z_1)\, C_Z(\mathbf{u}_\alpha - \mathbf{u}_\beta) + \sum_{\beta'=1}^{n'(\mathbf{u})} \lambda_{\beta'}(\mathbf{u}; z_1)\, C_{ZI}(\mathbf{u}_\alpha - \mathbf{u}_{\beta'}; z_1) \\
\qquad\qquad + \mu_Z(\mathbf{u}; z_1) \;=\; C_Z(\mathbf{u}_\alpha - \mathbf{u}) \qquad \alpha = 1, \ldots, n(\mathbf{u}) \\[2mm]
\displaystyle\sum_{\beta=1}^{n(\mathbf{u})} \lambda_\beta(\mathbf{u}; z_1)\, C_{IZ}(\mathbf{u}_{\alpha'} - \mathbf{u}_\beta; z_1) + \sum_{\beta'=1}^{n'(\mathbf{u})} \lambda_{\beta'}(\mathbf{u}; z_1)\, C_I(\mathbf{u}_{\alpha'} - \mathbf{u}_{\beta'}; z_1) \\
\qquad\qquad + \mu_I(\mathbf{u}; z_1) \;=\; C_{IZ}(\mathbf{u}_{\alpha'} - \mathbf{u}; z_1) \qquad \alpha' = 1, \ldots, n'(\mathbf{u}) \\[2mm]
\displaystyle\sum_{\beta=1}^{n(\mathbf{u})} \lambda_\beta(\mathbf{u}; z_1) \;=\; 1 \\[2mm]
\displaystyle\sum_{\beta'=1}^{n'(\mathbf{u})} \lambda_{\beta'}(\mathbf{u}; z_1) \;=\; 0
\end{cases}
$$

$$(6.53)$$

where $C_Z(\mathbf{h})$ and $C_I(\mathbf{h}; z_1)$ are the autocovariance functions of the primary variable $Z(\mathbf{u})$ and the indicator random function $I(\mathbf{u}; z_1)$, respectively. The cross covariance between $Z(\mathbf{u})$ and $I(\mathbf{u}; z_1)$ is denoted $C_{IZ}(\mathbf{h}; z_1)$. The generalization to $K \geq 2$ thresholds is straightforward using the formalism of equation (6.32). However, the modeling effort increases dramatically as more thresholds are considered.

Example

Figure 6.17 (top graph) shows 10 Cd concentrations (black dots) and 16 indicators of whether the tolerable maximum 0.8 ppm, denoted by the horizontal dashed line, is exceeded. A "$+$" sign means that the location is contaminated, whereas a "$-$" sign indicates that the location is safe. Indicator data are directly deduced from precise measurements at hard data locations where available (black dots). At the six other locations, the information is provided by some inexpensive device.

The two direct semivariograms $\widehat{\gamma}_Z(\mathbf{h})$ and $\widehat{\gamma}_I(\mathbf{h})$ and the cross semivariogram $\widehat{\gamma}_{ZI}(\mathbf{h})$ are computed using 259 Cd concentrations and indicator data shown at the top of Figures 2.5 and 2.7 (pages 23 and 25). Experimental values and the linear model of coregionalization fitted are depicted in Figure 6.17 (second row). Cadmium concentrations and indicator data are negatively correlated since a zero indicator value corresponds to a large Cd concentration, i.e., a concentration larger than the threshold 0.8 ppm.

Figure 6.17 (bottom graphs) shows the cokriging estimates (solid line) of Cd concentrations for two sampling densities: 16 soft data locations, and 10 and 6 hard data locations. The dashed line depicts ordinary kriging estimates using only the Cd data. Note the following:

- Both interpolators identify the hard data at their locations.

- The cokriging estimator need not honor the soft data. For example,

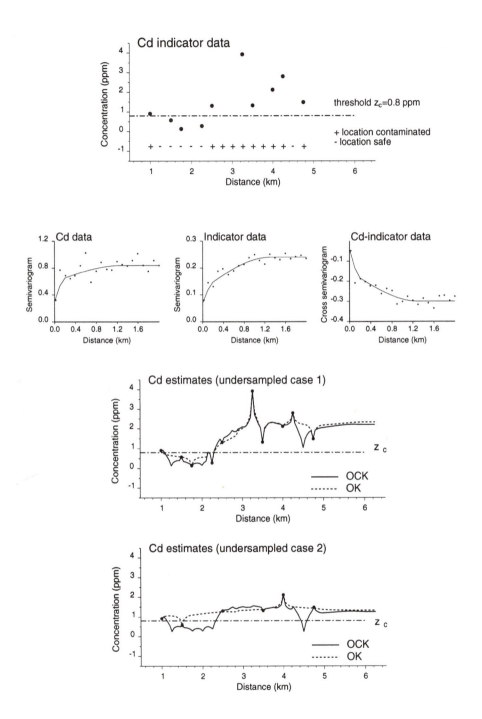

Figure 6.17: Accounting for soft information in the estimation of a continuous attribute. Soft data consist of 16 indicators of whether the tolerable maximum 0.8 ppm for Cd concentrations has been exceeded. Both hard Cd data and soft data are combined using ordinary cokriging and the direct and cross semivariogram models in the second row. The dashed line depicts the ordinary kriging estimate using only the Cd data.

the estimated OCK value at 4.75 km (undersampled case 1) exceeds the critical threshold (horizontal dashed line), although the colocated soft datum indicates that particular location as safe. At that location, the large neighboring hard data prevail on the soft (possibly inaccurate) colocated datum. Indicator kriging algorithms introduced in section 7.4.2 will allow both hard data and constraint interval data to be honored, provided they are consistent.

- Accounting for indicator data yields smaller Cd estimates at locations that are deemed safe than ordinary kriging using only hard data would. Discrepancies between kriging and cokriging estimates increase as hard data become sparser (undersampled case 2).

Categorical information

Consider a situation where primary data $\{z(\mathbf{u}_\alpha), \ \alpha = 1, \ldots, n\}$ are supplemented by exhaustively sampled secondary categorical information $\{s(\mathbf{u}), \ \forall \ \mathbf{u} \ \in \mathcal{A}\}$. The categorical attribute s can take K different states, say, K soil types or facies. For each state s_k, the proportion of z-values not exceeding a particular threshold z_c can be computed from those locations $\mathbf{u}_\alpha$ where both primary and secondary variables are known, see relation (2.10). Such calibration of soft data allows one to associate with each location $\mathbf{u}$ a conditional probability of type

$$
\begin{aligned}
y(\mathbf{u}; z_c) &= \text{Prob}\{Z(\mathbf{u}) \le z_c | S(\mathbf{u}) = s_k\} \\
&= F(z_c | s_k) \in [0, 1]
\end{aligned} \tag{6.54}
$$

Where a hard datum $z(\mathbf{u}_\alpha)$ is available, the soft datum $y(\mathbf{u}; z_c)$ is either zero or one, depending on whether the z-measurement exceeds the threshold z_c.

As with interval-type data, conditional probabilities of type (6.54) can be accounted for using the cokriging formalism. When the soft information relates to categorical variables, such as rock types or soil associations, the corresponding soft data are likely to be more continuous than hard data because all y-data in the same class s_k are constant. To avoid numerical problems caused by such highly redundant secondary information, the colocated standardized ordinary cokriging estimator is preferred:

$$
\frac{Z^*_{OCK}(\mathbf{u}; z_c) - m_Z}{\sigma_Z} = \sum_{\alpha=1}^{n(\mathbf{u})} \lambda_\alpha(\mathbf{u}; z_c) \left[\frac{Z(\mathbf{u}_\alpha) - m_Z}{\sigma_Z} \right]
$$

$$
+ \lambda(\mathbf{u}; z_c) \left[\frac{Y(\mathbf{u}; z_c) - m_Y(z_c)}{\sigma_Y(z_c)} \right] \tag{6.55}
$$

where the mean $m_Y(z_c)$ of variable $Y(\mathbf{u}; z_c)$ is the marginal probability that attribute z does not exceed the critical threshold z_c calculated across all K categories s_k. The variance $\sigma_Y^2(z_c)$ of the indicator variable is then

$m_Y(z_c)[1 - m_Y(z_c)]$. Note that both estimator and the cokriging weights depend on the threshold z_c.

The estimator (6.55) is unbiased provided all cokriging weights sum to one:

$$\sum_{\alpha=1}^{n(\mathbf{u})} \lambda_\alpha(\mathbf{u}; z_c) + \lambda(\mathbf{u}; z_c) = 1$$

The cokriging weights are obtained by solving the following system of $(n(\mathbf{u}) + 2)$ linear equations:

$$
\begin{cases}
\displaystyle\sum_{\beta=1}^{n(\mathbf{u})} \lambda_\beta(\mathbf{u}; z_c)\, \rho_Z(\mathbf{u}_\alpha - \mathbf{u}_\beta) + \lambda(\mathbf{u}; z_c)\, \rho_{ZY}(\mathbf{u}_\alpha - \mathbf{u}; z_c) \\
\qquad\qquad + \mu(\mathbf{u}; z_c) = \rho_Z(\mathbf{u}_\alpha - \mathbf{u}) \qquad \alpha = 1, \ldots, n(\mathbf{u}) \\
\displaystyle\sum_{\beta=1}^{n(\mathbf{u})} \lambda_\beta(\mathbf{u}; z_c)\, \rho_{YZ}(\mathbf{u} - \mathbf{u}_\beta; z_c) + \lambda(\mathbf{u}; z_c) + \mu(\mathbf{u}; z_c) = \rho_{YZ}(0; z_c) \\
\displaystyle\sum_{\beta=1}^{n(\mathbf{u})} \lambda_\beta(\mathbf{u}; z_c) + \lambda(\mathbf{u}; z_c) = 1
\end{cases}
$$

$$(6.56)$$

where $\rho_{ZY}(\mathbf{h}; z_c)$ is the cross correlogram between hard and soft data at threshold z_c.

Example
Using the conditional frequencies given in Table 2.5 (page 19), the geologic profile in Figure 6.1 (page 186, middle graph) is converted into a profile of probabilities of not exceeding the critical threshold for Cd concentrations (Figure 6.18, top graph). The largest probability is observed on Argovian rocks (1.25–2.75 km), whereas there is a zero probability of not exceeding the critical threshold on Portlandian rocks (4.2–4.8 km). Next, the 259 geologic data $s(\mathbf{u}_\alpha)$ available across the study area are converted into soft data (conditional probabilities) $y(\mathbf{u}_\alpha; z_c)$. Figure 6.18 (second row) shows the standardized semivariograms of soft data and Cd concentrations. The correlation between hard and soft data is much stronger along the NE-SW transect (10 locations) than over the study area (259 locations). Rather than modeling the experimental cross semivariogram of 259 data values depicting a negligible hard-soft correlation, for the sole purpose of this example a synthetic cross semivariogram model is built as a spherical model of range 1.2 km with a sill corresponding to a correlation coefficient $\rho_{ZY}(0; z_c)$ of 0.4. The linear model of coregionalization between hard and soft data is depicted by the solid line in Figure 6.18 (second row).

Figure 6.18 (two bottom graphs) shows the standardized cokriging (OCK) estimates (solid line) of Cd concentrations using the five closest Cd data and the colocated soft y-datum. The dashed line depicts ordinary kriging estimates using only the Cd data. For both sampling densities (10 or 6 Cd

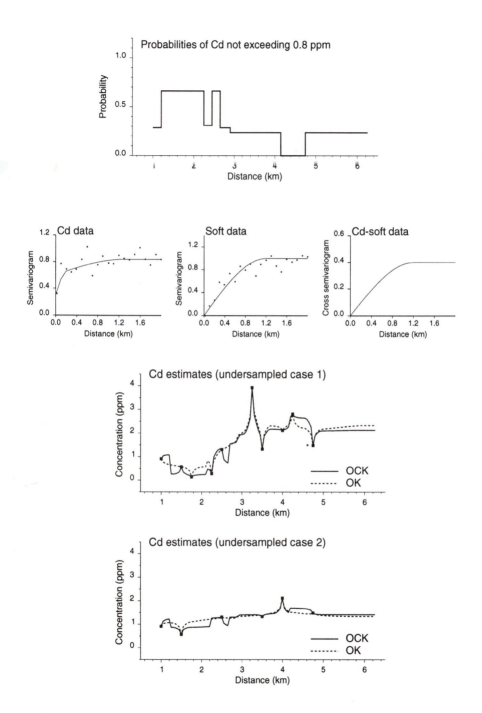

Figure 6.18: Accounting for soft information in the estimation of a continuous attribute. Soft data consist of probabilities of not exceeding the tolerable maximum 0.8 ppm for Cd concentrations as derived from geology. Both hard Cd data and soft data are combined using colocated standardized ordinary cokriging and the direct and cross semivariogram models in the second row. The dashed line depicts the ordinary kriging estimate using only the Cd data.

data), kriging and cokriging estimates depart most on rocks with very large or very small probabilities of contamination: accounting for soft information reduces the estimated Cd values on Argovian rocks and increases these estimates on Portlandian rocks. Kriging and cokriging estimates are similar on the other rocks where the conditional probability of contamination is close to the marginal probability $F(0.8 \text{ ppm}) = 0.65$.

6.2.8 Performance comparison

Cokriging algorithms are used to estimate metals with widespread contamination (Cd, Cu, Pb) and cobalt at 100 test locations. Table 6.2 gives, for each metal, the set of secondary metals retained in cokriging. Secondary information is available at 259 primary data locations (isotopic case) or at 259 primary data locations plus 100 test locations (heterotopic case). For both sampling densities, the 16 closest data locations of each primary/secondary variable are retained: $n_i(\mathbf{u}) = 16 \; \forall \; i$. In the heterotopic case, the situation where only the colocated secondary data are retained is also considered: $n_1(\mathbf{u}) = 16$ and $n_i(\mathbf{u}) = 1 \; \forall \; i > 1$. For the two sampling densities and the colocated case, two cokriging estimators, ordinary cokriging and standardized ordinary cokriging, are compared to the reference ordinary kriging estimator. The direct and cross semivariograms are modeled using the iterative procedure described in Appendix A. For example, Figures 4.18 and 4.19 show the linear model of coregionalization for cadmium, copper, and lead.

Figure 6.19 (left column) shows the rank correlation coefficient between true values and reference ordinary kriging (OK) estimates, then cokriging (CK) estimates for the isotopic (CK_{iso}), heterotopic (CK_{het}), and colocated case (CK_{col}). The corresponding mean absolute errors are displayed in the right column. Note the following:

- Kriging and cokriging scores are similar in the isotopic case. As discussed in section 6.2.2, primary data tend to screen the influence of colocated secondary data. Hence secondary information contributes little to the cokriging estimate when all metals are equally sampled.

Table 6.2: Secondary variables used to estimate primary metals at 100 test locations.

Primary variable	Secondary variables
Cd	Ni, Zn
Cu	Pb, Ni, Zn
Pb	Cu, Ni, Zn
Co	Ni, Zn

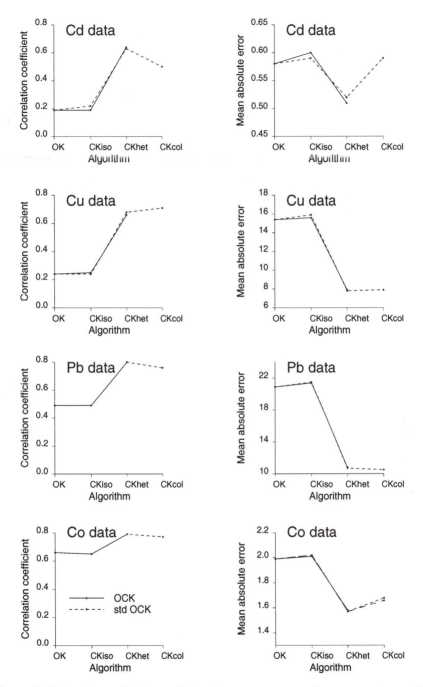

Figure 6.19: Rank correlation coefficients between true metal concentrations and cokriging estimates at 100 test locations (left column), and mean absolute errors (right column). Two cokriging estimators (ordinary cokriging, standardized ordinary cokriging) and three data configurations (isotopic, heterotopic, and colocated) are considered together with the reference ordinary kriging.

- Accounting for better sampled secondary metals (heterotopic case) significantly increases the rank correlation between true values and estimates, and reduces the mean absolute error. Traditional and standardized ordinary cokriging estimators perform equally.

- Retaining only the colocated secondary data (colocated cokriging) causes only a slight reduction of the cokriging performances.

Each test location is classified as contaminated or safe, depending on whether the cokriging estimate exceeds or not the critical threshold. Figure 6.20 shows the percentages of locations that are wrongly declared safe or contaminated using different estimators and data configurations. As with the rank correlation coefficients and the mean absolute errors, cokriging improves over kriging only when secondary metals are better sampled than the

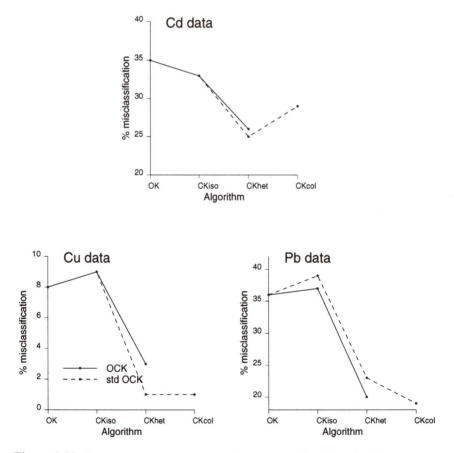

Figure 6.20: Proportion of test locations that are wrongly declared safe or contaminated according to cokriging estimates; as in Figure 6.19, two cokriging estimators and three data configurations are considered.

primary metal. Misclassification for Cd and Cu is further reduced by using
a single unbiasedness constraint (dashed line) rather than the traditional
constraints that call for the secondary data weights to sum to zero (solid
line). Retaining only the colocated secondary data slightly reduces cokriging
performances for cadmium in the sense that it increases the percentage of
locations misclassified.

6.2.9 Multivariate factorial kriging

As mentioned in section 5.6, many physical processes related to geology or
human activities control the spatial distribution of metal concentrations over
the study area. Observed relations between concentrations of different metals might then be connected with the occurrence of common sources of soil
contamination. For example, Figure 5.13 (middle graphs) showed that nickel
and cobalt concentrations have common regional features linked to the spatial distribution of rock types. The strong relation is also depicted by the
scattergram of Ni and Co regional components in Figure 6.21 (right top
graph). The two bottom scattergrams of Figure 6.21 show weaker relations
for local and micro-scale spatial components of both metals, which suggests

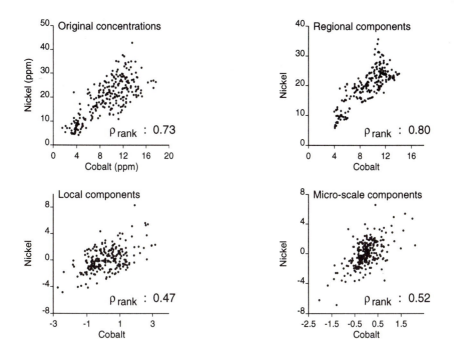

Figure 6.21: Scattergrams of nickel versus cobalt: original metal concentrations
(top left graph) and spatial components at the regional scale, local scale and micro-
scale.

that destructuring processes operate over shorter distances. Micro-scale variations arise partly from measurement errors that could be independent from one metal to another. The original nickel and cobalt concentrations result from a combination of all three different scales of variation, and their scattergram shown in Figure 6.21 (left top graph) depicts a relation intermediate between those observed at local and regional scales.

The relation between cobalt and nickel is said to be *scale-dependent* because it changes as a function of the spatial scale considered. Such attention to scale-dependence may enhance a relation between variables that is otherwise blurred in an approach where all different sources of variation are mixed, leading to a better understanding of the physical underlying mechanisms controlling spatial patterns (Goovaerts and Webster, 1994). Multivariate factorial kriging, also called factorial kriging analysis (Matheron, 1982; Wackernagel, 1988, 1995, p. 160–165), allows one to analyze relations between variables at the spatial scales detected and modeled from experimental semivariograms. The technique has been applied in various areas, such as geochemistry (Sandjivy, 1984; Bourgault and Marcotte, 1991; Wackernagel and Sanguinetti, 1993), soil science (Wackernagel et al., 1988; Goovaerts, 1992, 1994d; Goulard and Voltz, 1992), hydrogeology (Rouhani and Wackernagel, 1990; Goovaerts et al., 1993), mining (Sousa, 1989), and image processing (Daly et al., 1989).

Like factorial kriging, which is based on the linear model of regionalization (4.27), multivariate factorial kriging is based on the specific linear model of coregionalization (4.35) fitted to the experimental auto and cross covariance functions:

$$C_{ij}(\mathbf{h}) = \sum_{l=0}^{L} b_{ij}^l \, c_l(\mathbf{h}) \quad \forall \, i, j$$

Under that particular model, each RF $Z_i(\mathbf{u})$ can be interpreted as a linear combination of independent RFs $Y_k^l(\mathbf{u})$, each with zero mean and basic covariance function $c_l(\mathbf{h})$:

$$Z_i(\mathbf{u}) = \sum_{l=0}^{L} \sum_{k=1}^{N_v} a_{ik}^l Y_k^l(\mathbf{u}) + m_i(\mathbf{u})$$

$$= \sum_{l=0}^{L} Z_i^l(\mathbf{u}) + m_i(\mathbf{u}) \quad i = 1, \dots, N_v \quad (6.57)$$

where $Z_i^l(\mathbf{u}) = \sum_{k=1}^{N_v} a_{ik}^l Y_k^l(\mathbf{u})$ is the lth spatial component of the RF $Z_i(\mathbf{u})$, and the trend component $m_i(\mathbf{u})$ is assumed to be locally constant (decision of quasi-stationarity).

Recall from section 4.3.2 that the cross covariance between any two spatial components $Z_i^l(\mathbf{u})$ and $Z_j^{l'}(\mathbf{u})$ is a linear combination of cross covariances between RFs $Y_k^l(\mathbf{u})$:

$$
\begin{aligned}
\text{Cov}\{Z_i^l(\mathbf{u}), Z_j^{l'}(\mathbf{u}+\mathbf{h})\} &= \text{Cov}\{\sum_{k=1}^{N_v} a_{ik}^l Y_k^l(\mathbf{u}), \sum_{k'=1}^{N_v} a_{jk'}^{l'} Y_{k'}^{l'}(\mathbf{u}+\mathbf{h})\} \\
&= \sum_{k=1}^{N_v}\sum_{k'=1}^{N_v} a_{ik}^l\, a_{jk'}^{l'}\, \text{Cov}\{Y_k^l(\mathbf{u}), Y_{k'}^{l'}(\mathbf{u}+\mathbf{h})\}
\end{aligned}
$$

Because the RFs $Y_k^l(\mathbf{u})$ are, by construction, mutually orthogonal, that cross covariance reduces to

$$
\text{Cov}\{Z_i^l(\mathbf{u}), Z_j^{l'}(\mathbf{u}+\mathbf{h})\} = \begin{cases} b_{ij}^l c_l(\mathbf{h}) & \text{if } l=l' \\ 0 & \text{otherwise} \end{cases} \tag{6.58}
$$

At $|\mathbf{h}|=0$, the value of each basic covariance model $c_l(\mathbf{h})$ is 1. Thus, the coefficient b_{ij}^l is either the covariance at $|\mathbf{h}|=0$ between spatial components (case $i\neq j$) or the variance of the spatial component $Z_j^l(\mathbf{u})$ (case $j=i$).

For each covariance model $c_l(\mathbf{h})$, the coefficients b_{ij}^l can be arranged into an $N_v \times N_v$ coregionalization matrix $\mathbf{B}_l$:

$$
\mathbf{B}_l = \begin{bmatrix} b_{11}^l & \cdots & b_{1N_v}^l \\ \vdots & & \vdots \\ b_{N_v 1}^l & \cdots & b_{N_v N_v}^l \end{bmatrix}
$$

Accounting for relation (6.58), the matrix $\mathbf{B}_l$ is the variance–covariance matrix of the N_v spatial components $Z_i^l(\mathbf{u})$. The linear correlation between any two spatial components $Z_i^l(\mathbf{u})$ and $Z_j^l(\mathbf{u})$ is then measured by the structural correlation coefficient ρ_{ij}^l defined as

$$
\rho_{ij}^l = \frac{b_{ij}^l}{\sqrt{b_{ii}^l \cdot b_{jj}^l}}
$$

The matrix of structural correlation coefficients is denoted $\mathbf{R}_l = [\rho_{ij}^l]$.

In section 4.3.2, under second-order stationarity, the $(L+1)$ coregionalization matrices were shown to sum to the covariance function matrix at lag zero:

$$
\mathbf{C}(0) = \sum_{l=0}^{L} \mathbf{B}_l \tag{6.59}
$$

Multivariate analysis (Anderson, 1958) is traditionally conducted on the variance–covariance matrix $\mathbf{C}(0)$ or its standardized form, the correlation matrix $\mathbf{R}(0)$, thus ignoring the data coordinates. Factorial kriging accounts for the regionalized nature of variables by analyzing each coregionalization matrix $\mathbf{B}_l$ or matrix of structural correlation coefficients $\mathbf{R}_l$ separately. By so doing, each correlation structure is distinguished by filtering the structures belonging to other scales of spatial variation.

Multivariate factorial kriging proceeds in three steps:

1. The coregionalization matrices $\mathbf{B}_l$ are first estimated using the itera-tive procedure described in Appendix A. Recall that the decomposi-tion (6.59) and hence any subsequent interpretation of the coregional-ization matrices depend on the somewhat arbitrary decision of whether to include a particular component in the linear model of coregionaliza-tion. As in univariate factorial kriging, when modeling the direct and cross semivariograms, it is critical to account for any physical knowl-edge about the phenomenon and the study area; see related discussion in Goulard and Voltz (1992), Goovaerts (1992).

2. Multivariate methods, such as principal component analysis or discrim-inant analysis, are then applied to each matrix $\mathbf{B}_l$ or $\mathbf{R}_l$ (Wackernagel et al., 1989). To avoid results that are overinfluenced by the variables with the largest values, the N_v variables should be standardized to zero mean and unit variance at the beginning. Elements of each coregion-alization matrix $\mathbf{B}_l$ thus represent relative contributions of the basic model $g_l(\mathbf{h})$ to direct and cross semivariograms.

3. The variables $Y_k^l(\mathbf{u})$ in relation (6.57), called regionalized factors, are estimated at each location $\mathbf{u}$ as linear combinations of neighboring at-tribute values (see subsequent discussion). Again, beware that regional-ized factors are but mathematical constructions with no a priori physical meaning, and hence one should interpret these factors with caution.

Estimating regionalized factors

The ordinary cokriging[8] estimator of the kth regionalized factor at the lth spatial scale is

$$Y_k^{l^*}(\mathbf{u}) = \sum_{i=1}^{N_v} \sum_{\alpha_i=1}^{n_i(\mathbf{u})} \lambda_{\alpha_i k}^l(\mathbf{u}) \, Z_i(\mathbf{u}_{\alpha_i}) \tag{6.60}$$

where $\lambda_{\alpha_i k}^l(\mathbf{u})$ is the weight assigned to the datum $z_i(\mathbf{u}_{\alpha_i})$ interpreted as a realization of the RV $Z_i(\mathbf{u}_{\alpha_i})$. Since regionalized factors are built as RFs with zero mean, the estimator (6.60) is unbiased provided the N_v sets of cokriging weights sum to zero:

$$\sum_{\alpha_i=1}^{n_i(\mathbf{u})} \lambda_{\alpha_i k}^l(\mathbf{u}) = 0 \qquad i = 1, \ldots, N_v \tag{6.61}$$

---

[8] To avoid heavy notations, superscript and subscript "OCK" are removed from all notations hereafter.

The error variance $\sigma_E^2(\mathbf{u}) = \text{Var}\{Y_k^{l*}(\mathbf{u}) - Y_k^l(\mathbf{u})\}$ can be expressed as a linear combination of cross covariance values:

$$\sigma_E^2(\mathbf{u}) = c_l(0) + \sum_{i=1}^{N_v} \sum_{j=1}^{N_v} \sum_{\alpha_i=1}^{n_i(\mathbf{u})} \sum_{\alpha_j=1}^{n_j(\mathbf{u})} \lambda_{\alpha_i k}^l(\mathbf{u}) \, \lambda_{\alpha_j k}^l(\mathbf{u}) \, C_{ij}(\mathbf{u}_{\alpha_i} - \mathbf{u}_{\alpha_j})$$

$$-2\sum_{i=1}^{N_v} \sum_{\alpha_i=1}^{n_i(\mathbf{u})} \lambda_{\alpha_i k}^l(\mathbf{u}) \, \text{Cov}\{Z_i(\mathbf{u}_{\alpha_i}), Y_k^l(\mathbf{u})\}$$

Accounting for the decomposition (6.57), the cross covariance between the RV $Z_i(\mathbf{u}_{\alpha_i})$ and the regionalized factor $Y_k^l(\mathbf{u})$ is

$$\text{Cov}\{Z_i(\mathbf{u}_{\alpha_i}), Y_k^l(\mathbf{u})\} = \text{Cov}\{\sum_{l=0}^{L} \sum_{k=1}^{N_v} a_{ik}^l Y_k^l(\mathbf{u}_{\alpha_i}) + m_i(\mathbf{u}_{\alpha_i}), Y_k^l(\mathbf{u})\}$$

$$= a_{ik}^l \, c_l(\mathbf{u}_{\alpha_i} - \mathbf{u}),$$

because the regionalized factors are mutually independent. The linear model of coregionalization is very convenient: it allows one to infer the cross covariance between z_i-data and unavailable regionalized factors from components of the auto and cross covariance models of the original variables $Z_i(\mathbf{u})$.

By construction, the matrix $\mathbf{A}_l$ of coefficients a_{ik}^l is such that $\mathbf{A}_l \mathbf{A}_l^T = \mathbf{B}_l$; see expression (4.39). There are many ways to decompose a coregionalization matrix $\mathbf{B}_l$ into matrices $\mathbf{A}_l$. A principal component analysis would lead to the following spectral decomposition:

$$\mathbf{B}_l = \mathbf{Q}_l \, \mathbf{\Lambda}_l \, \mathbf{Q}_l^T = \mathbf{A}_l \, \mathbf{A}_l^T \quad \text{with} \quad \mathbf{A}_l = \mathbf{Q}_l \, \mathbf{\Lambda}_l^{1/2} \qquad (6.62)$$

where $\mathbf{Q}_l = [q_{ik}^l]$ is the orthogonal matrix of eigenvectors, and $\mathbf{\Lambda}_l = [\lambda_k^l]$ is the diagonal matrix of eigenvalues.

Minimizing the estimation variance $\sigma_E^2(\mathbf{u})$ under the N_v non-bias constraints (6.61) yields the following system of $(\sum_{i=1}^{N_v} n_i(\mathbf{u}) + N_v)$ linear equations:

$$\begin{cases} \sum_{j=1}^{N_v} \sum_{\beta_j=1}^{n_j(\mathbf{u})} \lambda_{\beta_j k}^l(\mathbf{u}) \, C_{ij}(\mathbf{u}_{\alpha_i} - \mathbf{u}_{\beta_j}) + \mu_{ik}^l(\mathbf{u}) = a_{ik}^l c_l(\mathbf{u}_{\alpha_i} - \mathbf{u}) \\[2mm] \qquad\qquad\qquad\qquad \alpha_i = 1, \ldots, n_i(\mathbf{u}) \quad i = 1, \ldots, N_v \qquad (6.63) \\[2mm] \sum_{\beta_i=1}^{n_i(\mathbf{u})} \lambda_{\beta_i k}^l(\mathbf{u}) = 0 \quad i = 1, \ldots, N_v \end{cases}$$

Apart from the right-hand-side terms and the unbiasedness constraints, that system is identical to the ordinary cokriging system (6.32) for attribute values.

Multivariate factorial kriging of metal concentrations

Multivariate factorial kriging is performed on the concentrations of the seven metals at all 259 locations (isotopic data set). All variables are standardized to zero mean and unit variance. The 28 experimental direct and cross semi-variograms are modeled as linear combinations of three basic structures: a nugget effect and two spherical models with ranges of 200 m and 1.3 km.

Principal component analysis is performed on the correlation matrix $\mathbf{R}(0)$ and the three coregionalization matrices $\mathbf{B}_l$ corresponding to the three basic semivariogram structures. One capitalizes on the property that the first few principal components account for most of the variance to display possible interrelations between variables. Start with the global interrelations between variables as described by the correlation matrix given in Table 2.6 (page 22); see Figure 6.22 (left top graph). The position of each variable (metal) Z_i in the plane of a pair of components Y_k and $Y_{k'}$ is given by the two correlation coefficients $(\rho_{ki}, \rho_{k'i})$ defined as

$$\rho_{ki} = q_{ki} \sqrt{\frac{\lambda_k}{\sigma_i^2}} \qquad (6.64)$$

Equation (6.64) is a mere rescaling of the loading q_{ki} by the variance λ_k of the kth principal component and the variance σ_i^2 of the ith original variable. A vector can be drawn from the origin in the plane $(0,0)$ to each plotted point $(\rho_{ki}, \rho_{k'i})$. The orientation of that vector with respect to the two axes reflects the correlation between the variable Z_i and the two principal components Y_k and $Y_{k'}$. The length of the vector, $\sqrt{\rho_{ki}^2 + \rho_{k'i}^2}$, measures the percentage of the variance σ_i^2 explained by the two components. If the two components completely account for the variance of Z_i, the point would lie on a circle of unit radius as drawn in Figure 6.22, hence the name *circle of correlations*. The closer the point is to the center, the smaller the proportion of variance accounted for by Y_k and $Y_{k'}$. Therefore, one should avoid interpreting relations between variables that plot near the center.

Figure 6.22 (left top graph) shows the circle of correlations defined by the first two components (Y_1, Y_2). The circle allows one to distinguish copper and lead, which are strongly correlated ($\rho = 0.78$), from a group of four metals (Cd, Cr, Co, and Ni). The isolated position of zinc expresses its equal correlation with variables of both groups (see Table 2.6, page 22).

Similarly, circles of correlations can be used to display spatial interrelations between variables as described by each coregionalization matrix $\mathbf{B}_l$. The position of the ith variable in the circle defined by the pair $(Y_k^l, Y_{k'}^l)$ is given by the pair of correlation coefficients between the spatial component Z_i^l and these two regionalized factors. By analogy with expression (6.64), the correlation coefficient between the spatial component Z_i^l and the regionalized factor Y_k^l is computed as

$$q_{ki}^l \sqrt{\frac{\lambda_k^l}{b_{ii}^l}}$$

where the coefficients q_{ki}^l and λ_k^l are derived from the spectral decomposition (6.62) of the coregionalization matrix $\mathbf{B}_l$, and b_{ii}^l is the sill of the basic structure $g_l(\mathbf{h})$ in the direct semivariogram model of variable Z_i.

The comparison of circles of correlations shown in Figure 6.22 indicates that interrelations between metals change as a function of the spatial scale. In particular, note:

- the stronger relation between copper and lead at the local scale, which probably reflects common sources of man-made pollution over short distances.

- the strong correlation between Cd, Co, and Ni at the regional scale that matches the scale of the stratigraphy, suggesting that the source of these metals is geochemical.

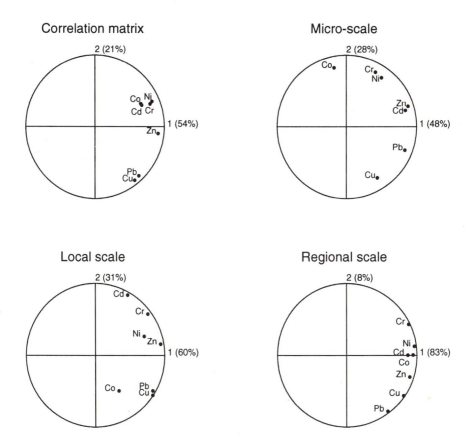

Figure 6.22: Circles of correlations computed from the correlation matrix of seven heavy metals and the corresponding three coregionalization matrices. The percentage of variance explained by each component or regionalized factor is given in parentheses.

Figure 6.23 shows the maps of the first regionalized factors at the local and regional scales. The map of the regional factor emphasizes the impact of the geology shown in Figure 5.13 (page 168, top graph) on the main regional features of metal concentrations.

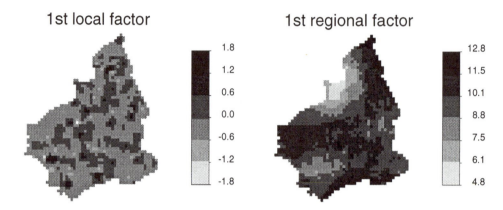

Figure 6.23: Maps of the first regionalized factor at local and regional scales.

Chapter 7

Assessment of Local Uncertainty

Unlike in Chapters 5 and 6, which focused on deriving an optimal estimate and the associated error variance, in this chapter priority is given to modeling the uncertainty about the unknown. Such a model of local uncertainty allows one to evaluate the risk involved in any decision-making process, such as delineation of contaminated areas where remedial measures should be taken. From the model of uncertainty, one can also derive different estimates optimal for different criteria, each customized to the specific problem at hand, instead of retaining the somewhat arbitrary least-squares error estimate.

Section 7.1 proposes modeling the local uncertainty through conditional probability distributions instead of confidence intervals. MultiGaussian- and indicator-based algorithms for determining such conditional distributions are introduced in sections 7.2 and 7.3. Section 7.4 presents tools for accounting for local uncertainty in risk analysis and decision-making processes. These tools are used in section 7.5 to classify test locations as safe or contaminated with respect to Cd, Cu, and Pb concentrations.

7.1 Two Models of Local Uncertainty

Consider the problem of modeling uncertainty about the concentration in cadmium at the two unsampled locations $\mathbf{u}'_1$ and $\mathbf{u}'_2$ shown in Figure 7.1 (top graph). The random function approach amounts to modeling jointly the two unknown values $z(\mathbf{u}'_1)$ and $z(\mathbf{u}'_2)$ as realizations of two spatially dependent random variables $Z(\mathbf{u}'_1)$ and $Z(\mathbf{u}'_2)$. Building on this random function model, one can assess uncertainty at these two locations in essentially two ways:

1. Compute minimum error variance (kriging) estimates, say, $z^*(\mathbf{u}'_1)$ and $z^*(\mathbf{u}'_2)$, of the unknown values and derive the corresponding confi-

dence intervals often assumed arbitrarily Gaussian (Figure 7.1, middle graphs).

2. Model the local probability distributions of the two RVs $Z(\mathbf{u}'_1)$ and $Z(\mathbf{u}'_2)$ (Figure 7.1, bottom graphs).

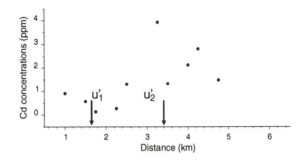

95% confidence interval

z($\mathbf{u}'_1$) ∈ [-0.9,1.9]   z($\mathbf{u}'_2$) ∈ [0.9,3.7]

Local probability distribution

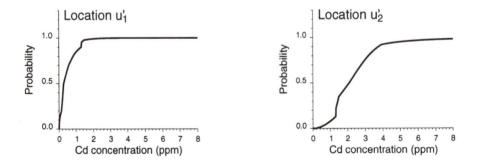

Figure 7.1: Modeling uncertainty about Cd concentrations at locations $\mathbf{u}'_1$ and $\mathbf{u}'_2$ (top graph). The traditional approach amounts to computing a minimum error variance estimate $z^*(\mathbf{u})$ and adopting Gaussian-type confidence intervals. A more rigorous approach calls for modeling the two local distributions of probability (bottom graphs).

7.1.1 Local confidence interval

The traditional approach for modeling local uncertainty at an unsampled location $\mathbf{u}$ consists of computing a minimum error variance (kriging) estimate, $z^*(\mathbf{u})$, of the unknown value $z(\mathbf{u})$ and the associated error variance $\sigma_E^2(\mathbf{u}) = Var\{Z^*(\mathbf{u}) - Z(\mathbf{u})\}$. The estimate and error variance are then typically combined to derive a Gaussian-type confidence interval centered on the estimated value. For example, the 95% confidence interval is taken as

$$\text{Prob}\{Z(\mathbf{u}) \in [z^*(\mathbf{u}) - 2\sigma_E(\mathbf{u}), z^*(\mathbf{u}) + 2\sigma_E(\mathbf{u})]\} = 0.95 \qquad (7.1)$$

where $\sigma_E^2(\mathbf{u})$ is the error (kriging) variance at $\mathbf{u}$.

The ordinary kriging estimate and error standard deviation at locations $\mathbf{u}_1'$ and $\mathbf{u}_2'$ shown in Figure 7.1 are, respectively:

$z^*(\mathbf{u}_1') = 0.5$ ppm, $\sigma_E(\mathbf{u}_1') = 0.7$ ppm
$z^*(\mathbf{u}_2') = 2.3$ ppm, $\sigma_E(\mathbf{u}_2') = 0.7$ ppm

Using relation (7.1), one deduces that the unknown Cd concentration has a 95% probability of lying in the interval $[-0.9, 1.9]$ ppm at $\mathbf{u}_1'$ and in the interval $[0.9, 3.7]$ ppm at $\mathbf{u}_2'$.

The derivation of a confidence interval of type (7.1) is straightforward in that only a measure of the correlation between z-values (semivariogram or covariance function) is required. However, the error model (7.1) calls for two stringent hypotheses (e.g., see Isaaks and Srivastava, 1989, p. 517–519):

1. The estimation error $z^*(\mathbf{u}) - z(\mathbf{u})$ is modeled as a realization of a Gaussian RV error.

2. The error variance $\sigma_E^2(\mathbf{u})$ is independent of the data values.

The first assumption implies the symmetry of the "local" distribution of errors. In practice, true values tend to be overestimated in low-valued areas and underestimated in high-valued areas (recall the discussion about the smoothing effect in section 5.8). Thus, local distributions of errors are generally positively or negatively skewed, the sign of the skewness depending on the data values retained at each location. Such asymmetry is more likely to occur when the sample histogram of the original data is also asymmetric. The "global" distribution of estimation errors, that is, the distribution of all local estimation errors pooled over the entire study area, may be symmetric, with the local overestimations balancing the local underestimations.

The second condition calls for the variance of the errors to be independent of the actual data values and to depend only on the data configuration, a situation referred to as *homoscedasticity* and rarely met in practice. In the example of Figure 7.1, ordinary kriging yields similar estimation variances $\sigma_E^2(\mathbf{u}_1') \simeq \sigma_E^2(\mathbf{u}_2')$ since the data configurations at locations $\mathbf{u}_1'$ and $\mathbf{u}_2'$ are similar: in both cases, the two closest data are approximately 250 m away. However, the potential for error is expected to be greater at $\mathbf{u}_2'$, which is surrounded by one very large and one small Cd concentration, than at $\mathbf{u}_1'$, which is surrounded by two consistently small values. Thus confidence intervals, such as that in equation (7.1), based on a mere estimation variance are generally not a satisfactory solution to the critical problem of assessing local uncertainty.

7.1.2 Local probability distributions

A more rigorous approach to estimation is to assess first the uncertainty about the unknown, then deduce an estimate optimal in some appropriate sense (see Srivastava, 1987a; Journel, 1989, Lesson 4). This is significantly different from the traditional approach of first deriving the estimate then attaching to it a confidence interval. Let $Z(\mathbf{u})$ be the RV modeling the uncertainty about $z(\mathbf{u})$. The distribution function $F(\mathbf{u}; z|(n)) = \text{Prob}\{Z(\mathbf{u}) \leq z|(n)\}$ made conditional to the information available (n) fully models that uncertainty in the sense that probability intervals can be derived, such as

$$\text{Prob}\{Z(\mathbf{u}) \in (a, b]|(n)\} = F(\mathbf{u}; b|(n)) - F(\mathbf{u}; a|(n))$$

or

$$\text{Prob}\{Z(\mathbf{u}) > b|(n)\} = 1 - F(\mathbf{u}; b|(n))$$

Note that these probability intervals are independent of any particular estimate $z^*(\mathbf{u})$ of the unknown value $z(\mathbf{u})$. Indeed, uncertainty depends on the information available (n), not on the particular optimality criterion retained to define an estimate. A lucid discussion of this important methodological priority is given in Srivastava (1987a).

Each conditional probability distribution function $F(\mathbf{u}; z|(n))$ provides a measure of *local* uncertainty in that it relates to a specific location $\mathbf{u}$. A series of single-point ccdfs do not provide any measure of multiple-point or *spatial* uncertainty, such as the probability that a string of locations jointly exceed a given threshold value. In Chapter 8, the concept of conditional simulation will allow the assessment of such spatial uncertainty from several realizations of the distribution in space of the attribute z.

The minimal information available about the z-value at any location $\mathbf{u}$ usually consists of a physical constraint interval $[z_{min}, z_{max}]$. For Cd values, the lower bound z_{min} of that interval would be zero since concentrations are necessarily non-negative. In addition, one may know that the Cd concentration cannot exceed 6 ppm. Thus, in the absence of any location-specific information, the unknown values $z(\mathbf{u}_1')$ and $z(\mathbf{u}_2')$ at any two locations $\mathbf{u}_1'$ and $\mathbf{u}_2'$ would be valued anywhere with equal probability in the interval $[0, 6$ ppm$]$. In other words, the cumulative distribution function (cdf) for both RVs $Z(\mathbf{u}_1')$ and $Z(\mathbf{u}_2')$ would be the uniform expression:

$$F(\mathbf{u}_1'; z) = F(\mathbf{u}_2'; z) = \begin{cases} 0 & \text{if } z \leq 0 \\ z/6 & \text{if } z \in (0, 6] \\ 1 & \text{if } z > 6 \end{cases} \quad \forall z \qquad (7.2)$$

The uniform cdf is depicted in Figure 7.2 (top graphs). Note the large

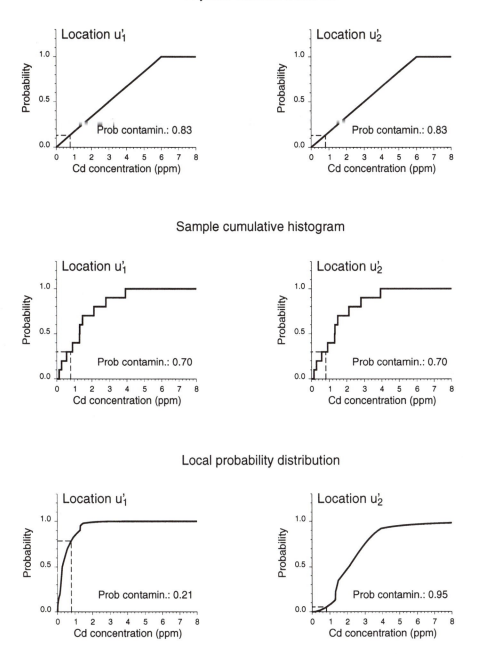

Figure 7.2: Reduction in the local uncertainty about Cd concentrations at locations u'_1 and u'_2 as more information is accounted for; minimal information: a physical constraint interval (top graphs), global information: cumulative histogram of the sample values (middle graphs), and its updating using local information (bottom graphs).

probability of exceeding the critical threshold at either locations $\mathbf{u}_1'$ and $\mathbf{u}_2'$:
$1 - F(\mathbf{u}; 0.8) = 1 - 0.8/6 = 0.83$.

The stationarity decision provides a first update of the uncertainty model (7.2). Indeed, the sample marginal distribution $F^*(z)$ can then be seen as global information available at any unsampled location:

$$F(\mathbf{u}; z) \;=\; \text{Prob}\{Z \le z\} = F^*(z) \qquad \forall\, \mathbf{u} \in A \qquad (7.3)$$

The uncertainty about the Cd concentration at any unsampled location $\mathbf{u}$ along the transect is then modeled from the cumulative histogram of the ten data values available (Figure 7.2, middle graphs). The probability for the unknown value to exceed the tolerable maximum 0.8 ppm would then be equal to the actual proportion of data that exceed that threshold, i.e., $1 - F^*(0.8\ \text{ppm}) = 0.70 = 7/10$.

Accounting for the global information specific to the study area (transect) reduces the range of possible values at the two unsampled locations. However, such a model of uncertainty (7.3) remains location-independent and does not account for the proximity of specific data values. Important information, such as that $\mathbf{u}_1'$ is in a low-valued area or that $\mathbf{u}_2'$ is in a high-valued area, is ignored. Intuitively, one would expect the probability of exceeding the tolerable maximum to be larger at location $\mathbf{u}_2'$ surrounded by contaminated locations than at location $\mathbf{u}_1'$ within a consistently low concentration area. The idea is to capitalize on the spatial dependence between any two Cd values to update the global model of uncertainty (7.3), that is, make it conditional to the local information available. This updating with the information available (n) would result in a usually different cdf, called the *posterior* distribution or conditional cumulative distribution function (ccdf):

$$F(\mathbf{u}; z|(n)) \;=\; \text{Prob}\{Z(\mathbf{u}) \le z|(n)\} \qquad (7.4)$$

where the notation "$|(n)$" expresses conditioning to the local information, say, $n(\mathbf{u})$ neighboring data $z(\mathbf{u}_\alpha)$. Since the local information retained depends on the location $\mathbf{u}$, a more rigorous but heavier notation would be $F(\mathbf{u}; z|\{n(\mathbf{u})\})$. In what follows, the more concise notation $F(\mathbf{u}; z|(n))$ is used regardless of the number or type of conditioning data retained.

Figure 7.2 (bottom graphs) shows the conditional distributions modeled at locations $\mathbf{u}_1'$ and $\mathbf{u}_2'$ using the multiGaussian approach introduced in the next section. Unlike the two uncertainty models (7.2) and (7.3), the ccdf model is location-dependent:

- There is a 21% probability that the unknown value $z(\mathbf{u}_1')$ exceeds the critical threshold 0.8 ppm depicted by the vertical dashed line. Accounting for small Cd concentrations around $\mathbf{u}_1'$ leads us to reduce the prior 70% probability of exceeding the tolerable maximum.

- On the other hand, the posterior probability for the unknown value $z(\mathbf{u}_2')$ to exceed 0.8 ppm is 0.95. Accounting for large Cd concentrations around $\mathbf{u}_2'$ leads us to increase the prior 70% probability of exceeding the tolerable maximum.

7.2 The MultiGaussian Approach

The easiest way to derive ccdfs consists of assuming a model for the entire spatial law (multivariate distribution) of the RF $Z(\mathbf{u})$. That law must be congenial enough so that all ccdfs $F(\mathbf{u}; z|(n))$, $\forall\ \mathbf{u} \in \mathcal{A}$, have the same analytical expression and are fully specified through a few parameters. The problem of determining the conditional cdf at location $\mathbf{u}$ thus reduces to that of estimating the few corresponding parameters, say, the mean and variance, hence the term *parametric approach*. The multivariate Gaussian RF model is by far the most widely used parametric model because its extremely congenial properties render the inference of the parameters of the ccdf straightforward.

7.2.1 The multiGaussian model

If $\{Y(\mathbf{u}), \mathbf{u} \in \mathcal{A}\}$ is a standard multivariate Gaussian RF with covariance function $C_Y(\mathbf{h})$, then the following are true (e.g., Anderson, 1958):

1. All subsets of that RF, e.g., $\{Y(\mathbf{u}), \mathbf{u} \in \mathcal{D} \subset \mathcal{A}\}$, are also multivariate normal.

2. The one-point[1] cdf of any linear combination of RV components is normal:

$$X \ = \ \sum_{\alpha=1}^{n} \lambda_\alpha\, Y(\mathbf{u}_\alpha)$$

is normally distributed, for any choice of n locations $\mathbf{u}_\alpha \in \mathcal{A}$ and any set of weights λ_α.

3. The two-point distribution of any pairs of RVs $Y(\mathbf{u})$ and $Y(\mathbf{u} + \mathbf{h})$ is normal and fully determined by the covariance function $C_Y(\mathbf{h})$ (Xiao, 1985):

$$
\begin{aligned}
G(\mathbf{h}; y_p, y_{p'}) \ &= \ \mathrm{Prob}\,\{Y(\mathbf{u}) \le y_p, Y(\mathbf{u} + \mathbf{h}) \le y_{p'}\} \\
&= \ p \cdot p' + \\
&\quad \frac{1}{2\pi} \int_{0}^{arcsin C_Y(\mathbf{h})} \exp\left[-\frac{y_p^2 + y_{p'}^2 - 2 y_p y_{p'} \sin\theta}{2\cos^2\theta}\right] d\theta
\end{aligned}
$$

$$\forall\ p, p' \in [0, 1] \qquad (7.5)$$

where $C_Y(\mathbf{h})$ is the correlogram of the standard RF $Y(\mathbf{u})$.

4. If two RVs $Y(\mathbf{u})$ and $Y(\mathbf{u}')$ are uncorrelated, i.e., if $\mathrm{Cov}\{Y(\mathbf{u}), Y(\mathbf{u}')\} = 0$, they are also independent.

---

[1] Because each RV $Y(\mathbf{u})$ relates to the same attribute at a particular point or location $\mathbf{u}$, the terms *one-point*, *two-point* and *multiple-point* are used hereafter; see related discussion in section 3.2.2. The term *multivariate* is reserved for cases involving several attributes of different types.

5. All conditional distributions of any subset of the RF $Y(\mathbf{u})$, given realizations of any other subsets of it, are (multiple-point) normal. In particular, the conditional distribution of the single variable $Y(\mathbf{u})$ given the $n(\mathbf{u})$ data $y(\mathbf{u}_\alpha)$ is normal and fully characterized by its two parameters, mean and variance, which are the conditional mean and conditional variance of the RV $Y(\mathbf{u})$ given the information (n):

$$G(\mathbf{u}; y|(n)) = G\left(\frac{y - E\{Y(\mathbf{u})|(n)\}}{\sqrt{\text{Var}\{Y(\mathbf{u})|(n)\}}}\right) \tag{7.6}$$

where $G(.)$ is the standard normal cdf (Abramovitz and Stegun, 1972, p. 932).

Under the multiGaussian model, the mean and variance of the ccdf at location $\mathbf{u}$ are identical to the simple kriging estimate $y_{SK}^*(\mathbf{u})$ and simple kriging variance $\sigma_{SK}^2(\mathbf{u})$ obtained from the $n(\mathbf{u})$ data $y(\mathbf{u}_\alpha)$ (Journel and Huijbregts, 1978, p. 566). The ccdf is then modeled as

$$[G(\mathbf{u}; y|(n))]_{SK}^* = G\left(\frac{y - y_{SK}^*(\mathbf{u})}{\sigma_{SK}(\mathbf{u})}\right) \tag{7.7}$$

with

$$y_{SK}^*(\mathbf{u}) = m(\mathbf{u}) + \sum_{\alpha=1}^{n(\mathbf{u})} \lambda_\alpha^{SK}(\mathbf{u})\,[y(\mathbf{u}_\alpha) - m(\mathbf{u}_\alpha)]$$

$$\sigma_{SK}^2(\mathbf{u}) = C_R(0) - \sum_{\alpha=1}^{n(\mathbf{u})} \lambda_\alpha^{SK}(\mathbf{u})\, C_R(\mathbf{u}_\alpha - \mathbf{u})$$

where the SK weights $\lambda_\alpha^{SK}(\mathbf{u})$ are provided by an SK system of type (5.10). The expected values of the $(n(\mathbf{u}) + 1)$ RVs $Y(\mathbf{u})$, $Y(\mathbf{u}_\alpha)$ are assumed known but not necessarily identical (non-stationary RF); see subsequent discussion in section 7.2.4.

Several attributes can be accounted for by considering a multiGaussian multivariate RF $\{Y_i(\mathbf{u}), i = 1, \ldots, N_v; \forall \mathbf{u} \in \mathcal{A}\}$. The parameters of the ccdf are then the simple cokriging estimate of the primary attribute at location $\mathbf{u}$ and the associated simple cokriging variance introduced in section 6.2.2.

7.2.2 Normal score transform

The multiGaussian (MG) approach is very convenient: the inference of the conditional cdf reduces to solving a simple kriging system at location $\mathbf{u}$. The trade-off cost is the assumption that data follow a multiGaussian distribution, which implies first that the one-point distribution of data (sample histogram) is normal.

Many variables in the earth sciences show an asymmetric distribution with a few very large values (positive skewness). Thus, the multiGaussian approach starts with an identification of the standard normal distribution and involves the following steps (Deutsch and Journel, 1992a, p. 138):

1. The original z-data are first transformed into y-values with a standard normal histogram. Such a transform is referred to as a *normal score transform*, and the y-values $y(\mathbf{u}_\alpha) = \phi(z(\mathbf{u}_\alpha))$ are called normal scores.

2. Provided the biGaussian assumption is not invalidated (see section 7.?.3), the multiGaussian model is applied to the normal scores, allowing the derivation of the Gaussian conditional cdf at any unsampled location $\mathbf{u}$:

$$G(\mathbf{u}; y|(n)) \;=\; \text{Prob}\,\{Y(\mathbf{u}) \le y|(n)\}$$

3. The conditional cdf of the original variable is then retrieved as

$$
\begin{aligned}
F(\mathbf{u}; z|(n)) \;&=\; \text{Prob}\,\{Z(\mathbf{u}) \le z|(n)\} \\
&=\; \text{Prob}\,\{Y(\mathbf{u}) \le y|(n)\} \\
&=\; G(\mathbf{u}; \phi(z)|(n))
\end{aligned}
$$

under the condition that the transform function $\phi(.)$ is monotonic increasing.

The normal score transform function $\phi(.)$ can be derived through a graphical correspondence between the cumulative one-point distributions of the original and standard normal variables; see Figure 7.3 and Journel and Huijbregts (1978), p. 478. Let $F(z)$ and $G(y)$ be the stationary one-point cdfs of the original RF $Z(\mathbf{u})$ and the standard normal RF $Y(\mathbf{u})$:

$$
\begin{aligned}
F(z) \;&=\; \text{Prob}\,\{Z(\mathbf{u}) \le z\} \\
G(y) \;&=\; \text{Prob}\,\{Y(\mathbf{u}) \le y\}
\end{aligned}
$$

The transform that allows one to go from a RF $Z(\mathbf{u})$ with any cdf $F(z)$ to a RF $Y(\mathbf{u})$ with standard Gaussian cdf $G(y)$ is depicted by the arrows in Figure 7.3 and is written as

$$Y(\mathbf{u}) \;=\; \phi(Z(\mathbf{u})) \;=\; G^{-1}\,[F(Z(\mathbf{u}))] \tag{7.8}$$

where $G^{-1}(.)$ is the inverse Gaussian cdf or quantile function of the RF $Y(\mathbf{u})$.

The normal score transform can be seen as a correspondence table between equal p-quantiles z_p and y_p of the two distributions. In other words, z_p and y_p correspond to the same cumulative probability p:

$$G(y_p) \;=\; G\left[G^{-1}(F(z_p))\right] \;=\; F(z_p) \;=\; p \qquad \forall\, p \in [0,1]$$

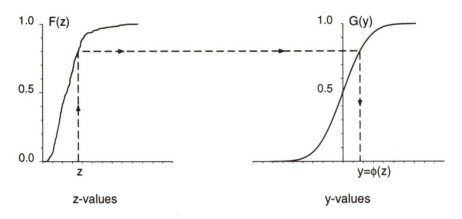

Figure 7.3: Graphical procedure for transforming the cumulative distribution of original z-values into the standard normal distribution of y-values called normal scores.

In practice, the normal score transform proceeds in three steps:

1. The n original data $z(\mathbf{u}_\alpha)$ are first ranked in ascending order:

$$[z(\mathbf{u}_{\alpha'})]^{(1)} \leq \ \cdots \ \leq [z(\mathbf{u}_\alpha)]^{(k)} \leq \ \cdots \ \leq [z(\mathbf{u}_{\alpha''})]^{(n)}$$

where the superscript k is the rank of datum $z(\mathbf{u}_\alpha)$ among all n data. Two or more data values may be identical, e.g., zero-valued data corresponding to concentrations below the detection limit. Since the normal score transform must be monotonic, ties in z-values must be broken. Such untying or despiking can be done randomly or, better, according to the local averages of the data surrounding each tied value (Verly, 1986). Tied values in high-valued areas would then get larger ranks than those located in low-valued areas.

2. The sample cumulative frequency of the datum $z(\mathbf{u}_\alpha)$ with rank k is then computed as

   - $p_k^* = k/n - 0.5/n$ if all data $z(\mathbf{u}_\alpha)$ receive the same weight $1/n$; that is, if the sample histogram is deemed representative of the study area.
   - $p_k^* = \sum_{i=1}^{k} w_i - 0.5 w_k \in [0,1]$ if declustering weights $w_i \neq 1/n$ are applied to data; refer to the discussion on declustering in section 4.1.2.

3. The normal score transform of the z-datum with rank k is matched to the p_k^*-quantile of the standard normal cdf:

$$y(\mathbf{u}_\alpha) \ = \ G^{-1}\left[F^*(z(\mathbf{u}_\alpha))\right] \ = \ G^{-1}(p_k^*) \tag{7.9}$$

The probability of being smaller than the minimum z-datum and the probability of being larger than the maximum z-datum are non-zero: $p_1^* \neq 0$ and $(1 - p_n^*) \neq 0$. Thus, the transform (7.9) never yields infinite normal scores. Other implementation details of the normal score transform are given in Deutsch and Journel (1992a, p. 210).

Example

Consider the normal score transform of the distribution of 10 Cd concentrations along the NE-SW transect. The three columns of Table 7.1 correspond to the three steps of the transformation, as follows:

1. The ten Cd concentrations are ranked from smallest to largest.

2. The ten cumulative frequencies p_k^* are computed as $(0.1k - 0.05)$ since data locations are evenly distributed along the transect (no declustering weight used).

3. The ten normal scores are computed according to relation (7.9).

The same procedure is applied to the sample set of 259 Cd concentrations. Figure 7.4 (top graphs) shows the histogram of the 259 Cd concentrations and the corresponding normal probability plot. A normal probability plot is a graph of the cumulative frequency, as introduced in section 2.1.2, with the ordinate scaled so that a normal distribution appears as a straight line. The convex shape of that curve indicates departure of the distribution of Cd data from the normal distribution model. Figure 7.4 (second row) shows the histogram of 259 normal scores and their cumulative distribution that now plots as a straight line on the normal probability graph.

Table 7.1: Cumulative frequency table of the ten Cd concentrations along the NE-SW transect and the corresponding normal scores.

Cd data	Cumul. freq.	Normal scores
0.135	0.05	-1.6449
0.275	0.15	-1.0364
0.570	0.25	-0.6745
0.910	0.35	-0.3853
1.310	0.45	-0.1257
1.325	0.55	0.1257
1.485	0.65	0.3853
2.120	0.75	0.6745
2.805	0.85	1.0364
3.925	0.95	1.6449

Original Cd data

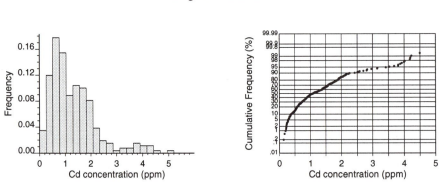

After normal score transform

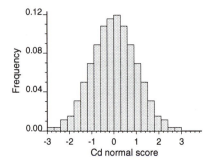
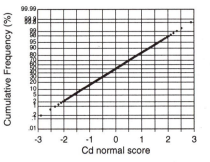

After logarithmic transform

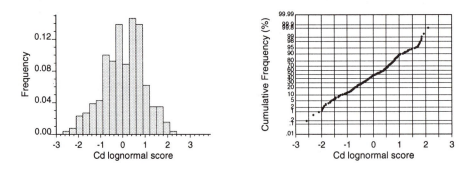

Figure 7.4: Histogram and normal probability plot of 259 Cd concentrations before (first row) and after normal score transform (second row) or logarithmic transform (third row).

A common approach for "normalizing" a positively skewed sample histogram is to take the logarithms of the original data, i.e., transform $y(\mathbf{u}_\alpha)$ as $\ln z(\mathbf{u}_\alpha)$. Figure 7.4 (bottom graphs) shows the histogram and normal probability plot of 259 Cd concentrations after such a transformation. The cumulative frequencies plot in a fairly straight line except at the two extremes of the distribution. Although of small magnitude, these deviations should not be disregarded since tail behaviors are generally critical. Thus, the normal score transform that ensures perfect reproduction of the normal distribution is the preferred transformation. Another shortcoming of the log transform is that it applies only to strictly positive variables (without zero-valued data).

7.2.3 Checking the multiGaussian assumption

The normality of the one-point cdf is a necessary but not sufficient condition to ensure that a RF model is multivariate Gaussian. One must also check the normality of the two-point,..., N-point cdfs of the normal score data.

Checking for two-point Gaussian distribution

Ideally one should define a transform similar to the one-point normal score transform, which would ensure that all two-point cdfs of the transformed RF $Y(\mathbf{u})$ are Gaussian. Unfortunately, such a transform is very difficult to determine in practice since there are as many two-point cdfs as there are different lag vectors $\mathbf{h}$ separating the two RVs $Y(\mathbf{u})$ and $Y(\mathbf{u}+\mathbf{h})$. Hence one should check whether the previously obtained data $y(\mathbf{u}_\alpha)$ are also reasonably bivariate Gaussian. If they are, then the multivariate Gaussian model can be adopted (it has been checked to the two-point level); if they are not, another model should be considered.

There are several ways to check that the two-point distribution of data $\{y(\mathbf{u}_\alpha), \alpha = 1,\ldots,n\}$ is normal. One method consists of verifying that the experimental two-point cdf values of any set of data pairs separated by the same vector $\mathbf{h}$ $\{(y(\mathbf{u}_\alpha), y(\mathbf{u}_\alpha + \mathbf{h})), \alpha = 1,\ldots,N(\mathbf{h})\}$ match the theoretical model given by the analytical expression (7.5). In practice, only the case $y_p = y_{p'}$ is considered, leading to the following expression for the two-point Gaussian cdf:

$$G(\mathbf{h}; y_p) = \text{Prob}\{Y(\mathbf{u}) \leq y_p, Y(\mathbf{u}+\mathbf{h}) \leq y_p\}$$

$$= p^2 + \frac{1}{2\pi} \int_0^{arcsin C_Y(\mathbf{h})} \exp\left[\frac{-y_p^2}{1 + sin\theta}\right] d\theta \qquad (7.10)$$

The check proceeds as follows:

1. The semivariogram $\gamma_Y(\mathbf{h})$ of the normal score data $y(\mathbf{u}_\alpha)$ is computed and modeled. The corresponding covariance model $C_Y(\mathbf{h})$ is then obtained as $1 - \gamma_Y(\mathbf{h})$.

2. For a series of p-quantile values y_p, the standard Gaussian two-point cdf values $G(\mathbf{h}; y_p)$ are computed using relation (7.10). The corresponding indicator semivariogram $\gamma_I(\mathbf{h}; y_p)$ for the p-quantile value y_p is then deduced as

$$\gamma_I(\mathbf{h}; y_p) \;=\; p \,-\, G(\mathbf{h}; y_p) \qquad (7.11)$$

where the indicator RF is defined as $I(\mathbf{u}; y_p) = 1$ if $Y(\mathbf{u}) \le y_p$, and equals zero otherwise. Indeed the two-point cdf $G(\mathbf{h}; y_p)$ is the non-centered indicator covariance for the threshold y_p.

3. For the same p-quantile values, the experimental indicator semivariograms of normal score data are computed as

$$\hat{\gamma}_I(\mathbf{h}; y_p) \;=\; \frac{1}{2N(\mathbf{h})} \sum_{\alpha=1}^{N(\mathbf{h})} [i(\mathbf{u}_\alpha; y_p) - i(\mathbf{u}_\alpha + \mathbf{h}; y_p)]^2$$

4. Experimental and Gaussian model-induced indicator semivariograms are compared graphically. Based on the quality of the fit, the user *decides* whether to reject the assumption of two-point normality. A goodness-of-fit criterion would consist of comparing the magnitude of deviations between experimental and model indicator semivariograms with those observed for simulated values that are known to be biGaussian.

Properties of indicator semivariograms
Expressions (7.10) and (7.11) point out two critical spatial properties of the multiGaussian RF model (Journel and Posa, 1990):

- The Gaussian model does not allow for any significant spatial correlation of extremely large or small values, a property known as *destructuration effect* and associated to the maximum entropy property of the Gaussian RF model. Indeed when the cumulative frequency p tends toward zero or one, $y_p^2 \to +\infty$, and the two-point cdf (7.10) tends toward the product p^2 of the two marginal probabilities (independence case). Hence, the indicator correlogram $\rho_I(\mathbf{h}; y_p)$ tends toward zero, and the indicator semivariogram $\gamma_I(\mathbf{h}; y_p)$ tends toward its sill $p(1 - p)$:

$$\left. \begin{array}{c} \rho_I(\mathbf{h}; y_p) \;\to\; 0 \\ \gamma_I(\mathbf{h}; y_p) \;\to\; p(1-p) \end{array} \right\} \text{ as } |y_p| \text{ increases}$$

for any vector $\mathbf{h}$. Figure 7.5 shows an example of the destructuration effect for a Gaussian RF $Y(\mathbf{u})$ with a spherical semivariogram model $\gamma(\mathbf{h})$ with a unit range: the standardized indicator semivariograms reach their unit sill faster for extreme threshold values.

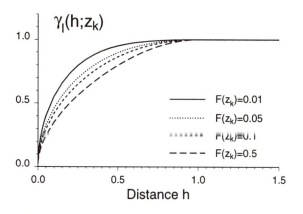

Figure 7.5: Standardized indicator semivariogram models implied by the multi-Gaussian model at four p-quantile values ranging from the median ($p = 0.5$) to the first percentile ($p = 0.01$). The semivariogram model of normal scores is spherical with a unit range. Note how the curves change shape and approach their sill more rapidly as the p-quantile value decreases.

- The pattern of indicator spatial correlation is symmetric with regard to the median threshold value $y_{0.5}$. Indeed, the median of the standard Gaussian distribution is zero, and p-quantiles are such that $y_p = -y_{p'}$ $\forall\, p' = 1 - p$, hence $y_p^2 = y_{p'}^2$. The indicator semivariogram models of any pair of symmetrical p-quantiles are thus identical:

$$\gamma_I(\mathbf{h}; y_p) \;=\; \gamma_I(\mathbf{h}; y_{p'}) \qquad \forall\, p' = 1 - p$$

For example, under the multiGaussian model the indicators of the 10% smallest y-values and those of the 10% largest y-values have a similar pattern of spatial variability, $\gamma_I(\mathbf{h}; y_{0.1}) \;=\; \gamma_I(\mathbf{h}; y_{0.9})$—a very specific property not necessarily shared by the actual data.

In the earth sciences, low entropy patterns, such as connected strings or patches of extreme values, often correspond to hazardous features that are worth specific attention. For example, strings of large permeability values can represent leakage conduits hazardous for a nuclear repository. Similarly, connected clusters of large metal concentrations above the tolerable maximum are critical for assessing the risk of soil pollution. Hence, and notwithstanding its analytical simplicity, the multiGaussian RF model may be inappropriate whenever the structural analysis or qualitative information indicates that extreme values are spatially correlated. Even in the absence of information about the connectivity of extreme values, the Gaussian model is not conservative in the sense that its maximum entropy character leads one to understate the potential for hazard (Gómez-Hernández and Wen, 1994). The more flexible but also more demanding indicator approach introduced in section 7.3 allows one to account for spatial correlation of extreme values.

Example

Figure 7.6 (top graph) shows the experimental semivariogram of Cd normal scores with the model fitted. Given that model, Gaussian-based indicator semivariogram models are deduced using expression (7.11) for the nine deciles of the sample distribution. The experimental indicator semivariograms (black dots) are seen to deviate from the Gaussian-induced models (solid line) for small threshold values. As already noticed in Figure 2.19 (page 45), small Cd concentrations are more connected in space than large concentrations. Thus,

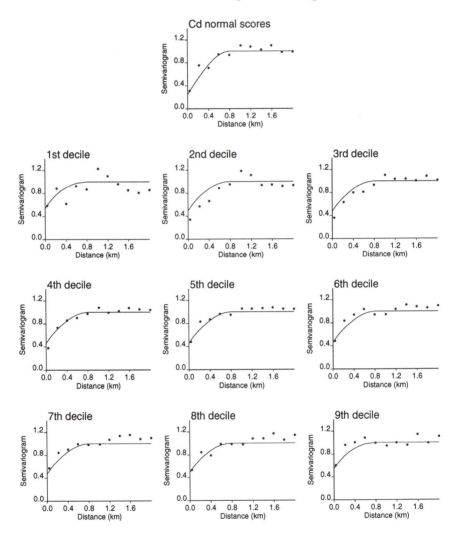

Figure 7.6: Experimental semivariogram of Cd normal scores (top graph) and experimental standardized indicator semivariograms for the nine deciles of the sample cumulative distribution. The solid lines depict the indicator semivariograms deduced from relation (7.11) for the multiGaussian model.

the pattern of correlation is not symmetric about the median and does not show the Gaussian-type destructuration of small values.

Checking for multiple-point Gaussian distribution

The normality of the one-point and two-point cdfs are both necessary but not sufficient conditions to ensure that a multivariate Gaussian RF model is appropriate for modeling the spatial distribution of normal score data $y(\mathbf{u}_\alpha)$. One should also check the normality of the three-point,..., N-point cdfs. As for the two point cdf, a check consists of comparing experimental multiple point frequencies with their theoretical Gaussian values obtained using an expression of type (7.10). Although such an expression has been established analytically, the main difficulty resides in the inference of such experimental frequencies. For example, the inference of a three-point frequency such that

$$\text{Prob}\{Y(\mathbf{u}) \le y_p, Y(\mathbf{u}') \le y_p, Y(\mathbf{u}'') \le y_p\}$$

requires the availability of a series of triplet values with the same geometric configuration as the triplet $(\mathbf{u}, \mathbf{u}', \mathbf{u}'')$.

Non-regular gridding and data sparsity prevent us from computing sample statistics involving more than two locations at a time. Therefore, in practice, if indicator semivariograms or ancillary information do not invalidate the biGaussian assumption, the multiGaussian formalism is adopted.

7.2.4 Estimating the Gaussian ccdf parameters

Once the multiGaussian model is adopted, inference of the normal ccdf $G(\mathbf{u}; y|(n))$ reduces to estimating its two parameters (mean and variance) at any unsampled location $\mathbf{u}$. Consider the decomposition of the multiGaussian RF $Y(\mathbf{u})$ into a residual component $R(\mathbf{u})$ and a trend component $m_Y(\mathbf{u})$:

$$Y(\mathbf{u}) \;=\; m_Y(\mathbf{u}) \,+\, R(\mathbf{u})$$

The strict multiGaussian formalism requires strict stationarity, i.e., $m_Y(\mathbf{u}) = m \; \forall \; \mathbf{u} \in \mathcal{A}$, if only to be able to use the original sample histogram for normal score transform and back-transform. In practice, non-stationary behaviors are accounted for using algorithms other than simple kriging to estimate the mean of the Gaussian ccdf $G(\mathbf{u}; y|(n))$. For example, ordinary kriging allows one to account for significant changes in the local mean of normal scores over the study area. An alternative consists of modeling the trend component $m_Y(\mathbf{u})$ in one of two ways:

1. A low-order polynomial of the vector of coordinates, say, a two-dimensional linear trend:

$$m_Y(\mathbf{u}) = m_Y(x, y) = a_0(\mathbf{u}) + a_1(\mathbf{u})x + a_2(\mathbf{u})y$$

The mean of the Gaussian ccdf is then provided by a KT estimator of type (5.25).

2. A linear rescaling of a smoothly varying secondary variable $d(\mathbf{u})$:

$$m_Y(\mathbf{u}) = a_0(\mathbf{u}) + a_1(\mathbf{u})\, d(\mathbf{u})$$

where d-values are normally distributed with zero mean and unit variance (normal scores). The mean of the Gaussian ccdf is then provided by a KED estimator of type (6.4).

Beware that whatever the kriging algorithm (OK, KT, KED) considered, the variance of the ccdf must be identified with the theoretically correct simple kriging variance (Journel, 1980). An alternative to non-stationary kriging of the normal score y-data consists of detrending the original z-data prior to a normal score transformation of the corresponding z-residual data.

Example

Figure 7.7 (top graphs) shows the simple kriging estimate $y_{SK}^*(\mathbf{u})$ and variance $\sigma_{SK}^2(\mathbf{u})$ computed using the ten Cd normal scores given in Table 7.1 (page 269, third column). According to relation (7.7), the Gaussian conditional cdfs at locations $\mathbf{u}_1'$ to $\mathbf{u}_4'$ are modeled as

$$[G(\mathbf{u}_\beta'; y|(n))]_{SK}^* = G\left(\frac{y - y_{SK}^*(\mathbf{u}_\beta')}{\sigma_{SK}(\mathbf{u}_\beta')}\right) \qquad \beta = 1, \ldots, 4,$$

and are displayed at the bottom of Figure 7.7, which shows the following:

- The ccdf at the datum location $\mathbf{u}_3'$ is a unit-step function defined as

$$[G(\mathbf{u}_3'; y|(n))]_{SK}^* = \begin{cases} 1 & \text{if } y \geq y(\mathbf{u}_3') \\ 0 & \text{otherwise} \end{cases}$$

  Since the SK estimator is an exact interpolator, the ccdf mean at datum location $\mathbf{u}_3'$ is the normal score $y(\mathbf{u}_3')$ itself, and the ccdf variance is zero (no uncertainty).

- The ccdf at location $\mathbf{u}_4'$, which is beyond the correlation range of normal scores, is the standard normal cdf (marginal cdf):

$$[G(\mathbf{u}_4'; y|(n))]_{SK}^* = G(y)$$

  Since the information available is beyond the correlation range of the y-data, the global (location-independent) model of uncertainty is not updated. The SK estimate and variance are then equal to the stationary zero mean and unit variance (no reduction in uncertainty).

- The shape (steepness) of the Gaussian ccdf at locations $\mathbf{u}_1'$ and $\mathbf{u}_2'$ is intermediate between the unit-step function at $\mathbf{u}_3'$ and the marginal cdf at $\mathbf{u}_4'$. Indeed, the local uncertainty at these two locations, measured

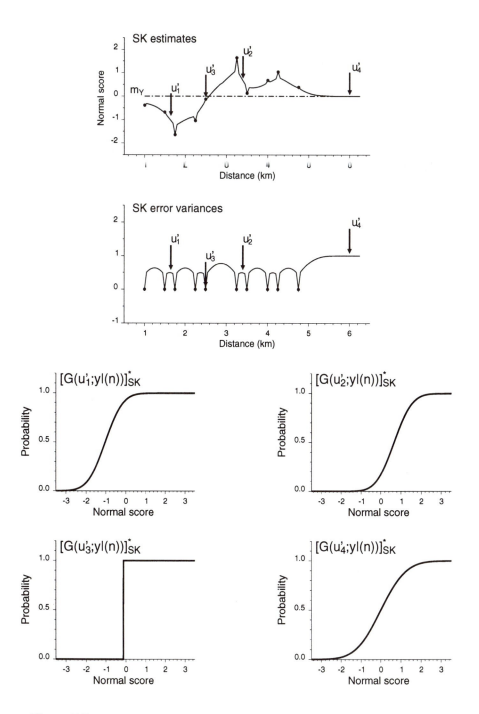

Figure 7.7: SK estimates of Cd normal scores and the corresponding error variances. These values are used as the mean and variance of the Gaussian conditional cdf at locations u_1' to u_4'.

by the SK variance, is larger than at datum location $\mathbf{u}'_3$ but smaller than at $\mathbf{u}'_4$, which is beyond the correlation range:

$$\sigma^2_{SK}(\mathbf{u}'_3) = 0 \; < \; \sigma^2_{SK}(\mathbf{u}'_1) = \sigma^2_{SK}(\mathbf{u}'_2) = 0.8 \; < \; \sigma^2_{SK}(\mathbf{u}'_4) = 1$$

Because the kriging variance is independent of the data values, the uncertainty at both locations $\mathbf{u}'_1$ and $\mathbf{u}'_2$ with similar data configurations is modeled as similar. As discussed in section 7.1.1, the potential for error is, however, expected to be greater at location $\mathbf{u}'_2$, which is surrounded by one very large and one small normal score, than at $\mathbf{u}'_1$, which is surrounded by two consistently small normal scores.

7.2.5 Increasing the resolution of the sample cdf

The local uncertainty model is needed in the space of the original variable Z, not in the space of the normal score variable Y. Because the normal score transform $\phi(.)$ is monotonic increasing, the z-ccdf value at any threshold z' can be retrieved from the Gaussian ccdf in two steps:

1. The z-threshold z' is first transformed into a normal score or y-threshold y':

$$y' \; = \; \phi(z') \; = \; G^{-1}[F^*(z')] \tag{7.12}$$

where $G^{-1}(.)$ is the inverse Gaussian cdf and $F^*(.)$ is the original sample cdf.

2. The z-ccdf value $F(\mathbf{u}; z'|(n))$ is then read from the Gaussian ccdf using that y-threshold:

$$F(\mathbf{u}; z'|(n)) \; = \; G(\mathbf{u}; y'|(n)) \tag{7.13}$$

Consider, for example, the Gaussian ccdf at location $\mathbf{u}'_2$ along the NE-SW transect (Figure 7.8, right bottom graph). Using the transformation Table 7.1 also depicted at the top of Figure 7.8, the ten z-ccdf values corresponding to the ten original data values $z(\mathbf{u}_\alpha)$ are readily retrieved from the Gaussian ccdf as:

$$F(\mathbf{u}; z(\mathbf{u}_\alpha)|(n)) \; = \; G(\mathbf{u}; y(\mathbf{u}_\alpha)|(n))$$

where $y(\mathbf{u}_\alpha)$ is the normal score transform of $z(\mathbf{u}_\alpha)$; see the black dots in Figure 7.8 (left bottom graph).

In the example of Figure 7.8, it is important to know the conditional probability that the Cd concentration does not exceed the tolerable maximum 0.8 ppm, $F(\mathbf{u}; 0.8|(n))$. Inference of such a z-ccdf value from the Gaussian ccdf calls for the normal score threshold $\phi(0.8) = G^{-1}[F^*(0.8)]$, which is unknown since no original data value in Table 7.1 (first column) is exactly 0.8 ppm. Thus the resolution of the discrete sample cdf in Figure 7.8 (left top graph) must be increased to provide the marginal cumulative probability associated with the z-threshold 0.8 ppm necessary to determine the normal score threshold $\phi(0.8)$.

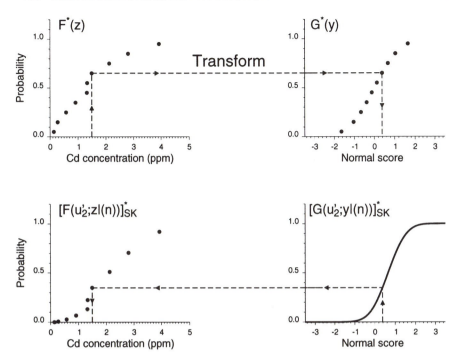

Figure 7.8: Graphical transform of ten Cd data values into ten normal scores (top graph). Accounting for that correspondence, the ccdf value of the original variable at each of the ten threshold Cd data values is identified with the Gaussian ccdf value at the corresponding normal score.

Piecewise interpolation/extrapolation

The first approach for increasing the resolution of any sample cdf consists of interpolating cdf values within each class of thresholds $(z_{k-1}, z_k]$ and extrapolating these cdf values beyond the smallest z-data value z_1 (lower tail) and the largest z-data value z_K (upper tail). Several interpolation and extrapolation models (Journel, 1987; Deutsch and Journel, 1992a, p. 131–135) can be used, depending on the part of the cdf that is modeled: the lower tail, middle classes $(z_{k-1}, z_k]$, or the upper tail.

Linear cdf model
A linear model is generally adopted for interpolation within classes of threshold values $(z_{k-1}, z_k]$. Such a linear model amounts to assuming a uniform distribution within that class, that is,

$$[F(z)]_{Lin.} = F^*(z_{k-1}) + \left[\frac{z - z_{k-1}}{z_k - z_{k-1}} \right] \cdot [F^*(z_k) - F^*(z_{k-1})]$$

$$\forall z \in (z_{k-1}, z_k] \qquad (7.14)$$

At class bounds, $[F(z_k)]_{Lin} = F^*(z_k)$ as it should.

Power cdf model
The linear model (7.14) is a particular case of the power model:

$$[F(z)]_{Pow.} = F^*(z_{k-1}) + \left[\frac{z - z_{k-1}}{z_k - z_{k-1}}\right]^{\omega} \cdot [F^*(z_k) - F^*(z_{k-1})]$$
$$\forall z \in (z_{k-1}, z_k] \qquad (7.15)$$

where the parameter (the power) ω is strictly positive, $\omega > 0$. Different interpolation models are obtained by varying ω (Figure 7.9):

- $\omega = 1$ corresponds to the linear model (uniform distribution).

- Distributions with $\omega < 1$ are positively skewed.

- Distributions with $\omega > 1$ are negatively skewed.

The positive (negative) skewness of the distribution increases as the parameter ω decreases (increases).

The power model can be used also to model the lower and upper tails of the distribution. A negatively skewed lower tail could be modeled as

$$[F(z)]_{Pow.} = \left[\frac{z - z_{min}}{z_1 - z_{min}}\right]^{\omega > 1} \cdot F^*(z_1) \quad \forall z \in (z_{min}, z_1]$$

where z_1 is the smallest z-data value and z_{min} is the minimum z-value fixed

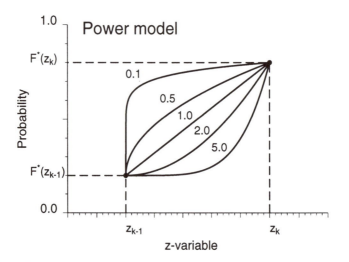

Figure 7.9: Power models for interpolating within-class cdf values. The shape of the distribution is controlled by the parameter ω: positive skewness for ω less than 1.0, uniform distribution for $\omega = 1$ (linear model), and negative skewness for ω greater than 1.0.

by the user. Conversely, a power model for a positively skewed upper tail
could be

$$[F(z)]_{Pow} = F^*(z_K) + \left[\frac{z - z_K}{z_{max} - z_K}\right]^{\omega < 1} \cdot [1 - F^*(z_K)]$$

$$\forall\, z \in (z_K, z_{max}]$$

where z_K is the largest z-data value and z_{max} is the maximum z-value fixed
by the user.

Hyperbolic cdf model
The power model for an upper tail calls for the sometimes arbitrary choice
of a maximum z-value. The hyperbolic model allows one to extrapolate the
upper tail of a positively skewed distribution toward an infinite upper bound.
The hyperbolic model is

$$[F(z)]_{Hyp.} = 1 - \frac{\lambda}{z^\omega} \quad \forall\, z > z_K \tag{7.16}$$

with the parameter $\omega \geq 1$ controling how fast the cdf model reaches its
limiting value 1: the smaller is ω, the longer is the tail of the distribution
(Figure 7.10). The parameter λ is such that the hyperbolic model (7.16)
identifies the sample cumulative frequency $F^*(z_K)$:

$$\lambda = z_K^\omega \cdot [1 - F^*(z_K)]$$

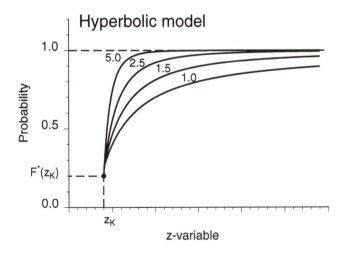

Figure 7.10: Hyperbolic models for extrapolating cdf values beyond the largest
threshold value z_K (upper tail). The parameter ω controls how fast the positively
skewed cdf model reaches the upper limit value 1: the smaller is ω, the longer is
the tail of the distribution.

When modeling an upper tail, one should avoid understating the probability of occurrence of very large values, which are often the most consequential for risk analysis. Practice has shown that a hyperbolic upper tail distribution with $\omega = 1.5$ yields acceptable results in a wide variety of applications.

Example
Figure 7.11 (left top graph) shows the class interpolation and tail extrapolation applied to the sample cdf of the original ten Cd concentrations along the NE-SW transect. The steps are as follows:

1. The lower tail is extrapolated toward a zero minimum using a negatively skewed power model with $\omega = 2.5$.

2. A linear interpolation is performed separately within each of the nine middle classes $(z_1, z_2]$ to $(z_9, z_{10}]$.

3. The upper tail is extrapolated toward an infinite upper bound using a hyperbolic model with $\omega = 1.5$.

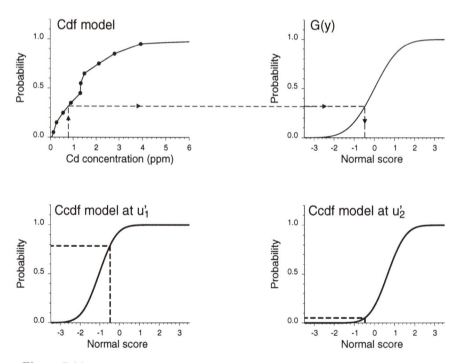

Figure 7.11: Interpolation and extrapolation models applied to the sample cumulative distribution of the ten Cd concentrations (right top graph). That cdf model is then used for a normal score transformation of the critical threshold 0.8 ppm into a normal score threshold depicted by the vertical dashed line on the Gaussian ccdfs at locations $\mathbf{u}_1'$ and $\mathbf{u}_2'$ (bottom graph).

Using that z-cdf model and the standard normal cdf shown in Figure 7.11 (right top graph), the z-threshold 0.8 ppm is transformed into a normal score threshold $y = -0.47$. That threshold is depicted by the vertical dashed line on the two Gaussian ccdfs at the bottom of Figure 7.11. The posterior probability of exceeding 0.8 ppm is 0.21 and 0.95, respectively, at locations $\mathbf{u}_1'$ and $\mathbf{u}_2'$. As mentioned in section 7.1, the probability of contamination is much larger at $\mathbf{u}_2'$ in the high-valued part of the transect than at $\mathbf{u}_1'$.

Smoothing the sample cdf

Because data are sparse, sample histograms generally show multiple sawtooth-like spikes, which would not be present if the entire population had been exhaustively sampled. Thus, a good alternative to using a series of interpolation and extrapolation models consists of smoothing the sample cdf. Such an approach allows one to increase the resolution of the sample z-cdf while smoothing out artificial fluctuations.

Traditional kernel-based smoothing (Silverman, 1986) does not allow the sample mean, variance, and quantile values to be honored simultaneously, and may lead to spurious results, such as negative quantiles for positive variables. Several algorithms have been recently developed to smooth a sample histogram while ensuring reproduction of sample statistics deemed representative of the underlying population. These smoothing algorithms are based on either quadratic programming (Xu and Journel, 1995) or simulated annealing (Deutsch, 1994b).

Example
A quadratic programming-based approach is applied to smoothing the cumulative distribution of the ten Cd concentrations along the the NE-SW transect (Figure 7.12). The resolution of the sample cdf is increased by considering 100 classes of probability values instead of ten classes for the original distribution.

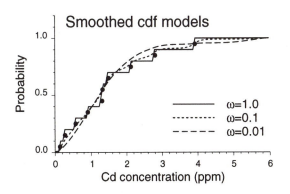

Figure 7.12: Cdf models provided by a smoothing interpolation algorithm using 100 classes of probability values. As ω decreases, the cdf model becomes increasingly smoother.

A weight ω allows a balance between smoothness and closeness to the original distribution: as ω decreases, the cdf model becomes increasingly smoother but deviates more from the original distribution. This particular smoothing algorithm requires prior determination of the minimum and maximum z-values.

7.3   The Indicator Approach

The congeniality of the multiGaussian approach presented in the previous section is balanced by several shortcomings:

1. Strong assumptions must be made about the multiple-point distribution of data. Unfortunately, the normality of three-point,..., multiple-point experimental cdfs cannot be checked in practice (see section 7.2.3).

2. Under the multiGaussian model, extremely large and small values are spatially uncorrelated, an assumption often invalidated in practice or a non-conservative model for applications where connectivity of extreme values are dangerous features.

3. The variance of the conditional cdf in the normal space depends only on the data configuration, not on the data themselves (homoscedasticity).

When sample indicator semivariograms do not support the biGaussian assumption or when critical information cannot be incorporated using the multi-Gaussian framework, a non-parametric approach should be adopted. That second class of algorithms does not assume any particular shape or analytical expression for the conditional distributions. Instead, the function $F(\mathbf{u}; z|(n))$ is modeled through a series of K threshold values z_k discretizing the range of variation of z:

$$F(\mathbf{u}; z_k|(n)) \;=\; \text{Prob} \left\{ Z(\mathbf{u}) \leq z_k|(n) \right\} \quad k = 1, \dots, K \qquad (7.17)$$

Like the cdf values in section 7.2.5, the K ccdf values are then interpolated within each class $(z_k, z_{k+1}]$ and extrapolated beyond the two extreme threshold values z_1 and z_K.

The indicator approach is based on the interpretation of the conditional probability (7.17) as the conditional expectation of an indicator RV $I(\mathbf{u}; z_k)$ given the information (n):

$$F(\mathbf{u}; z_k|(n)) \;=\; \text{E} \left\{ I(\mathbf{u}; z_k)|(n) \right\} \qquad (7.18)$$

with $I(\mathbf{u}; z_k)=1$ if $Z(\mathbf{u}) \leq z_k$ and zero otherwise.

According to the projection theorem (Luenberger, 1969, p. 49), the least-squares (kriging) estimate of the indicator $i(\mathbf{u}; z_k)$ is also the least-squares estimate of its conditional expectation. Thus, the ccdf value $F(\mathbf{u}; z_k|(n))$ can be obtained by (co)kriging the unknown indicator $i(\mathbf{u}; z_k)$ using indicator transforms of the neighboring information.

7.3.1   Indicator coding of information

The indicator approach starts with a selection of the number of thresholds and their values. To alleviate computation and inference efforts and reduce the occurrence of order relation problems (see discussion in section 7.3.4), the number of thresholds should rarely exceed 15. On the other hand, the number of thresholds should not be smaller than five to provide a reasonable discretization of the local distribution. The set of K threshold values is typically chosen such that the range of z-values is split into $(K+1)$ classes of approximately equal frequency, e.g., the nine deciles of the sample cumulative distribution. Note the following guidelines:

- It is good practice to identify the threshold values with any critical z-value, such as the concentration above which remedial action should be taken. Thus, the ccdf value at this threshold will not have to be interpolated or extrapolated later.

- More threshold values should be chosen within the part of the distribution that is of greater interest than the rest, e.g., the lower or upper tail.

- Sample indicator semivariograms at extreme threshold values are not well defined because they depend on the spatial distribution of a few pairs of indicator data (see section 2.3.4). Therefore, threshold values beyond the first or ninth decile of the sample cdf may be inappropriate. Indicator semivariogram models at extreme threshold values could, however, be built by extrapolating parameters of the indicator semivariograms available at intermediate thresholds or from soft structural information (see discussion in section 7.3.5).

Once the K threshold values have been chosen, each piece of information (e.g., metal concentration, soil, or rock type) is coded into a vector of K cumulative probabilities of the type

$$\text{Prob}\,\{Z(\mathbf{u}) \le z_k | \text{ specific local information at } \mathbf{u}\} \quad k = 1, \dots, K \quad (7.19)$$

The discrete cdf (7.19) represents the local information about the z-value at $\mathbf{u}$ prior to any correction or updating based on neighboring data, hence the term *local prior*. Different types of local prior cdfs can be distinguished, depending on the nature of the local information available.

Hard data

A hard datum $z(\mathbf{u}_\alpha)$ is a precise measurement of the attribute of interest. There is no uncertainty at the datum location $\mathbf{u}_\alpha$, hence the local prior probabilities are binary (hard) indicator data defined as

$$i(\mathbf{u}_\alpha; z_k) = \begin{cases} 1 & \text{if } z(\mathbf{u}_\alpha) \le z_k \\ 0 & \text{otherwise} \end{cases} \quad k = 1, \dots, K \quad (7.20)$$

Consider the indicator coding of the ten Cd concentrations shown in Figure 7.13 (top graph). The horizontal dashed lines depict the four threshold values retained, $z_k = 0.80$, 1.38, 1.88, and 2.26 ppm, corresponding to the tolerable maximum and three upper quantiles ($p = 0.6$, 0.8, and 0.9) of the distribution of 259 Cd concentrations. Indicator coding yields a vector of four indicator values at each datum location (Figure 7.13, bottom graph). For example, the vector of hard indicator data related to the measurement $z(\mathbf{u}_5) = 1.31$ ppm is (fifth column of bottom Figure 7.13):

$$\begin{bmatrix} 1 \\ 1 \\ 1 \\ 0 \end{bmatrix} = \begin{bmatrix} i(\mathbf{u}_5; 2.26) \\ i(\mathbf{u}_5; 1.88) \\ i(\mathbf{u}_5; 1.38) \\ i(\mathbf{u}_5; 0.80) \end{bmatrix} = \begin{bmatrix} \text{Prob}\,\{Z(\mathbf{u}_5) \le 2.26 | z(\mathbf{u}_5) = 1.31\} \\ \text{Prob}\,\{Z(\mathbf{u}_5) \le 1.88 | z(\mathbf{u}_5) = 1.31\} \\ \text{Prob}\,\{Z(\mathbf{u}_5) \le 1.38 | z(\mathbf{u}_5) = 1.31\} \\ \text{Prob}\,\{Z(\mathbf{u}_5) \le 0.80 | z(\mathbf{u}_5) = 1.31\} \end{bmatrix}$$

$$(7.21)$$

The prior probability that the Cd concentration at $\mathbf{u}_5$ is no greater than 0.8 ppm is zero because the measured concentration is 1.31 ppm. That probability is 1 for the three larger threshold values 1.38, 1.88, and 2.26 ppm.

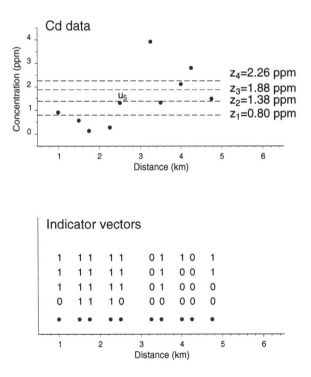

Figure 7.13: Coding of ten Cd data values into ten vectors of four indicators of non-exceedence of threshold values depicted by the four horizontal dashed lines.

Constraint intervals

There is always some uncertainty attached to any measurement $z(\mathbf{u}_\alpha)$ because of measurement errors. The coding (7.20) into hard indicator data (0 or 1 only) amounts to treating that error as negligible. When imprecision of measurement cannot be ignored, the information at location $\mathbf{u}_\alpha$ can be modeled into an interval of possible values for z, $z(\mathbf{u}_\alpha) \in (a_\alpha, b_\alpha]$, equivalent to $a_\alpha < z(\mathbf{u}_\alpha) \leq b_\alpha$. Such interval-type data are typically provided by inexpensive measurement devices, such as colorimetric papers to evaluate pollution levels. The prior cdf is then modeled as an incomplete vector of hard indicator data:

$$i(\mathbf{u}_\alpha; z_k) = \begin{cases} 1 & \text{if } b_\alpha \leq z_k \\ \text{undefined (missing)} & \text{if } z_k \in (a_\alpha, b_\alpha] \\ 0 & \text{if } a_\alpha > z_k \end{cases} \quad k = 1, \ldots, K \quad (7.22)$$

Consider, for example, the constraint interval $(a_0, b_0] = (1.0, 2.0]$, which indicates that the Cd concentration at location $\mathbf{u}_0$ is larger than 1.0 and smaller than or equal to 2.0 ppm. The vector of prior probabilities corresponding to the four above threshold values is

$$\begin{bmatrix} 1 \\ ? \\ ? \\ 0 \end{bmatrix} = \begin{bmatrix} \text{Prob}\{Z(\mathbf{u}_0) \leq 2.26 | z(\mathbf{u}_0) \in (1.0, 2.0]\} \\ \text{Prob}\{Z(\mathbf{u}_0) \leq 1.88 | z(\mathbf{u}_0) \in (1.0, 2.0]\} \\ \text{Prob}\{Z(\mathbf{u}_0) \leq 1.38 | z(\mathbf{u}_0) \in (1.0, 2.0]\} \\ \text{Prob}\{Z(\mathbf{u}_0) \leq 0.80 | z(\mathbf{u}_0) \in (1.0, 2.0]\} \end{bmatrix} \quad (7.23)$$

The two missing indicator data reflect the uncertainty about whether the Cd concentration exceeds 1.38 and 1.88 ppm, respectively.

Soft categorical data

The local information need not relate directly to the attribute of interest. For example, one may know that a particular state s_l of the categorical attribute s, say, a particular facies or soil type, prevails at $\mathbf{u}'_\alpha$. From the n' locations where both z and s are known, the proportions of z-data belonging to that state s_l while not exceeding any particular threshold value z_k can be computed as

$$F^*(z_k | s_l) = \frac{1}{\sum_{\alpha=1}^{n'} i(\mathbf{u}'_\alpha; s_l)} \sum_{\alpha=1}^{n'} i(\mathbf{u}'_\alpha; z_k) \cdot i(\mathbf{u}'_\alpha; s_l) \quad k = 1, \ldots, K$$

where $i(\mathbf{u}'_\alpha; s_l)$ is the indicator of occurrence of the state s_l; that is, $i(\mathbf{u}'_\alpha; s_l) = 1$ if $s(\mathbf{u}'_\alpha) = s_l$ and zero otherwise.

The set of K local prior probabilities at location $\mathbf{u}'_\alpha$ are then identified with the K sample conditional frequencies $F^*(z_k | s_l)$ corresponding to the soil type $s(\mathbf{u}'_\alpha) = s_l$ at $\mathbf{u}'_\alpha$:

$$y(\mathbf{u}'_\alpha; z_k) = F^*(z_k | s_l) \in [0, 1] \quad k = 1, \ldots, K \quad (7.24)$$

Whereas the hard indicator data $i(\mathbf{u}_\alpha; z_k)$ are valued 0 or 1 (no uncertainty), the data $y(\mathbf{u}'_\alpha; z_k)$ are valued between 0 and 1, thereby expressing the uncertainty about the actual z-value at $\mathbf{u}'_\alpha$. The information $y(\mathbf{u}'_\alpha; z_k)$ is a "soft indicator" of the continuous value $z(\mathbf{u}'_\alpha)$ being no greater than the cutoff z_k.

Let L be the number of states s_l of the categorical attribute s. When the set of calibration data, of size n', is small in comparison to the number of frequency classes, $K \times L$, the resulting local prior cdfs of type (7.24) might be erratic and might need to be smoothed or interpolated (Deutsch, 1994b; Xu and Journel, 1995). Another solution consists of importing such calibration from a better sampled area where the same attributes are measured.

Suppose that soft information along the NE-SW transect relates to the profile of two rock categories shown in Figure 7.14 (top graph). The category s_1 corresponds to Argovian rocks, whereas the category s_2 includes the four other rock types. Figure 7.14 (middle graph) shows, for each category, the cumulative distribution of Cd data as derived from the sample data set ($n' = 259$). The calibration step amounts to computing the proportions of Cd data not exceeding the threshold values depicted by the vertical dashed lines. The profile of rock categories is then converted into four profiles of local prior probabilities, one for each threshold (Figure 7.14, bottom graph). For example, the vector of local prior probabilities at any location belonging to rock category s_1, say, at $\mathbf{u}'_3$, is

$$
\begin{bmatrix} 0.85 \\ 0.83 \\ 0.74 \\ 0.66 \end{bmatrix}
=
\begin{bmatrix} \text{Prob}\,\{Z(\mathbf{u}'_3) \leq 2.26 | s(\mathbf{u}'_3) = s_1\} \\ \text{Prob}\,\{Z(\mathbf{u}'_3) \leq 1.88 | s(\mathbf{u}'_3) = s_1\} \\ \text{Prob}\,\{Z(\mathbf{u}'_3) \leq 1.38 | s(\mathbf{u}'_3) = s_1\} \\ \text{Prob}\,\{Z(\mathbf{u}'_3) \leq 0.80 | s(\mathbf{u}'_3) = s_1\} \end{bmatrix}
$$

Similarly, the local prior information at any location belonging to rock category s_2, say, at $\mathbf{u}_0$, is

$$
\begin{bmatrix} 0.91 \\ 0.79 \\ 0.53 \\ 0.26 \end{bmatrix}
=
\begin{bmatrix} \text{Prob}\,\{Z(\mathbf{u}_0) \leq 2.26 | s(\mathbf{u}_0) = s_2\} \\ \text{Prob}\,\{Z(\mathbf{u}_0) \leq 1.88 | s(\mathbf{u}_0) = s_2\} \\ \text{Prob}\,\{Z(\mathbf{u}_0) \leq 1.38 | s(\mathbf{u}_0) = s_2\} \\ \text{Prob}\,\{Z(\mathbf{u}_0) \leq 0.80 | s(\mathbf{u}_0) = s_2\} \end{bmatrix}
$$

Prior to any updating based on neighboring data, the probability of exceeding the critical threshold 0.8 ppm is about twice as large at $\mathbf{u}_0$ in category s_2 as it is at $\mathbf{u}'_3$ on Argovian rocks.

Soft continuous data

Soft information about the z-value at a location $\mathbf{u}'_\alpha$ may also originate from a related continuous attribute v, such as Ni concentration. The indicator coding of a soft datum $v(\mathbf{u}'_\alpha)$ proceeds as follows:

1. The range of variation of the secondary attribute v is discretized into L classes $(v_{l-1}, v_l]$.

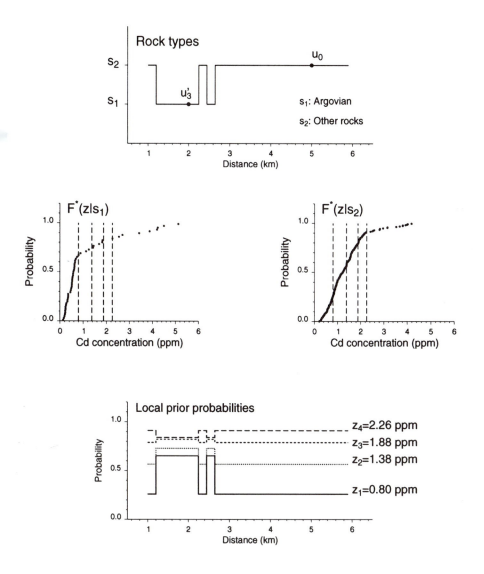

Figure 7.14: Coding of the profile of two rock categories into four profiles of local prior probabilities of not exceeding Cd threshold values. The prior probabilities at a location u belonging to rock category s_l are inferred from the cumulative distribution of Cd data belonging to that category (middle graph). The four threshold values are depicted by the vertical dashed lines.

2. For each class of v-values $(v_{l-1}, v_l]$, the proportion of z-data not exceeding a particular threshold value z_k is computed as

$$F^*(z_k | v_l) = \frac{1}{\sum_{\alpha=1}^{n'} i(\mathbf{u}'_\alpha; v_l)} \sum_{\alpha=1}^{n'} i(\mathbf{u}'_\alpha; z_k) \cdot i(\mathbf{u}'_\alpha; v_l)$$

where n' is the number of locations $\mathbf{u}'_\alpha$ where both primary and secondary attributes are known. The indicator variable $i(\mathbf{u}'_\alpha; v_l)$ is defined as

$$i(\mathbf{u}'_\alpha; v_l) = \begin{cases} 1 & \text{if } v(\mathbf{u}'_\alpha) \in (v_{l-1}, v_l] \\ 0 & \text{otherwise} \end{cases}$$

As mentioned for soft categorical information, the conditional frequencies $F^*(z_k | v_l)$ can be smoothed prior to being used or they can be borrowed from another better sampled field.

3. The set of K local prior probabilities at $\mathbf{u}'_\alpha$ are then identified with the K sample conditional frequencies $F^*(z_k | v_l)$ corresponding to the prevailing class of v-values:

$$y(\mathbf{u}'_\alpha; z_k) = F^*(z_k | v_l) \in [0, 1] \qquad k = 1, \dots, K \qquad (7.25)$$

if $v(\mathbf{u}'_\alpha) \in (v_{l-1}, v_l]$.

Consider the indicator coding of five Ni concentrations along the NE-SW transect (Figure 7.15, top graph). The middle graph shows the calibration scattergram of Ni-values versus Cd-values ($n = 259$). Three classes of Ni concentrations depicted by the vertical dashed lines are considered: $(0, 15]$ ppm, $(15, 25]$ ppm, and $(25, 50]$ ppm. For each class, the cumulative distribution of Cd data is plotted, and the proportions of data not exceeding the threshold values z_k=0.80, 1.38, 1.88, or 2.26 ppm are computed. These calibration results are then used to code the prior information provided by each Ni datum (Figure 7.15, bottom graph). For example, the vector of soft indicator data at location $\mathbf{u}'_3$ where the Ni concentration is valued within the interval $(0,15]$ ppm is (second column at bottom Figure 7.15)

$$\begin{bmatrix} 0.97 \\ 0.96 \\ 0.85 \\ 0.81 \end{bmatrix} = \begin{bmatrix} \text{Prob}\,\{Z(\mathbf{u}'_3) \le 2.26 | v(\mathbf{u}'_3) \in (0, 15]\} \\ \text{Prob}\,\{Z(\mathbf{u}'_3) \le 1.88 | v(\mathbf{u}'_3) \in (0, 15]\} \\ \text{Prob}\,\{Z(\mathbf{u}'_3) \le 1.38 | v(\mathbf{u}'_3) \in (0, 15]\} \\ \text{Prob}\,\{Z(\mathbf{u}'_3) \le 0.80 | v(\mathbf{u}'_3) \in (0, 15]\} \end{bmatrix}$$

Colocated sources of information

When a soft datum location $\mathbf{u}'_\alpha$ coincides with a hard datum location $\mathbf{u}_\alpha$, the set of hard indicator data $i(\mathbf{u}_\alpha; z_k)$ prevails over the colocated soft indicator data $y(\mathbf{u}'_\alpha; z_k)$. For example, the hard datum $z(\mathbf{u}_5)$=1.31 ppm prevails over the soft information that $\mathbf{u}_5$ belongs to rock category s_1. If the primary

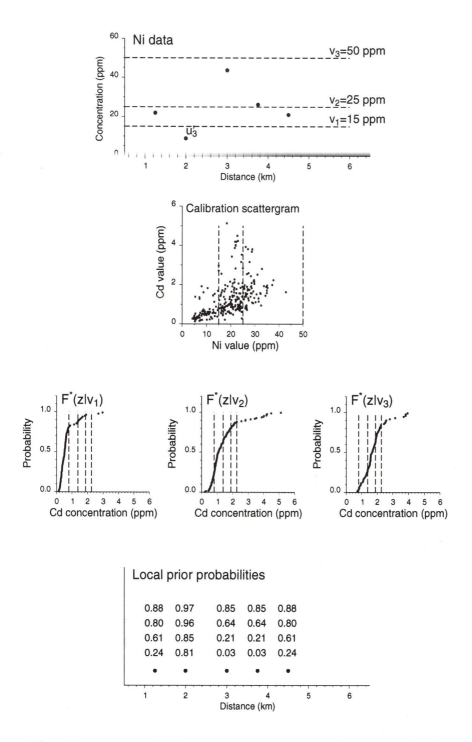

Figure 7.15: Coding of five Ni concentrations into vectors of local prior probabilities of not exceeding Cd threshold values. The prior probabilities at location $\mathbf{u}$ where the Ni-concentration is within the interval $(v_{l-1}, v_l]$ are inferred from the cumulative distribution of Cd data within that class of Ni-concentrations (middle graph).

datum is a constraint interval $(a_\alpha, b_\alpha]$, the corresponding missing indicator data are replaced by soft indicator data at the corresponding thresholds, i.e., the local prior cdf (7.22) becomes:

$$i(\mathbf{u}_\alpha; z_k) = \begin{cases} 1 & \text{if } b_\alpha \le z_k \\ F^*(z_k | s_l) & \text{if } z_k \in (a_\alpha, b_\alpha] \\ 0 & \text{if } a_\alpha > z_k \end{cases} \quad k = 1, \dots, K \qquad (7.26)$$

if $s(\mathbf{u}_\alpha) = s_l$.

For example, at location $\mathbf{u}_0$ belonging to the rock category s_2 (Figure 7.14), the vector of local prior probabilities (7.23) is completed as

$$\begin{bmatrix} 1 \\ 0.79 \\ 0.53 \\ 0 \end{bmatrix} = \begin{bmatrix} \text{Prob}\{Z(\mathbf{u}_0) \le 2.26 | z(\mathbf{u}_0) \in (1.0, 2.0]\} \\ \text{Prob}\{Z(\mathbf{u}_0) \le 1.58 | s(\mathbf{u}_0) = s_2\} \\ \text{Prob}\{Z(\mathbf{u}_0) \le 1.38 | s(\mathbf{u}_0) = s_2\} \\ \text{Prob}\{Z(\mathbf{u}_0) \le 0.80 | z(\mathbf{u}_0) \in (1.0, 2.0]\} \end{bmatrix}$$

When several soft data are available at the same location $\mathbf{u}'_\alpha$, the local prior information consists of several prior cdfs of type (7.24) or (7.25). For example, one may know that the concentration of the secondary variable Ni is valued within $(0,15]$ ppm at location $\mathbf{u}'_3$ belonging to rock category s_1. Thus, there are two colocated vectors of local prior probabilities at $\mathbf{u}'_3$. The first vector contains the probabilities for the Cd concentration to be smaller than each of the four threshold values given that Ni concentration is valued within $(0,15]$ ppm:

$$\begin{bmatrix} y_1(\mathbf{u}'_3; 2.26) \\ y_1(\mathbf{u}'_3; 1.88) \\ y_1(\mathbf{u}'_3; 1.38) \\ y_1(\mathbf{u}'_3; 0.80) \end{bmatrix} = \begin{bmatrix} 0.97 \\ 0.96 \\ 0.85 \\ 0.81 \end{bmatrix}$$

The second vector consists of local prior probabilities derived from the calibration of the rock category s_1:

$$\begin{bmatrix} y_2(\mathbf{u}'_3; 2.26) \\ y_2(\mathbf{u}'_3; 1.88) \\ y_2(\mathbf{u}'_3; 1.38) \\ y_2(\mathbf{u}'_3; 0.80) \end{bmatrix} = \begin{bmatrix} 0.85 \\ 0.83 \\ 0.74 \\ 0.66 \end{bmatrix}$$

Different sources of soft information could be ranked according to their ability to predict hard data; see the discussion related to Figure 7.23 (page 316). The vector of local prior probabilities corresponding to the most accurate soft information would then prevail over the other vectors.

7.3.2   Updating into ccdf values

Indicator coding allows different types of information (hard and soft data) to be processed together, regardless of their origins. The objective is to evaluate at any location $\mathbf{u}$ the set of K ccdf values or posterior probabilities:

$$F(\mathbf{u}; z_k|(n)) \;=\; \text{Prob}\,\{Z(\mathbf{u}) \le z_k|(n)\} \qquad k = 1, \ldots, K \qquad (7.27)$$

where the conditioning information (n) consists of hard and soft data retained in a search neighborhood $W(\mathbf{u})$ centered at $\mathbf{u}$. As mentioned previously, the least-squares (kriging) estimate of the indicator $i(\mathbf{u}, z_k)$ can be used as a model for the ccdf value of $z(\mathbf{u})$ at a particular threshold value z_k:

$$[F(\mathbf{u}; z_k|(n))]^* \;=\; [i(\mathbf{u}; z_k)]^*_{krig.}$$

Each ccdf value can thus be derived as a linear combination of neighboring hard and soft indicator data, $i(\mathbf{u}_\alpha; z_k)$ and $y(\mathbf{u}'_\alpha; z_k)$, using kriging algorithms similar to those introduced in Chapters 5 and 6. Note that K (co)kriging systems, one for each threshold value z_k, must be solved at any location $\mathbf{u}$.

   This section presents kriging algorithms that account for hard indicator transforms of the sole attribute of interest. Algorithms for incorporating soft information are introduced in section 7.3.3.

Indicator kriging

First, consider the problem of estimating the indicator value $i(\mathbf{u}; z_k)$ at any location $\mathbf{u}$ using only hard indicator data defined at the same threshold value z_k. The linear estimator (5.1) is expressed in terms of indicator RVs as

$$[I(\mathbf{u}; z_k)]^* - \text{E}\,\{I(\mathbf{u}; z_k)\} \;=\; \sum_{\alpha=1}^{n(\mathbf{u})} \lambda_\alpha(\mathbf{u}; z_k)\;[I(\mathbf{u}_\alpha; z_k) - \text{E}\,\{I(\mathbf{u}_\alpha; z_k)\}]$$

$$(7.28)$$

where $\lambda_\alpha(\mathbf{u}; z_k)$ is the weight assigned to the indicator datum $i(\mathbf{u}_\alpha; z_k)$ interpreted as a realization of the indicator RV $I(\mathbf{u}_\alpha; z_k)$. As with kriging estimators of the z-attribute values, two indicator kriging (IK) variants are distinguished, depending on whether the indicator mean is considered constant within the study area $\mathcal{A}$.

Simple indicator kriging
The simple indicator kriging (sIK) estimator considers the indicator mean known and constant throughout $\mathcal{A}$:

$$\text{E}\,\{I(\mathbf{u}; z_k)\} \;=\; F(z_k), \text{ known } \forall\, \mathbf{u} \in \mathcal{A}$$

The linear estimator (7.28) is thus written as a linear combination of $(n(\mathbf{u})+1)$ pieces of information, the $n(\mathbf{u})$ indicator RVs $I(\mathbf{u}_\alpha; z_k)$ and the cdf value $F(z_k)$:

$$[F(\mathbf{u}; z_k|(n))]^*_{sIK} = [I(\mathbf{u}; z_k)]^*_{SK}$$

$$= F(z_k) + \sum_{\alpha=1}^{n(\mathbf{u})} \lambda_\alpha^{SK}(\mathbf{u}; z_k) \, [I(\mathbf{u}_\alpha; z_k) - F(z_k)]$$

$$= \sum_{\alpha=1}^{n(\mathbf{u})} \lambda_\alpha^{SK}(\mathbf{u}; z_k) \, I(\mathbf{u}_\alpha; z_k) + \lambda_m^{SK}(\mathbf{u}; z_k) \, F(z_k)$$

$$(7.29)$$

where the weight of the mean is defined as

$$\lambda_m^{SK}(\mathbf{u}; z_k) = 1 - \sum_{\alpha=1}^{n(\mathbf{u})} \lambda_\alpha^{SK}(\mathbf{u}; z_k)$$

The indicator mean could be estimated by the sample cumulative frequency $F^*(z_k)$ after possible correction for preferential sampling (see section 4.1.2).

The simple IK weights are provided by a simple kriging system of type (5.10):

$$\sum_{\beta=1}^{n(\mathbf{u})} \lambda_\beta^{SK}(\mathbf{u}; z_k) \, C_I(\mathbf{u}_\alpha - \mathbf{u}_\beta; z_k) = C_I(\mathbf{u}_\alpha - \mathbf{u}; z_k)$$

$$\alpha = 1, \dots, n(\mathbf{u}) \qquad (7.30)$$

where $C_I(\mathbf{h}; z_k)$ is the covariance function of the indicator RF $I(\mathbf{u}; z_k)$ at threshold z_k.

Ordinary indicator kriging
Ordinary indicator kriging (oIK) allows one to account for local fluctuations of the indicator mean by limiting the domain of stationarity of that mean to the local neighborhood $W(\mathbf{u})$:

$$E\{I(\mathbf{u}'; z_k)\} = \text{constant but unknown } \forall \, \mathbf{u}' \in W(\mathbf{u})$$

The ordinary IK estimator is a linear combination of the $n(\mathbf{u})$ indicator RVs $I(\mathbf{u}_\alpha; z_k)$ in the neighborhood $W(\mathbf{u})$:

$$[F(\mathbf{u}; z_k|(n))]^*_{oIK} = [I(\mathbf{u}; z_k)]^*_{OK}$$

$$= \sum_{\alpha=1}^{n(\mathbf{u})} \lambda_\alpha^{OK}(\mathbf{u}; z_k) \, I(\mathbf{u}_\alpha; z_k) \qquad (7.31)$$

where the weights are given by an ordinary kriging system of type (5.18):

$$\begin{cases} \displaystyle\sum_{\beta=1}^{n(\mathbf{u})} \lambda_\beta^{OK}(\mathbf{u}; z_k) \, C_I(\mathbf{u}_\alpha - \mathbf{u}_\beta; z_k) + \mu_{OK}(\mathbf{u}; z_k) = C_I(\mathbf{u}_\alpha - \mathbf{u}; z_k) \\[4pt] \qquad\qquad\qquad\qquad\qquad\qquad \alpha = 1, \dots, n(\mathbf{u}) \\[6pt] \displaystyle\sum_{\beta=1}^{n(\mathbf{u})} \lambda_\beta^{OK}(\mathbf{u}; z_k) = 1 \end{cases}$$

$$(7.32)$$

As with ordinary kriging of a continuous variable, ordinary IK with local search neighborhood amounts to:

1. estimating the local indicator mean, say, $F_{oIK}^*(\mathbf{u}; z_k)$, at each location $\mathbf{u}$ using the hard indicator data $i(\mathbf{u}_\alpha; z_k)$ specific to the neighborhood of $\mathbf{u}$, then

2. applying the simple IK estimator (7.29) using that estimate of the mean rather than the sample cumulative frequency $F^*(z_k)$:

$$[I(\mathbf{u}; z_k)]_{OK}^* = \sum_{\alpha=1}^{n(\mathbf{u})} \lambda_\alpha^{SK}(\mathbf{u}; z_k)\, I(\mathbf{u}_\alpha; z_k) + \lambda_m^{SK}(\mathbf{u}; z_k)\, F_{oIK}^*(\mathbf{u}; z_k)$$

Exactitude property
Both simple and ordinary IK estimators are exact interpolators because they honor hard indicator data at their locations:

$$[i(\mathbf{u}; z_k)]^* = i(\mathbf{u}_\alpha; z_k) \qquad \forall\, \mathbf{u} = \mathbf{u}_\alpha,\ \alpha = 1, \ldots, n$$

Where the z-value is known exactly, the posterior cdf is the unit-step prior cdf (no updating):

$$[F(\mathbf{u}_\alpha; z_k|(n))]^* = \begin{cases} 1 & \text{if } z(\mathbf{u}_\alpha) \le z_k \\ 0 & \text{otherwise} \end{cases} \qquad k = 1, \ldots, K$$

Where the prior information is a constraint interval $(a_\alpha, b_\alpha]$, only the missing indicator values are updated by posterior probabilities:

$$[F(\mathbf{u}_\alpha; z_k|(n))]^* = \begin{cases} 1 & \text{if } b_\alpha \le z_k \\ [i(\mathbf{u}_\alpha; z_k)]^* & \text{if } z_k \in (a_\alpha, b_\alpha] \\ 0 & \text{if } a_\alpha > z_k \end{cases} \qquad k = 1, \ldots, K$$

Example
Simple and ordinary indicator kriging are used along the NE-SW transect to determine ccdf values at the four threshold values $z_k = 0.80$, 1.38, 1.88, and 2.26 ppm. The information available consists of ten Cd concentrations at locations $\mathbf{u}_1$ to $\mathbf{u}_{10}$ (black dots) plus the constraint interval $(1.0, 2.0$ ppm] at location $\mathbf{u}_0$ (Figure 7.16, top graph). For each threshold value, the indicator semivariogram is inferred from all data available over the study area (Figure 7.16, middle graphs). Estimation is performed every 50 m using at each location the five closest indicator data related only to the threshold z_k being considered. The stationary indicator means $F(z_k)$ required by simple IK are estimated from the cumulative histogram of the ten Cd values.

Figure 7.16 (bottom graph) shows both oIK (solid line) and sIK (small dashed line) estimates of the probability of exceeding the critical threshold

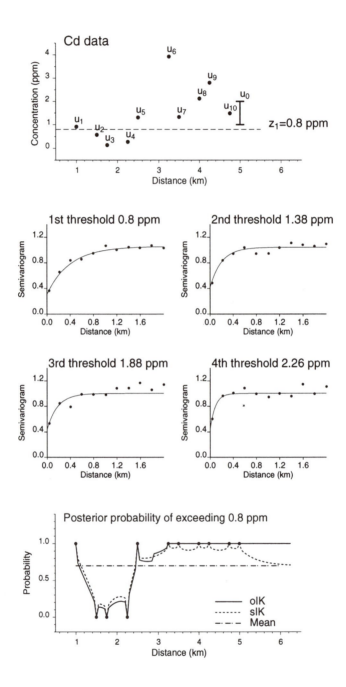

Figure 7.16: Simple and ordinary indicator kriging estimates of the probability of exceeding the critical threshold value 0.8 ppm. The information available consists of ten Cd concentrations and the constraint interval $(1.0, 2.0$ ppm] at u_0, plus the indicator semivariogram modeled from 259 indicator data.

$z_1 = 0.8$ ppm computed as $1 - [F(\mathbf{u}; z_1|(n))]^*$. The horizontal dashed line depicts the marginal probability $[1 - F(z_1)]$ of exceeding the threshold value z_1, estimated by the proportion of Cd values greater than 0.8 ppm: $7/10 = 0.7$. Note the following:

- Both IK estimators are exact. The posterior probability of exceeding the critical threshold is zero at data locations $\mathbf{u}_2$ to $\mathbf{u}_4$, where the Cd concentration is smaller than 0.8 ppm, and that probability is 1 at other data locations known to be contaminated.

- The local estimation of the indicator mean within each search neighborhood yields ordinary IK estimates that follow the data trends better than the simple IK estimates. The posterior probabilities of contamination are smaller in the low-valued (left) part of the transect and are larger in the high-valued (right) part.

- Beyond location $\mathbf{u}_6$, ordinary IK yields a unit probability of exceeding the critical threshold, which corresponds to a zero cumulative probability $[F(\mathbf{u}; z_1|(n))]^*$. In this part of the transect, all data locations within the search neighborhood $W(\mathbf{u})$ are contaminated, which entails that all indicator data $i(\mathbf{u}_\alpha; z_1)$ are zero and so is their estimated mean (7.31). In contrast, away from the data (extrapolation situation), the SK estimate approaches the marginal probability 0.7.

At locations $\mathbf{u}_5$ and $\mathbf{u}_0$, ordinary IK yields the following vectors of posterior probabilities: $\mathbf{u}_5$: $[0\ 1\ 1\ 1]$, $\mathbf{u}_0$: $[0.0\ 0.16\ 0.54\ 1.0]$. The exactitude property of the IK estimator entails that the ccdf at the datum location $\mathbf{u}_5$ is identical to the prior cdf (7.21) (no updating). Similarly, the ccdf at location $\mathbf{u}_0$ honors the constraint interval information (7.23); only the two missing indicator values at thresholds z_2 and z_3 are updated into the posterior probabilities 0.16 and 0.54, respectively.

Indicator cokriging

Neither IK estimators (7.29) and (7.31) make full use of the information available in that they ignore indicator data at thresholds different from that being estimated. Information from all K thresholds can be accounted for using the cokriging formalism introduced in section 6.2. For example, the ordinary indicator cokriging (oICK) estimator of ccdf value at threshold z_{k_0} would be written

$$
\begin{aligned}
[F(\mathbf{u}; z_{k_0}|(n))]^*_{oICK} &= [I(\mathbf{u}; z_{k_0})]^*_{OCK} \\
&= \sum_{k=1}^{K} \sum_{\alpha_k=1}^{n_k(\mathbf{u})} \lambda^{OCK}_{\alpha_k}(\mathbf{u}; z_{k_0})\, I(\mathbf{u}_{\alpha_k}; z_k) \qquad (7.33)
\end{aligned}
$$

where $\lambda^{OCK}_{\alpha_k}(\mathbf{u}; z_{k_0})$ is the weight assigned to indicator datum $i(\mathbf{u}_{\alpha_k}; z_k)$ at location $\mathbf{u}_{\alpha_k}$. These cokriging weights are solutions of an ordinary cokriging system of type (6.32):

$$
\begin{cases}
\displaystyle\sum_{k'=1}^{K}\sum_{\beta_{k'}=1}^{n_{k'}(\mathbf{u})} \lambda_{\beta_{k'}}^{OCK}(\mathbf{u}; z_{k_0})\, C_I(\mathbf{u}_{\alpha_k} - \mathbf{u}_{\beta_{k'}}; z_k, z_{k'}) \;+\; \mu_k^{OCK}(\mathbf{u}; z_{k_0}) \\[2mm]
\qquad\qquad = C_I(\mathbf{u}_{\alpha_k} - \mathbf{u}; z_k; z_{k_0}) \qquad \alpha_k = 1, \ldots, n_k(\mathbf{u}) \quad k = 1, \ldots, K \\[2mm]
\displaystyle\sum_{\beta_k=1}^{n_k(\mathbf{u})} \lambda_{\beta_k}^{OCK}(\mathbf{u}; z_{k_0}) \;=\; \delta_{kk_0} \qquad k = 1, \ldots, K
\end{cases}
$$

$$(7.34)$$

where $C_I(\mathbf{h}; z_k, z_{k'})$ is the cross covariance between any two indicator RVs $I(\mathbf{u}; z_k)$ and $I(\mathbf{u} + \mathbf{h}; z_{k'})$, and $\delta_{kk_0} = 1$ for $k = k_0$ and $\delta_{kk_0} = 0$ otherwise.

As mentioned in section 6.2.4, ordinary cokriging constraints that call for the secondary data weights to sum to zero tend to limit artificially the influence of these secondary data. These constraints also increase the occurrence of negative weights hence the risk of getting unacceptable estimates such as negative posterior probabilities (Goovaerts, 1994a). Therefore, one might consider rescaling each secondary indicator variable $I(\mathbf{u}; z_k)$ to the mean of the primary indicator variable, then solving the ordinary indicator cokriging system (7.34) with the single unbiasedness constraint that calls for all primary and secondary data weights to sum to 1.

Indicator kriging versus indicator cokriging
Indicator cokriging is much more demanding than indicator kriging for two reasons:

1. $K(K+1)/2$ direct and cross indicator semivariograms must be inferred and jointly modeled (e.g., 45 semivariograms for 9 threshold values).

2. K larger cokriging matrices must be inverted at each location $\mathbf{u}$.

Note that the modeling and computational effort necessitated by indicator cokriging may be alleviated by kriging principal indicator components; see section 6.2.5 and Suro-Pérez and Journel (1991).

In theory, the indicator cokriging estimator is better (in a least-squares sense) than the IK estimator because it accounts for additional information available across all thresholds. Practice has shown, however, that indicator cokriging improves little over indicator kriging (Goovaerts, 1994a) for the following reasons:

• Cumulative indicator data, as defined in relation (7.20), already carry substantial information from one threshold to the next, which diminishes the actual loss of information from disregarding other thresholds.

• A complete vector of hard indicator data is usually available at each datum location (equally sampled case). As shown in section 6.2.2, the primary datum thus tends to screen the influence of colocated secondary data; that is, the indicator value at the threshold z_{k_0} being estimated screens the influence of colocated indicator values at other thresholds $z_k \neq z_{k_0}$.

- Indicator cokriging creates many more order relation problems, which must be corrected a posteriori; see Figure 7.18 and section 7.3.4.

Example
Figure 7.17 shows the ten omnidirectional standardized indicator direct and cross semivariograms for the four threshold Cd values $z_k = 0.80$, 1.38, 1.88, and 2.26 ppm. The iterative procedure described in Appendix A is used to fit a linear model of coregionalization, including three basic structures:

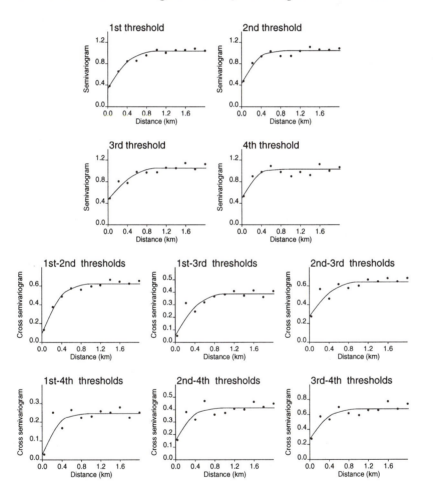

Figure 7.17: Experimental standardized indicator direct and cross semivariograms corresponding to four different Cd threshold values, with the linear model of coregionalization fitted.

a nugget effect and two spherical models with ranges of 500 m and 1.0 km, respectively.

Figure 7.18 shows the four profiles of posterior probabilities estimated using ordinary indicator kriging (dashed line) and ordinary indicator cokriging with a single unbiasedness constraint of the type described in section 6.2.4 (solid line). Note the following:

- Except for the second threshold, kriging and cokriging yield fairly similar probabilities. Differences between estimates arise from the binary nature of indicator data. Consider, for example, the first threshold 0.8 ppm. In the right part of the transect where all Cd data exceed that threshold, there is only one possible kriging estimate: 1.0, since the closest indicator data $i(\mathbf{u}_\alpha; 0.8)$ are all equal to 1 (Figure 7.13, page 286). Accounting for indicator values from other thresholds increases the number of possible estimated values, hence the profile of oICK estimates is more variable.

- Cokriging produces probabilities outside the interval $[0, 1]$ at the two extremes of the transect depicted by arrows. Though only a little outside $[0, 1]$, such inconsistent probabilities must be corrected. As discussed subsequently, the correction amounts to resetting these probabilities to the nearest bound, 0 or 1.

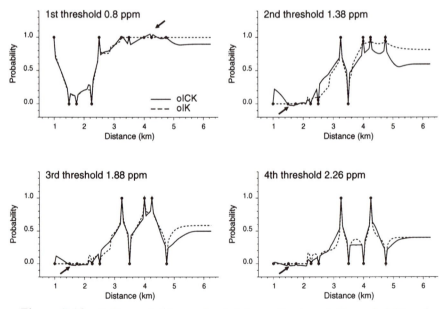

Figure 7.18: Ordinary kriging and cokriging estimates of the probability of exceeding four different threshold values. Arrows indicate inconsistent probabilities outside the interval $[0, 1]$ produced by indicator cokriging.

Probability kriging

A shortcut to indicator cokriging consists of using the original data values $z(\mathbf{u}_\alpha)$ instead of their indicator transforms at thresholds different from that being estimated. The resulting cokriging estimator is less cumbersome because it accounts for a single secondary variable:

$$[F(\mathbf{u}; z_k|(n))]^* = \sum_{\alpha=1}^{n(\mathbf{u})} \lambda_\alpha(\mathbf{u}; z_k)\, I(\mathbf{u}_\alpha; z_k) + \sum_{\alpha=1}^{n(\mathbf{u})} \nu_\alpha(\mathbf{u}; z_k)\, Z(\mathbf{u}_\alpha)$$

Large differences between the units of measurement of z and its indicator transforms may cause instability problems when solving the cokriging system. One solution consists of replacing the z-values by their standardized ranks $x(\mathbf{u}_\alpha) = r(\mathbf{u}_\alpha)/n$, where $r(\mathbf{u}_\alpha) \in [1, n]$ is the rank of the datum $z(\mathbf{u}_\alpha)$ in the sample cumulative distribution (see section 2.1.2). The indicator data $i(\mathbf{u}_\alpha; z_k)$ are valued either 0 or 1, whereas the rank-order transforms $x(\mathbf{u}_\alpha)$ are uniformly distributed in $[0, 1]$.

The cokriging of the indicator transform $i(\mathbf{u}; z_k)$ using the rank-order transform $x(\mathbf{u})$ as a secondary variable is referred to as *probability kriging* (Isaaks, 1984; Journel, 1984b; Sullivan, 1984, 1985). The probability kriging (PK) estimator is

$$[F(\mathbf{u}; z_k|(n))]^*_{PK} = [I(\mathbf{u}; z_k)]^*_{PK}$$

$$= \sum_{\alpha=1}^{n(\mathbf{u})} \lambda_\alpha^{PK}(\mathbf{u}; z_k)\, I(\mathbf{u}_\alpha; z_k) + \sum_{\alpha=1}^{n(\mathbf{u})} \nu_\alpha^{PK}(\mathbf{u}; z_k)\, X(\mathbf{u}_\alpha)$$

$$(7.35)$$

The PK weights $\lambda_\alpha^{PK}(\mathbf{u}; z_k)$ and $\nu_\alpha^{PK}(\mathbf{u}; z_k)$ are obtained by solving the following ordinary cokriging system of $(2n(\mathbf{u}) + 2)$ equations:

$$\begin{cases} \sum_{\beta=1}^{n(\mathbf{u})} \lambda_\beta^{PK}(\mathbf{u}; z_k)\, C_I(\mathbf{u}_\alpha - \mathbf{u}_\beta; z_k) + \sum_{\beta=1}^{n(\mathbf{u})} \nu_\beta^{PK}(\mathbf{u}; z_k)\, C_{IX}(\mathbf{u}_\alpha - \mathbf{u}_\beta; z_k) \\ \qquad\qquad + \mu_I^{PK}(\mathbf{u}; z_k) = C_I(\mathbf{u}_\alpha - \mathbf{u}; z_k) \qquad \alpha = 1, \ldots, n(\mathbf{u}) \\ \sum_{\beta=1}^{n(\mathbf{u})} \lambda_\beta^{PK}(\mathbf{u}; z_k)\, C_{XI}(\mathbf{u}_\alpha - \mathbf{u}_\beta; z_k) + \sum_{\beta=1}^{n(\mathbf{u})} \nu_\beta^{PK}(\mathbf{u}; z_k)\, C_X(\mathbf{u}_\alpha - \mathbf{u}_\beta) \\ \qquad\qquad + \mu_X^{PK}(\mathbf{u}; z_k) = C_{XI}(\mathbf{u}_\alpha - \mathbf{u}; z_k) \qquad \alpha = 1, \ldots, n(\mathbf{u}) \\ \sum_{\beta=1}^{n(\mathbf{u})} \lambda_\beta^{PK}(\mathbf{u}; z_k) = 1 \\ \sum_{\beta=1}^{n(\mathbf{u})} \nu_\beta^{PK}(\mathbf{u}; z_k) = 0 \end{cases}$$

$$(7.36)$$

where $C_X(\mathbf{h})$ is the covariance function of the uniform RF $X(\mathbf{u})$, and $C_{XI}(\mathbf{h}; z_k)$ is the cross covariance function between $X(\mathbf{u})$ and the indicator RF $I(\mathbf{u}; z_k)$.

The PK estimator (7.35) uses more information than the ordinary IK estimator (7.31) because the rank of the datum $z(\mathbf{u}_\alpha)$ in the sample cdf is taken into account in addition to its indicator of exceeding the threshold value z_k. The trade-off cost for this better conditioning of the posterior cdf is the inference of the semivariogram $\gamma_X(\mathbf{h})$ and the K cross semivariograms $\gamma_{XI}(\mathbf{h}; z_k)$, and the modeling of the coregionalization between indicator and uniform transforms at each threshold.

To reduce the proportion of negative cokriging weights and enhance the influence of rank-order transforms in the PK estimator, the two unbiasedness constraints in the cokriging system (7.36) can be replaced by the single constraint that all the cokriging weights must sum to 1:

$$\sum_{\alpha=1}^{n(\mathbf{u})} \lambda_\alpha^{PK}(\mathbf{u}; z_k) + \sum_{\alpha=1}^{n(\mathbf{u})} \nu_\alpha^{PK}(\mathbf{u}; z_k) \;\; = \;\; 1$$

The PK estimator is thus rewritten

$$[F(\mathbf{u}; z_k | (n))]_{PK}^* \;\; = \;\; \sum_{\alpha=1}^{n(\mathbf{u})} \lambda_\alpha^{PK}(\mathbf{u}; z_k)\, I(\mathbf{u}_\alpha; z_k)$$

$$+ \sum_{\alpha=1}^{n(\mathbf{u})} \nu_\alpha^{PK}(\mathbf{u}; z_k)\, [X(\mathbf{u}_\alpha) - 0.5 + F(z_k)]$$

$$(7.37)$$

since the stationary mean of the uniform transform $X(\mathbf{u})$ is 0.5.

Example

Figure 7.19 (top graph) shows the uniform transforms (open circles) and indicator transforms for the threshold value 0.8 ppm (closed circles) of ten Cd concentrations. The rank-order transform is based on only the ten Cd data values, hence the two extreme uniform data values are 0.1 for the smallest Cd concentration at $\mathbf{u}_3$ and 1.0 for the largest concentration at $\mathbf{u}_6$. Both data sets are combined using the PK estimator (7.37) with a single unbiasedness constraint. The semivariograms of indicator and uniform transforms and their cross semivariogram are inferred from all data available over the study area (Figure 7.19, middle graph). By construction, indicator and uniform data are negatively correlated since a zero indicator datum corresponds to a large Cd concentration and, therefore, to a high rank in the sample cumulative distribution.

Figure 7.19 (bottom graph) shows both PK (solid line) and oIK (dashed line) estimates of the probability that Cd concentration exceeds 0.8 ppm:

- Like the cokriging in Figure 7.18, probability kriging yields an interpolation profile more variable in space, particularly in the right part of the transect. Indeed, the data ranks valued in $[0, 1]$ allow one to discriminate Cd concentrations with similar indicator transforms 0 or 1. In this

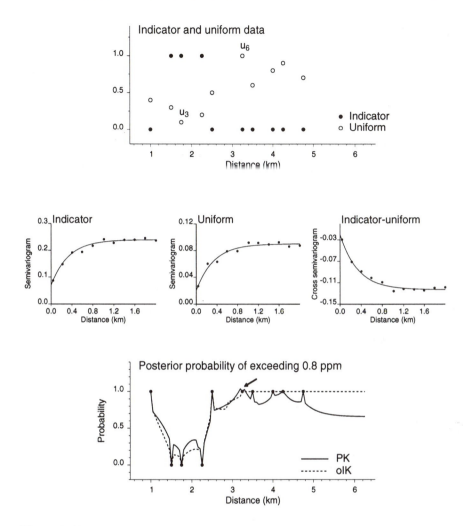

Figure 7.19: Probability kriging estimates of the probability of exceeding 0.8 ppm. Indicator and uniform transforms are combined using ordinary cokriging with a single unbiasedness constraint and the semivariogram models shown in the second row. The dashed line depicts the ordinary kriging estimate using only indicator data. The arrow indicates two inconsistent probabilities produced by probability kriging.

way, probability kriging corrects for the loss of resolution caused by the use of a single threshold in indicator kriging.

- Unlike ordinary cokriging, probability kriging produces only two probabilities outside the interval $[0, 1]$. The use of a single secondary variable in the PK estimator lessens the screening effect between variables, thereby reducing the risk of getting negative cokriging weights and inconsistent probabilities.

Median indicator kriging

Though less demanding than indicator cokriging or probability kriging, indicator kriging still calls for estimating and modeling K indicator semivariograms and solving K kriging systems at each location $\mathbf{u}$. The modeling and computational effort can be substantially alleviated if the two following conditions are jointly met:

1. The K indicator RFs $I(\mathbf{u}; z_k)$ are intrinsically correlated; recall expression (4.41). All the $K(K+1)/2$ indicator direct and cross semivariogram models are then proportional to a common semivariogram model $\gamma_{mI}(\mathbf{h})$:

$$\gamma_I(\mathbf{h}; z_k, z_{k'}) \ = \ \varphi_{kk'} \cdot \gamma_{mI}(\mathbf{h}) \qquad \forall \, k, k' \qquad (7.38)$$

The corresponding RF model $Z(\mathbf{u})$ is called the mosaic model (Journel, 1984b).

2. All vectors of hard indicator data retained in the estimation are complete (equally sampled case); there are no missing indicator values such as implied by constraint intervals of type (7.22).

As shown in section 6.2.2, kriging and cokriging estimators are then identical; for example, for the ordinary (co)kriging case:

$$
\begin{aligned}
[F(\mathbf{u}; z_k|(n))]^*_{mIK} \ &= \ \sum_{\alpha=1}^{n(\mathbf{u})} \lambda_\alpha^{OK}(\mathbf{u}) \, I(\mathbf{u}_\alpha; z_k) \qquad\qquad (7.39) \\
&\equiv \ [F(\mathbf{u}; z_k|(n))]^*_{oICK} \quad k = 1, \dots, K
\end{aligned}
$$

The estimator (7.39) is called median indicator kriging (mIK) since the common model $\gamma_{mI}(\mathbf{h})$ is usually inferred from the indicator semivariogram at the median threshold value $z_M = F^{-1}(0.5)$ (Journel, 1984b). Indeed, median indicator data $i(\mathbf{u}_\alpha; z_M)$ are evenly distributed as 0 and 1 values, which usually renders the experimental indicator semivariogram $\widehat{\gamma}_I(\mathbf{h}; z_M)$ better defined than at other threshold values.

Since the indicator data configuration is the same for all threshold values, the kriging weights $\lambda_\alpha^{OK}(\mathbf{u})$ do not depend on the threshold being considered; therefore, only one IK system needs to be solved at each location $\mathbf{u}$:

$$
\begin{cases}
\displaystyle\sum_{\beta=1}^{n(\mathbf{u})} \lambda_\beta^{OK}(\mathbf{u}) \, C_{mI}(\mathbf{u}_\alpha - \mathbf{u}_\beta) + \mu_{OK}(\mathbf{u}) \ = \ C_{mI}(\mathbf{u}_\alpha - \mathbf{u}) \\
\qquad\qquad\qquad\qquad\qquad\qquad\qquad \alpha = 1, \dots, n(\mathbf{u}) \qquad (7.40) \\
\displaystyle\sum_{\beta=1}^{n(\mathbf{u})} \lambda_\beta^{OK}(\mathbf{u}) \ = \ 1
\end{cases}
$$

where the indicator covariance function $C_{mI}(\mathbf{h})$ is deduced as $0.25 - \gamma_{mI}(\mathbf{h})$ if the model $\gamma_{mI}(\mathbf{h})$ relates to the median threshold value z_M.

Median IK is very fast because it requires only one indicator semivariogram (median) to be modeled and a single IK system to be solved at each location $\mathbf{u}$. However, such an approach calls for the particular coregionalization model (7.38) that does not allow different shapes or anisotropy patterns for the K indicator semivariograms. When sample indicator semivariograms appear not to be proportional to each other, as in the case of the Cd semivariograms in Figure 7.17 (page 299), the more flexible and still reasonably fast indicator kriging should be used.

Block versus composite ccdfs

Let $z_V(\mathbf{u})$ be the linear average value of an attribute z over a block V of any specific dimensions:

$$z_V(\mathbf{u}) = \frac{1}{|V|} \int_{V(\mathbf{u})} z(\mathbf{u}')d\mathbf{u}' \simeq \frac{1}{N} \sum_{i=1}^{N} z(\mathbf{u}'_i) \qquad (7.41)$$

where $|V|$ is the measure (length, area, volume) of block V. The integral is, in practice, approximated by a discrete sum of z-values defined at N points $\mathbf{u}'_i$ discretizing the block $V(\mathbf{u})$. For the level of information (n), the uncertainty about the block value $z_V(\mathbf{u})$ is modeled by the "block" posterior cdf:

$$\begin{aligned} F_V(\mathbf{u}; z|(n)) &= \text{Prob}\{Z_V(\mathbf{u}) \leq z|(n)\} \\ &= \text{E}\{I_V(\mathbf{u}; z)|(n)\} \qquad (7.42) \end{aligned}$$

with the block indicator RV defined as $I_V(\mathbf{u}; z)=1$ if $Z_V(\mathbf{u}) \leq z$ and equal to zero otherwise.

If block data $z_V(\mathbf{u}_\alpha)$ were available, the posterior cdf (7.42) at any threshold z_k could be obtained by indicator kriging from block indicator data defined as

$$i_V(\mathbf{u}_\alpha; z_k) = \begin{cases} 1 & \text{if } z_V(\mathbf{u}_\alpha) \leq z_k \\ 0 & \text{otherwise} \end{cases}$$

Unfortunately, block data $z_V(\mathbf{u}_\alpha)$ do not usually exist, hence the block ccdf must be modeled from the point-data $z(\mathbf{u}_\alpha)$ alone.

Because the indicator variable $i(\mathbf{u}; z)$ is a non-linear transform of the original variable $z(\mathbf{u})$, the block indicator $i_V(\mathbf{u}; z)$ is not a linear average of point indicators $i(\mathbf{u}; z)$:

$$i_V(\mathbf{u}; z) \neq \frac{1}{|V|} \int_{V(\mathbf{u})} i(\mathbf{u}'; z)d\mathbf{u}'$$

Therefore, the block ccdf $F_V(\mathbf{u}; z|(n))$ cannot be derived as a linear average of point ccdfs:

$$[F_V(\mathbf{u}; z|(n))]^* \neq \frac{1}{N} \sum_{i=1}^{N} [F(\mathbf{u}'_i; z|(n))]^* = [F_N(\mathbf{u}; z|(n))]^*$$

The "composite" ccdf $[F_N(\mathbf{u}; z|(n))]^*$ is an estimate of the proportion of point values within $V(\mathbf{u})$ that do not exceed the threshold value z, whereas the block ccdf gives the probability that the average z-value is no greater than z. A simulation-based approach for deriving block ccdfs from point data is introduced in section 8.5.

In many environmental applications, risks relate to the occurrence of large concentrations over small volumes. Such risks are underestimated by linear averaging (7.41) over the block $V(\mathbf{u})$ because extreme values are smoothed out. Hence, it is often the composite ccdf that is of interest, not the block ccdf. Block kriging formalism allows one to estimate the composite probability at threshold z_k as a linear combination of point indicator data $i(\mathbf{u}_\alpha; z_k)$. For example, the block oIK estimator is written

$$[F_N(\mathbf{u}; z_k|(n))]^*_{oIK} \;=\; \sum_{\alpha=1}^{n(\mathbf{u})} \lambda^{OK}_{\alpha V}(\mathbf{u}; z_k)\, I(\mathbf{u}_\alpha; z_k)$$

where the weights are given by a block OK system of type (5.44):

$$\begin{cases} \displaystyle\sum_{\beta=1}^{n(\mathbf{u})} \lambda^{OK}_{\beta V}(\mathbf{u}; z_k)\, C_I(\mathbf{u}_\alpha - \mathbf{u}_\beta; z_k) \;+\; \mu^{OK}_V(\mathbf{u}; z_k) \;=\; \overline{C}_I(\mathbf{u}_\alpha, V(\mathbf{u}); z_k) \\ \qquad\qquad\qquad\qquad\qquad\qquad\qquad \alpha = 1, \ldots, n(\mathbf{u}) \\ \displaystyle\sum_{\beta=1}^{n(\mathbf{u})} \lambda^{OK}_{\beta V}(\mathbf{u}; z_k) \;=\; 1 \end{cases}$$

The average indicator covariance $\overline{C}_I(\mathbf{u}_\alpha, V(\mathbf{u}); z_k)$ is approximated by the arithmetic average of the point-support indicator covariances $C_I(\mathbf{u}_\alpha - \mathbf{u}'_i; z_k)$ defined between $\mathbf{u}_\alpha$ and any of the N points $\mathbf{u}'_i$ discretizing the block $V(\mathbf{u})$:

$$\overline{C}_I(\mathbf{u}_\alpha, V(\mathbf{u}); z_k) \;\approx\; \frac{1}{N}\sum_{i=1}^{N} C_I(\mathbf{u}_\alpha - \mathbf{u}'_i; z_k)$$

7.3.3 Accounting for secondary information

The major advantage of the indicator approach is its ability to incorporate soft information of various types in addition to direct measurements on the attribute of interest. Once soft data have been coded into local prior probabilities of type (7.24) or (7.25), they can be processed with hard data using kriging algorithms introduced in Chapter 6. This section presents two other indicator algorithms for incorporating soft data:

1. The most straightforward method consists of a simple IK of the hard indicator data using the soft prior probabilities as local indicator means.

2. The second method is a form of indicator cokriging of hard indicator data using the soft prior probabilities as secondary data.

Simple indicator kriging with local prior means

Consider the situation where primary data $\{z(\mathbf{u}_\alpha),\ \alpha = 1,\ldots,n\}$ are supplemented by an exhaustively sampled[2] secondary information that may relate to either a categorical attribute s or a continuous attribute v. Indicator coding yields, for each threshold value z_k, a set of hard indicator data $\{i(\mathbf{u}_\alpha; z_k),\ \alpha = 1,\ldots,n\}$ and an exhaustive set of soft indicator data $\{y(\mathbf{u}; z_k),\ \mathbf{u} \in \mathcal{A}\}$ defined as:

$$y(\mathbf{u}; z_k) \quad = \quad \text{Prob}\,\{Z(\mathbf{u}) \leq z_k \mid \text{secondary information at } \mathbf{u}\}$$

Recall the simple IK estimator (7.29):

$$[F(\mathbf{u}; z_k | (n))]^*_{sIK} \quad = \quad F(z_k) + \sum_{\alpha=1}^{n(\mathbf{u})} \nu_\alpha^{SK}(\mathbf{u}; z_k)\ [I(\mathbf{u}_\alpha; z_k) - F(z_k)]$$

In simple kriging, the marginal probability $F(z_k)$ does not depend on the location $\mathbf{u}$ and represents the global prior information common to all unsampled locations under the decision of stationarity. To account for the soft datum available at each location, the marginal probability is replaced by the soft prior probability $y(\mathbf{u}; z_k)$ (Goovaerts and Journel, 1995). The simple IK estimator is then rewritten:

$$[F(\mathbf{u}; z_k | (n))]^*_{sIK} \quad = \quad y(\mathbf{u}; z_k) + \sum_{\alpha=1}^{n(\mathbf{u})} \lambda_\alpha^{SK}(\mathbf{u}; z_k)\ [I(\mathbf{u}_\alpha; z_k) - y(\mathbf{u}_\alpha; z_k)]$$

$$(7.43)$$

where $y(\mathbf{u}_\alpha; z_k)$ is identified with the sample conditional frequency $F^*(z_k | s_l)$ or $F^*(z_k | v_l)$, depending on whether the soft information relates to a categorical or a continuous attribute (recall section 7.3.1). The kriging weights are obtained by solving a simple IK system:

$$\sum_{\beta=1}^{n(\mathbf{u})} \lambda_\beta^{SK}(\mathbf{u}; z_k)\ C_R(\mathbf{u}_\alpha - \mathbf{u}_\beta; z_k) \quad = \quad C_R(\mathbf{u}_\alpha - \mathbf{u}; z_k) \qquad \alpha = 1,\ldots,n(\mathbf{u})$$

where $C_R(\mathbf{h}; z_k)$ is the covariance function of the residual RF $R(\mathbf{u}; z_k) = I(\mathbf{u}; z_k) - y(\mathbf{u}; z_k)$ at threshold value z_k.

If the secondary data do not allow a significant differentiation of z-values, the L prior probabilities $F^*(z_k | s_l)$ or $F^*(z_k | v_l)$ would be similar and close to the sample marginal probability $F^*(z_k)$. The estimate would then revert to the simple IK estimate with constant indicator mean. The estimator (7.43) is exact because it honors hard indicator data $i(\mathbf{u}; z_k)$ at their locations.

---

[2] "Exhaustively sampled" refers to a secondary information that is available at all data locations $\mathbf{u}_\alpha$ and at all locations $\mathbf{u}$ being estimated.

Example

Consider that the information available along the NE-SW transect consists of ten Cd concentrations and the profile of five rock categories, as shown in Figure 6.1 (middle graph). The objective is to model the posterior probability that the unknown Cd concentration does not exceed any given threshold value, say, the critical threshold $z_c = 0.8$ ppm. Simple IK with local prior means involves the following steps:

1. The proportion of Cd data not exceeding z_c is first computed for each rock category s_l (Figure 7.20, left top graph).

2. These calibration results are then used to convert the profile of rock types into a profile of local prior probabilities $y(\mathbf{u}; z_c)$ of not exceeding the critical threshold (Figure 7.20, second row). The ten Cd concentrations are transformed into ten hard indicator data $i(\mathbf{u}_\alpha; z_c)$ using relation (7.20).

3. At each hard datum location $\mathbf{u}_\alpha$, the residual value $r(\mathbf{u}_\alpha; z_c)$ is computed by subtracting the soft indicator datum $y(\mathbf{u}_\alpha; z_c)$ from the colocated hard indicator datum $i(\mathbf{u}_\alpha; z_c)$. The semivariogram of residuals is then computed and modeled. Figure 7.20 (right top graph) shows the experimental residual semivariogram inferred from the residual data set (259 data), with the model fitted.

4. The residual values are estimated along the transect using simple kriging and the five closest residual data $r(\mathbf{u}_\alpha; z_c^*)$ (third row). The posterior probability $[F(\mathbf{u}; z_c|(n))]^*_{sIK}$ is obtained by adding the soft indicator datum $y(\mathbf{u}; z_c)$ to the SK estimate $r^*_{SK}(\mathbf{u}; z_c)$ (Figure 7.20, bottom graph).

The profile of simple IK estimates is converted into a profile of probabilities of exceeding the critical threshold 0.8 ppm (Figure 7.21, solid line). As discussed in previous sections and shown in Figure 7.21 (dashed line), ordinary indicator kriging yields a unit probability of exceeding 0.8 ppm in the right part of the transect where all data locations are contaminated. Calibration of geologic information indicates, however, that Cd concentration has a smaller probability (76.5%) of exceeding that threshold on the rock category s_2 prevailing in this part of the transect. Accounting for this local prior information reduces the probability of contamination. Similarly, accounting for the nonzero (33.9%) prior probability of contamination on Argovian rocks increases the probability of contamination in the low-valued part of the transect.

Soft cokriging

Rather than using the soft indicator data $y(\mathbf{u}; z_k)$ as local indicator means, these data can be interpreted as a realization of a RF $Y(\mathbf{u}; z_k)$ correlated with the indicator RF $I(\mathbf{u}; z_k)$. Hard and soft indicator data are then combined using indicator cokriging where $I(\mathbf{u}; z_k)$ and $Y(\mathbf{u}; z_k)$ are the primary

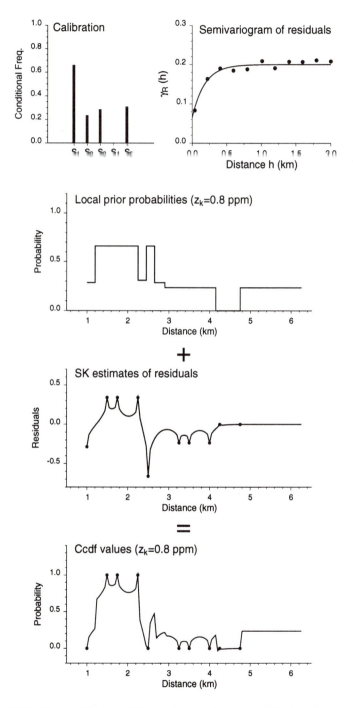

Figure 7.20: Simple kriging with varying local means. The trend component at location **u** is identified with the probability of not exceeding the critical threshold of 0.8 ppm for the rock type prevailing there.

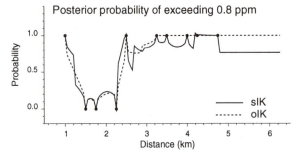

Figure 7.21: Probability of exceeding the critical threshold 0.8 ppm obtained using simple kriging with varying local means, as shown in Figure 7.20 (bottom graph). The dashed line depicts the ordinary kriging estimate using only indicator data.

and secondary variables, respectively. For example, the ordinary indicator cokriging estimator at threshold z_k is

$$[F(\mathbf{u}; z_k | (n))]^*_{oICK} = \sum_{\alpha_1=1}^{n_1(\mathbf{u})} \lambda_{\alpha_1}^{OCK}(\mathbf{u}; z_k) \, I(\mathbf{u}_{\alpha_1}; z_k)$$

$$+ \sum_{\alpha_2=1}^{n_2(\mathbf{u})} \lambda_{\alpha_2}^{OCK}(\mathbf{u}; z_k) \, Y(\mathbf{u}'_{\alpha_2}; z_k) \quad (7.44)$$

where $\lambda_{\alpha_1}^{OCK}(\mathbf{u}; z_k)$ and $\lambda_{\alpha_2}^{OCK}(\mathbf{u}; z_k)$ are the cokriging weights of hard and soft indicator data at locations $\mathbf{u}_{\alpha_1}$ and $\mathbf{u}'_{\alpha_2}$.

Unlike the simple IK estimator (7.43), the soft information need no longer be exhaustive. Note that the "soft" cokriging estimator (7.44) retains only hard and soft indicator data at the threshold z_k being considered. As discussed in section 7.3.2, it is generally not worth accounting for information at other thresholds when indicator vectors are complete (equally sampled case).

The cokriging weights are obtained by solving the following ordinary cokriging system of $(n_1(\mathbf{u}) + n_2(\mathbf{u}) + 1)$ equations:

$$\begin{cases} \displaystyle\sum_{\beta_1=1}^{n_1(\mathbf{u})} \lambda_{\beta_1}^{OCK}(\mathbf{u}; z_k) \, C_I(\mathbf{u}_{\alpha_1} - \mathbf{u}_{\beta_1}; z_k) + \sum_{\beta_2=1}^{n_2(\mathbf{u})} \lambda_{\beta_2}^{OCK}(\mathbf{u}; z_k) \, C_{IY}(\mathbf{u}_{\alpha_1} - \mathbf{u}'_{\beta_2}; z_k) \\[2mm] \qquad + \mu^{OCK}(\mathbf{u}; z_k) = C_I(\mathbf{u}_{\alpha_1} - \mathbf{u}; z_k) \qquad \alpha_1 = 1, \ldots, n_1(\mathbf{u}) \\[2mm] \displaystyle\sum_{\beta_1=1}^{n_1(\mathbf{u})} \lambda_{\beta_1}^{OCK}(\mathbf{u}; z_k) \, C_{YI}(\mathbf{u}'_{\alpha_2} - \mathbf{u}_{\beta_1}; z_k) + \sum_{\beta_2=1}^{n_2(\mathbf{u})} \lambda_{\beta_2}^{OCK}(\mathbf{u}; z_k) \, C_Y(\mathbf{u}'_{\alpha_2} - \mathbf{u}'_{\beta_2}; z_k \\[2mm] \qquad + \mu^{OCK}(\mathbf{u}; z_k) = C_{YI}(\mathbf{u}'_{\alpha_2} - \mathbf{u}; z_k) \qquad \alpha_2 = 1, \ldots, n_2(\mathbf{u}) \\[2mm] \displaystyle\sum_{\beta_1=1}^{n_1(\mathbf{u})} \lambda_{\beta_1}^{OCK}(\mathbf{u}; z_k) + \sum_{\beta_2=1}^{n_2(\mathbf{u})} \lambda_{\beta_2}^{OCK}(\mathbf{u}; z_k) = 1 \end{cases}$$

$$(7.45)$$

where $\mu^{OCK}(\mathbf{u}; z_k)$ is a Lagrange parameter, $C_I(\mathbf{h}; z_k)$ and $C_Y(\mathbf{h}; z_k)$ are the autocovariance functions of the hard and soft indicator RFs, and $C_{IY}(\mathbf{h}; z_k)$ is their cross covariance function. Note the following:

- Only one unbiasedness condition is needed if one can assume that the primary and secondary indicator variables have the same mean within each search neighborhood $W(\mathbf{u})$:

$$E\left[Y(\mathbf{u}'; z_k)\right] \;=\; E\left[I(\mathbf{u}'; z_k)\right] \qquad \forall\, \mathbf{u}' \in W(\mathbf{u})$$

  Unbiasedness is then ensured by constraining all the weights to sum to 1, see the last equation of system (7.45). If this assumption cannot be made, the two traditional unbiasedness constraints of type (6.28) must be used:

$$\sum_{\alpha_1=1}^{n_1(\mathbf{u})} \lambda_{\alpha_1}^{OCK}(\mathbf{u}; z_k) \;=\; 1 \qquad \sum_{\alpha_2=1}^{n_2(\mathbf{u})} \lambda_{\alpha_2}^{OCK}(\mathbf{u}; z_k) \;=\; 0$$

- The cokriging estimator (7.44) can be readily extended to incorporate several different soft indicator variables.

- Two autocovariance functions $C_I(\mathbf{h}; z_k)$ and $C_Y(\mathbf{h}; z_k)$ and a cross covariance $C_{IY}(\mathbf{h}; z_k)$ must be inferred and jointly modeled at each threshold z_k.

Example
Consider that the ten Cd concentrations along the NE-SW transect are supplemented by the five Ni concentrations shown at the top of Figure 7.15. Accounting for the calibration performed in section 7.3.1, hard and soft information are coded into local prior probabilities of not exceeding the critical threshold 0.8 ppm (Figure 7.22, top graph). The direct and cross indicator semivariograms required by the cokriging system (7.45) are inferred from all data available over the study area. The model of coregionalization between hard and soft indicator data for threshold 0.8 ppm is depicted by the solid line in Figure 7.22 (middle graph).

Figure 7.22 (bottom graph) shows the soft cokriging estimate (solid line) of the probability that Cd concentration exceeds 0.8 ppm. The dashed line depicts the ordinary IK estimate using only the ten hard indicator data. As for the example of Figure 7.21, accounting for soft information reduces the probability of contamination in the high-valued (right) part of the transect.

Colocated indicator cokriging

Soft data are usually more numerous than hard data. As a result, the cokriging system (7.45) may be unstable because the correlation between close soft data is much greater than the correlation between more distant hard data. As mentioned in section 6.2.6, one solution to instability problems caused by

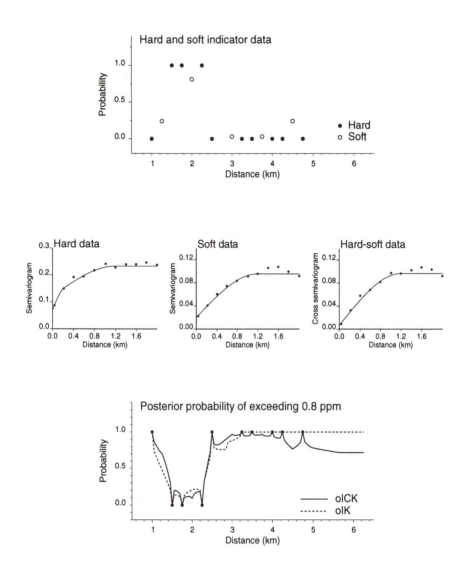

Figure 7.22: Soft cokriging estimate of the probability of exceeding 0.8 ppm. The soft information consists of five prior probabilities of not exceeding the critical threshold as derived from the calibration of Ni concentrations in Figure 7.15. Hard and soft data are combined using ordinary cokriging and the semivariogram models shown in the second row. The dashed line depicts the ordinary kriging estimate using only hard indicator data.

highly redundant secondary information consists of retaining only the secondary datum closest to the location $\mathbf{u}$ being estimated, e.g., the colocated soft indicator datum $y(\mathbf{u}; z_k)$. The indicator cokriging estimator (7.44) is then rewritten

$$[F(\mathbf{u}; z_k|(n))]^*_{oICK} = \sum_{\alpha_1=1}^{n_1(\mathbf{u})} \lambda_{\alpha_1}^{OCK}(\mathbf{u}; z_k) \, I(\mathbf{u}_{\alpha_1}; z_k)$$
$$+ \lambda_2^{OCK}(\mathbf{u}; z_k) \, Y(\mathbf{u}; z_k) \qquad (7.46)$$

where the cokriging weights are solutions of the following system of $(n_1(\mathbf{u})+2)$ equations:

$$\begin{cases} \displaystyle\sum_{\beta_1=1}^{n_1(\mathbf{u})} \lambda_{\beta_1}^{OCK}(\mathbf{u}; z_k) \, C_I(\mathbf{u}_{\alpha_1} - \mathbf{u}_{\beta_1}; z_k) + \lambda_2^{OCK}(\mathbf{u}; z_k) \, C_{IY}(\mathbf{u}_{\alpha_1} - \mathbf{u}; z_k) \\ \qquad\qquad + \mu^{OCK}(\mathbf{u}; z_k) = C_I(\mathbf{u}_{\alpha_1} - \mathbf{u}; z_k) \qquad \alpha_1 = 1, \dots, n_1(\mathbf{u}) \\ \displaystyle\sum_{\beta_1=1}^{n_1(\mathbf{u})} \lambda_{\beta_1}^{OCK}(\mathbf{u}; z_k) \, C_{YI}(\mathbf{u} - \mathbf{u}_{\beta_1}; z_k) + \lambda_2^{OCK}(\mathbf{u}; z_k) \, C_Y(0; z_k) \\ \qquad\qquad + \mu^{OCK}(\mathbf{u}; z_k) = C_{YI}(0; z_k) \\ \displaystyle\sum_{\beta_1=1}^{n_1(\mathbf{u})} \lambda_{\beta_1}^{OCK}(\mathbf{u}; z_k) + \lambda_2^{OCK}(\mathbf{u}; z_k) = 1 \end{cases}$$
$$(7.47)$$

If hard and soft indicator means are different, the soft indicator variable $Y(\mathbf{u}; z_k)$ must be rescaled so that its mean equals the hard indicator mean $F(z_k)$. The cokriging estimator (7.46) is then written

$$[F(\mathbf{u}; z_k|(n))]^*_{oICK} = \sum_{\alpha_1=1}^{n_1(\mathbf{u})} \lambda_{\alpha_1}^{OCK}(\mathbf{u}; z_k) \, I(\mathbf{u}_{\alpha_1}; z_k)$$
$$+ \lambda_2^{OCK}(\mathbf{u}; z_k) \, [Y(\mathbf{u}; z_k) - m_Y(z_k) + F(z_k)]$$

where $m_Y(z_k) = E\{Y(\mathbf{u}; z_k)\}$.

Colocated indicator cokriging is faster than the full indicator cokriging and avoids instability problems caused by densely sampled soft information. Moreover, the colocated cokriging system (7.47) does not require the soft autocovariance model, except at $\mathbf{h} = 0$ where it is $m_Y(z_k)[1 - m_Y(z_k)]$.

Markov–Bayes algorithm

The cokriging system (7.45) calls for the joint modeling of two autocovariance functions and one cross covariance function at each threshold z_k. This modeling of the hard-soft coregionalization may be alleviated by a Markov-type hypothesis stating that "a hard indicator datum at $\mathbf{u}$, $i(\mathbf{u}; z_k)$, screens the influence of any colocated soft indicator data $y(\mathbf{u}; z_k)$ on the estimation of the primary variable at any other location $\mathbf{u}'''$, that is,

$$\text{Prob}\left\{Z(\mathbf{u}') \leq z_k \mid i(\mathbf{u}; z_k), y(\mathbf{u}; z_k)\right\} = \text{Prob}\left\{Z(\mathbf{u}') \leq z_k \mid i(\mathbf{u}; z_k)\right\}$$
$$\forall \, \mathbf{u}, \mathbf{u}', z_k \quad (7.48)$$

Building on this relation, Zhu and Journel (1993) established the following relations between indicator auto and cross covariance functions at any threshold z_k:

$$\begin{aligned} C_Y(\mathbf{h}; z_k) &= |B(z_k)| \, C_I(\mathbf{h}; z_k) \quad \mathbf{h} = 0 \\ &= B^2(z_k) \, C_I(\mathbf{h}; z_k) \quad \forall \, \mathbf{h} > 0 \end{aligned} \quad (7.49)$$

$$C_{IY}(\mathbf{h}; z_k) = B(z_k) \, C_I(\mathbf{h}; z_k) \quad \forall \, \mathbf{h} \quad (7.50)$$

where each coefficient $B(z_k)$ is defined as the difference between the two conditional expectations:

$$B(z_k) = m^{(1)}(z_k) - m^{(0)}(z_k) \in [-1, 1] \quad k = 1, \ldots, K$$

with

$$\begin{aligned} m^{(1)}(z_k) &= \text{E}\left[Y(\mathbf{u}; z_k) \mid I(\mathbf{u}; z_k) = 1\right] \in [0, 1] \\ m^{(0)}(z_k) &= \text{E}\left[Y(\mathbf{u}; z_k) \mid I(\mathbf{u}; z_k) = 0\right] \in [0, 1] \end{aligned}$$

Under the Markov hypothesis (7.48), the autocovariance model $C_Y(\mathbf{h}; z_k)$ at $\mathbf{h} > 0$ and the cross covariance model $C_{IY}(\mathbf{h}; z_k)$ are deduced simply by linear rescaling of the hard autocovariance model, $C_I(\mathbf{h}; z_k)$. Therefore, modeling the hard-soft coregionalization requires the modeling of only a single covariance function per threshold z_k. The Bayesian updating of local prior probabilities by indicator cokriging under the Markov-type relations (7.49) and (7.50) is referred to as the *Markov–Bayes algorithm*.

Remark:
The Markov statement "hard data screen the influence of any colocated soft data" sounds misleadingly trivial. Indeed, if the soft datum refers to a volume larger than the colocated hard datum, it may carry valuable additional information. The resulting covariance models (7.49) and (7.50) must be checked against experimental covariance functions (see hereafter and Figure 7.24).

Estimating the coefficients $B(z_k)$
The determination of coefficients $B(z_k)$ requires the estimation of the two conditional expectations $m^{(1)}(z_k)$ and $m^{(0)}(z_k)$ at each threshold z_k. The quantity $m^{(1)}(z_k)$ is estimated by the arithmetic average of the soft indicator data $y(\mathbf{u}_\alpha; z_k)$, where $i(\mathbf{u}_\alpha; z_k) = 1$:

$$\widehat{m}^{(1)}(z_k) = \frac{1}{\sum_{\alpha=1}^{n_{IY}} i(\mathbf{u}_\alpha; z_k)} \sum_{\alpha=1}^{n_{IY}} y(\mathbf{u}_\alpha; z_k) \cdot i(\mathbf{u}_\alpha; z_k)$$

where n_{IY} is the number of locations where both hard and soft data are known. The best situation is when $\widehat{m}^{(1)}(z_k) = 1$, which means that the soft information y exactly predicts that the value $z(\mathbf{u}_\alpha)$, with $z(\mathbf{u}_\alpha) \leq z_k$, is no greater than the threshold value z_k.

Conversely, $m^{(0)}(z_k)$ is estimated by the arithmetic average of the soft indicator data at locations where $i(\mathbf{u}_\alpha; z_k) = 0$:

$$\widehat{m}^{(0)}(z_k) \;=\; \frac{1}{\sum_{\alpha=1}^{n_{IY}}[1 - i(\mathbf{u}_\alpha; z_k)]} \sum_{\alpha=1}^{n_{IY}} y(\mathbf{u}_\alpha; z_k) \cdot [1 - i(\mathbf{u}_\alpha; z_k)]$$

Here, the best situation is when $\widehat{m}^{(0)}(z_k) = 0$, which indicates that the soft information exactly predicts that the value $z(\mathbf{u}_\alpha)$, with $z(\mathbf{u}_\alpha) > z_k$, exceeds the threshold value z_k.

The difference $\widehat{B}(z_k) = \widehat{m}^{(1)}(z_k) - \widehat{m}^{(0)}(z_k)$ measures the ability of the soft information y to separate the two cases $i(\mathbf{u}; z_k) = 1$ and $i(\mathbf{u}; z_k) = 0$. In other words, $\widehat{B}(z_k)$ is an accuracy index for the soft information:

1. If $\widehat{B}(z_k) = 1$, the autocovariance model of hard and soft indicator RFs, as well as their cross covariance model given by relations (7.49) and (7.50), are identical:

$$C_I(\mathbf{h}; z_k) \;=\; C_Y(\mathbf{h}; z_k) \;=\; C_{IY}(\mathbf{h}; z_k) \qquad \forall\, \mathbf{h}$$

   The indicator cokriging system (7.45) then reverts to an IK system of type (7.32). The local soft prior probabilities $y(\mathbf{u}_\alpha'; z_k)$ are treated as hard indicator data and are not updated.

2. If $\widehat{B}(z_k) = 0$, relation (7.50) yields a zero cross covariance model between hard and soft indicator RFs:

$$C_{IY}(\mathbf{h}; z_k) \;=\; 0 \qquad \forall\, \mathbf{h}$$

   All soft data weights $\lambda_{\alpha_2}^{OCK}(\mathbf{u}; z_k)$ in the cokriging system (7.45) are then negligible.[3]

Example

Consider the estimation of B at the threshold Cd value $z_k = 0.8$ ppm. The soft information consists of either five different rock types s_l or nine classes of Ni concentrations $(v_{l-1}, v_l]$. A calibration like the one performed in section 7.3.1 yields two sets of local prior probabilities: $y_1(\mathbf{u}_\alpha; 0.8) = F^*(0.8|s_l)$ and $y_2(\mathbf{u}_\alpha; 0.8) = F^*(0.8|v_l)$. Figure 7.23 (top graphs) shows the scattergrams of 259 Cd concentrations versus colocated local prior probabilities derived from rock types (left graph) or Ni concentrations (right graph). In

---

[3] Soft data weights are exactly equal to zero if the two traditional unbiasedness constraints of type (6.28) are used in the cokriging system.

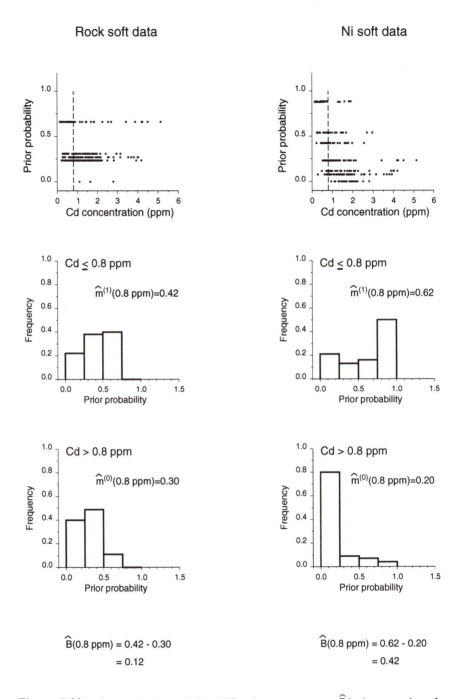

Figure 7.23:   Determination of the calibration parameters $\widehat{B}(z_k)$ measuring the ability of the soft information (rock type on left or Ni concentration on right) to predict whether the Cd concentration exceeds the critical threshold $z_k=0.8$ ppm.

both cases, the 259 prior probabilities are split into two groups, depending on whether the colocated Cd concentration exceeds the critical threshold 0.8 ppm depicted by the vertical dashed line. The histograms of each subset of local prior probabilities at the bottom of Figure 7.23 show the following:

- For Ni data (right column), the two histograms have very different shapes. The prior probabilities of not exceeding 0.8 ppm are large when the actual Cd concentration is no greater than that threshold. Conversely, these prior probabilities are small when the actual Cd concentration exceeds 0.8 ppm. The reasonably large B-value (0.42) obtained as the difference between the means of the two distributions reflects the ability of Ni concentrations to predict whether the Cd concentration exceeds 0.8 ppm.

- For rock types (left column), the contrast between the two histograms of prior probabilities is less apparent. The small B-value (0.12) indicates that geology provides little information on whether the Cd concentration exceeds 0.8 ppm.

Table 7.2 gives the calibration parameters for eight other thresholds of the cumulative distribution of Cd data. The decrease in B with increasing threshold value indicates that the soft information is most valuable for small Cd concentrations. Indeed, very small Cd concentrations are mainly on Argovian rocks, whereas medium and large concentrations, which are evenly distributed on all rock types, are poorly discriminated. Similarly, the calibration scattergram in Figure 7.15 (page 291, second row) shows a tighter relation between small Ni and Cd concentrations.

Table 7.2: Measures of the ability of rock types or Ni concentrations to predict whether the Cd concentration exceeds a given threshold value; prediction is best if $\widehat{B}(z_k) = 1$.

		$\widehat{B}(z_k)$	
z_k	$F^*(z_k)$	Rock type	Ni conc.
0.37	0.10	0.12	0.44
0.56	0.20	0.17	0.44
0.73	0.30	0.16	0.46
0.80	0.35	0.12	0.42
1.06	0.50	0.06	0.30
1.38	0.60	0.02	0.26
1.57	0.70	0.02	0.21
1.88	0.80	0.02	0.15
2.26	0.90	0.04	0.07

Checking the Markov hypothesis
The Markov hypothesis (7.48) and resulting expressions (7.49) and (7.50) are very congenial in that only the autocovariance function or semivariogram of hard indicator data must be modeled. However, that hypothesis should be checked, particularly relation (7.50), which is the most critical in the ensuing cokriging process. If that relation is invalidated, the three auto and cross covariance functions must be jointly modeled using the linear model of coregionalization (4.37) as in the example of Figure 7.22.

Checking the Markov hypothesis at threshold z_k involves the following three steps:

1. Compute and model the autocovariance function $C_I(\mathbf{h}; z_k)$ of hard indicator data.

2. Use that model and the B-value $\widehat{B}(z_k)$ to deduce the auto and cross covariance models $C_Y(\mathbf{h}; z_k)$, $C_{IY}(\mathbf{h}; z_k)$ through relations (7.49) and (7.50),

3. Compare these models with the experimental auto and cross covariance values $\widehat{C}_Y(\mathbf{h}; z_k)$ and $\widehat{C}_{IY}(\mathbf{h}; z_k)$.

Figure 7.24 (top graph) shows the experimental omnidirectional standardized indicator covariance function of Cd at threshold $z_k = 0.8$ ppm with the model fitted. Using that model and the B-values computed in Figure 7.23, the soft autocovariance model and the hard-soft cross covariance model are derived in both cases where the soft information originates from rock types or Ni concentrations. The resulting models are shown as the continuous curves in Figure 7.24 (bottom graph).

From expression (7.49), the Markov-derived autocovariance model $C_Y(\mathbf{h}; z_k)$ shows a nugget effect the size of which increases with decreasing B-value. Therefore, the discontinuity at the origin of the soft autocovariance model is larger for rock type data than for Ni data. In both cases, the Markov-related model severely overestimates the nugget effect of soft indicator data. Such a poor match of the covariance model $C_Y(\mathbf{h}; z_k)$ is of no consequence as long as only the colocated secondary datum $y(\mathbf{u}; z_k)$ is used in the cokriging estimator (7.44). Indeed, the corresponding cokriging system (7.45) requires only the soft autocovariance value at $|\mathbf{h}| = 0$.

Better fits are obtained for the cross covariance functions between hard and soft data. The Markov-derived model still underestimates the short-range continuity for the pair Cd-Ni. In this case, the linear model of coregionalization shown in Figure 7.22 is preferred.

Example
As in the example of Figure 7.20, the information available along the NE-SW transect consists of ten Cd concentrations and the profile of local prior probabilities of not exceeding the threshold $z_c = 0.8$ ppm as derived from five rock types. Rather than interpreting the soft data as local indicator means, these

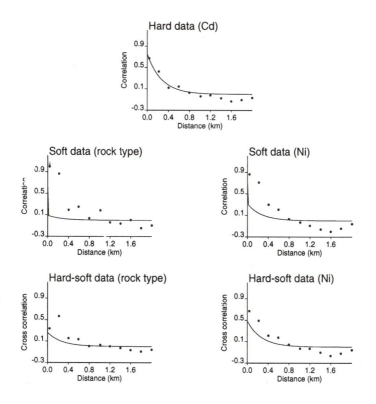

Figure 7.24: Markov-derived models for the indicator auto and cross correlograms of the pairs Cd-rock type and Cd-Ni at the threshold value 0.8 ppm. Models (continuous line) are deduced from the autocovariance model of hard indicator data (top graph) using the Markov-type approximations (7.49) and (7.50) and the calibration parameters obtained in Figure 7.23.

data are here accounted for using the colocated cokriging estimator (7.46) and the Markov-related models of Figure 7.24.

Figure 7.25 shows both colocated ordinary indicator cokriging (solid line) and ordinary indicator kriging (dashed line) estimates of the probability of exceeding 0.8 ppm. Results are similar to those obtained using simple kriging with varying local means in Figure 7.21.

7.3.4 Correcting for order relation deviations

At any location $\mathbf{u}$, each estimated posterior probability $[F(\mathbf{u}; z_k|(n))]^*$ must lie in the interval $[0, 1]$ and the series of such K estimates must be a non-decreasing function of the threshold value z_k:

$$[F(\mathbf{u}; z_k|(n))]^* \in [0, 1] \tag{7.51}$$

$$[F(\mathbf{u}; z_k|(n))]^* \leq [F(\mathbf{u}; z_{k'}|(n))]^* \qquad \forall\, z_{k'} > z_k \tag{7.52}$$

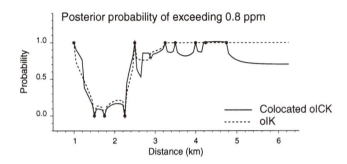

Figure 7.25: Soft cokriging estimate of the probability of exceeding 0.8 ppm. The soft information is the profile of prior probabilities shown in Figure 7.20 (second row). Hard and soft data are combined using ordinary cokriging and the Markov-related models of Figure 7.24. The dashed line depicts the ordinary kriging estimate using only hard indicator data.

The first order relation may not be satisfied because the (co)kriging estimate is a non-convex linear combination of the conditioning data, i.e., the (co)kriging weights can be negative (see section 5.8.1). Because the K probabilities are not estimated jointly, the second condition may not be met either. The posterior cdf in Figure 7.26 shows both types of order relation deviations:

$$[F(\mathbf{u}; z_2|(n))]^* > [F(\mathbf{u}; z_3|(n))]^*$$
$$[F(\mathbf{u}; z_6|(n))]^* > [F(\mathbf{u}; z_7|(n))]^*$$
$$[F(\mathbf{u}; z_8|(n))]^* > 1$$

All order relation deviations in this example have been exaggerated for better illustration. In practice, both types of deviations are generally small, around 0.01–0.03 (Goovaerts, 1994a).

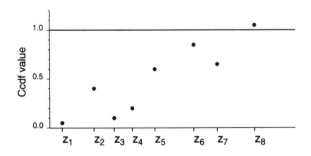

Figure 7.26: Examples of order relation problems shown by ccdf values (black dots) estimated by an indicator approach. The magnitude of order relation deviations is, in practice, much smaller than in this fictitious example.

The following are two major causes of order relations problems (Deutsch and Journel, 1992a, p. 78):

1. The occurrence of negative (co)kriging weights.
   Practice has shown that ordinary indicator kriging and cokriging algorithms produce many more and larger order relation deviations than simple indicator kriging and cokriging (Goovaerts, 1994a). Indeed, the constraints on the weights, particularly the constraint that the secondary data weights must sum to zero, increase the possibility of getting negative weights with a resulting increase in inconsistent estimated probabilities.

2. The lack of z-data in some classes of threshold values.
   Suppose, for example, that the class $(z_6, z_7]$ contains no z-data. The two IK estimates at thresholds z_6 and z_7 are then based on the same indicator data set, since

$$i(\mathbf{u}_\alpha; z_6) \;=\; i(\mathbf{u}_\alpha; z_7) \qquad \forall\, \alpha = 1, \dots, n(\mathbf{u})$$

The difference between the two IK estimates is thus a linear combination of differences between IK weights at the two thresholds z_6 and z_7:

$$
\begin{aligned}
\Delta_{IK}^*(\mathbf{u}; z_6, z_7) \;&=\; [F(\mathbf{u}; z_7|(n))]_{IK}^* \;-\; [F(\mathbf{u}; z_6|(n))]_{IK}^* \\
&=\; \sum_{\alpha=1}^{n(\mathbf{u})} i(\mathbf{u}_\alpha; z_7) \cdot [\lambda_\alpha^{IK}(\mathbf{u}; z_7) - \lambda_\alpha^{IK}(\mathbf{u}; z_6)]
\end{aligned}
$$

$$(7.53)$$

A negative value for the difference $\Delta_{IK}^*(\mathbf{u}; z_6, z_7)$ entails violation of the order relation (7.52). Were the indicator semivariogram models $\gamma_I(\mathbf{h}, z_6)$ and $\gamma_I(\mathbf{h}, z_7)$ the same, the two sets of IK weights would be identical since the same data locations are retained at both thresholds:

$$\lambda_\alpha^{IK}(\mathbf{u}; z_6) \;=\; \lambda_\alpha^{IK}(\mathbf{u}; z_7) \qquad \alpha = 1, \dots, n(\mathbf{u})$$

The difference (7.53) is thus zero, hence there is no order relation deviation of type (7.52). In contrast, a sudden change in two consecutive indicator semivariogram models, say, from z_6 to z_7, leads to different IK weights with an increasing risk of order relation problems.

Implementation tips

Inconsistent probabilities could be avoided by imposing the order relations (7.51) and (7.52) as constraints in the kriging algorithm. Such a solution is, however, computationally expensive. Instead, the common practice is to correct a posteriori for order relation deviations, see subsequent discussion. The proportion and magnitude of these deviations and the required corrections can be reduced using the following implementation tips (see also Deutsch and Journel, 1992a, p. 79–80):

1. Apply oICK/PK algorithms with a single unbiasedness constraint for all cokriging weights so as to reduce the occurrence of negative cokriging weights.

2. Avoid sudden changes in indicator semivariogram parameters from one threshold to the next. One solution consists of modeling all indicator semivariograms using different linear combinations of the same set of basic structures, e.g., a nugget effect and an exponential model:

$$\gamma_I(\mathbf{h}; z_k) \; = \; b^0(z_k) \; + \; b^1(z_k) \, \mathrm{Exp}\,(|\mathbf{h}|; a(z_k)) \qquad k = 1, \ldots, K \tag{7.54}$$

The indicator semivariogram parameters (sill, range, anisotropy direction, and anisotropy ratio) should vary smoothly from one threshold to the next so that:

- There is a continuum in the spatial variability with increasing (or decreasing) thresholds.

- Semivariogram parameters are easily interpolated or extrapolated beyond the initial thresholds z_k. This allows one to retain more thresholds without increasing the inference and modeling effort.

For example, the nine Cd indicator semivariograms in Figure 7.27 are all modeled as the sum of a nugget effect and an exponential model. Note the continuum in the relative nugget effects and effective range values with increasing threshold value (Figure 7.27, bottom graphs).

Relation (7.53) shows that all order relation problems of type (7.52) caused by the lack of data in some classes would be eliminated if the same indicator semivariogram model is used at all thresholds z_k (median indicator kriging). The trade-off cost is the lack of flexibility to model changes in the pattern of spatial continuity from one threshold to another.

3. Select thresholds z_k so that within each search neighborhood $W(\mathbf{u})$ there is at least one datum from each class $(z_{k-1}, z_k]$. Sampling sparsity may dramatically reduce the number of such thresholds. Therefore, rather than using the same set of thresholds z_k over the study area, the thresholds can be made dependent on the local information available within each neighborhood $W(\mathbf{u})$:

- Thresholds z_k that are upper bounds of classes $(z_{k-1}, z_k]$ with no z-data are ignored. In the previous example of Figure 7.26 where the class $(z_6, z_7]$ was assumed empty, the ccdf value at threshold z_7 would not be inferred.

- The threshold values are identified with the $n(\mathbf{u})$ data values $z(\mathbf{u}_\alpha)$ retained in each local neighborhood $W(\mathbf{u})$. The indicator semivariogram model at any threshold value $z_\alpha = z(\mathbf{u}_\alpha)$ is then deduced

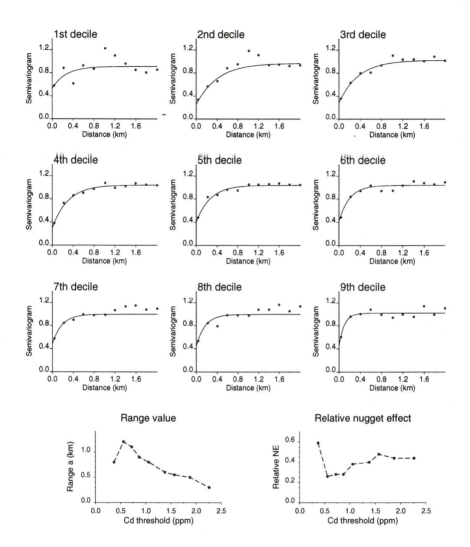

Figure 7.27: Experimental indicator semivariograms for the nine deciles of the sample distribution of Cd data with models fitted. Note the decrease in range value and the increase in relative nugget effect as the threshold value increases.

as

$$\gamma_I(\mathbf{h}; z_\alpha) = b^0(z_\alpha) + b^1(z_\alpha) \, \mathrm{Exp}\left(|\mathbf{h}|; a(z_\alpha)\right)$$

where the parameters $b^0(z_\alpha)$, $b^1(z_\alpha)$, and $a(z_\alpha)$ are interpolated from the parameters of the available indicator semivariogram models $\gamma_I(\mathbf{h}; z_k)$; see relation (7.54) and Figure 7.27 (bottom graphs).

A posteriori correction of ccdf values

In the presence of order relation problems, the original series of ccdf values $\{[F(\mathbf{u}; z_k|(n))]^*, \ k = 1, \ldots, K\}$ must be corrected to a new set of posterior probabilities $\{[F(\mathbf{u}; z_k|(n))]^{**}, \ k = 1, \ldots, K\}$ honoring the two order relations (7.51) and (7.52). Several correction algorithms are available.

One algorithm (Journel, 1984b; Sullivan, 1984) determines ccdf values $[F(\mathbf{u}; z_k|(n))]^{**}$ so as to minimize the weighted sum of squares of corrections over the K thresholds; that is,

$$\sum_{k=1}^{K} \omega_k \cdot \left([F(\mathbf{u}; z_k|(n))]^{**} - [F(\mathbf{u}; z_k|(n))]^* \right)^2$$

is minimized under the $(K + 1)$ linear constraints:

$$\begin{cases} [F(\mathbf{u}; z_{k-1}|(n))]^{**} \leq [F(\mathbf{u}; z_k|(n))]^{**} & k = 2, \ldots, K \\ [F(\mathbf{u}; z_1|(n))]^{**} \geq 0 \\ [F(\mathbf{u}; z_K|(n))]^{**} \leq 1 \end{cases}$$

The weights ω_k allow one to give more importance to critical parts of the ccdf, e.g., lower or upper tails. The solution of such a system can be obtained through quadratic programming.

A more straightforward approach (Deutsch and Journel, 1992a, p. 81) consists of averaging the results of an upward and downward correction of ccdf values; see Figure 7.28 (open circles) and Table 7.3. The correction proceeds in three steps:

1. An upward correction is first performed resulting in a set of K ccdf values $[F(\mathbf{u}; z_k|(n))]_U^{**}$ (Table 7.3, third column):

   - Reset all ccdf values that are not within $[0, 1]$ to the closest bound, 0 or 1:

$$\begin{aligned} [F(\mathbf{u}; z_k|(n))]_U^{**} &= 0 \quad \text{if } [F(\mathbf{u}; z_k|(n))]^* < 0 \\ [F(\mathbf{u}; z_k|(n))]_U^{**} &= 1 \quad \text{if } [F(\mathbf{u}; z_k|(n))]^* > 1 \end{aligned}$$

   - Loop upward through all thresholds applying the correction:

$$\begin{aligned} [F(\mathbf{u}; z_k|(n))]_U^{**} &= [F(\mathbf{u}; z_k|(n))]^* \\ &\quad \text{if } [F(\mathbf{u}; z_k|(n))]^* \geq [F(\mathbf{u}; z_{k-1}|(n))]^* \\ &= [F(\mathbf{u}; z_{k-1}|(n))]^* \quad \text{otherwise} \end{aligned}$$

2. The same approach is used to perform a downward correction resulting in a set of K ccdf values $[F(\mathbf{u}; z_k|(n))]_D^{**}$ (Table 7.3, fourth column):

   - Reset all ccdf values that are not within $[0, 1]$ to the closest bound, 0 or 1:

$$\begin{aligned} [F(\mathbf{u}; z_k|(n))]_D^{**} &= 0 \quad \text{if } [F(\mathbf{u}; z_k|(n))]^* < 0 \\ [F(\mathbf{u}; z_k|(n))]_D^{**} &= 1 \quad \text{if } [F(\mathbf{u}; z_k|(n))]^* > 1 \end{aligned}$$

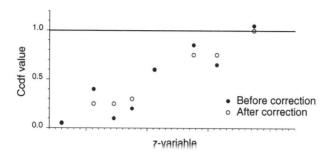

Figure 7.28: Correction of order relation problems of Figure 7.26. The corrected probabilities (open circles) are obtained by averaging results of an upward and downward corrections (Table 7.3).

- Loop downward through all thresholds applying the correction:

$$[F(\mathbf{u}; z_k|(n))]_D^{**} = [F(\mathbf{u}; z_k|(n))]^*$$
$$\text{if } [F(\mathbf{u}; z_k|(n))]^* \leq [F(\mathbf{u}; z_{k+1}|(n))]^*$$
$$= [F(\mathbf{u}; z_{k+1}|(n))]^* \quad \text{otherwise}$$

3. The final set of ccdf values (Table 7.3, fifth column) is the average of the two sets of corrected ccdf values:

$$[F(\mathbf{u}; z_k|(n))]^{**} = \frac{[F(\mathbf{u}; z_k|(n))]_U^{**} + [F(\mathbf{u}; z_k|(n))]_D^{**}}{2} \quad k = 1, \ldots, K$$

To alleviate notations, the corrected ccdf values are hereafter denoted $[F(\mathbf{u}; z_k|(n))]^*$.

Table 7.3: Inconsistent ccdf values shown in Figure 7.28 and their correction using the average of an upward and downward corrections.

		Corrected ccdf values		
Threshold	Faulty prob.	Upward	Downward	Average
z_1	0.05	0.05	0.05	0.05
z_2	0.40	0.40	0.10	0.25
z_3	0.10	0.40	0.10	0.25
z_4	0.20	0.40	0.20	0.30
z_5	0.60	0.60	0.60	0.60
z_6	0.85	0.85	0.65	0.75
z_7	0.65	0.85	0.65	0.75
z_8	1.05	1.00	1.00	1.00

7.3.5 Interpolating/extrapolating ccdf values

Several factors control the number of thresholds that can be handled in the practice of indicator kriging:

1. The sampling density.
   Increasing the number of thresholds enhances the risk of occurrence of empty classes $(z_{k-1}, z_k]$ and the resulting order relation problems, particularly in sparsely sampled areas.

2. The size of the calibration data set.
   For the indicator algorithms that incorporate soft data, the number of local prior probabilities $F^*(z_k|s_l)$ or $F^*(z_k|v_l)$ to infer increases as $K \times L$. When calibration data are sparse, reliable statistics can be obtained for only a few thresholds. One solution consists of a prior smoothing of the sample scattergram, followed by extracting as many conditional cdfs as needed from the smoothed scattergram (Deutsch, 1994b; Xu and Journel, 1995).

3. Inference and modeling resources.
   Except for median IK, the inference and modeling effort increases dramatically as more thresholds are considered; for example, for K thresholds:

   - When performing IK: K direct indicator semivariograms
   - When performing oICK: $K(K + 1)/2$ direct and cross indicator semivariograms
   - When performing PK: $(1 + 2K)$ direct and cross indicator semivariograms

   As mentioned previously, the inference and modeling effort could be alleviated by interpolating semivariogram parameters from the parameters of fewer indicator semivariograms.

The usually poor resolution of the posterior cdf renders critical the interpolation of ccdf values within each class of threshold values $(z_{k-1}, z_k]$ and, more importantly, their extrapolation beyond the smallest threshold z_1 (lower tail) and the largest threshold z_K (upper tail). The three interpolation/extrapolation cdf models (linear, power, hyperbolic) introduced in section 7.2.5 can be applied to ccdf values. For example, Figure 7.29 (top graph) shows the model fitted to the four ccdf values provided by probability kriging (PK) at location $\mathbf{u}_2'$ shown at the top of Figure 7.7 (page 277):

1. The lower tail is extrapolated toward a zero minimum concentration using a negatively skewed power model with $\omega = 2.5$.

2. A linear interpolation is performed separately within each of the three middle classes $(0.56, 0.80]$, $(0.80, 1.38]$, and $(1.38, 1.88]$.

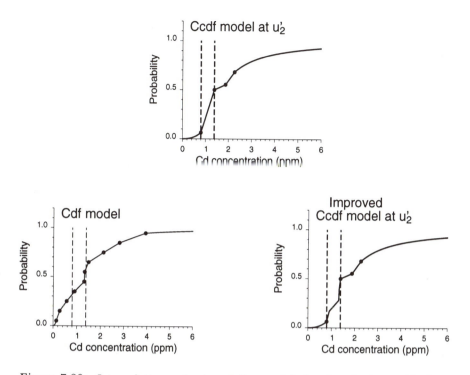

Figure 7.29: Interpolation and extrapolation models fitted to the four ccdf values at location $\mathbf{u}'_2$. The resolution within the class $(0.8, 1.38]$ can be improved by accounting for the more detailed non-conditional cdf model (see bottom graphs).

3. The upper tail is extrapolated toward an infinite upper bound using a hyperbolic model with $\omega = 1.5$.

Because of the limited number of thresholds z_k, the posterior cdf model is much less detailed than the sample cdf model, which is based on all n data $z(\mathbf{u}_\alpha)$ available, say, the ten Cd concentrations along the NE-SW transect (Figure 7.29, left bottom graph). The idea is to capitalize on the higher level of discretization of the cdf to improve the within-class resolution of the ccdf (Figure 7.29, right bottom graph). Interpolation within any class of threshold values $(z_{k-1}, z_k]$ would then proceed as follows:

1. The class $(z_{k-1}, z_k]$ is first split into $L^{(k)}$ subclasses $(z_k^{(l-1)}, z_k^{(l)}]$; for example, for $L^{(k)} = 3$, the three subclasses are $(z_k^{(0)} = z_{k-1}, z_k^{(1)}]$, $(z_k^{(1)}, z_k^{(2)}]$, and $(z_k^{(2)}, z_k^{(3)} = z_k]$. The bound values $z_k^{(l)}$ of the subclasses can be identified with the sample data values falling within the class $(z_{k-1}, z_k]$. Consider, for example, the ccdf class $(0.8, 1.38]$ depicted by the vertical dashed lines in Figure 7.29 (right bottom graph). Four subclasses are defined using as bound values the three Cd data values falling within that class of the sample cumulative distribution (Figure 7.29, left bottom graph).

2. A linear cdf interpolation of type (7.14) is then performed separately within each subclass:

$$[F(z)]_{Lin} \; = \; F^*(z_k^{(l-1)}) \; + \; \left[\frac{z - z_k^{(l-1)}}{z_k^{(l)} - z_k^{(l-1)}} \right] \cdot \left[F^*(z_k^{(l)}) \; - \; F^*(z_k^{(l-1)}) \right]$$

$$\forall \, z \in (z_k^{(l-1)}, z_k^{(l)}] \quad l = 1, \ldots, L^{(k)}$$

3. The series of $L^{(k)}$ cdf models are rescaled linearly, such that the ccdf values at thresholds z_{k-1} and z_k are honored:

$$[F(\mathbf{u}; z|(n))]_{Lin} \; = \; [F(\mathbf{u}; z_{k-1}|(n))]^* \; + \; \varphi \cdot ([F(z)]_{Lin} \; - \; F^*(z_{k-1}))$$

$$\forall \, z \in (z_{k-1}, z_k]$$

$$\text{with} \quad \varphi \; = \; \frac{[F(\mathbf{u}; z_k|(n))]^* \; - \; [F(\mathbf{u}; z_{k-1}|(n))]^*}{F^*(z_k) \; - \; F^*(z_{k-1})}$$

Such an interpolation amounts to using the same intraclass distribution model $[F(z)]_{Lin}$ at all locations $\mathbf{u}$; in other words, the intraclass distribution is non-conditional. A similar approach allows one to increase the resolution of the lower and upper tail classes $(z_{min}, z_1]$ and $(z_K, z_{max}]$.

A good alternative to the piecewise interpolation/extrapolation of the sample cdf consists of smoothing and extrapolating the cdf using the algorithms introduced in section 7.2.5.

7.3.6   Modeling uncertainty for categorical attributes

Consider the problem of modeling the uncertainty about the state s_k of the categorical attribute s at the unsampled location $\mathbf{u}$. For the level of information (n), that uncertainty is modeled by the conditional probability distribution function (cpdf) of the discrete RV $S(\mathbf{u})$:

$$p(\mathbf{u}; s_k|(n)) \; = \; \text{Prob}\,\{S(\mathbf{u}) = s_k|(n)\} \qquad k = 1, \ldots, K \qquad (7.55)$$

where the conditioning information may consist of n neighboring categorical data $s(\mathbf{u}_\alpha)$.

Each conditional (posterior) probability $p(\mathbf{u}; s_k|(n))$ is also the conditional expectation of the class-indicator RV $I(\mathbf{u}; s_k)$ given the information (n):

$$p(\mathbf{u}; s_k|(n)) \; = \; \text{E}\,\{I(\mathbf{u}; s_k)|(n)\} \qquad (7.56)$$

with $I(\mathbf{u}; s_k)=1$ if $S(\mathbf{u}) = s_k$ and zero otherwise. Thus, the conditional pdf of the categorical variable $S(\mathbf{u})$ can be modeled using an indicator approach similar to that introduced to model the conditional cdf of the continuous variable $Z(\mathbf{u})$.

Indicator coding of information

The indicator approach requires a preliminary coding of each piece of information into a vector of K local prior probabilities corresponding to the K states s_k:

$$\text{Prob} \{S(\mathbf{u}) = s_k | \text{ local information at } \mathbf{u}\} \quad k = 1, \ldots, K$$

Different types of local prior p̄dfs can be distinguished, depending on the nature of the local information available:

- A hard datum $s(\mathbf{u}_\alpha)$ is a precise measurement of the state s_k at location $\mathbf{u}_\alpha$ (no uncertainty). The local prior probabilities are then binary (hard) indicator data defined as

$$i(\mathbf{u}_\alpha; s_k) = \begin{cases} 1 & \text{if } s(\mathbf{u}_\alpha) = s_k \\ 0 & \text{otherwise} \end{cases} \quad k = 1, \ldots, K$$

- The local information may consist of a zero probability of occurrence of one or more states $s_{k'}$; for example, a particular land use is known to be absent in a given geologic environment. The local prior pdf is then an incomplete vector of hard indicator data:

$$i(\mathbf{u}_\alpha; s_k) = \begin{cases} 0 & \text{if } k = k' \\ (\text{missing}) & \text{otherwise} \end{cases} \quad k = 1, \ldots, K$$

- Ancillary information (e.g.,calibration of continuous v-data) may provide prior probabilities of occurrence for the K states s_k at location $\mathbf{u}'_\alpha$. The set of local soft indicator data is then defined as

$$y(\mathbf{u}'_\alpha; s_k) \quad = \quad F^*(s_k | v_l) \in [0, 1] \quad k = 1, \ldots, K$$

where $F^*(s_k | v_l)$ is the sample frequency of occurrence of state s_k given that the v-value falls into the class $(v_{l-1}, v_l]$.

Updating into posterior pdf values

The indicator algorithms introduced in sections 7.3.2 and 7.3.3 can be used to estimate each of the K conditional probability values $p(\mathbf{u}; s_k | (n))$ as a linear combination of neighboring hard and soft indicator data.

If a single category s_k prevails at each location $\mathbf{u}$ (mutually exclusive categories), the K hard indicator data $i(\mathbf{u}_\alpha; s_k)$ sum to 1 at any location $\mathbf{u}$. Thus, the K class-indicator RVs $I(\mathbf{u}; s_k)$ are linearly related, leading to linear dependence in the rows and columns of the experimental matrix of indicator covariance functions and a risk of numerical instabilities if all K categories are accounted for in a cokriging system (recall discussion in section 6.2.2). As with the cokriging of linearly related continuous variables, two solutions are:

1. Estimate each posterior probability one at a time, using a cokriging system of type (7.34) and discarding any category that is weakly correlated with the category being estimated.

2. Estimate all posterior probabilities but one, say, the probability $p(\mathbf{u}; s_{k_0}|(n))$ of the category s_{k_0} with the largest global proportion p_{k_0}. The posterior probability of that category at $\mathbf{u}$ is then computed as the complement:

$$[p(\mathbf{u}; s_{k_0}|(n))]^*_{CK} \;=\; 1 \;-\; \sum_{\substack{k=1 \\ k \neq k_0}}^{K} [p(\mathbf{u}; s_k|(n))]^*_{CK}$$

For categorical attributes, the common indicator experimental semivariogram $\widehat{\gamma}_{mIK}(\mathbf{h})$ required by median indicator kriging can be computed as the mean of the K rescaled experimental indicator semivariograms:

$$\widehat{\gamma}_{mI}(\mathbf{h}) = \frac{1}{K} \sum_{k=1}^{K} \frac{\widehat{\gamma}_I(\mathbf{h}; s_k)}{p_k^*(1 - p_k^*)},$$

where $\widehat{\gamma}_I(\mathbf{h}; s_k)$ is the indicator semivariogram (2.24) of category s_k with global proportion p_k^*. Each indicator semivariogram is rescaled by the indicator variance $p_k^*(1 - p_k^*)$. Posterior probabilities are then obtained using an indicator kriging system of type (7.40).

Correcting for order relation deviations

At each location $\mathbf{u}$, the K estimated probabilities $[p(\mathbf{u}; s_k|(n))]^*$ must be valued within $[0, 1]$ and must sum to 1:

$$[p(\mathbf{u}; s_k|(n))]^* \in [0, 1] \qquad k = 1, \ldots, K \qquad\qquad (7.57)$$

$$\sum_{k=1}^{K} [p(\mathbf{u}; s_k|(n))]^* \;=\; 1 \qquad\qquad\qquad (7.58)$$

The non-convexity of the kriging estimator entails that an estimated probability may be negative or greater than 1. Again, the ordinary cokriging constraints on the weights of secondary data increase the risk of getting negative weights, with a resulting increase in faulty probabilities. The second condition (7.58) is rarely satisfied when the K probabilites are estimated separately. However, practice has shown that the magnitude of both types of order relation deviation is usually small, around 0.01–0.03 (Goovaerts, 1994b).

The correction procedure is more straightforward than for conditional cdfs of continuous variables and typically proceeds in two steps:

1. Any posterior probability outside the interval $[0, 1]$ is first reset to the closest bound, 0 or 1.

2. The K estimates $[p(\mathbf{u}; s_k|(n))]^* \in [0, 1]$ are then standardized by the sum $\sum_{k=1}^{K}[p(\mathbf{u}; s_k|(n))]^*$ to meet the second condition (7.58):

$$[p(\mathbf{u}; s_k|(n))]^{**} = \frac{[p(\mathbf{u}; s_k|(n))]^*}{\sum_{k=1}^{K}[p(\mathbf{u}; s_k|(n))]^*}$$

An alternative consists of estimating all probabilities but one, say, the posterior probability $p(\mathbf{u}; s_{k_0}|(n))$ related to the most frequent category s_{k_0}. The estimates outside the interval $[0, 1]$ are then reset to the nearest bound, and the remaining probability $[p(\mathbf{u}; s_{k_0}|(n))]^*$ is computed as

$$[p(\mathbf{u}; s_{k_0}|(n))]^* = 1 - \sum_{\substack{k=1 \\ k \neq k_0}}^{K} [p(\mathbf{u}; s_k|(n))]^*$$

One drawback of this approach is that the posterior probability $[p(\mathbf{u}; s_{k_0}|(n))]^*$ tends to accumulate all errors affecting the $(K-1)$ other estimates $[p(\mathbf{u}; s_k|(n))]^*$. Therefore, the category s_{k_0} should be the one of least interest or the one with a large global proportion p_{k_0}, or both.

Example

Consider the problem of modeling the uncertainty about the prevailing land use along the NE-SW transect. Figure 7.30 (top graphs) shows the ten data available and their coding into indicators of presence/absence of each land use. Standardized indicator semivariograms are inferred from all data available over the study area (Figure 7.30, bottom graphs). All semivariograms have a small nugget effect and, except for forests, reach a sill at around 300 m. The probability of occurrence of each land use is determined every 50 m using ordinary indicator kriging and the five closest indicator data. The four profiles of corrected probabilities are shown in Figure 7.31.

7.4   Using Local Uncertainty Models

Let $\{F(\mathbf{u}; z|(n)), \mathbf{u} \in \mathcal{A}\}$ be the set of conditional distributions (ccdfs) defined over the study area $\mathcal{A}$. Each function $F(\mathbf{u}; z|(n))$ fully models the uncertainty at location $\mathbf{u}$ in that it gives, for a continuous variable, the probability that the unknown is no greater than any given threshold z:

$$F(\mathbf{u}; z|(n)) = \text{Prob}\{Z(\mathbf{u}) \leq z|(n)\}$$

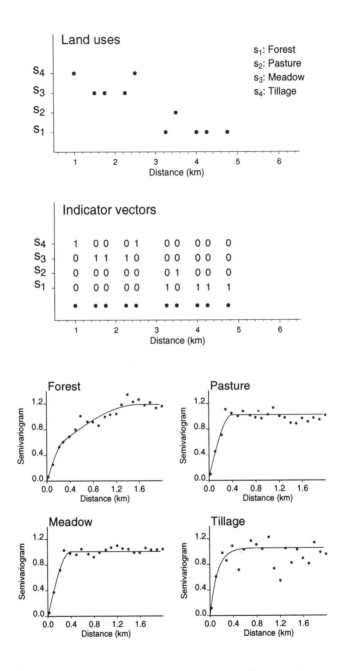

Figure 7.30:   Coding of ten categorical data into indicators of presence/absence of each land use. Standardized indicator semivariograms are inferred from all data in the study area.

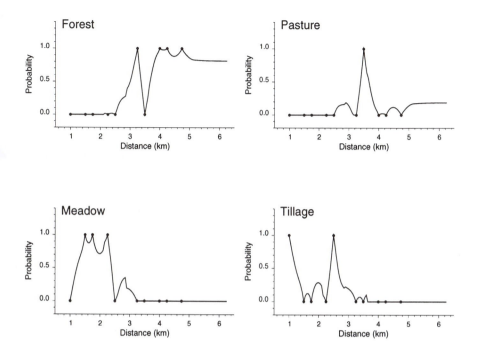

Figure 7.31: Ordinary kriging estimate of the probability that a given land use prevails along the NE-SW transect. The information available consists of ten data and the indicator semivariogram models shown in Figure 7.30.

The model of local uncertainty is typically post-processed to retrieve single values such as

1. the probability of exceeding a critical threshold at $\mathbf{u}$,

2. an estimate of the unknown value $z(\mathbf{u})$, which is optimal for a given criterion which need not be least-squares, and

3. the expected cost of classifying the location $\mathbf{u}$ as, for example, safe or contaminated.

Maps of these different quantities are then used in decision-making processes, such as the delineation of candidate areas for remediation or additional sampling.

7.4.1   Measures of local uncertainty

Knowledge of the conditional cdf model at location $\mathbf{u}$ allows a straightforward assessment of the uncertainty about the unknown $z(\mathbf{u})$ before and independently of the choice of a particular estimate for that unknown.

Probability interval

The probability that the unknown is valued within an interval $(a, b]$, called a probability interval, is computed as the difference between ccdf values for thresholds b and a:

$$\text{Prob}\{Z(\mathbf{u}) \in (a, b]|(n)\} \;=\; F(\mathbf{u}; b|(n)) - F(\mathbf{u}; a|(n)) \qquad (7.59)$$

A 50% probability interval means that the unknown has equal probability to lie inside or outside the interval $(a, b]$.

Setting the upper bound b of the interval to $+\infty$ provides the probability of exceeding the threshold a:

$$
\begin{aligned}
\text{Prob}\{Z(\mathbf{u}) \in (a, +\infty]|(n)\} \;&=\; \text{Prob}\{Z(\mathbf{u}) > a|(n)\} \\
&=\; 1 - F(\mathbf{u}; a|(n)) \qquad (7.60)
\end{aligned}
$$

This latter probability is of particular importance for environmental applications, where the focus is on the risk of exceeding regulatory limits.

Consider the two ccdf models $F(\mathbf{u}'_1; z|(n))$ and $F(\mathbf{u}'_2; z|(n))$ in Figure 7.32:

- The probability that the Cd concentration is valued between 1.5 and 4 ppm, depicted between the two vertical dashed lines, is negligible at $\mathbf{u}'_1$ in the low-valued part of the transect.

- The uncertainty is much larger at $\mathbf{u}'_2$, with the unknown Cd concentration having similar probabilities to be valued inside or outside the interval $(1.5, 4 \text{ ppm}]$.

Unlike the local confidence intervals shown in Figure 7.1 (page 260), probability intervals do not require the prior determination of particular estimates $z^*(\mathbf{u}'_1)$ and $z^*(\mathbf{u}'_2)$.

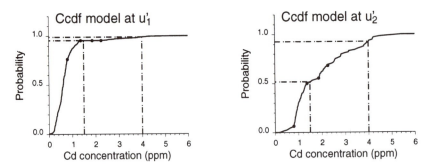

Figure 7.32: Conditional cdf models provided by probability kriging at locations $\mathbf{u}'_1$ and $\mathbf{u}'_2$. In both cases, a linear interpolation is performed between tabulated bounds identified with the 259 Cd concentrations over the study area. The dashed lines depict the probability interval corresponding to the interval $(1.5, 4.0]$ ppm.

Local entropy

A measure of local uncertainty, not specific to any particular interval $(a, b]$, is provided by the entropy of the local probability density function (Shannon, 1948; Christakos, 1990; Journel and Deutsch, 1993):

$$H(\mathbf{u}) = \int_{-\infty}^{+\infty} -[\ln f(\mathbf{u}; z|(n))] \cdot f(\mathbf{u}; z|(n)) \, dz \qquad (7.61)$$

where $f(\mathbf{u}; z|(n)) = \partial F(\mathbf{u}; z|(n))/\partial z$ is the conditional pdf, and all zero pdf values are excluded from the integral.

The bounded pdf with maximum entropy is the uniform distribution (3.17): all outcomes have the same probability of occurrence. The entropy, and hence uncertainty, decreases as the probability distribution focuses more toward a single value.

In practice, the range of variation of z-values is discretized into K non-overlapping classes $(z_{k-1}, z_k]$, and the corresponding K probability intervals are computed as

$$p_k(\mathbf{u}) = F(\mathbf{u}; z_k|(n)) - F(\mathbf{u}; z_{k-1}|(n))$$

There are typically many more discretization values z_k than original threshold values used for indicator coding.

The entropy of the conditional pdf at location $\mathbf{u}$ is then computed as

$$H(\mathbf{u}) \simeq -\sum_{k=1}^{K'} [\ln p'_k(\mathbf{u})] \cdot p'_k(\mathbf{u}) \geq 0 \qquad (7.62)$$

where $K' \leq K$ is the number of non-zero probability intervals $p'_k(\mathbf{u})$. Beware that a measure of local uncertainty depends on the particular ccdf model $F(\mathbf{u}; z|(n))$, hence better but heavier notations would be $p'_k(\mathbf{u}|(n))$ and $H(\mathbf{u}|(n))$.

For a given level of discretization, extreme entropy values are

- zero   if $p_{k'}(\mathbf{u}) = 1$ and $p_k(\mathbf{u}) = 0$ $\forall \ k \neq k'$
  The unknown is certainly valued in the interval $(z_{k'-1}, z_{k'}]$
  (minimum uncertainty).

- $\ln K$ if $p_k(\mathbf{u}) = 1/K$ $\forall \ k$
  Each interval $(z_{k-1}, z_k]$ is equally likely to include the unknown (maximum uncertainty).

Thus, for a discretization level K, a standardized measure of the entropy is

$$H_R(\mathbf{u}) = \frac{H(\mathbf{u})}{\ln K} \in [0, 1] \qquad (7.63)$$

The two conditional cdfs of Figure 7.32 are converted into pdfs using 30 discretization classes of equal amplitude 0.2 ppm (Figure 7.33). The entropy

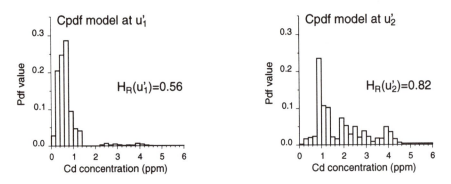

Figure 7.33:   Conditional pdf models deduced from a discretization of the ccdf models in Figure 7.32 using 30 classes of equal amplitude 0.2 ppm. The smaller entropy $H_R(\mathbf{u}'_1)$ expresses the greater certainty about the Cd concentration as modeled by the narrower cpdf at location $\mathbf{u}'_1$.

of each distribution is then computed using relation (7.63). The smaller entropy at $\mathbf{u}'_1$ reflects the greater certainty associated with that location. Such a result agrees with the intuitive feeling that the uncertainty should be smaller at $\mathbf{u}'_1$, which is surrounded by two similar data values, than at $\mathbf{u}'_2$, which is surrounded by two extreme data values. Recall that the kriging variance, which is data-independent, would not differentiate the uncertainty prevailing at these two locations (Figure 7.1, page 260).

Conditional variance

Other statistics can be used to measure the spread of the conditional pdf at location $\mathbf{u}$. For example, the conditional variance $\sigma^2(\mathbf{u})$ measures the spread of the conditional probability distribution around its mean $z_E^*(\mathbf{u})$:

$$\sigma^2(\mathbf{u}) \quad = \quad \int_{-\infty}^{+\infty} [z - z_E^*(\mathbf{u})]^2 \cdot f(\mathbf{u}; z|(n)) \, dz$$

As with the entropy measure, this integral is, in practice, approximated by the discrete sum:

$$\sigma^2(\mathbf{u}) \quad \simeq \quad \sum_{k=1}^{K+1} [\bar{z}_k - z_E^*(\mathbf{u})]^2 \cdot [F(\mathbf{u}; z_k|(n)) - F(\mathbf{u}; z_{k-1}|(n))] \qquad (7.64)$$

where

- z_k, $k = 1, \ldots, K$, are K threshold values discretizing the range of variation of z-values. By convention, $F(\mathbf{u}; z_0|(n)) = 0$ and $F(\mathbf{u}; z_{K+1}|(n)) = 1$. Other thresholds z_k could be identified with p-quantiles corresponding to regularly spaced ccdf increments, $z_k = F^{-1}(\mathbf{u}; k/[K + 1]|(n))$.

- $\bar{z}_k$ is the mean of the class $(z_{k-1}, z_k]$, which depends on the within-class interpolation model; e.g., for the linear model, $\bar{z}_k = (z_{k-1} + z_k)/2$.

- $z_E^*(\mathbf{u})$ is the expected value of the ccdf approximated by the discrete sum:

$$z_E^*(\mathbf{u}) \simeq \sum_{k=1}^{K+1} \bar{z}_k \cdot [F(\mathbf{u}; z_k|(n)) - F(\mathbf{u}; z_{k-1}|(n))] \qquad (7.65)$$

Unlike the entropy measure (7.63), the variance is defined around a specific central value, the mean of the conditional distribution, and it depends on the K within-class means $\bar{z}_k$. Beware that both the ccdf mean and its upper tail mean can be very sensitive to the choice of the extrapolation model, as would be the conditional variance $\sigma^2(\mathbf{u})$ (see subsequent example).

For highly asymmetric distributions, a more robust measure of spread is the interquartile range, defined as the difference between the upper and lower quartiles of the distribution:

$$\begin{aligned} q_R(\mathbf{u}) &= q_{0.75}(\mathbf{u}) - q_{0.25}(\mathbf{u}) \\ &= F^{-1}(\mathbf{u}; 0.75|(n)) - F^{-1}(\mathbf{u}; 0.25|(n)) \end{aligned} \qquad (7.66)$$

Because it does not use means of extreme classes, the interquartile range is less affected by the choice of a particular extrapolation model for the upper tail.

Consider the following three different extrapolation models for the upper tail of the two ccdfs at locations $\mathbf{u}_1'$ and $\mathbf{u}_2'$ (Figure 7.34, top graphs):

1. Linear interpolation between tabulated bounds provided by the sample cdf (259 Cd data),

2. Hyperbolic model with a short tail ($\omega = 5$), and

3. Hyperbolic model with a long tail ($\omega = 1.5$).

The three models provide similar fits at location $\mathbf{u}_1'$ because of the small contribution of the upper tail to the ccdf $F(\mathbf{u}_1'; z|(n))$. The impact of the upper tail model is much greater at $\mathbf{u}_2'$ in the high-valued part of the transect. In this example, the sample-interpolated distribution model (solid line) is intermediate between the two hyperbolic models depicted by dashed lines. The short-tail model (small dashed line) yields larger probabilities for intermediate Cd concentrations (2.5–3 ppm), whereas the long-tail model (large dashed line) increases the probability of occurrence of large Cd concentrations (> 5.4 ppm) (Figure 7.34, bottom graphs). Note the following:

- The choice of the upper tail extrapolation model greatly influences the ccdf mean and variance at location $\mathbf{u}_2'$.

- Differences between ccdf models at $\mathbf{u}_1'$, though of small magnitude, lead to means $z_E^*(\mathbf{u}_1')$ that fluctuate above or below the critical threshold 0.8 ppm. Decision-making based on that threshold would then dramatically depend on the upper tail model.

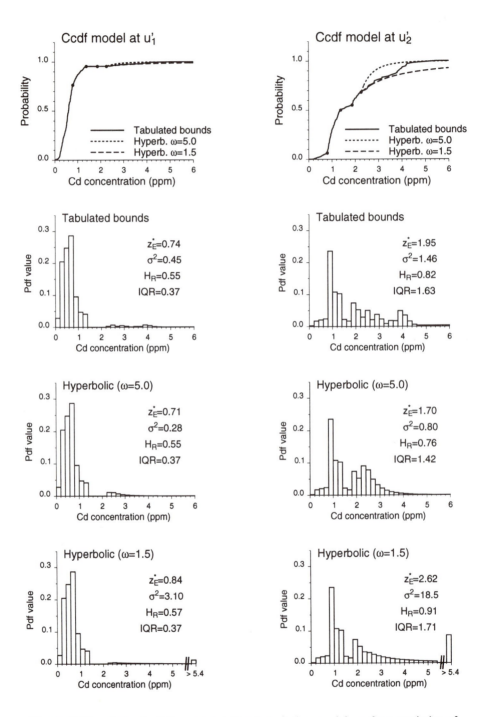

Figure 7.34: Impact of the upper tail extrapolation model on four statistics of the conditional pdfs at locations $\mathbf{u}_1'$ and $\mathbf{u}_2'$: mean, variance, local entropy, and interquartile range.

- Other measures of spread (entropy, interquartile range) are less sensitive
  to the choice of the upper tail extrapolation model.

Example

Figure 7.35 shows the ordinary kriging variance and three statistics (variance, entropy, and interquartile range) of the ccdf models obtained along the NE-SW transect using probability kriging. The upper tail is linearly interpolated between tabulated bounds (first option). At all locations with the same data configuration, the ordinary kriging variance is the same whatever the

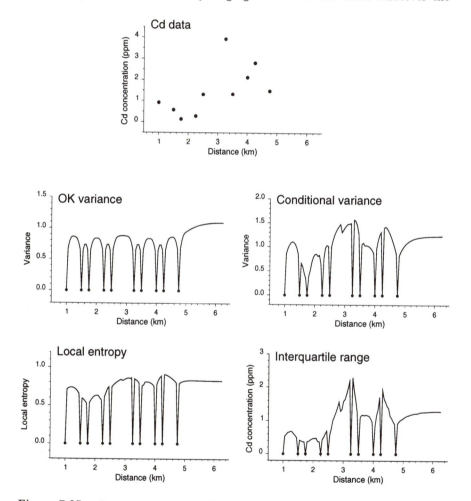

Figure 7.35:   Four measures of local uncertainty along the NE-SW transect. As opposed to the kriging variance, ccdf statistics (conditional variance, entropy, and interquartile range) account for data values and indicate a larger uncertainty in the right part of the transect where the range of Cd concentrations is larger.

surrounding Cd concentrations. The three other measures, which account for data values, indicate that the uncertainty about Cd concentration is greater where the range of data is larger, that is, in the central and right parts of the transect (Figure 7.35). The contrast between the two parts of the transect is most apparent when using the interquartile range.

7.4.2   Optimal estimates

Probability or local entropy maps allow locations to be ranked according to their level of uncertainty about the unknown value $z(\mathbf{u})$. Such a ranking could suffice for decision-making such as delineation of areas where remedial measures should be taken. Areas with the largest probabilities of exceeding the tolerable maximum would be cleaned first, followed by others in decreasing order of their probabilities of exceedence. However, measures of local uncertainty must be typically supplemented by an estimate $z^*(\mathbf{u})$ of the unknown value because decision makers rarely think only in terms of probability.

Loss function

The selection of a unique estimate within the range of possible z-values requires an optimality criterion. One common criterion is the minimization of the impact attached to the estimation error $e(\mathbf{u}) = z^*(\mathbf{u}) - z(\mathbf{u})$ that is likely to occur. Consider, for example, the estimation of a toxic concentration. Underestimation of that concentration (negative estimation error) may cause ill health and lead to insurance claims and lawsuits. Conversely, overestimation of the toxic concentration (positive estimation error) may cause costly and unnecessary cleaning.

Evaluate the impact or loss associated with any error as a function $L(.)$ of that error, e.g., $L(e(\mathbf{u})) = [e(\mathbf{u})]^2$. Given that particular loss function, the estimate $z_L^*(\mathbf{u})$ should be chosen so as to minimize the resulting loss; that is,

$z_L^*(\mathbf{u})$ should be such that $L(z^*(\mathbf{u}) - z(\mathbf{u}))$ is minimal.

Determination of the actual loss $L(z^*(\mathbf{u}) - z(\mathbf{u}))$ requires the actual value $z(\mathbf{u})$, which is unknown in practice. However, the uncertainty about $z(\mathbf{u})$ is modeled by the ccdf $F(\mathbf{u}; z|(n))$, which is available. The idea is to use this model of uncertainty to determine the expected loss:

$$\varphi_L(z^*(\mathbf{u})|(n)) = \mathrm{E}\{L(z^*(\mathbf{u}) - Z(\mathbf{u}))|(n)\}$$
$$= \int_{-\infty}^{+\infty} L(z^*(\mathbf{u}) - z) \, dF(\mathbf{u}; z|(n))$$

In practice, the following discrete approximation is used:

$$\varphi_L(z^*(\mathbf{u})|(n)) \simeq \sum_{k=1}^{K+1} L(z^*(\mathbf{u}) - \bar{z}_k) \cdot [F(\mathbf{u}; z_k|(n)) - F(\mathbf{u}; z_{k-1}|(n))]$$

$$(7.67)$$

where z_k and $\bar{z}_k$ are defined as in equation (7.64). The "L-optimal" estimate
for the loss function $L(.)$ is then the z-value that minimizes the expected loss:

$$z_L^*(\mathbf{u}) \text{ is such that } \varphi_L(z^*(\mathbf{u})|(n)) \text{ is minimal.}$$

Unlike interpolation algorithms introduced in Chapters 5 and 6, here the
determination of an optimal estimate proceeds in two steps:

1. The uncertainty about the unknown value $z(\mathbf{u})$ is first modeled by the
   conditional cdf $F(\mathbf{u}; z|(n))$.

2. From that model, an estimate $z_L^*(\mathbf{u})$ is deduced according to a specific
   optimality criterion.

Such a dichotomy between assessment of uncertainty and estimation clearly
shows that there is no best estimate for all situations. Rather, the goodness
of an estimate depends on the intended use of that estimate. For a given
model of uncertainty, different estimates can be obtained depending on the
loss function chosen (Journel, 1984a; Srivastava, 1987a).

E-type estimate

Consider first the common least-squares criterion; that is, the loss is modeled
as a quadratic function of either underestimation or overestimation error:

$$L(e(\mathbf{u})) \quad = \quad [e(\mathbf{u})]^2 \tag{7.68}$$

see Figure 7.36 (left top graph). The optimal estimate is shown to be the
expected value of the ccdf at location $\mathbf{u}$, also called the E-type estimate, recall
relation (7.65):

$$z_L^*(\mathbf{u}) \quad = \quad z_E^*(\mathbf{u}) \simeq \sum_{k=1}^{K+1} \bar{z}_k \cdot [F(\mathbf{u}; z_k|(n)) \; - \; F(\mathbf{u}; z_{k-1}|(n))]$$

For example, the E-type estimate at $\mathbf{u}_2'$, $z_E^*(\mathbf{u}_2')$, depicted by the vertical
dashed line in Figure 7.36 (right top graph), is 1.95 ppm.

E-type and (co)kriging estimates are usually different, although both are
optimal for the least-squares criterion. They differ in that the ccdf from which
the E-type estimate derives depends on data values. One exception occurs
when the original z-values are normally distributed and the ccdf is modeled
using a multiGaussian approach. In all other situations, the advantage of the
E-type estimate lies in the availability of a model of uncertainty much richer
than a mere kriging variance (recall sections 7.1.1 and 7.4.1).

Because of the squaring of the estimation error in expression (7.68), extreme error values tend to have a great impact on the expected loss. Thus,
the E-type estimate may overly depend on the modeling of the upper and
lower tails of the distribution $F(\mathbf{u}; z|(n))$. For the positively skewed ccdf at
location $\mathbf{u}_2'$, the E-type estimate varies from 1.70 to 2.62 ppm, depending on
whether the ω-parameter of the hyperbolic tail model is set to 5 (short tail)
or 1.5 (long tail) (Table 7.4).

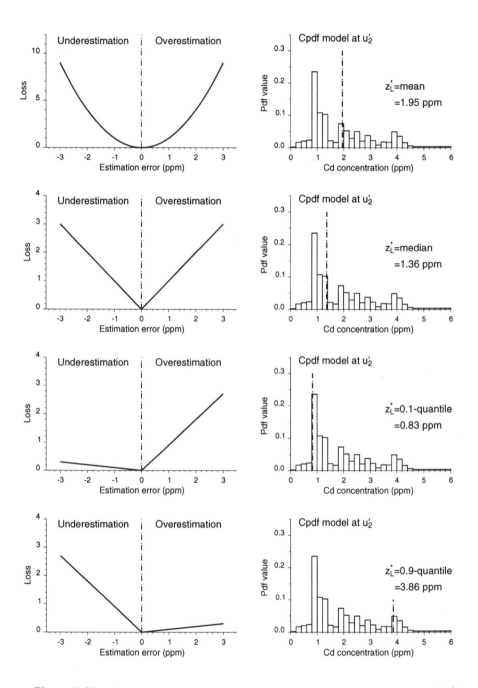

Figure 7.36:   Four loss functions and the corresponding optimal estimates $z_L^*(u_2')$ depicted by the vertical dashed line on the conditional probability distribution.

Table 7.4: Impact of the upper tail extrapolation model on various optimal estimates at location $\mathbf{u}_1'$ and $\mathbf{u}_2'$.

Optimal estimates		Extrapolation options	
	Tabulated bounds	Hyperbolic ($\omega = 5$)	Hyperbolic ($\omega = 1.5$)
Location $\mathbf{u}_1'$			
Mean (E-type)	0.74	0.71	0.84
Median	0.61	0.61	0.61
0.1-quantile	0.28	0.28	0.28
0.9-quantile	1.12	1.12	1.12
Location $\mathbf{u}_2'$			
Mean (E-type)	1.95	1.70	2.62
Median	1.36	1.36	1.36
0.1-quantile	0.83	0.83	0.83
0.9-quantile	3.86	2.85	4.90

Median estimate

Using a linear function of the estimation error rather than the quadratic function (7.68) allows one to reduce the impact of tail models on the expected loss. For example, the loss can be modeled as proportional to the absolute estimation error:

$$L(e(\mathbf{u})) \;=\; |e(\mathbf{u})| \tag{7.69}$$

see Figure 7.36 (second row, left graph). The optimal estimate for this mean absolute deviation criterion is the median of the ccdf:

$$z_L^*(\mathbf{u}) \;=\; q_{0.5}(\mathbf{u}) \;=\; F^{-1}(\mathbf{u}; 0.5|(n))$$

The median of the ccdf at location $\mathbf{u}_2'$ is 1.36 ppm, a value much smaller than the E-type estimate $z_E^*(\mathbf{u}_2')$=1.95 ppm. This difference arises because of the positive skewness of the conditional pdf at $\mathbf{u}_2'$. The median value is usually not affected by the choice of a particular extrapolation model for the upper tail (Table 7.4).

Quantile estimates

In the two previous examples, the loss was modeled as a function of the absolute magnitude of the estimation error regardless of sign. Such symmetric loss functions amount to penalizing overestimation and underestimation equally. In most applications, the actual loss is asymmetric. For decisions about cleaning, the impact of underestimating a toxic concentration (ill health and lawsuit) is likely larger than that of an overestimation of equal amplitude (cost of unnecessary cleaning). Conversely, when predicting deficiencies in soil nutrients, the impact of underestimation (undue application of corrective treatments) is likely to be smaller than that of overestimation (risk of retarded growth or premature death). Thus, it is often critical to distinguish underestimation from overestimation when modeling the impact of estimation errors.

Consider the following asymmetric linear loss function:

$$L(e(\mathbf{u})) = \begin{cases} \omega_1 \cdot e(\mathbf{u}) & \text{for } e(\mathbf{u}) \geq 0 \text{ (overestimation)} \\ \omega_2 \cdot |e(\mathbf{u})| & \text{for } e(\mathbf{u}) < 0 \text{ (underestimation)} \end{cases} \qquad (7.70)$$

where the non-negative parameters ω_1 and ω_2 are the relative impacts attached to overestimation and underestimation, respectively. The optimal estimate is shown to be the p-quantile of the ccdf (Journel, 1984a):

$$z_L^*(\mathbf{u}) = q_p(\mathbf{u}) = F^{-1}(\mathbf{u}; p|(n))$$

$$\text{with} \qquad p = \frac{\omega_2}{\omega_1 + \omega_2} \in [0, 1]$$

Three cases can be distinguished:

1. $\omega_1 = \omega_2$
   The linear loss function (7.70) is then symmetric, and the optimal estimate is the 0.5-quantile, i.e., the median previously discussed.

2. $\omega_1 > \omega_2$
   The impact of overestimation is larger than that of underestimation of the same magnitude. Thus, $p < 0.5$, and the optimal estimate is smaller than the median (a conservative choice for detecting deficiencies). Consider the asymmetric loss function in Figure 7.36 (third row, left graph), where $\omega_1 = 0.9$ and $\omega_2 = 0.1$. The optimal estimate is the 0.1-quantile of the ccdf depicted by the vertical dashed line in Figure 7.36 (third row, right graph). That quantile estimate can be interpreted as the threshold value that has a 90% probability of being exceeded by the unknown value:

$$\text{Prob} \{Z(\mathbf{u}) > q_{0.1}(\mathbf{u})|(n)\} = 0.9$$

   Thus a large estimate $q_{0.1}(\mathbf{u})$ indicates that the unknown value is certainly large at location $\mathbf{u}$.

3. $\omega_1 < \omega_2$
   The impact of overestimation is smaller than that of underestimation of the same magnitude. Thus, $p > 0.5$, and the optimal estimate is larger than the median (a conservative choice for detecting pollution). Consider the asymmetric loss function in Figure 7.36 (left bottom graph), where $\omega_1 = 0.1$ and $\omega_2 = 0.9$. The optimal estimate is the 0.9-quantile

of the ccdf depicted by the vertical dashed line in Figure 7.36 (right bottom graph). That quantile estimate can be interpreted as the threshold value that has only a 10% probability of being exceeded by the unknown value:

$$\text{Prob}\,\{Z(\mathbf{u}) > q_{0.9}(\mathbf{u})|(n)\} \;=\; 0.1$$

A small estimate $q_{0.9}(\mathbf{u})$ thus indicates that the unknown value is certainly small at location $\mathbf{u}$.

Increasing the contrast between the relative impacts w_1 and w_2 yields quantile estimates corresponding to very small or very large cumulative probabilities p. Beware that any quantile value outside the range of threshold values, i.e., $q_p(\mathbf{u}) < z_1$ or $q_p(\mathbf{u}) > z_K$, would depend strongly on the model used to extrapolate the lower or the upper tail. For example, the 0.9-quantile estimate at location $\mathbf{u}_2'$, $q_{0.1}(\mathbf{u}_2') = 3.86$ ppm, is larger than the maximum threshold $z_4=2.26$ ppm. Such an estimate is thus very sensitive to the upper tail model (Table 7.4, last row).

Other L-optimal estimates

The three types of loss functions shown in Figure 7.36 (left column) allow a straightforward (analytical) determination of the optimal estimate. In absence of such an analytical solution, the expected loss can be computed for a series of z-values, and the one yielding the smallest expected loss is retained as the optimal estimate. Consider, for example, the following asymmetric loss function, which severely penalizes underestimation:

$$L(e(\mathbf{u})) \;=\; \begin{cases} e(\mathbf{u}) & \text{for } e(\mathbf{u}) \geq 0 \text{ (overestimation)} \\ [e(\mathbf{u})]^2 & \text{for } e(\mathbf{u}) < 0 \text{ (underestimation)} \end{cases} \qquad (7.71)$$

see Figure 7.37 (left top graph). Knowledge of the ccdf model $F(\mathbf{u}_2'; z|(n))$ allows one to compute the expected loss for a series of Cd concentrations ranging from 0 to 6 ppm (Figure 7.37, bottom graph). The optimal estimate is 2.4 ppm, which corresponds to the minimum expected loss.

Remarks

1. The impact of the estimation error may vary from one area to another; for example, the underestimation of a toxic concentration is likely to be more prejudicial in residential areas than in industrial yards. One should then define a loss function $L(e(\mathbf{u}); \mathbf{u})$ that changes locally, depending on whether $\mathbf{u}$ is in an industrial or residential area.

2. In practice, the losses associated with both types of estimation error may be difficult to evaluate precisely. However, the user generally knows which type of error would be most prejudicial in the situation at hand. Simple asymmetric loss functions, such as expressions (7.70) or (7.71),

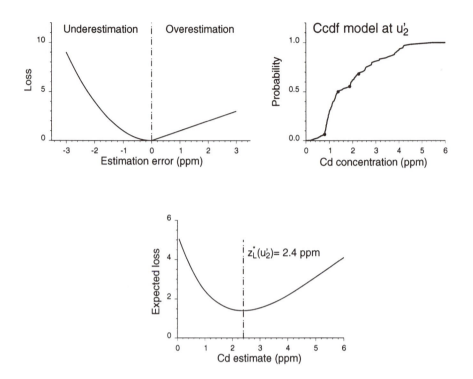

Figure 7.37:   Expected loss for various possible estimates of the Cd concentration at location u'_2 (bottom graph). The optimal estimate $z^*_L(u'_2)$ is the concentration that minimizes the expected loss assessed from the ccdf model and the asymmetric loss function shown at the top of the figure.

should then be preferred to the usual least-squares criterion that penalizes equally underestimation and overestimation.

3. Whatever the optimality criterion retained, the resulting estimate $z^*_L(\mathbf{u})$ honors the soft information available at location $\mathbf{u}$. Consider, for example, that the soft information consists of a constraint interval $(a, b]$ for the unknown value $z(\mathbf{u})$. The posterior cdf at $\mathbf{u}$ then takes the form

$$F(\mathbf{u}; z|(n)) = \begin{cases} 1 & \text{if } b \le z \\ \in [0, 1] & \text{if } z \in (a, b] \\ 0 & \text{if } a > z \end{cases}$$

The optimal estimate $z^*_L(\mathbf{u})$ necessarily lies within the interval $(a, b]$ since there is a zero probability for it being outside that interval.

7.4.3 Decision making in the face of uncertainty

Many investigations lead to important decisions, such as cleaning hazardous areas or correcting for soil deficiencies. Decisions are most often made in the face of uncertainty because concentrations in toxic or nutrient elements are rarely known with certainty. Given a conditional cdf model, there are several ways to account for such uncertainty in the decision-making process.

Exceeding a probability threshold

Consider the decision of cleaning locations contaminated by cadmium. A straightforward approach consists of declaring contaminated all locations where the probability of exceeding the tolerable maximum 0.8 ppm is larger than a given probability threshold. There is no doubt that location $\mathbf{u}_2'$, with a 95% probability of exceeding 0.8 ppm, is a prime candidate for remediation. Decision making is much more difficult at location $\mathbf{u}_1'$, which has a 34% probability of exceeding the tolerable maximum. In such a case, the choice of a probability threshold is subjective and depends on political and social decisions; for example, a 34% probability of contamination may be unacceptable for residential areas, yet tolerable for industrial yards.

In practice, remedial measures are applied to an area or block $V(\mathbf{u})$, not to a single location $\mathbf{u}$. Decision making can then proceed in many different ways, depending on whether the regulatory threshold applies to a block, say, z_c^V for a truckload-sized volume, or to a single location, say, z_c^p for an auger-sampled size volume. Examples of different approaches follow:

- Model the ccdf at N locations $\mathbf{u}_j'$ discretizing the block $V(\mathbf{u})$, and retrieve the N corresponding probabilities of exceeding the critical threshold, $1 - \left[F(\mathbf{u}_j'; z_c^p|(n))\right]^*$. Then, decide to clean the block if the probability threshold p_c is exceeded by a given proportion of locations.

- Model the composite ccdf, which gives the average probability that a location $\mathbf{u}$ within $V(\mathbf{u})$ is no greater than any threshold value z, as $\left[F_N(\mathbf{u}; z|(n))\right]^* = 1/N \sum_{j=1}^{N} \left[F(\mathbf{u}_j'; z|(n))\right]^*$. Retrieve the composite probability of contamination, $1 - \left[F_N(\mathbf{u}; z_c^p|(n))\right]^*$, then decide to clean the block if that probability exceeds the probability threshold p_c.

- Model the block ccdf $\left[F_V(\mathbf{u}; z|(n))\right]^*$, which gives the probability that the average z-value over the block $V(\mathbf{u})$ is no larger than any threshold value z, and retrieve the block probability of contamination, $1 - \left[F_V(\mathbf{u}; z_c^V|(n))\right]^*$. Then, decide to clean the block if that block probability is larger than the probability threshold p_c.

The first approach does not call for averaging probabilities or z-estimated values. Thus, a block could be declared contaminated if the probability threshold is exceeded at a few locations within that block. In contrast, the use of a block ccdf may lead to overly optimistic decisions because the average z-value smooths out extreme values.

Exceeding a physical threshold

Another approach consists of declaring contaminated all locations where the estimated Cd concentration exceeds the tolerable maximum $z_c = 0.8$ ppm. This approach requires the prior determination of an estimate for the unknown concentration. As shown for location $\mathbf{u}'_1$ in Table 7.4 (first row, page 343), different optimality criteria and interpolation ccdf models yield different estimates, which may or may not exceed the critical threshold 0.8 ppm. Therefore, there is a risk of declaring contaminated a safe location. Conversely, one might declare safe a contaminated location. These two misclassification risks can be assessed from the conditional cdf model $F(\mathbf{u}; z|(n))$:

1. The risk $\alpha(\mathbf{u})$ of wrongly classifying a location $\mathbf{u}$ as hazardous (false positive) is

$$\begin{aligned} \alpha(\mathbf{u}) &= \text{Prob}\left\{Z(\mathbf{u}) \le z_c | z_L^*(\mathbf{u}) > z_c, (n)\right\} \\ &= F(\mathbf{u}; z_c|(n)) \end{aligned}$$

for all locations $\mathbf{u}$ such that the estimate $z_L^*(\mathbf{u}) > z_c$.

2. The risk $\beta(\mathbf{u})$ of wrongly classifying a location $\mathbf{u}$ as safe (false negative) is

$$\begin{aligned} \beta(\mathbf{u}) &= \text{Prob}\left\{Z(\mathbf{u}) > z_c | z_L^*(\mathbf{u}) \le z_c, (n)\right\} \\ &= 1 - F(\mathbf{u}; z_c|(n)) \end{aligned}$$

for all locations $\mathbf{u}$ such that the estimate $z_L^*(\mathbf{u}) \le z_c$.

Note that the risk $\beta(\mathbf{u})$ is not defined where the risk $\alpha(\mathbf{u})$ is, and conversely. Again, one faces the difficult problem of choosing a probability threshold for each misclassification risk.

Figure 7.38 (top graph) shows the median and 0.9-quantile estimates retrieved from the ccdf models obtained using probability kriging. In both cases, a location $\mathbf{u}$ is declared hazardous with a risk $\alpha(\mathbf{u})$ of false positive if the estimate exceeds the critical threshold depicted by the horizontal dashed line (Figure 7.38, middle graphs). Conversely, if the estimate at $\mathbf{u}$ is smaller than 0.8 ppm, that location is declared safe with a risk $\beta(\mathbf{u})$ of false negative (Figure 7.38, bottom graphs). Note the following:

- At any particular location $\mathbf{u}$, the magnitude of the misclassification risk depends on the ccdf model, not on the particular estimate (median or 0.9-quantile) retained.

- The type of misclassification risk at $\mathbf{u}$ depends, however, on the estimate $z_L^*(\mathbf{u})$, that is, on the optimality criterion. As the loss function penalizes underestimation more severely, estimates tend to be larger, thus limiting the areas with false negative risk $\beta(\mathbf{u})$ at the expense of potential large

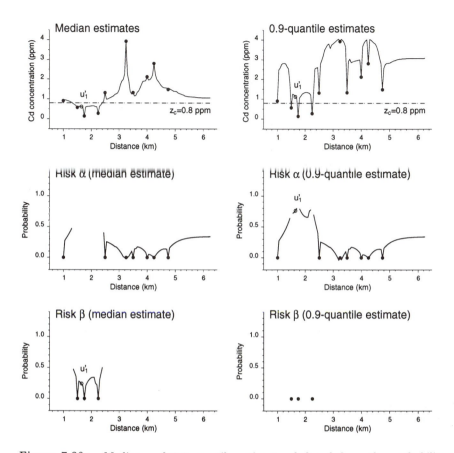

Figure 7.38: Median and 0.9-quantile estimates deduced from the probability kriging ccdf models, and the corresponding misclassification risks: α is the risk of wrongly declaring that a location is hazardous on the basis that the estimate $z_L^*(\mathbf{u})$ exceeds 0.8 ppm, whereas β is the risk of wrongly declaring that a location is safe on the basis that the estimate $z_L^*(\mathbf{u})$ is smaller than 0.8 ppm.

risks $\alpha(\mathbf{u})$. For example, location $\mathbf{u}_1'$ is classified as safe with a risk $\beta(\mathbf{u}_1') = 0.24$ on the basis of the median estimate (Figure 7.38, bottom left graph). The same location $\mathbf{u}_1'$ is classified as hazardous with a risk $\alpha(\mathbf{u}_1') = 0.76$ on the basis of the 0.9-quantile estimate (Figure 7.38, middle right graph). In the latter case, all unsampled locations are declared contaminated with a large misclassification risk α in the low-valued (left) part of the transect.

- Misclassification risks can be used to rank locations candidates for additional sampling: locations with the highest risks $\beta(\mathbf{u})$ are preferentially sampled. Such a criterion would lead one to locate additional samples in the low-valued part of the transect.

Beware that a classification based on an estimate is no more objective than a classification based on a probability of exceeding a critical threshold. Indeed, an optimality criterion must be chosen, and the resulting estimate may depend on somewhat subjective decisions about models for extrapolating ccdf tails.

Minimization of the expected loss

A third approach consists of evaluating the economic impact of the two possible decisions using the concept of loss functions, introduced in section 7.4.2. Each location is classified as safe or contaminated so as to minimize the resulting expected loss (Journel, 1987). Unlike previous approaches based on probabilities of exceeding critical thresholds or misclassification risks, here the decision is based on financial costs.

As in the determination of optimal estimates, the key step is the specification of economic functions that measure the impact of the two types of misclassification:

1. The loss associated with classifying a location $\mathbf{u}$ as safe could be modeled as

$$
L_1(z(\mathbf{u})) = \begin{cases} 0 & \text{if } z(\mathbf{u}) \leq z_c \\ w_1 \left[z(\mathbf{u}) - z_c \right] & \text{otherwise} \end{cases} \tag{7.72}
$$

   where w_1 is the relative cost of underestimating the toxic concentration, e.g., potential ill health (w_1 is in units of money/concentration, say, dollar / ppm), and z_c is the critical threshold. If the location $\mathbf{u}$ is actually safe, $z(\mathbf{u}) \leq z_c$, then the classification is correct and there is no loss. If the location is actually contaminated, $z(\mathbf{u}) > z_c$, the misclassification cost is modeled as proportional to the actual contamination $[z(\mathbf{u}) - z_c]$.

2. The loss associated with classifying a location $\mathbf{u}$ as contaminated could be modeled as

$$
L_2(z(\mathbf{u})) = \begin{cases} 0 & \text{if } z(\mathbf{u}) > z_c \\ w_2 & \text{otherwise} \end{cases} \tag{7.73}
$$

   The remediation cost is here modeled as a constant value w_2. For example, the cleaning procedure amounts to removing the upper layer of soil, hence the cost is independent of the actual concentration and w_2 is in units of money only. An alternative consists of modeling the remediation cost as proportional to the estimated contamination $[z_L^*(\mathbf{u}) - z_c]$, which calls for a prior estimate of the unknown concentration at $\mathbf{u}$.

The conditional cdf model $F(\mathbf{u}; z|(n))$ allows one to determine the expected loss attached to the two types of classification as

$$
\varphi_1(\mathbf{u}) \;=\; E\left[L_1(Z(\mathbf{u}))|(n)\right] \;=\; \int_{-\infty}^{+\infty} L_1(z(\mathbf{u})) \, dF(\mathbf{u}; z|(n))
$$

$$\varphi_2(\mathbf{u}) \quad = \quad E\left[L_2(Z(\mathbf{u}))|(n)\right] \quad = \quad \int_{-\infty}^{+\infty} L_2(z(\mathbf{u})) \, dF(\mathbf{u}; z|(n))$$

which are, in practice, approximated as

$$\varphi_1(\mathbf{u}) \quad \simeq \quad \sum_{k=1}^{K+1} L_1(\bar{z}_k) \, [F(\mathbf{u}; z_k|(n)) - F(\mathbf{u}; z_{k-1}|(n))] \qquad (7.74)$$

$$\varphi_2(\mathbf{u}) \quad \simeq \quad \sum_{k=1}^{K+1} L_2(\bar{z}_k) \, [F(\mathbf{u}; z_k|(n)) - F(\mathbf{u}; z_{k-1}|(n))] \qquad (7.75)$$

where $\bar{z}_k$ and K are defined as in equation (7.64). The location $\mathbf{u}$ is then declared safe or contaminated so as to minimize the resulting expected loss:

$$\varphi_1(\mathbf{u}) \quad > \quad \varphi_2(\mathbf{u}) \quad \Rightarrow \quad \mathbf{u} \text{ is classified as contaminated.}$$
$$\varphi_1(\mathbf{u}) \quad < \quad \varphi_2(\mathbf{u}) \quad \Rightarrow \quad \mathbf{u} \text{ is classified as safe.}$$

Consider the problem of whether to classify locations along the NE-SW transect as contaminated by cadmium. The loss attached to the two types of classification is modeled by functions (7.72) and (7.73) with two sets of w-costs: (1) $w_1=1$, $w_2=0.5$ and (2) $w_1=1$, $w_2=2.5$ (Figure 7.39, top graphs). Figure 7.39 (middle graphs) shows the expected losses computed using the ccdf models provided by probability kriging. The expected loss associated with declaring a location safe (solid line) is larger in the high-valued (right) part of the transect, where the unknown Cd concentration more likely exceeds the critical threshold. Conversely, the expected loss associated with declaring a location contaminated (dashed line) is larger in the low-valued (left) part of the transect, where there is a small probability of exceeding the critical threshold. The minimization of the expected loss yields the classifications shown at the bottom of Figure 7.39. As the remediation cost w_2 increases, the impact of a false positive (unnecessary cleaning) increases, hence more locations are declared safe (Figure 7.39, right bottom graph).

7.4.4   Simulation

Let $F(\mathbf{u}; z|(n))$ be the conditional distribution (ccdf) modeling the uncertainty about the unknown $z(\mathbf{u})$. Rather than deriving a single estimated value $z^*(\mathbf{u})$ from that ccdf, one may draw from it a series of L simulated values $z^{(l)}(\mathbf{u})$, $l = 1, \dots, L$. Each value $z^{(l)}(\mathbf{u})$ represents a possible outcome or realization of the RV $Z(\mathbf{u})$ modeling the uncertainty at location $\mathbf{u}$.

The "Monte-Carlo" simulation proceeds in two steps:

1. A series of L independent random numbers $p^{(l)}$, $l = 1, \dots, L$, uniformly distributed in $[0, 1]$, is drawn.

2. The lth simulated value $z^{(l)}(\mathbf{u})$ is identified with the $p^{(l)}$-quantile of the ccdf (Figure 7.40):

$$z^{(l)}(\mathbf{u}) \quad = \quad F^{-1}(\mathbf{u}; p^{(l)}|(n)) \qquad l = 1, \dots, L \qquad (7.76)$$

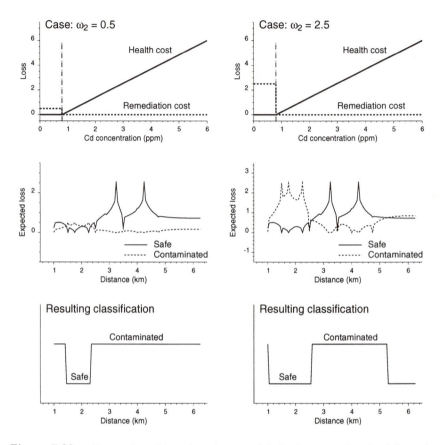

Figure 7.39:   Two pairs of loss functions model the loss associated with an incorrect classification of a location as safe (ill health) or contaminated (unnecessary cleaning). As the remediation cost ω_2 increases, the expected loss attached to undue cleaning increases relatively to the expected health costs, leading to more locations being classified as safe.

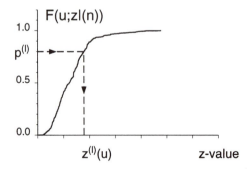

Figure 7.40:   Monte-Carlo simulation from a conditional cdf $F(\mathbf{u}; z|(n))$. The lth simulated value $z^{(l)}(\mathbf{u})$ is obtained as the $p^{(l)}$-quantile of the conditional distribution where $p^{(l)}$ is a random number uniformly distributed in $[0, 1]$.

The L simulated values $z^{(l)}(\mathbf{u})$ are distributed according to the conditional cdf. Indeed,

$$\text{Prob}\{Z^{(l)}(\mathbf{u}) \le z\} = \text{Prob}\{F^{-1}(\mathbf{u}; p^{(l)}|(n)) \le z\}$$

from the definition (7.76),

$$= \text{Prob}\{p^{(l)} \le F(\mathbf{u}; z|(n))\}$$

since $F(\mathbf{u}; z|(n))$ is monotonic nondecreasing,

$$- F(\mathbf{u}; z|(n))$$

since the random numbers $p^{(l)}$ are uniformly distributed in $[0, 1]$.

This property of ccdf reproduction allows one to approximate any moment or quantile of the conditional distribution by the corresponding moment or quantile of the histogram of many realizations $z^{(l)}(\mathbf{u})$. Thus, Monte-Carlo simulation provides an alternative to the approximations of type (7.64) or (7.65) for computing the conditional variance or E-type estimate. Note though that determination of the quantile value (7.76) still requires interpolation and extrapolation from calculated ccdf values.

Stochastic simulation can be extended to the modeling of uncertainty about the output value of any complex transfer function at any location $\mathbf{u}$. Consider, for example, the problem of modeling the uncertainty about the loss associated with classifying the location $\mathbf{u}_2$ as safe. Figure 7.41 (left graph) shows the histogram of 1,000 simulated Cd-values drawn from the ccdf model depicted in Figure 7.32 (page 334, right graph). The simulated histogram is close to the conditional pdf model shown in Figure 7.33 (right graph), as it should be. The 1,000 realizations $z^{(l)}(\mathbf{u}_2)$ can be fed into the loss function (7.72) to yield a simulated distribution of costs at location $\mathbf{u}_2$, $y^{(l)}(\mathbf{u}_2) = L_1[z^{(l)}(\mathbf{u}_2)]$, $l = 1, \dots, L$ (Figure 7.41, right graph). In

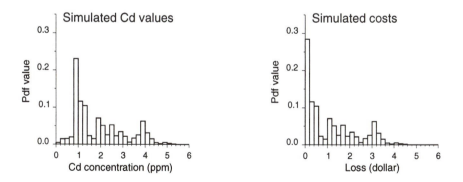

Figure 7.41: Histogram of 1,000 Cd values simulated at location $\mathbf{u}_2$ using the quantile algorithm (left graph). These 1,000 realizations are fed into the loss function (7.72), yielding the output distribution of costs on the right.

contrast, the approximation (7.74) provides only an estimate of the mean of the distribution (expected cost) without any information on the spread of that distribution.

7.4.5   Classification of categorical attributes

The previous analysis of local uncertainty can be extended to a categorical attribute s with K mutually exclusive categories s_k. Consider, for example, the land use data at the top of Figure 7.42. The uncertainty about whether land use s_k prevails at any unsampled location $\mathbf{u}$ is modeled by the vector of K conditional probabilities $[p(\mathbf{u}; s_k|(n))]^*$. Figure 7.42 (middle graphs) shows the probability distributions of four land uses at locations $\mathbf{u}'_3$ and $\mathbf{u}'_4$. Intuitively, the uncertainty is smaller at $\mathbf{u}'_3$ where the category s_3 (meadow) has a much larger probability of occurrence than any other category. In contrast, three categories have non-negligible probabilities to prevail at $\mathbf{u}'_4$. The post-processing of conditional probability distributions allows one to quantify such local uncertainty about the prevailing category $s(\mathbf{u})$ and derive an optimal estimate for that category.

Measures of local uncertainty

The uncertainty attached to a particular location $\mathbf{u}$ could be measured as 1 minus the largest conditional probability at this location:

$$\phi(\mathbf{u}) \quad = \quad 1 - [p(\mathbf{u}; s_{k_{max}}(\mathbf{u})|(n))]^* \tag{7.77}$$

where $s_{k_{max}}$ is the category with the largest conditional probability at $\mathbf{u}$. At any datum location, the probability is 1 for the prevailing category, hence measure (7.77) equals zero. At other locations, the probability $\phi(\mathbf{u})$ is valued in $(0, 1]$. The probability $\phi(\mathbf{u})$ at locations $\mathbf{u}'_3$ and $\mathbf{u}'_4$ is 0.10 and 0.53, respectively, which reflects the larger uncertainty about the land use at $\mathbf{u}'_4$.

Another measure of local uncertainty is the entropy of the conditional probability distribution:

$$H(\mathbf{u}) = - \sum_{k=1}^{K} \ln[p(\mathbf{u}; s_k|(n))]^* \ [p(\mathbf{u}; s_k|(n))]^*$$

As with the measure (7.77), the local entropy equals zero at any datum location (no uncertainty). At other locations, the local entropy is valued in $(0, \ln K]$. The upper bound $(\ln K)$ is the maximum entropy associated with the uniform distribution, $[p(\mathbf{u}; s_k|(n))]^* = 1/K, \ \forall \ k$. Thus, a standardized measure of the local entropy is

$$H_R(\mathbf{u}) = \frac{H(\mathbf{u})}{\ln K} \quad \in \ [0, 1] \tag{7.78}$$

The local entropy at locations $\mathbf{u}'_3$ and $\mathbf{u}'_4$ is 0.28 and 0.87, respectively. Again, the uncertainty is larger at $\mathbf{u}'_4$.

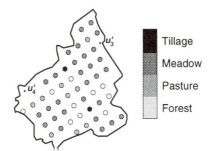

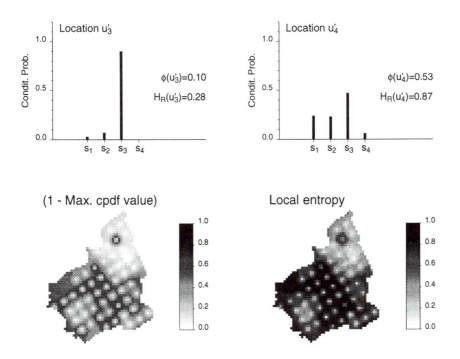

Figure 7.42: Probabilities of occurrence of the four land uses as determined at locations u'_3 and u'_4 using ordinary indicator kriging and 50 categorical data. The bottom maps depict two measures of local uncertainty over the study area: one minus the largest probability of occurrence, and the entropy of the conditional probability distribution.

Figure 7.42 (bottom graphs) shows the maps of the two measures of local uncertainty over the study area. Both maps look alike and indicate greater uncertainty in the central part where all four land uses are recorded. The uncertainty is smallest in the northern part of the area where almost all data belong to the same category, meadow.

Classification procedures

Unlike continuous attributes, categorical attributes, such as rock types or land uses, cannot be estimated as a mere linear combination of neighboring data. A straightforward technique of estimation or classification would amount to allocating the unsampled location **u** to the category of the nearest datum, a technique known as Thiessen polygons or polygonal method (Figure 7.43, right top graph). The polygonal method has two shortcomings: (1) it ignores important information, such as patterns of spatial continuity and transition probabilities among categories, and (2) it provides no measure of the reliability of the classification.

Another approach, which accounts for spatial information, consists of allocating the location **u** to the category with the largest conditional probability of occurrence (Figure 7.43, left bottom graph). The risk of misclassification is then depicted by the ϕ-probability map shown in Figure 7.42 (left bottom graph). Because a conditional probability $[p(\mathbf{u}; s_k|(n))]^*$ is likely to be large if the corresponding marginal probability p_k is already large, most of the locations tend to be allocated to the most frequent category. For example, 74% of the grid nodes in Figure 7.43 (left bottom graph) are allocated to meadow. Conversely, the less frequent category, in this case tillage, tends to be underrepresented since its conditional probability of occurrence is generally much smaller than that of other categories. Such a classification is thus inadequate if one aims at reproducing sample proportions, which are deemed representative of the entire area.

Soares (1992) developed a classification algorithm that preferentially allocates locations to the category with the largest conditional probability of occurrence under the constraint of reproduction of global proportions. The classification algorithm is dynamic in the sense that the allocation rule changes as the classification progresses. The algorithm proceeds as follows:

1. For each category s_k, the N grid locations are ranked according to decreasing conditional probabilities $[p(\mathbf{u}; s_k|(n))]^*$.

2. The n_k locations ($n_k = N\ p_k$) with the largest conditional probabilities $[p(\mathbf{u}; s_k|(n))]^*$ are allocated to the category s_k.

3. If a location **u** is classified into two or more categories, e.g., s_k and $s_{k'}$, it is allocated to the category s_k with the largest conditional probability of occurrence, i.e., $[p(\mathbf{u}; s_k|(n))]^* > [p(\mathbf{u}; s_{k'}|(n))]^*$. The $(n_{k'} + 1)$th location with the largest conditional probability $[p(\mathbf{u}; s_{k'}|(n))]^*$ is then allocated to category $s_{k'}$, so that the global proportion $p_{k'}$ is reproduced. The procedure is repeated until each grid node belongs to a single category.

Such a classification maximizes the average conditional probability π_S of occurrence over the study area $\mathcal{A}$ while ensuring an exact reproduction of the

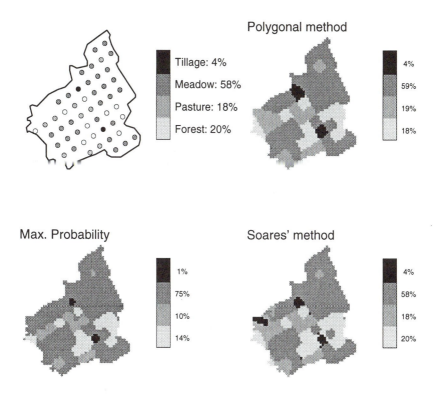

Figure 7.43:   Sample data and maps of land uses obtained using the polygonal method, the maximization of the local conditional probability, and Soares' algorithm. Note the overrepresentation of meadow and the underrepresentation of tillage produced by maximization of the local conditional probability.

global proportions p_k; that is,

$$\pi_S = \frac{1}{N} \sum_{\beta=1}^{N} \sum_{k=1}^{K} i_S(\mathbf{u}_\beta; k) \left[p(\mathbf{u}_\beta; s_k | (n)) \right]^*$$

is maximum under the K constraints:

$$\frac{1}{N} \sum_{\beta=1}^{N} i_S(\mathbf{u}_\beta; s_k) = p_k \qquad k = 1, \dots, K$$

where $i_S(\mathbf{u}_\beta; s_k) = 1$ if location $\mathbf{u}_\beta$ is allocated to class s_k, and $i_S(\mathbf{u}_\beta; s_k) = 0$ otherwise. Figure 7.43 (right bottom graph) shows the map produced by Soares' algorithm. In this particular example, the polygonal method and Soares' algorithm yield similar results.

7.5 Performance Comparison

The risk of the Cd, Cu, or Pb concentrations exceeding the tolerable maximum within the study area is modeled using different multiGaussian- and indicator-based algorithms. Next, regions where remedial measures should be taken are delineated by applying classification criteria introduced in section 7.4 to the local ccdf models. Classification results are compared with actual metal concentrations known at test locations.

Probability maps

Each of the five probability maps shown in Figure 7.44 depicts the probability that Cd concentration exceeds the tolerable maximum 0.8 ppm across the study area. These probabilities are determined using several algorithms that differ in their underlying assumptions and the information accounted for:

1. The first algorithm requires the multivariate Gaussian RF model. Under that model, the mean and variance of each Gaussian ccdf $G(\mathbf{u}; y|(n))$ are estimated from Cd normal scores using simple kriging. The probability of exceeding the critical threshold $z_c = 0.8$ ppm is then retrieved as the probability $G(\mathbf{u}; y_c|(n))$ of exceeding the normal score $y_c = \phi(z_c)$ (Figure 7.44, left top graph).

2. The four other algorithms require a prior indicator transform of the Cd data. Whereas ordinary indicator kriging (oIK) uses only indicator data defined at the threshold being estimated, 0.8 ppm, ordinary indicator cokriging (oICK) accounts for three additional thresholds 1.38, 1.88, and 2.26 ppm. Instead of indicator transforms at thresholds different from 0.8 ppm, probability kriging (PK) uses uniform transforms of Cd data as secondary information.

3. The last indicator algorithm incorporates soft information related to geology in addition to the indicator data at threshold 0.8 ppm. A calibration of rock types allows one to convert the geologic map of Figure 6.6 (left top graph) into a map of local prior probabilities of exceeding 0.8 ppm, shown in Figure 7.44 (left bottom graph). These probabilities are then used as local means in the simple IK estimator (7.43); see the resulting probability map in Figure 7.44 (right bottom graph).

The first four algorithms generate probability maps with very similar spatial patterns. Ordinary indicator kriging provides results similar to those of the more demanding indicator cokriging and probability kriging. Accounting for secondary information (rock types) yields a more detailed probability map (Figure 7.44, right bottom graph). In particular, the contrast between low-valued Argovian rocks with smaller probabilities of exceeding the critical threshold and other rocks is enhanced.

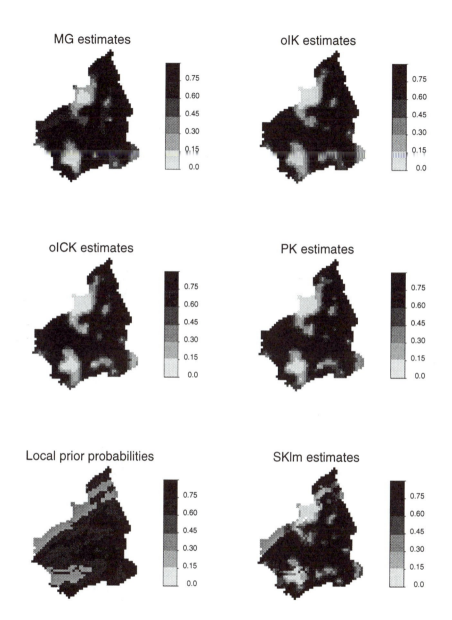

Figure 7.44: Probability maps of exceeding the critical Cd concentration 0.8 ppm produced using the multiGaussian approach (MG), ordinary indicator kriging (oIK), ordinary indicator cokriging (oICK), probability kriging (PK), simple kriging with varying local probabilities (SKlm) as derived from the geologic map (left bottom graph).

Optimal estimates and measures of local uncertainty

Figure 7.45 shows the maps of ordinary kriging Cd estimates (top graph) and four estimates retrieved from the oIK-related ccdfs. Although both OK and E-type estimates are optimal for the least-squares criterion, they are not identical. As mentioned in section 7.4.2, the two estimates differ in that the ccdf from which the E-type estimate derives is dependent on the data values. That difference is further illustrated in Figure 7.46, which shows the maps of the OK variance and three measures of spread of the conditional distributions

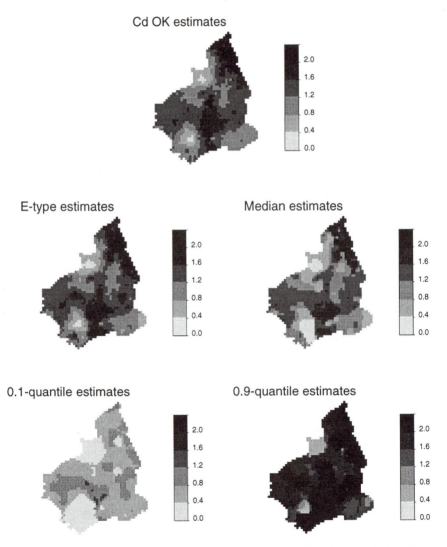

Figure 7.45:   Maps of ordinary kriging Cd estimates and various statistics of the local ccdfs: mean, median, 0.1-quantile, and 0.9-quantile.

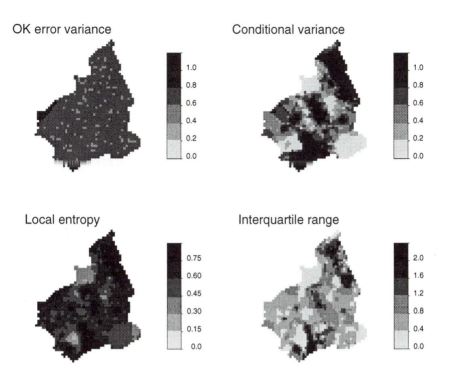

Figure 7.46:   Four measures of local uncertainty over the study area: ordinary kriging variance and three ccdf statistics (conditional variance, entropy, and interquartile range).

(variance, entropy, interquartile range). The map of OK variance indicates greater uncertainty in the extreme west corner of the study area where data are sparse, whereas the uncertainty is smallest near data locations. Elsewhere the kriging variance is about the same whatever the surrounding data values. In contrast, measures of spread deduced from the ccdf models all indicate that uncertainty is larger in the high Cd-valued parts of the study area where the data variance is also the largest (recall the proportional effect shown in Figure 4.4). The uncertainty is smaller on Argovian rocks where Cd concentrations are consistently small.

In addition to the means of the conditional cdfs, three p-quantiles of these distributions are mapped in Figure 7.45: the median, the 0.1-quantile, and the 0.9-quantile. These three conditional quantiles map the threshold values that have, respectively, a 50%, 90%, and 10% probability of being exceeded by the unknown Cd values. High-valued parts (dark zones) of the 0.1-quantile map indicate areas where the unknown Cd concentrations are certainly large, whereas low-valued parts (light grey zones) of the 0.9-quantile map correspond to areas where the concentrations are certainly small. As with the marginal distribution (histogram), the local distributions of Cd data tend to be positively skewed, hence the median estimate is generally smaller than the E-type estimate at the same location.

Delimitation of contaminated areas

Figures 7.47–7.49 show three subdivisions of the study area into safe regions and regions where remedial measures should be taken. In all three cases, the conditional cdfs were established using ordinary indicator kriging. The three criteria for classifying a location as contaminated follow:

1. The probability of exceeding the tolerable maximum 0.8 ppm is larger than the marginal probability of contamination estimated in section 2.1.2 (Figure 7.47).

2. The E-type estimate exceeds the critical threshold (Figure 7.48, top graphs). The bottom maps depict the risks attached to each classification. For example, the risk α of wrongly declaring that a location is hazardous (false positive) is larger around low-valued areas (Figure 7.48, left bottom graph).

3. The expected loss associated with wrongly declaring a location contaminated (undue cleaning of a safe location) is smaller than the cost associated with wrongly classifying a location as safe (ill health) (Figure 7.49). These two losses were modeled using functions (7.72) and (7.73) with for ω-costs: $\omega_1=1$, $\omega_2=2.5$.

These different criteria lead to classifications that may greatly differ. In particular, the E-type estimate, which is optimal for a least-squares criterion, tends to overestimate the Cd concentration, leading to most locations being classified as contaminated (Figure 7.48, right top graph).

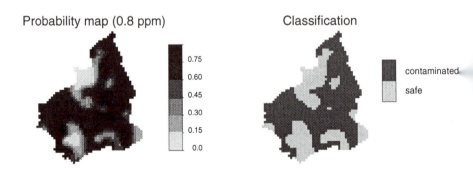

Figure 7.47: Classification of locations as contaminated by cadmium on the basis that the probability of exceeding the critical threshold 0.8 ppm is larger than the marginal probability of contamination (0.653).

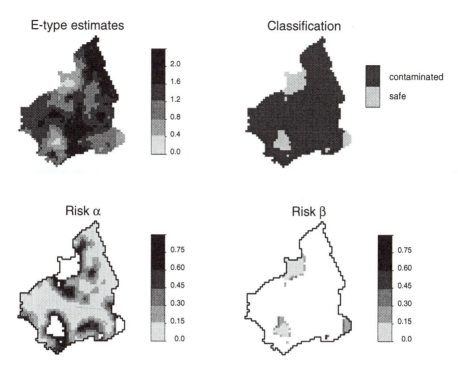

Figure 7.48: Classification of locations as contaminated by cadmium on the basis that the E-type estimate exceeds the critical threshold 0.8 ppm. Bottom graphs show the corresponding risks of wrongly declaring that a location is hazardous (risk α) or safe (risk β).

Classification of test locations

The 100 test locations were classified as contaminated or safe with respect to Cd concentration, using the following three criteria:

1. The probability of exceeding the tolerable maximum 0.8 ppm is larger than a probability threshold p.

2. The tolerable maximum 0.8 ppm is exceeded by a p-quantile estimate retrieved from the local ccdf.

3. The expected loss associated with wrongly declaring a location contaminated (undue cleaning of a safe location) is smaller than the cost associated with wrongly declaring it safe (ill health).

A range of values for the probability p and the ω-costs was considered to investigate the sensitivity of classification results to the probability threshold p, the p-quantile estimate, or the loss functions retained.

Figure 7.50 (left column) shows the proportion of locations declared contaminated versus the probability p or the cost ratio ω_1/ω_2, with ω_2 set to 1.

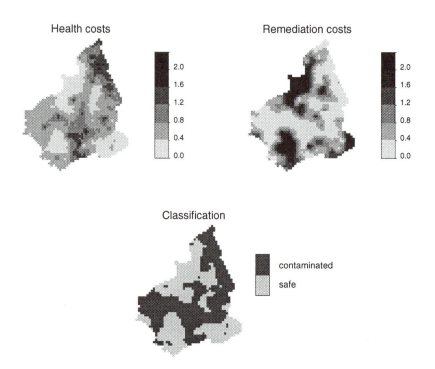

Figure 7.49:   Classification of locations as contaminated by cadmium on the basis that the resulting expected cost (unnecessary cleaning) is smaller than the cost associated with wrongly classifying a location as safe (potential ill health).

Ccdf models are determined using two different algorithms:

1. **oIK**: ccdf values are determined using ordinary indicator kriging at five different threshold values corresponding to the first, third, fifth, seventh, and ninth deciles of the marginal distribution. The resolution of the ccdf is increased by linear interpolation between tabulated bounds provided by the sample cdf (259 data).

2. **oICK**: ccdf values are determined using ordinary indicator cokriging at the same five thresholds used for oIK. The secondary (soft) information consists of the colocated local prior probability of type (7.25) derived from a calibration of the relation between Cd and Ni values.

   Whatever the algorithm used to determine the ccdf values, the proportion of locations declared contaminated decreases as the probability threshold p increases (Figure 7.50, left top graph). Conversely, as p increases, the corresponding p-quantile estimate increases, hence more locations are declared contaminated (Figure 7.50, left middle graph). As the relative cost ω_1 of declaring a location safe increases, the proportion of locations classified as contaminated increases (Figure 7.50, left bottom graph).

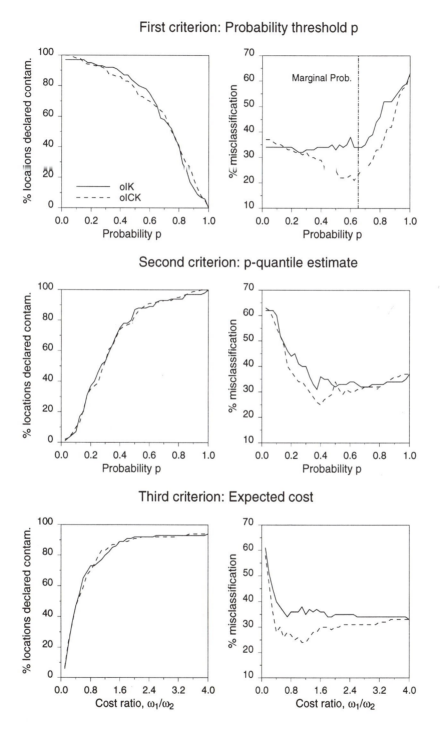

Figure 7.50:   Proportion of test locations that are declared contaminated with respect to Cd concentration (left column) or that are wrongly classified (right column), using three different classification criteria and two algorithms for determining the ccdf models.

Figure 7.50 (right column) shows the proportion of test locations that are wrongly classified versus the probability p or the cost ratio w_1/w_2. This proportion is generally minimal for a probability threshold close to the marginal probability of contamination, $(1-F^*(0.8 \text{ ppm})) = 0.653$, depicted by the vertical dashed line on the right top graph. The proportion of locations wrongly classified using the second criterion is minimal when the p-quantile estimate corresponds to a probability close to $p=F^*(0.8 \text{ ppm}) = 0.347$ (Figure 7.50, right middle graph). Using the first criterion with a probability p yields a classification similar to that provided by the second criterion with a probability $(1-p)$, as expected. For the third criterion, the proportion of misclassified locations is minimal close to a unit cost ratio corresponding to about 70% of locations classified as contaminated (Figure 7.50, bottom graphs). For all three criteria, accounting for secondary information (oICK) reduces the proportion of misclassified locations.

Because the actual Cd concentration $z(\mathbf{u}_t)$ is known at each test location $\mathbf{u}_t$, the actual cost of wrongly classifying that location as safe or contaminated can be computed as

$$C(\mathbf{u}_t) \quad = \quad i(\mathbf{u}_t)L_1(z(\mathbf{u}_t)) + [1 - i(\mathbf{u}_t)]L_2(z(\mathbf{u}_t))$$

where $L_1(.)$ and $L_2(.)$ are the loss functions (7.72) and (7.73), $i(\mathbf{u}_t) = 1$ if the location $\mathbf{u}_t$ is classified as safe and $i(\mathbf{u}_t) = 0$ otherwise. For example, the cost of classifying a contaminated location $\mathbf{u}_0$ as safe is

$$C(\mathbf{u}_0) \quad = \quad L_1(z(\mathbf{u}_0)) \quad = \quad w_1[z(\mathbf{u}_0) - z_c]$$

The total misclassification cost C is the sum of the local costs $C(\mathbf{u}_t)$ over the 100 test locations:

$$C \quad = \quad \sum_{t=1}^{100} C(\mathbf{u}_t)$$

In the following discussion, the cost C is expressed as a percentage of the cost resulting from the extreme decision of declaring all locations safe, $i(\mathbf{u}_t) = 1 \; \forall \; t$. In this example, the relative costs w_1 and w_2 were arbitrarily chosen such that the actual cost of declaring all locations safe is twice the actual cost of declaring all locations contaminated.

Figure 7.51 (top graphs) shows the misclassification costs corresponding to the first and third classification criteria considered in Figure 7.50. The Cd cost curves look like the proportion curves in Figure 7.50 (top and bottom right graphs) because the misclassification cost increases with the proportion of misclassified locations. As for cadmium, the misclassification costs for copper and lead are minimal for a probability threshold close to the marginal probability of contamination (Figure 7.51, left column). Accounting for secondary information (oICK) reduces the cost. Regardless of the algorithm or criterion used, the resulting classification generally costs less than the extreme decision of taking no remedial measure (percentage is smaller than 100%).

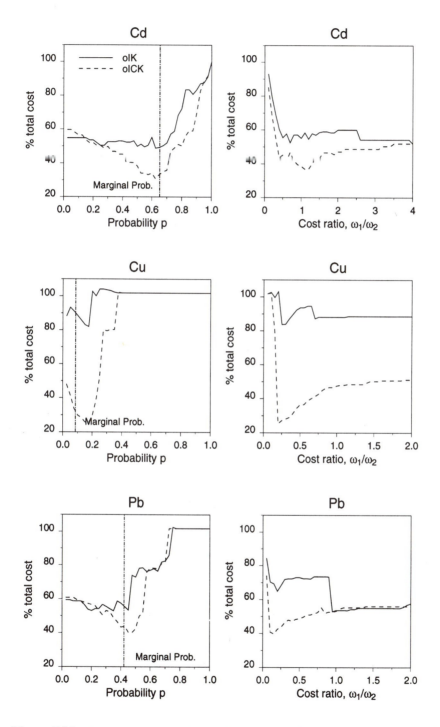

Figure 7.51: Economic impact of wrongly classifying the 100 test locations as safe or contaminated using two different criteria: probability threshold (left column) and expected cost (right column). Ccdf models are determined using two different algorithms. Results are expressed as a percentage of the cost of classifying all locations as safe.

Chapter 8

Assessment of Spatial Uncertainty

The multiGaussian and indicator-based algorithms introduced in Chapter 7 provide only models of local uncertainty in that each conditional cdf is specific to one single location. Most applications require a measure of the joint uncertainty about attribute values at several locations taken together, for example, the probability of occurrence of a string of large or small values. Such spatial uncertainty is modeled by generating multiple realizations of the joint distribution of attribute values in space, a process known as stochastic simulation. Then, transfer functions, such as flow simulators, can be applied to the set of alternative representations, yielding a distribution of response values, such as the time for a fluid to travel from one location to another, used in subsequent risk analysis.

The conceptual difference between stochastic simulation and estimation is discussed in section 8.1. Sections 8.2–8.7 present various algorithms for simulating one or several interdependent continuous variables: sequential Gaussian and indicator simulation techniques, the LU decomposition algorithm, the p-field simulation algorithm, and simulated annealing. Algorithms for simulating categorical variables are introduced in section 8.8. Important topics such as reproduction of model statistics, visualization of spatial uncertainty, and accuracy and precision of response distributions are addressed in section 8.9.

8.1 Estimation versus Simulation

Let $\{z^*(\mathbf{u}), \mathbf{u} \in \mathcal{A}\}$ be the set of kriging estimates of attribute z over the study area $\mathcal{A}$. Each estimate $z^*(\mathbf{u})$ taken separately, i.e., independently of neighboring estimates $z^*(\mathbf{u}')$, is "best" in the least-squares sense because the local error variance $\text{Var}\{Z^*(\mathbf{u}) - Z(\mathbf{u})\}$ is minimum. The map of such best local estimates, however, may not be best as a whole. As shown in Figure 5.19 and Table 5.2 (pages 181 and 182), interpolation algorithms tend to smooth out local details of the spatial variation of the attribute. Typically, small values are overestimated, whereas large values are underestimated. Such conditional bias is a serious shortcoming when trying to detect patterns of extreme attribute values, such as zones of high permeability values or zones rich in a metal. Another drawback of estimation is that the smoothing is not uniform. Rather, it depends on the local data configuration: smoothing is minimal close to the data locations and increases as the location being estimated gets farther away from data locations. A map of kriging estimates appears more variable in densely sampled areas than in sparsely sampled areas. Thus the kriged map may display artifact structures.

Smooth interpolated maps should not be used for applications sensitive to the presence of extreme values and their patterns of continuity. Consider, for example, the problem of assessing groundwater-travel times from a nuclear repository to the surface. A smooth map of estimated transmissivities would fail to reproduce critical features, such as strings of large or small values that form flow paths or barriers. The processing of a kriged transmissivity map through a flow simulator may yield inaccurate travel times. Similarly, the risk of soil pollution by heavy metals would be underestimated by a kriged map of metal concentrations that fails to reproduce clusters of large concentrations above the tolerable maximum.

Reproducing model statistics

Instead of a map of local best estimates, stochastic simulation generates a map or a realization of z-values, say, $\{z^{(l)}(\mathbf{u}), \mathbf{u} \in \mathcal{A}\}$ with l denoting the lth realization, which reproduces statistics deemed most consequential for the problem in hand. Typical requisites for such simulated map are as follows:

1. Data values are honored at their locations:

$$z^{(l)}(\mathbf{u}) = z(\mathbf{u}_\alpha) \qquad \forall\, \mathbf{u} = \mathbf{u}_\alpha,\ \alpha = 1, \ldots, n$$

   The realization is then said to be conditional (to the data values).

2. The histogram of simulated values reproduces closely the declustered sample histogram.

3. The covariance model $C(\mathbf{h})$ or, better, the set of indicator covariance models $C_I(\mathbf{h}; z_k)$ for various thresholds z_k are reproduced.

As shown subsequently, more complex features, such as spatial correlation with a secondary attribute or multiple-point statistics, may also be reproduced.

Figure 8.1 (top graphs) shows the maps of estimated and simulated Cd values over the study area. In both cases, the 259 Cd data values are honored at their locations. Unlike simulation, ordinary kriging yields a smooth map of estimated values:

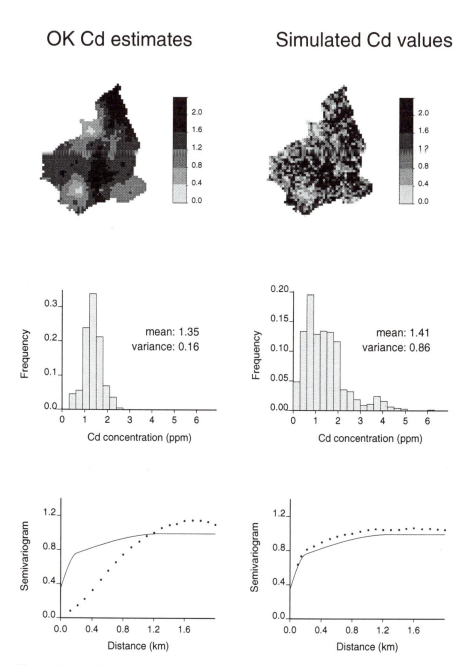

Figure 8.1: Ordinary kriging estimates and simulated values of Cd concentrations over the study area. Bottom graphs show the corresponding histograms and standardized experimental and model (solid line) semivariograms. Note the smoothing effect of kriging that leads to underestimation of the short-range variability of Cd values.

- The variance of kriged estimates is much smaller than the sample variance $\hat{\sigma}^2 = 0.83$ (Figure 8.1, left middle graph).

- The semivariogram of kriged estimates has a much smaller relative nugget effect than the semivariogram model depicted by the solid line in Figure 8.1 (left bottom graph), which reflects the underestimation of the short-range variability of Cd values.

The reproduction of the sample histogram and semivariogram model by the simulated values (right column) is much better, yet it is not exact—nor should it be. Reasons for departures between model and realization statistics are discussed in section 8.9.1, and algorithms for improving reproduction of model statistics when deemed necessary will be introduced.

Modeling spatial uncertainty

One may generate multiple realizations that all reasonably match the same sample statistics and exactly match the conditioning data. For example, the three realizations shown in Figure 8.2 all honor the Cd data values while approximately reproducing the sample histogram and semivariogram model. The set of alternative realizations $\{z^{(l)}(\mathbf{u}), \mathbf{u} \in \mathcal{A}\}$, $l = 1, \ldots, L$, provides a visual and quantitative measure (actually a model) of *spatial uncertainty*. Spatial features, such as specific strings of large values, are deemed certain if seen on most of the L simulated maps. Conversely, a feature is deemed uncertain if seen only on a few simulated maps.

The multiGaussian and indicator algorithms introduced in the previous chapter provide a measure only of *local uncertainty* because each conditional cdf relates to a single location $\mathbf{u}$. Recall that a series of single-point ccdfs $F(\mathbf{u}_j; z_c|(n))$, $j = 1, \ldots, J$, do not provide any measure of multiple-point or spatial uncertainty, such as the probability that the z-values at J locations $\mathbf{u}_j$ are jointly no greater than a critical threshold z_c:

$$\text{Prob}\,\{Z(\mathbf{u}_j) \leq z_c, j = 1, \ldots, J|(n)\} \quad \neq \quad \prod_{j=1}^{J} F(\mathbf{u}_j; z_c|(n))$$

where the symbol $\prod$ denotes a product.

There is equality only if the J RVs $Z(\mathbf{u}_j)$ are independent—a case of little interest. The joint probability may be assessed numerically from a set of L realizations of the spatial distribution of z-values over the J locations $\mathbf{u}_j$:

$$\text{Prob}\,\{Z(\mathbf{u}_j) \leq z_c, j = 1, \ldots, J|(n)\} \quad \approx \quad \frac{1}{L}\sum_{l=1}^{L}\prod_{j=1}^{J} i^{(l)}(\mathbf{u}_j; z_c) \qquad (8.1)$$

where the indicator value $i^{(l)}(\mathbf{u}_j; z_c)$ is 1 if the simulated z-value at $\mathbf{u}_j$ does not exceed the threshold z_c, and zero otherwise.

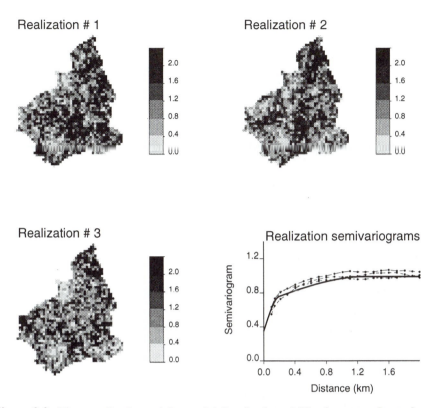

Figure 8.2: Three realizations of the spatial distribution of Cd values over the study area and the corresponding standardized semivariograms. Differences between realizations provide a model for the uncertainty about the distribution in space of Cd values.

Using the spatial uncertainty model

Generating alternative realizations of the spatial distribution of an attribute is rarely a goal per se. Rather, these realizations serve as input to complex transfer functions, such as flow simulators in reservoir engineering or remediation processes in pollution control. Transfer functions differ from the loss functions introduced in section 7.4 in that they consider all locations simultaneously rather than one at a time. The processing of each input realization yields a unique value for each response, for example, a unique value for groundwater travel time or remediation cost. The distribution (histogram) of the L response values corresponding to the L input realizations provides a measure of response uncertainty resulting from our imperfect knowledge of the distribution in space of z. That measure can then be used in risk analysis and decision making.

Consider the problem of assessing the economic impact of declaring the study area safe with respect to Cd. One wants to evaluate the cost that might result from taking no remedial measure. That cost can be computed from any realization $\{z^{(l)}(\mathbf{u}'_j), j = 1, \ldots, N\}$ as the sum of local costs defined at the N locations $\mathbf{u}'_j$ discretizing the study area $\mathcal{A}$:

$$\mathcal{C}(l) = \sum_{j=1}^{N} L_1(z^{(l)}(\mathbf{u}'_j)) \tag{8.2}$$

Utilizing expression (7.72), one can model the cost associated with declaring safe a location $\mathbf{u}'_j$ as

$$L_1(z^{(l)}(\mathbf{u}'_j)) = \begin{cases} 0 & \text{if } z^{(l)}(\mathbf{u}'_j) \leq z_c \\ \omega_1(\mathbf{u}'_j)\,[z^{(l)}(\mathbf{u}'_j) - z_c] & \text{otherwise} \end{cases}$$

where z_c is the critical threshold, and $\omega_1(\mathbf{u}'_j)$ is the relative cost of underestimating the toxic concentration at $\mathbf{u}'_j$, e.g., the cost associated with ill health (ω_1-unit is cost/ppm). The cost $\omega_1(\mathbf{u}'_j)$ could be modeled as a function of the land use prevailing at $\mathbf{u}'_j$, say, $\omega_1 = 1$ for forest, $\omega_1 = 5$ for meadow and pasture, and $\omega_1 = 10$ for tillage. If the simulated value $z^{(l)}(\mathbf{u}'_j)$ does not exceed z_c, the classification is correct and there is no cost. Conversely, if the simulated location is contaminated, $z^{(l)}(\mathbf{u}'_j) > z_c$, the misclassification cost is proportional to the contamination $[z^{(l)}(\mathbf{u}'_j) - z_c]$.

That operation can be repeated for many, say, $L = 100$, realizations that all honor Cd data and reproduce the sample histogram and semivariogram model. The histogram of the 100 costs $\mathcal{C}(l)$, $l = 1, \ldots, L$, provides an assessment of the risk involved with declaring the area safe (Figure 8.3, top graph). The worst scenario corresponds to realization # 10, which shows an important contamination of agricultural land that leads to the maximum cost $\mathcal{C}_{max} = 5187$. The minimum cost associated with this simulation model is obtained for realization # 41, $\mathcal{C}_{min} = 3517$. These two extreme realizations are shown at the bottom of Figure 8.3.

The same cost function was applied to the kriged map shown at the top of Figure 8.1 (left graph), yielding a cost $\mathcal{C}^* = 3854$ denoted by a vertical arrow on the histogram of Figure 8.3. In addition to providing no measure of response uncertainty, the use of the smooth kriged map would most likely have led to an underestimation of the cost associated with declaring the study area safe.

Simulation algorithms

In section 7.4.4, the quantile algorithm for generating a set of L realizations $z^{(l)}(\mathbf{u})$, $l = 1, \ldots, L$, at any specific location $\mathbf{u}$ was introduced. This was done by sampling the one-point ccdf that models uncertainty at that location:

$$F(\mathbf{u}; z|(n)) = \text{Prob}\{Z(\mathbf{u}) \leq z|(n)\}$$

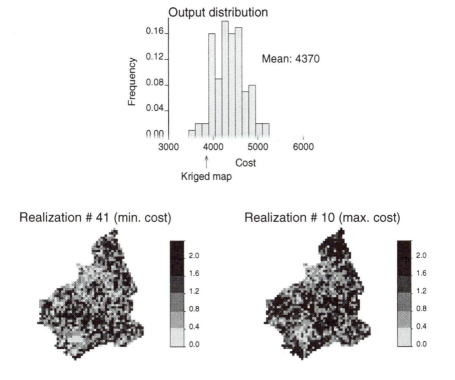

Figure 8.3: The distribution of costs resulting from a wrong decision to declare the study area safe with respect to Cd (top graph). This distribution is obtained by post-processing 100 realizations of the spatial distribution of Cd values; the two realizations yielding the smallest and the largest costs are shown at the bottom of the figure.

Similarly, a set of simulated maps $\{z^{(l)}(\mathbf{u}'_j), j = 1, \ldots, N\}, l = 1, \ldots, L$, can be generated by sampling the N-variate or N-point ccdf that models the joint uncertainty at the N locations $\mathbf{u}'_j$:

$$F(\mathbf{u}'_1, \ldots, \mathbf{u}'_N; z_1, \ldots, z_N | (n)) = \text{Prob}\{Z(\mathbf{u}'_1) \leq z_1, \ldots, Z(\mathbf{u}'_N) \leq z_N | (n)\} \qquad (8.3)$$

Inference of the conditional cdf (8.3) requires knowledge or stringent hypotheses about the spatial law of the RF $Z(\mathbf{u})$. The multivariate Gaussian RF model is one model whose spatial law is fully determined by the sole z-covariance function; it underlies several simulation algorithms, such as the *LU decomposition* algorithm introduced in section 8.5. In the following presentation, the focus is on three classes of simulation algorithms that are not limited to the Gaussian formalism:

1. The wide class of simulation algorithms known under the generic name of *sequential simulation* algorithms: instead of modeling the N-point ccdf (8.3), a one-point ccdf is modeled and sampled at each of the N nodes visited along a random sequence. To ensure reproduction of the z-covariance model, each one-point ccdf is made conditional not only to the original n data but also to all values simulated at previously visited locations.

2. The *p-field approach* also trades sampling of the N-point ccdf (8.3) for sampling of N successive one-point ccdfs. Unlike the sequential approach, all one-point ccdfs are conditioned only to the original n data. Reproduction of the z-covariance model is here approximated by imposing an autocorrelation pattern on the probability values used for sampling these ccdfs.

3. Unlike the two previous simulation algorithms, in *simulated annealing* the creation of a stochastic image is formulated as an optimization problem without any reference to a RF model. The basic idea is to perturb an initial (seed) image gradually so as to match target constraints, such as reproduction of the z-covariance model. Different realizations are obtained by considering different initial images.

8.2   The Sequential Simulation Paradigm

Let $\{Z(\mathbf{u}_j'), j = 1, \ldots, N\}$ be a set of random variables defined at N locations $\mathbf{u}_j'$ within the study area $\mathcal{A}$; these locations need not be gridded. The objective is to generate several joint realizations of these N RVs:

$$\{z^{(l)}(\mathbf{u}_j'), j = 1, \ldots, N\}   l = 1, \ldots, L$$

conditional to the data set $\{z(\mathbf{u}_\alpha), \alpha = 1, \ldots, n\}$.

Consider first the joint simulation of z-values at two locations only, say, $\mathbf{u}_1'$ and $\mathbf{u}_2'$. A set of pairs of realizations $\{z^{(l)}(\mathbf{u}_1'), z^{(l)}(\mathbf{u}_2')\}, l = 1, \ldots, L$, can be generated by sampling the bivariate or two-point conditional cdf:

$$F(\mathbf{u}_1', \mathbf{u}_2'; z_1, z_2|(n)) = \text{Prob}\,\{Z(\mathbf{u}_1') \leq z_1, Z(\mathbf{u}_2') \leq z_2|(n)\}   (8.4)$$

An alternative approach is provided by Bayes' axiom, whereby any two-point ccdf can be expressed as a product of one-point ccdfs:

$$F(\mathbf{u}_1', \mathbf{u}_2'; z_1, z_2|(n)) = F(\mathbf{u}_2'; z_2|(n+1)) \cdot F(\mathbf{u}_1'; z_1|(n))   (8.5)$$

where "$|(n+1)$" denotes conditioning to the n data values $z(\mathbf{u}_\alpha)$ and to the realization $Z(\mathbf{u}_1') = z^{(l)}(\mathbf{u}_1')$. The decomposition (8.5) allows one to generate the pair $\{z^{(l)}(\mathbf{u}_1'), z^{(l)}(\mathbf{u}_2')\}$ in two steps: the value $z^{(l)}(\mathbf{u}_1')$ is first drawn from the ccdf $F(\mathbf{u}_1'; z_1|(n))$, then the ccdf at location $\mathbf{u}_2'$ is conditioned to the realization $z^{(l)}(\mathbf{u}_1')$ in addition to original data (n) and its sampling yields the correlated value $z^{(l)}(\mathbf{u}'_2)$. The idea is to trade the sampling hence modeling of the two-point ccdf (8.4) for the sequential sampling of two one-point ccdfs easier to infer, hence the generic name *sequential* simulation algorithm.

The sequential principle can be generalized to more than two locations. By recursive application of Bayes' axiom (8.5), the N-point conditional cdf (8.3) can be written as the product of N one-point conditional cdfs:

$$
\begin{aligned}
F(\mathbf{u}'_1, \ldots, \mathbf{u}'_N; z_1, \ldots, z_N | (n)) \;=\; & F(\mathbf{u}'_N; z_N | (n + N - 1)) \\
& \cdot F(\mathbf{u}'_{N-1}; z_{N-1} | (n + N - 2)) \cdot \; \ldots \\
& \cdot F(\mathbf{u}'_2; z_2 | (n + 1)) \cdot F(\mathbf{u}'_1; z_1 | (n))
\end{aligned}
$$
$$(8.6)$$

where, for example, $F(\mathbf{u}'_N; z_N | (n + N - 1))$ is the conditional cdf of $Z(\mathbf{u}'_N)$ given the set of n original data values and the $(N - 1)$ realizations $Z(\mathbf{u}'_j) = z^{(l)}(\mathbf{u}'_j), j = 1, \ldots, N - 1$. The decomposition (8.6) allows one to generate a realization of the random vector $\{Z(\mathbf{u}'_j), j = 1, \ldots, N\}$ in N successive steps:

- Model the cdf at the first location $\mathbf{u}'_1$ conditional to the n original data $z(\mathbf{u}_\alpha)$:

$$
F(\mathbf{u}'_1; z | (n)) \;=\; \text{Prob}\,\{Z(\mathbf{u}'_1) \leq z | (n)\}
$$

- Draw from that ccdf a realization $z^{(l)}(\mathbf{u}'_1)$, which becomes a conditioning datum for all subsequent drawings.

$$\vdots$$

- At the ith node $\mathbf{u}'_i$ visited, model the conditional cdf of $Z(\mathbf{u}'_i)$ given the n original data and all $(i-1)$ values $z^{(l)}(\mathbf{u}'_j)$ simulated at the previously visited locations $\mathbf{u}'_j, j = 1, \ldots, i - 1$:

$$
F(\mathbf{u}'_i; z | (n + i - 1)) \;=\; \text{Prob}\,\{Z(\mathbf{u}'_i) \leq z | (n + i - 1)\}
$$

- Draw from that ccdf a realization $z^{(l)}(\mathbf{u}'_i)$, which becomes a conditioning datum for all subsequent drawings.

- Repeat the two previous steps until all N nodes are visited and each has been given a simulated value.

The resulting set of simulated values $\{z^{(l)}(\mathbf{u}'_j), j = 1, \ldots, N\}$ represents one realization of the RF $\{Z(\mathbf{u}), \mathbf{u} \in \mathcal{A}\}$ over the N nodes $\mathbf{u}'_j$. Any number L of such realizations, $\{z^{(l)}(\mathbf{u}'_j), j = 1, \ldots, N\}, l = 1, \ldots, L$, can be obtained by repeating L times the entire sequential process with possibly different paths to visit the N nodes.

Remarks

1. The sequential simulation algorithm requires the determination of a conditional cdf at each location being simulated. Two major classes of sequential simulation algorithms can be distinguished, depending on whether the series of conditional cdfs are determined using the multi-Gaussian or the indicator formalisms introduced in the previous chapter.

2. Sequential simulation ensures that data are honored at their locations (conditional realizations). Indeed, at any datum location $\mathbf{u}_\alpha$, the simulated value is drawn from a zero-variance, unit-step conditional cdf with mean equal to the z-datum $z(\mathbf{u}_\alpha)$ itself. If large measurement errors render questionable the exact matching of data values, one should allow the simulated values to deviate somewhat from data at their locations. If the errors are normally distributed, the simulated value could be drawn from a Gaussian ccdf centered on the datum value and with a variance equal to the error variance.

3. The sequential principle can be extended to simulate several continuous or categorical attributes (see subsequent sections).

Implementation

As in many applied fields, successful application of a basic principle relies on experience and a few critical implementation tips; see Deutsch and Journel, 1992a, p. 30–34 and 124–125; Gómez-Hernández and Cassiraga (1994).

Search strategies
The sequential simulation algorithm requires the determination of N successive ccdfs $F(\mathbf{u}'_1; z|(n)), \ldots, F(\mathbf{u}'_N; z|(n + N - 1))$, with an increasing level of conditioning information. Correspondingly, the size of the kriging system(s) to be solved to determine these ccdfs increases and becomes quickly prohibitive as the simulation progresses. As shown in section 5.8.2, the data closest to the location being estimated tend to screen the influence of more distant data. Thus, in the practice of sequential simulation, only the original data and those previously simulated values closest to the location $\mathbf{u}'$ being simulated are retained. Good practice consists of using the semivariogram distance $\gamma(\mathbf{u}' - \mathbf{u}_\alpha)$ rather than the Euclidian distance $|\mathbf{u}' - \mathbf{u}_\alpha|$ so that conditioning data are preferentially selected along the direction of maximum continuity.

As the simulation progresses, the original data tend to be overwhelmed by the large number of previously simulated values, particularly when the simulation grid is dense. A balance between the two types of conditioning information can be preserved by separately searching the original data and the previously simulated values (two-part search): at each location $\mathbf{u}'$, a fixed number $n(\mathbf{u}')$ of closest original data are retained no matter how many previously simulated values are in the neighborhood of $\mathbf{u}'$.

Visiting sequence

In theory, the N nodes can be simulated in any sequence as long as all data and all previously simulated values are used in the determination of ccdfs. However, because only neighboring data are retained, artificial continuity may be generated along a deterministic path visiting the N nodes. Hence a random sequence or path is recommended (Isaaks, 1990). An exception to this practice is the multiple-grid simulation procedure described subsequently.

When generating several realizations, the computational time can be reduced considerably by keeping the same random path for all realizations. Indeed, the N kriging systems, one for each node $\mathbf{u}'_j$, need be solved only once since the N conditioning data configurations remain the same from one realization to another. The trade-off cost is the risk of generating realizations that are too similar. Therefore, it is better to use a different random path for each realization.

Multiple-grid simulation

The use of a search neighborhood limits reproduction of the input covariance model to the radius of that neighborhood. Another obstacle to reproduction of long-range structures is the screening of distant data by too many data closer to the location being simulated. The multiple-grid concept (Gómez-Hernández, 1991; Tran, 1994) allows one to reproduce long-range correlation structures without having to consider large search neighborhoods with too many conditioning data. For example, a two-step simulation of a square grid 500×500 could proceed as follows:

1. The attribute values are first simulated on a coarse grid (e.g., 25×25) using a large search neighborhood so as to reproduce long-range correlation structures. Because the grid is coarse, each neighborhood contains few data, which reduces the screening effect.

2. Once the coarse grid has been completed, the simulation continues on the finer grid 500×500 using a smaller search neighborhood so as to reproduce short-range correlation structures. The previously simulated values on the coarse grid are used as data for the simulation on the fine grid.

A random path is followed within each grid.

The procedure can be generalized to any number of intermediate grids; this number depends on the number of structures with different ranges to be reproduced and the final grid spacing.

8.3   Sequential Gaussian Simulation

Implementation of the sequential principle under the multiGaussian RF model is referred to as sequential Gaussian simulation (sGs). Algorithms for simulating a single attribute using only values of that attribute then accounting for secondary information are first introduced. The joint simulation of several correlated attributes is then addressed.

8.3.1   Accounting for a single attribute

Consider the simulation of the continuous attribute z at N nodes $\mathbf{u}'_j$ of a grid (not necessarily regular) conditional to the data set $\{z(\mathbf{u}_\alpha), \alpha = 1, \ldots, n\}$. Sequential Gaussian simulation proceeds as follows:

1. The first step is to check the appropriateness of the multiGaussian RF model, which calls for a prior transform of z-data into y-data with a standard normal cdf using the normal score transform (7.8). Normality of the two-point distribution of the resulting normal score variable $Y(\mathbf{u}) = \phi(Z(\mathbf{u}))$ is then checked (recall section 7.2.3). If the biGaussian assumption is invalidated, other procedures for determination of the local ccdfs must be considered, for example, indicator-based sequential simulation algorithms presented subsequently.

2. If the multiGaussian RF model is retained for the y-variable, sequential simulation is performed on the y-data:

   - Define a random path visiting each node of the grid only once.
   - At each node $\mathbf{u}'$, determine the parameters (mean and variance) of the Gaussian ccdf $G(\mathbf{u}'; y|(n))$ using SK with the normal score semivariogram model $\gamma_Y(\mathbf{h})$. The conditioning information (n) consists of a specified number $n(\mathbf{u}')$ of both normal score data $y(\mathbf{u}_\alpha)$ and values $y^{(l)}(\mathbf{u}'_j)$ simulated at previously visited grid nodes.
   - Draw a simulated value $y^{(l)}(\mathbf{u}')$ from that ccdf, and add it to the data set.
   - Proceed to the next node along the random path, and repeat the two previous steps.
   - Loop until all N nodes are simulated.

3. The final step consists of back-transforming the simulated normal scores $\{y^{(l)}(\mathbf{u}'_j), j = 1, \ldots, N\}$ into simulated values for the original variable, which amounts to applying the inverse of the normal score transform (7.8) to the simulated y-values:

$$z^{(l)}(\mathbf{u}'_j) \;=\; \phi^{-1}(y^{(l)}(\mathbf{u}'_j)) \qquad j = 1, \ldots, N \qquad (8.7)$$

with $\phi^{-1}(.) = F^{-1}(G(.))$, where $F^{-1}(.)$ is the inverse cdf or quantile function of the variable Z, and $G(.)$ is the standard Gaussian cdf. That back-transform allows one to identify the original z-histogram $F(z)$. Indeed,

$$\text{Prob}\left\{Z^{(l)}(\mathbf{u}) \leq z\right\} = \text{Prob}\left\{\phi^{-1}(Y^{(l)}(\mathbf{u})) \leq z\right\}$$

from the definition (8.7) of the back-transform,

$$= \text{Prob}\{Y^{(l)}(\mathbf{u}) \leq \phi(z)\}$$

since the transform function $\phi(.)$ is monotonic increasing,

$$= G[\phi(z)] = F(z)$$

from the definition (7.8) of the normal score transform.

Other realizations $\{z^{(l')}(\mathbf{u}'_j), j = 1, \ldots, N\}$, $l' \neq l$, are obtained by repeating steps 2 and 3 with a different random path.

As mentioned in section 7.2.4, non-stationary behaviors could be accounted for using algorithms other than simple kriging to estimate the mean of the Gaussian ccdf: ordinary kriging or kriging with a trend model. However, Gaussian theory requires that the simple (co)kriging variance of normal scores be used for variance of the Gaussian ccdf (see Journel, 1980). Consider, for example, that the mean and variance of the Gaussian ccdf at $\mathbf{u}$ are estimated using kriging with a trend model. The simulation model is

$$Y^{(l)}(\mathbf{u}) = Y^*_{KT}(\mathbf{u}) + E(\mathbf{u}) \tag{8.8}$$

where the error component $E(\mathbf{u})$ is independent of $Y^*_{KT}(\mathbf{u})$, $E\{E(\mathbf{u})\} = 0$ and $\text{Var}\{E(\mathbf{u})\} = \sigma^2_{KT}(\mathbf{u})$. Recall the KT model $Y(\mathbf{u}) = m(\mathbf{u}) + R(\mathbf{u})$, where the trend component $m(\mathbf{u})$ is modeled as a linear combination of $(K+1)$ functions of the coordinates. Accounting for the KT system (5.26), one can show that the simulation model (8.8) does not reproduce the residual covariance $C_R(\mathbf{h})$ unless $K = 0$ (simple kriging case):

$$\text{Cov}\{Y(\mathbf{u}_\alpha), Y^{(l)}(\mathbf{u})\} = \sum_{\beta=1}^{n(\mathbf{u})} \lambda^{KT}_\beta(\mathbf{u}) \, C_R(\mathbf{u}_\alpha - \mathbf{u}_\beta)$$

$$\neq C_R(\mathbf{u}_\alpha - \mathbf{u}) \quad \forall \, \alpha$$

Consider now the situation where the unknown trend $m(\mathbf{u})$ is identified with its KT estimate $m^*_{KT}(\mathbf{u})$. Since $m^*_{KT}(\mathbf{u})$ is known everywhere, simulation of the normal score variable Y amounts to simulating the residual $R(\mathbf{u})$ using simple kriging and the residual covariance $C_R(\mathbf{h})$. The simulation model (8.8) becomes

$$Y^{(l)}(\mathbf{u}) = m^*_{KT}(\mathbf{u}) + R^*_{SK}(\mathbf{u}) + E(\mathbf{u})$$

with

- $E(\mathbf{u})$ independent of $R_{SK}^*(\mathbf{u})$

- $E\{E(\mathbf{u})\} = 0$, thus $E\{Y^{(l)}(\mathbf{u})\} = m_{KT}^*(\mathbf{u})$

- $\mathrm{Var}\{E(\mathbf{u})\} = \sigma_{SK}^2(\mathbf{u}) = C_R(0) - \sum_{\alpha=1}^{n(\mathbf{u})} \lambda_\alpha^{SK}(\mathbf{u})\, C_R(\mathbf{u}_\alpha - \mathbf{u})$

where the SK weights $\lambda_\alpha^{SK}(\mathbf{u})$ are provided by an SK system of type (5.10). Accounting for the relation $m_{KT}^*(\mathbf{u}) + R_{SK}^*(\mathbf{u}) = Y_{KT}^*(\mathbf{u})$, one can write the simulation model (8.9) as

$$Y^{(l)}(\mathbf{u}) \;=\; Y_{KT}^*(\mathbf{u}) + E(\mathbf{u})$$

One can show that residual covariance $C_R(\mathbf{h})$ is reproduced:

$$
\begin{aligned}
\mathrm{Cov}\{Y(\mathbf{u}_\alpha), Y^{(l)}(\mathbf{u})\} \;&=\; E\left\{[Y(\mathbf{u}_\alpha) - m(\mathbf{u}_\alpha)] \cdot \left[Y^{(l)}(\mathbf{u}) - m_{KT}^*(\mathbf{u})\right]\right\} \\
&=\; \mathrm{Cov}\{R(\mathbf{u}_\alpha), R_{SK}^*(\mathbf{u})\}, \text{ since } m(\mathbf{u}) = m_{KT}^*(\mathbf{u}) \\
&=\; \sum_{\beta=1}^{n(\mathbf{u})} \lambda_\beta^{SK}(\mathbf{u})\, C_R(\mathbf{u}_\alpha - \mathbf{u}_\beta) \\
&=\; C_R(\mathbf{u}_\alpha - \mathbf{u}) \qquad \forall\, \alpha
\end{aligned}
$$

as in the SK system (5.9).

Determining a kriging estimate and a different kriging variance would require solving two systems at each location $\mathbf{u}$, hence it is not practical. The solution adopted by program *sgsim* in GSLIB2.0 is to allow the user to input a constant variance correction factor multiplying all kriging variances (of normal scores). This factor is to be determined by trial and error to ensure that the variance of all simulated normal scores for any particular realization is indeed 1 as required by the theory.

Note

Actually, reproduction of the covariance model $C_Y(\mathbf{h})$ does not require the successive ccdf models to be Gaussian; they can be of any type as long as their means and variances are determined by simple kriging (Journel, 1994a). This result leads to an important theoretical extension of the sequential simulation paradigm whereby original z-attribute values are simulated directly without any prior normal score transform. The algorithm is then called direct sequential simulation (dssim), see Xu and Journel (1994), Xu (1995b). In the absence of a normal score transform and back-transform, there is, however, no control on the histogram of simulated values. Reproduction of a target histogram can be achieved by post-processing the dssim realization using the algorithms introduced in section 8.9.1.

Example

Consider the conditional simulation of Cd concentrations along the NE-SW transect. Figure 8.4 shows the main steps of the sGs approach:

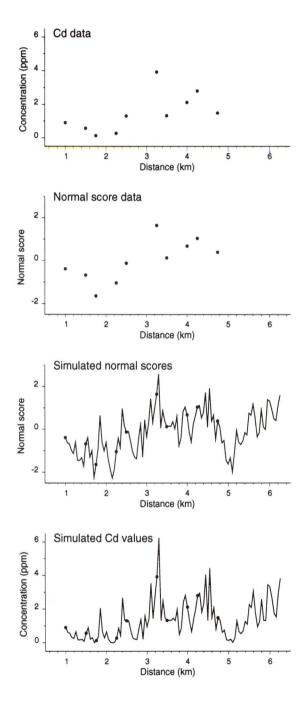

Figure 8.4: Sequential Gaussian simulation. The ten original Cd data are first transformed into ten normal score data, then conditional simulation is performed in the normal space (third row). Finally, the simulated normal scores are back-transformed into simulated Cd values.

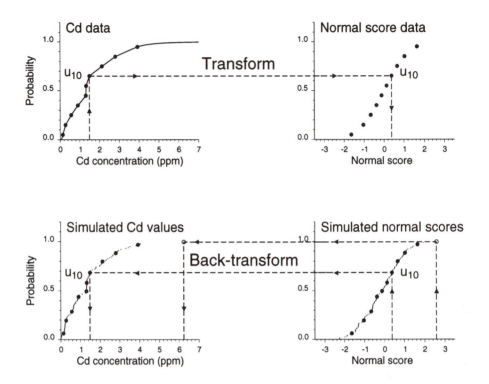

Figure 8.5: Graphical transform of ten Cd data into ten normal scores (top graph). Using this correspondence in reverse, the 106 simulated normal scores shown in Figure 8.4 (third row) are back-transformed into simulated Cd values. The ten original Cd data depicted by the large black dots are retrieved exactly.

1 The ten Cd data are first transformed into ten normal score data using the transformation Table 7.1 on page 269 (Figure 8.4, top graphs).

2. Sequential simulation is then performed in the normal space using the semivariogram model of Cd normal scores shown at the top of Figure 7.6 (page 274). At each simulated node, the five closest Cd normal score data and the five closest previously simulated values are retained in the SK system (two-part search). Figure 8.4 (third row) shows the profile of simulated normal scores.

3. Last, the simulated normal scores are back-transformed into simulated Cd values using expression (8.7) (Figure 8.4, bottom graph).

Figure 8.5 (bottom graph) shows the graphical back-transform of the 106 normal scores simulated along the NE-SW transect. Like the normal score transform displayed at the top of the figure, the back-transform can be seen as a correspondence table between equal p-quantiles of the standard Gaussian cdf $G(.)$ and the sample z-cdf $F^*(.)$. More precisely, a simulated value $y^{(l)}(\mathbf{u}'_j)$ and its back-transform $z^{(l)}(\mathbf{u}'_j) = \phi^{-1}(y^{(l)}(\mathbf{u}'_j))$ correspond to the same cumulative probability p_j:

$$G\left[y^{(l)}(\mathbf{u}'_j)\right] = F^*\left[z^{(l)}(\mathbf{u}'_j)\right] = p_j$$

which amounts to identifying $z^{(l)}(\mathbf{u}'_j)$ with the p_j-quantile of the sample z-cdf:

$$z^{(l)}(\mathbf{u}'_i) = F^{*-1}(p_j) \quad \text{with } p_j \in [0,1].$$

Using the transformation Table 7.1, which is displayed at the top of Figure 8.5, the ten normal score data $y(\mathbf{u}_\alpha)$ depicted by the large black dots are readily back-transformed into the ten original values $z(\mathbf{u}_\alpha)$. For example, the back-transform of the normal score at datum location $\mathbf{u}_{10}$ (4th largest Cd datum) yields the original Cd concentration $z(\mathbf{u}_{10})=1.49$ ppm (Figure 8.5, left bottom graph).

The back-transform of simulated normal scores $y^{(l)}(\mathbf{u}'_j) \neq y(\mathbf{u}_\alpha)$, say, the largest normal score $y^{(l)}_{max} = 2.58$ depicted by an open circle in Figure 8.5 (right bottom graph), is written

$$z^{(l)}(\mathbf{u}'_j) = F^{*-1}(p_j) \quad \text{with } p_j \neq p_\alpha$$

The resolution of the sample z-cdf shown at the top of Figure 8.5 (left graph, black dots) must be increased to provide quantile z-values corresponding to cumulative frequencies $p_j \neq p_\alpha$. In this example, cdf values are interpolated and extrapolated using the following piecewise model: power model ($\omega = 2.5$) for the lower tail, linear model for the nine middle classes, and hyperbolic model ($\omega = 5$) for the upper tail, see Figure 8.5 (left top graph, solid line) and section 7.2.5. Using that model, the largest simulated normal score is back-transformed into a simulated Cd concentration of 6.24 ppm.

Beware that the back-transform of extreme simulated values depends completely on the modeling of the upper and lower tails of the sample z-cdf. In the example of Figure 8.5, had the ω-parameter of the hyperbolic tail model been set to 1.5 (long tail) instead of 5 (short tail), the back-transform of the largest normal score would yield a Cd concentration of 18.36 ppm instead of 6.24 ppm. Whenever the back-transform yields unrealistic z-values (e.g., values outside some physical constraint interval $[z_{min}, z_{max}]$), one should question the pertinence of the model adopted for extrapolating the lower or the upper tail.

8.3.2 Accounting for secondary information

Consider the situation where primary data $\{z_1(\mathbf{u}_{\alpha_1}), \alpha_1 = 1, \ldots, n_1\}$ are supplemented by secondary data related to $(N_v - 1)$ different continuous attributes z_i, $\{z_i(\mathbf{u}_{\alpha_i}), \alpha_i = 1, \ldots, n_i, i = 2 \ldots, N_v\}$ possibly at different locations. The objective for now is to simulate only the primary variable Z_1 conditional to both primary and secondary information.

Sequential Gaussian simulation starts with transforming each set of z_i-data $\{z_i(\mathbf{u}_{\alpha_i}), \alpha_i = 1, \ldots, n_i\}$ into a set of normal score y_i-data $\{y_i(\mathbf{u}_{\alpha_i}), \alpha_i = 1, \ldots, n_i\}$ using the transform (7.8):

$$
\begin{aligned}
y_i(\mathbf{u}_{\alpha_i}) &= \phi_i(z_i(\mathbf{u}_{\alpha_i})) \\
&= G^{-1}[F_i^*(z_i(\mathbf{u}_{\alpha_i}))] \quad \alpha_i = 1, \ldots, n_i \quad i = 1, \ldots, N_v
\end{aligned}
$$

where $F_i^*(.)$ is the marginal cdf of the variable Z_i.

The next step is to check whether the auto and joint two-point distributions of the normal score RFs $\{Y_1(\mathbf{u}), \ldots, Y_{N_v}(\mathbf{u})\}$ are reasonably normal. In practice, the biGaussian assumption is checked only for each variable separately using the procedure described in section 7.2.3.

If the two-point distribution of each normal score variable appears reasonably normal, a multivariate multiple-point Gaussian RF model is then adopted. Under that model, the ccdf parameters (mean and variance) of the normal score variable $Y_1(\mathbf{u})$ are equal to the simple cokriging estimate $y_{SCK}^{(1)*}(\mathbf{u})$ and corresponding simple cokriging variance $(\sigma_{SCK}^{(1)}(\mathbf{u}))^2$ obtained from all normal score data $y_i(\mathbf{u}_{\alpha_i})$. The ccdf is then modeled as

$$
[G(\mathbf{u}; y_1|(n))]_{SCK}^* = G\left(\frac{y_1 - y_{SCK}^{(1)*}(\mathbf{u})}{\sigma_{SCK}^{(1)}(\mathbf{u})}\right) \tag{8.9}
$$

with

$$
y_{SCK}^{(1)*}(\mathbf{u}) = \sum_{i=1}^{N_v} \sum_{\alpha_i=1}^{n_i(\mathbf{u})} \lambda_{\alpha_i}^{SCK}(\mathbf{u}) \, y_i(\mathbf{u}_{\alpha_i})
$$

$$
(\sigma_{SCK}^{(1)}(\mathbf{u}))^2 = C_{11}(0) - \sum_{i=1}^{N_v} \sum_{\alpha_i=1}^{n_i(\mathbf{u})} \lambda_{\alpha_i}^{SCK}(\mathbf{u}) \, C_{i1}(\mathbf{u}_{\alpha_i} - \mathbf{u})
$$

where the simple cokriging (SCK) weights $\lambda_{\alpha_i}^{SCK}(\mathbf{u})$ are provided by an SCK system of type (6.16).

The sequential Gaussian simulation of the primary variable Z_1 proceeds as follows:

1. Define a random path visiting each node of the grid only once.

2. At each node $\mathbf{u}'$, determine the parameters (mean and variance) of the Gaussian ccdf (8.9) using simple cokriging with the direct and cross semivariogram models of normal score variables. The conditioning information consists of neighboring primary and secondary normal score data and previously simulated values of Y_1.

3. Draw a simulated value $y_1^{(l)}(\mathbf{u}')$ from that ccdf, and add it to the data set.

4. Proceed to the next node along the random path, and repeat the two previous steps.

5. Loop until all N nodes are simulated.

6. Back-transform the simulated normal scores $\{y_1^{(l)}(\mathbf{u}_j'), j = 1, \ldots, N\}$ into simulated values for the primary variable $\{z_1^{(l)}(\mathbf{u}_j') = \phi_1^{-1}(y_1^{(l)}(\mathbf{u}_j')), j = 1, \ldots, N\}$.

Other realizations $\{z_1^{(l')}(\mathbf{u}_j'), j = 1, \ldots, N\}$, $l' \neq l$, are obtained by repeating the entire process with a different random path for each realization.

Colocated cokriging and the Markov model

The multivariate sGs algorithm is demanding in the sense that a full cokriging system must be solved at each grid node being simulated. Implementation of the algorithm is alleviated by using colocated cokriging and the Markov-type model introduced in section 6.2.6.

To reduce the computational time and avoid instability problems caused by possibly highly redundant secondary information, only the secondary data colocated with the node being simulated could be retained in the simple cokriging system. The trade-off costs of such an approach are the following:

- There is no control on reproduction of the correlation between primary and secondary variables at lags $|\mathbf{h}| \neq 0$.

- The secondary variables must be available at all simulated grid nodes. If this is not the case and the secondary information is dense, nodal values of the secondary variables could be interpolated or better simulated prior to simulation of the primary variable.

Unlike full cokriging, colocated cokriging requires only the inference and modeling of the primary covariance function $C_{Y_1,Y_1}(\mathbf{h})$ and the cross covariance functions $C_{Y_1,Y_i}(\mathbf{h})$, $i = 2, \ldots, N_v$, defined as

$$\begin{aligned} C_{Y_1,Y_1}(\mathbf{h}) &= E\{Y_1(\mathbf{u}) \cdot Y_1(\mathbf{u}+\mathbf{h})\} = \rho_{Y_1,Y_1}(\mathbf{h}) \\ C_{Y_1,Y_i}(\mathbf{h}) &= E\{Y_1(\mathbf{u}) \cdot Y_i(\mathbf{u}+\mathbf{h})\} = \rho_{Y_1,Y_i}(\mathbf{h}) \qquad i = 2, \ldots, N_v \end{aligned}$$

since $E\{Y_i(\mathbf{u})\} = 0$ and $\mathrm{Var}\{Y_i(\mathbf{u})\} = 1$, $\forall\ i$, by definition of the normal score transform. The secondary autocovariance functions $C_{Y_i,Y_i}(\mathbf{h})$, $i \geq 2$, are required only at lag zero since a single datum of each secondary variable is retained in the cokriging system. The modeling effort implied by colocated cokriging can be further alleviated by using the Markov-type model (6.47):

$$\rho_{Y_1,Y_i}(\mathbf{h}) = \rho_{Y_1,Y_i}(0) \cdot \rho_{Y_1,Y_1}(\mathbf{h}) \qquad i = 2, \ldots, N_v$$

or, in terms of semivariogram models

$$\gamma_{Y_1,Y_i}(\mathbf{h}) = \rho_{Y_1,Y_i}(0) \cdot \gamma_{Y_1,Y_1}(\mathbf{h}) \qquad i = 2, \ldots, N_v \qquad (8.10)$$

Each cross semivariogram model $\gamma_{Y_1,Y_i}(\mathbf{h})$ is then derived as a mere linear rescaling of the primary semivariogram model $\gamma_{Y_1,Y_1}(\mathbf{h})$ by the correlation coefficient $\rho_{Y_1,Y_i}(0)$. One should check whether the proportionality relation (8.10) actually applies to the experimental cross semivariograms.

Example

Consider the conditional simulation of Cd concentrations along the NE-SW transect using as exhaustively sampled secondary information the Ni block estimates shown at the bottom of Figure 6.1. The multivariate sGs algorithm proceeds as follows:

1. The ten primary Cd data and the profile of secondary Ni data are first transformed into normal score data (Figure 8.6, top graph).

2. Sequential simulation is then performed in the normal space. At each simulated node, the conditioning information consists of the five closest normal score Cd data, the colocated normal score Ni datum, and the five closest previously simulated normal scores. The colocated simple cokriging system is solved using the direct and cross semivariogram models inferred from all 259 normal scores available over the study area (Figure 8.6, second row). In this example, the Markov-type approximation (8.10) was not used because the experimental cross semivariogram between Cd and Ni normal scores is not proportional to the semivariogram of Cd normal scores.

3. Finally, the simulated normal scores are back-transformed into simulated Cd values (Figure 8.6, bottom graphs).

Kriging with an external drift (KED)

An alternative to colocated cokriging for incorporating exhaustively sampled secondary information is provided by kriging with an external drift. The key assumption here is that, in the normal space, the trend of the primary variable Y_1 is a linear function of the secondary variable Y_2, which must therefore vary smoothly in space:

$$m_{Y_1}(\mathbf{u}) = a_0(\mathbf{u}) + a_1(\mathbf{u}) \cdot y_2(\mathbf{u}) \quad \forall\, \mathbf{u} \in \mathcal{A} \qquad (8.11)$$

As mentioned in section 6.1.3, it is critical to validate this assumption either from calibration data or by using some physical rationale.

Under the trend model (8.11), the mean of the Gaussian ccdf at any grid node $\mathbf{u}$ is identified with the KED estimate:

$$y_{KED}^{(1)^*}(\mathbf{u}) = \sum_{\alpha=1}^{n(\mathbf{u})} \lambda_\alpha^{KED}(\mathbf{u})\, y_1(\mathbf{u}_\alpha) + \sum_{j=1}^{n'(\mathbf{u})} \lambda_j^{KED}(\mathbf{u})\, y_1^{(l)}(\mathbf{u}_j')$$

where the weights of normal score data and previously simulated values, $\lambda_\alpha^{KED}(\mathbf{u})$ and $\lambda_j^{KED}(\mathbf{u})$, are solutions of a KED system of type (6.5). As in kriging with a trend, the conditional variance must be identified with the simple kriging variance, not the KED variance.

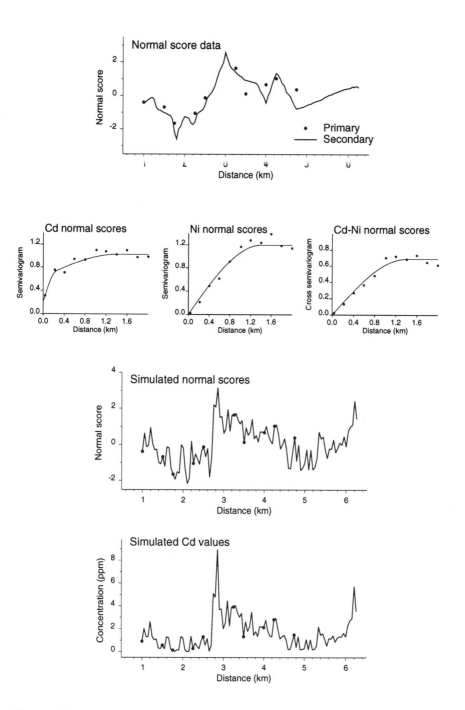

Figure 8.6: Accounting for exhaustive secondary information in sequential Gaussian simulation. Both primary (ten Cd data) and secondary data (Ni block estimates) are transformed into normal score data and combined using colocated simple cokriging and the semivariogram models shown in the second row. The simulated normal scores are then back-transformed into simulated Cd values.

8.3.3   Joint simulation of multiple variables

Many applications require the joint simulation of interdependent attributes; for example, permeability and porosity values for reservoir modeling, or several metal concentrations for modeling a contaminated area. Using the same sample information as in section 8.3.2, $\{z_i(\mathbf{u}_{\alpha_i}), \alpha_i = 1, \ldots, n_i, \ i = 1 \ldots, N_v\}$, the objective is now to simulate the vector RF $[Z_1(\mathbf{u}), \ldots, Z_{N_v}(\mathbf{u})]$ rather than the sole RF $Z_1(\mathbf{u})$.

Vectorial simulation

The first approach consists of simulating directly the vector RF using sequential Gaussian cosimulation (Verly, 1993).

As with the procedure introduced in section 8.3.2, such joint simulation starts with a normal score transform of each variable Z_i, then the two-point normality of each normal score variable $Y_i = \phi(Z_i)$ is checked. If the multivariate multiple-point Gaussian RF model is retained, the simulation proceeds as follows:

- Define a random path visiting each node of the grid only once.

- At each node $\mathbf{u}'$, build the covariance matrix $\mathbf{C}$ of the random vector $\mathbf{Y} = (\mathbf{Y}_c, \mathbf{Y}_v)$, where $\mathbf{Y}_c$ includes the N_c random variables surrounding $\mathbf{u}'$ (conditioning normal score data and previously simulated normal scores of all N_v types), and $\mathbf{Y}_v$ is the vector of random variables to be simulated at $\mathbf{u}'$, $[Y_1(\mathbf{u}'), \ldots, Y_{N_v}(\mathbf{u}')]$:

$$\mathbf{C} = \begin{bmatrix} \mathbf{C}_{11} & \mathbf{C}_{12} \\ \mathbf{C}_{21} & \mathbf{C}_{22} \end{bmatrix}$$

where $\mathbf{C}_{11}$ is the $N_c \times N_c$ covariance matrix of the vector $\mathbf{Y}_c$, $\mathbf{C}_{22}$ is the $N_v \times N_v$ covariance matrix of the vector $\mathbf{Y}_v$, and $\mathbf{C}_{12} = \mathbf{C}_{21}^T$ is the $N_c \times N_v$ covariance matrix between the two vectors $\mathbf{Y}_c$ and $\mathbf{Y}_v$. The random vector $\mathbf{Y}$ has a multiGaussian distribution.

- Decompose the matrix $\mathbf{C}$ into the product of a lower and an upper triangular matrix (LU decomposition):

$$\mathbf{C} = \mathbf{L} \cdot \mathbf{U} = \begin{bmatrix} \mathbf{L}_{11} & \mathbf{0} \\ \mathbf{L}_{21} & \mathbf{L}_{22} \end{bmatrix} \cdot \begin{bmatrix} \mathbf{U}_{11} & \mathbf{U}_{12} \\ \mathbf{0} & \mathbf{U}_{22} \end{bmatrix}$$

- Generate a realization $[y_1^{(l)}(\mathbf{u}'), \ldots, y_{N_v}^{(l)}(\mathbf{u}')]$ conditional to the "data" vector $\mathbf{y}_c$ as the linear combination:

$$\mathbf{y}^{(l)}(\mathbf{u}') = \mathbf{L}_{21} \cdot \mathbf{L}_{11}^{-1} \cdot \mathbf{y}_c + \mathbf{L}_{22} \cdot \boldsymbol{\omega}^{(l)}$$

where $\boldsymbol{\omega}^{(l)}$ is a vector of N_v independent standard normal deviates.

- Add the vector of simulated values to the conditioning data set.

- Proceed to the next node along the random path, and repeat the four previous steps.

- Loop until all N nodes are simulated.

- Back-transform the realizations $\{y_i^{(l)}(\mathbf{u}_j'), j = 1, \ldots, N\}$, $i = 1, \ldots, N_v$, into simulated values for the original variables Z_i.

Repeat the procedure with another random path to generate another set of joint realizations.

Defining a hierarchy of variables

Instead of simulating the N_v variables simultaneously, each variable can be simulated in turn as long as it is done conditionally to the previously simulated values (Gómez-Hernández and Journel, 1993; Almeida and Journel, 1994). In particular, Almeida and Journel (1994) proposed to simulate each variable in turn according to a predefined hierarchy, which allows the implementation of the colocated cokriging approximation.

The simulation algorithm proceeds as follows:

1. Define a hierarchy of variables, starting with the most important or better auto-correlated variable Z_1.

2. Transform all variables Z_i into their normal scores Y_i.

3. Define a random path visiting each node of the grid only once.

4. At each node $\mathbf{u}'$:

   - Use simple kriging to determine the parameters of the Gaussian ccdf of the first variable $Y_1(\mathbf{u}')$. The conditioning information consists of neighboring normal score data $y_1(\mathbf{u}_{\alpha_1})$ and previously simulated values $y_1^{(l)}(\mathbf{u}_j')$ of the first variable. Then, draw a simulated value $y_1^{(l)}(\mathbf{u}')$ from that ccdf, and add it to the conditioning data set.

   - Use colocated simple cokriging to determine the parameters of the Gaussian ccdf of the second variable $Y_2(\mathbf{u}')$. In addition to neighboring normal score data $y_2(\mathbf{u}_{\alpha_2})$ and previously simulated values $y_2^{(l)}(\mathbf{u}_j')$, the previously simulated colocated value $y_1^{(l)}(\mathbf{u}')$ is retained as a datum. Then, draw a simulated value $y_2^{(l)}(\mathbf{u}')$ from that ccdf, and add it to the conditioning data set.

     $\vdots$

   - Use colocated simple cokriging to determine the parameters of the Gaussian ccdf of the last variable $Y_{N_v}(\mathbf{u}')$. The conditioning information consists of the neighboring normal score data $y_{N_v}(\mathbf{u}_{\alpha_i})$ and previously simulated values $y_{N_v}^{(l)}(\mathbf{u}_j')$ of that variable, plus all previously simulated colocated values $y_1^{(l)}(\mathbf{u}'), \ldots, y_{N_v-1}^{(l)}(\mathbf{u}')$. Then, draw a simulated value $y_{N_v}^{(l)}(\mathbf{u}')$ from that ccdf, and add it to the conditioning data set.

5. Loop until all N nodes are simulated.

6. Back-transform the N_v realizations into simulated values for the original variables.

Unlike the vectorial simulation, this procedure does not require modeling the full matrix of cross covariance functions $\mathbf{C}(\mathbf{h}) = [C_{Y_i,Y_j}(\mathbf{h})]$. The trade-off cost is that the cross covariance functions may be poorly reproduced at lags $|\mathbf{h}| \neq 0$. As already mentioned in section 8.3.2, a further approximation consists of adopting a Markov-type model to alleviate the modeling of the matrix of cross covariance functions.

Transformation into independent vectors

An alternative to the direct simulation of the vector $[Z_1(\mathbf{u}), \ldots, Z_{N_v}(\mathbf{u})]$ is to simulate separately a set of independent factors $[X_1(\mathbf{u}), \ldots, X_{N_v}(\mathbf{u})]$ from which the original Z-variables can be reconstituted (Luster, 1985).

The simulation proceeds in three steps:

1. The N_v original variables $Z_i(\mathbf{u})$ are first decomposed into N_v orthogonal factors $X_k(\mathbf{u})$, e.g., the N_v principal components of the Z-correlation matrix at $|\mathbf{h}| = 0$, $\mathbf{R}(0)$ (recall section 6.2.5 on principal component kriging). The critical assumption is that the orthogonality of the principal components at lag zero extends to all other separation vectors.

2. Each set of x_k-data $\{x_k(\mathbf{u}_\alpha), \alpha = 1, \ldots, n\}$ is then transformed into the normal score data $\{y_k(\mathbf{u}_\alpha) = \phi_k(x_k(\mathbf{u}_\alpha)), \alpha = 1, \ldots, n\}$. Sequential Gaussian simulation is performed for each factor independently of the others, and the resulting simulated normal scores are back-transformed into simulated x_k-values, $\{x_k^{(l)}(\mathbf{u}_j') = \phi_k^{-1}(y_k^{(l)}(\mathbf{u}_j')), j = 1, \ldots, N\}$.

3. At each grid node $\mathbf{u}_j'$, the simulated value of the variable Z_i is then retrieved as a linear combination of the N_v independently simulated x_k-values at that location.

The method is fast because no cokriging system must be solved. Thus there is no need to infer and jointly model the $N_v(N_v - 1)/2$ cross semivariograms. The method has the following drawbacks:

• Only those data locations where all N_v attributes z_i are simultaneously measured can be considered.

• There is no control on reproduction of the spatial cross correlations between variables at lag $|\mathbf{h}| \neq 0$.

- The modeling of the semivariograms of factor normal scores cannot capitalize on ancillary information available directly from either the z_i-data or their normal scores.

8.4 Sequential Indicator Simulation

Sequential Gaussian simulation is very fast and straightforward because the modeling of the Gaussian ccdf at each location **u** requires the solution of only a single (co)kriging system at that location. An implicit assumption is that the spatial variability of the attribute values can be fully characterized by a single covariance function. In particular, this precludes modeling patterns of spatial continuity specific to different classes of values. Possibly more critical, the maximum entropy property of the multiGaussian RF model (recall section 7.2.3) does not allow for any significant correlation of extreme values (Journel and Alabert, 1988; Journel and Deutsch, 1993) and for a given covariance maximizes their scattering in space (destructuration effect). For example, the sGs realization shown at the top of Figure 8.7 (left graph) reproduces the Cd semivariogram fairly well, but the spatial continuity of small Cd values is underestimated: the experimental indicator semivariogram at the second decile has a nugget effect that is too large; see dots on Figure 8.7 (second row).

In the earth sciences, connected strings of large or small values are common and are critical for many applications. Consider the problem of assessing ground-water travel times from a nuclear repository to the surface. Sequential Gaussian simulation generates realizations that minimize connectivity of high permeability values leading to a possible understatement of the risks of leakage. More conservative results, that is, shorter travel times, are provided by non-Gaussian simulation algorithms that allow for continuity of large extreme values yet reproduce the same across-all-classes semivariogram (Gómez-Hernández and Wen, 1994). Similarly, the spatial distribution of ore grades in mineral deposits often displays rich zones oriented preferentially within a more homogeneous isotropic background. In such cases, modeling the specific continuity of high grades using a non-Gaussian technique would yield a better prediction of the recovered grade (Bourgault and Journel, 1994).

The multiGaussian RF model is inappropriate whenever the structural analysis or qualitative information indicates that extreme values could be better correlated in space than medium values. Even in the absence of information about connectivity of extreme values, the user must be aware that the analytical simplicity of sequential Gaussian simulation is balanced by the risk of understating the potential for critical features, such as strings of small or large values.

Sequential indicator simulation (sis) is the most widely used non-Gaussian simulation technique. The indicator formalism introduced in section 7.3 is used to model the sequence of conditional cdfs from which simulated values

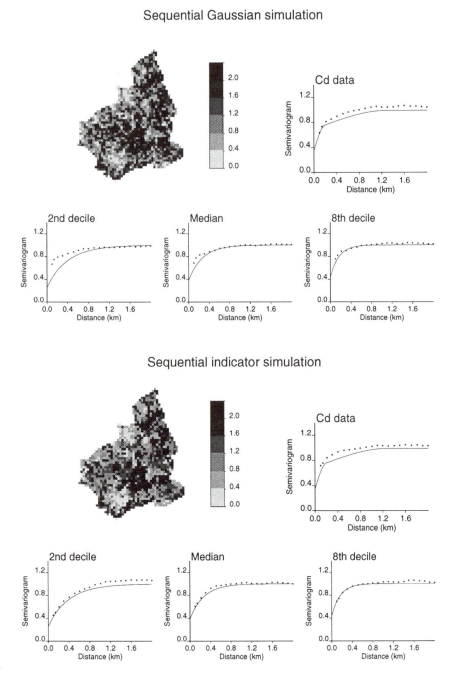

Figure 8.7: Gaussian versus indicator sequential simulation algorithms. Indicator-based algorithms allow one to account for class-specific models of spatial continuity (indicator semivariograms); hence the sis realization (third row) better reproduces the spatial continuity of small Cd values as modeled by the indicator semivariogram at the second decile.

are drawn (Alabert, 1987b; Journel and Alabert, 1988; Gómez-Hernández and Srivastava, 1990). Unlike sequential Gaussian simulation, the indicator approach allows one to account for class-specific patterns of spatial continuity through different indicator semivariogram models. For example, the realization in Figure 8.7 (third row) shows a better continuity of small Cd values relatively to median and large values as modeled by their standardized indicator semivariograms (bottom graphs). Another advantage of indicator-based simulation techniques is their flexibility in incorporating soft information coded under the format of local prior probabilities.

8.4.1 Accounting for a single attribute

Consider first the simulation of a single continuous attribute z at N grid nodes $\mathbf{u}'_j$ conditional only to the z-data $\{z(\mathbf{u}_\alpha), \alpha = 1, \ldots, n\}$. Sequential indicator simulation proceeds as follows:

- Discretize the range of variation of z into $(K + 1)$ classes using K threshold values z_k. Then, transform each datum $z(\mathbf{u}_\alpha)$ into a vector of hard indicator data, defined as

$$i(\mathbf{u}_\alpha; z_k) = \begin{cases} 1 & \text{if } z(\mathbf{u}_\alpha) \leq z_k \\ 0 & \text{otherwise} \end{cases} \qquad k = 1, \ldots, K$$

- Define a random path visiting each node of the grid only once.

- At each node $\mathbf{u}'$:

  1. Determine the K ccdf values $[F(\mathbf{u}'; z_k|(n))]^*$ using any one of the indicator kriging algorithms introduced in section 7.3.2: simple or ordinary indicator kriging, median indicator kriging, indicator cokriging or probability kriging. The conditioning information consists of indicator transforms (and uniform transforms for probability kriging) of neighboring original z-data and previously simulated z-values.

  2. Correct for any order relation deviations (recall section 7.3.4). Then, build a complete ccdf model $F(\mathbf{u}'; z|(n))$, $\forall z$, using the interpolation/extrapolation algorithms introduced in section 7.3.5.

  3. Draw a simulated value $z^{(l)}(\mathbf{u}')$ from that ccdf.

  4. Add the simulated value to the conditioning data set.

  5. Proceed to the next node along the random path, and repeat steps 1 to 4.

Repeat the entire procedure with a different random path to generate another realization $\{z^{(l')}(\mathbf{u}'_j), j = 1, \ldots, N\}$, $l' \neq l$.

Example

Consider the conditional simulation of Cd concentrations along the NE-SW transect. Figure 8.8 depicts the main steps of the sequential indicator approach, as follows:

1. The ten Cd data are first transformed into ten vectors of indicator data using the three threshold values $z_k = 0.8, 1.38$, and 2.26 ppm (Figure 8.8, top graphs).

2. For each threshold value, the indicator semivariogram is inferred from all data over the study area (Figure 8.8, third row).

3. Ccdfs are determined using ordinary indicator kriging. At each simulated location, the conditioning information consists of the indicator transforms of the five closest Cd data and of the five closest previously simulated Cd values (two-part search). The resolution of the discrete ccdf model is increased by linear interpolation between tabulated bounds obtained from the ten Cd data. Lower and upper tails are modeled using, respectively, a power model ($\omega = 2.5$) and a hyperbolic model ($\omega = 5$).

The resulting profile of simulated values is shown at the bottom of Figure 8.8.

Reproduction of model statistics

At each grid node, the indicator-based simulation can be viewed as a two-step procedure:

1. A simulated class-value is first assigned to the grid node $\mathbf{u}'$, say, $\mathbf{u}' \in (z_{k'-1}, z_{k'}]$.

2. A simulated z-value is then drawn from that class $(z_{k'-1}, z_{k'}]$ using some within-class distribution model (e.g., a uniform distribution).

Consequently, indicator-based algorithms guarantee (approximate) reproduction of only the K class proportions and corresponding indicator semivariograms $\gamma_I(\mathbf{h}; z_k)$, not reproduction of the cdf and semivariogram of the continuous z-values.

The actual approximation of one-point and two-point z-statistics by a sequential indicator realization thus depends on several factors, such as the discretization level (number of thresholds), the information accounted for when performing indicator kriging, and the interpolation/extrapolation models used for increasing the resolution of the discrete ccdf.

Reproduction of the z-cdf

The fewer the thresholds (classes) retained, the greater the impact of interpolation/extrapolation ccdf models on the resulting marginal distribution of simulated z-values. Consider, for example, the simulation of Cd values over

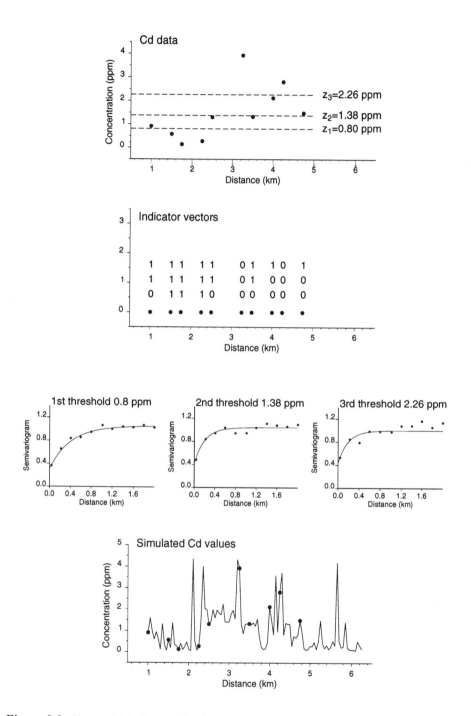

Figure 8.8: Sequential indicator simulation. After coding the ten original Cd data into indicator vectors, conditional simulation is performed using ordinary indicator kriging and the three indicator semivariograms shown in the third row. The bottom graph shows the resulting simulated profile.

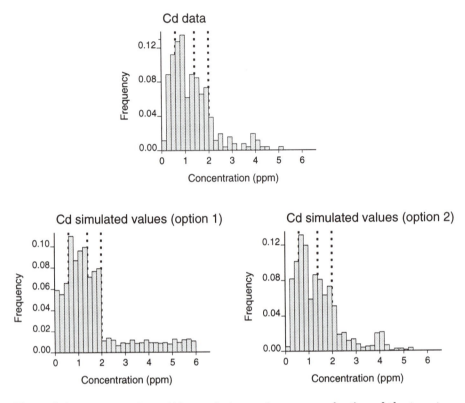

Figure 8.9: Impact of the ccdf interpolation option on reproduction of the target histogram (top graph) by a sequential indicator realization: (1) within-class linear interpolation, (2) linear interpolation between tabulated bounds identified with the 259 Cd data values.

the study area using only three threshold values: $z_k = 0.6$, 1.4, and 2.0 ppm. Figure 8.9 shows the target sample histogram (top graph) and the histograms of simulated values resulting from using two different types of interpolation and extrapolation:

1. The first model considers a series of linear interpolations within each of the four classes $(z_{k-1}, z_k]$. The resulting distribution of simulated values is approximately uniform within each class, which is delineated by the vertical dashed lines on the histogram of Figure 8.9 (left bottom graph).

2. In the second model, a linear interpolation is performed between tabulated bounds identified with the 259 Cd data values. The resulting ccdf model has a higher level of discretization, leading to a more detailed histogram of simulated values that is closer to the target histogram (Figure 8.9, right bottom graph).

The normal score back-transform, which is an integral part of the sGs algorithm, generally leads to a better reproduction of the z-cdf than do indicator-based techniques. If that reproduction is deemed necessary, the realizations can be post-processed using the algorithms introduced later in section 8.9.1.

Reproduction of the z-semivariogram
Because the interpolation/extrapolation of ccdf values is done independently from one location to another, simulated z-values within the same class $(z_{k-1}, z_k]$ are spatially independent. Consequently, whenever the number of classes (thresholds) is small, the sequential indicator realization may appear noisy as a result of this artificial noise within classes.

Implementation tips
A better reproduction of marginal target statistics can be achieved by increasing the number, K, of thresholds. In theory, the z-semivariogram is reproduced exactly if indicator cokriging with an infinite number of thresholds is used; if only indicator kriging is used, again with an infinite number of thresholds, it is the z-madogram (2.18) that is reproduced (Alabert, 1987b).

As discussed in section 7.3.5, several factors limit, in practice, the number of thresholds used in indicator kriging. Too many thresholds drastically increase computational time, inference and modeling effort, and the risk of order relation deviations. Various implementations reduce order relation problems while providing a reasonable discretization of the local ccdfs (see Deutsch and Journel, 1992a, p. 77–80; Chu, 1996).

One approach consists of using the same semivariogram model at all thresholds (median indicator kriging). No matter how many thresholds are retained, only one (median) indicator semivariogram is retained; hence only a single IK system need be solved at each grid node. Consequently, all order relation deviations caused by lack of data in some classes are eliminated (see section 7.3.4). One loses, however, the flexibility to model class-specific patterns of spatial continuity.

Another possibility is to interpolate the parameters of many indicator semivariograms from the parameters of a few explicitly modeled ones (recall section 7.3.4 and Figure 7.27, bottom graphs). More thresholds could then be retained for indicator kriging without increasing the inference and modeling effort.

Instead of modeling the whole ccdf before sampling it, another approach consists of modeling only that part of the ccdf that is sampled by the random probability value p (Deutsch and Journel, 1992a, p. 186; Chu, 1996). At each grid node $\mathbf{u}'$, the simulation would then proceed as follows:

1. Draw the random number $p \in [0, 1]$, a cumulative probability value.

2. Determine iteratively the two bounding ccdf values between which the probability p lies, say,

$$F(\mathbf{u}'; z_{k'-1}|(n)) \leq p \leq F(\mathbf{u}'; z_{k'}|(n))$$

3. Draw the simulated value $z^{(l)}(\mathbf{u}')$ in the class $(z_{k'-1}, z_{k'}]$ using some within-class distribution model.

This limited modeling of the ccdf accelerates the simulation: the maximum number of kriging systems to be solved at each grid node is $\log_2(K+1)$, the worst score of the bisection algorithm (Press et al., 1986, p. 117) used for determining the bounding ccdf values, instead of K systems for a "full" ccdf modeling.

8.4.2   Accounting for secondary information

An advantage of sequential indicator simulation is its flexibility in incorporating soft information complementing the usually sparse direct measurements of the primary attribute of interest. The key step in data integration is to code any soft information into local prior probabilities for the primary variable. The resulting "soft" indicator data can then be used in the determination of conditional cdfs through any one of the algorithms introduced in section 7.3.3: simple kriging with varying local means, kriging with an external drift, soft (colocated) cokriging, or the Markov–Bayes algorithm (Zhu and Journel, 1993; Bourgault et al., 1995).

Example

Consider the simulation of Cd values along the NE-SW transect conditional to the ten Cd data and three profiles of local prior probabilities derived from calibration of two rock categories (Figure 8.10, top two rows). At each grid node, the conditioning information consists of the five closest Cd data, the five closest previously simulated Cd values, and the colocated soft datum. Three thresholds $z_k = 0.8$, 1.38, and 2.26 ppm are considered. The corresponding hard-soft indicator cross semivariogram models are deduced from the three hard indicator semivariogram models of Figure 7.27 using the Markov-type relation (7.50) and the calibration parameters of Table 7.2 (page 317, third column).

Figure 8.10 (bottom graph) shows the profile of simulated Cd values obtained using the Markov–Bayes algorithm (solid line) and ordinary indicator kriging not accounting for soft information (dashed line). The same random path and series of random numbers are used in both cases, so differences between the two simulated profiles originate only from the additional soft information. Accounting for geology yields smaller simulated values on Argovian rocks, which have smaller probabilities of exceeding the different thresholds.

8.4.3   Joint simulation of multiple variables

The Gaussian-based cosimulation paradigm introduced in section 8.3.3 can be extended to indicators.

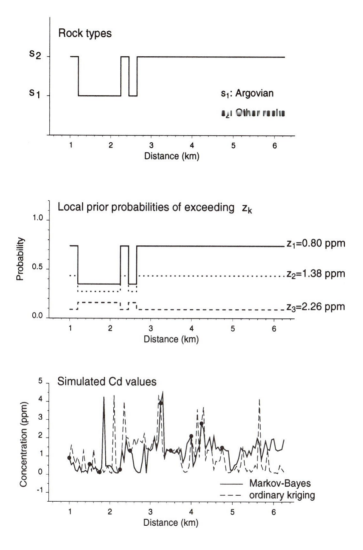

Figure 8.10: Accounting for soft information in sequential indicator simulation. The soft information consists of three profiles of local prior probabilities of exceeding given thresholds z_k, as derived from the profile of rock types shown at the top of the figure. Hard and soft secondary indicator data are combined using colocated indicator cokriging and Markov-type semivariogram models, yielding the bottom simulated profile (solid line). The dashed line depicts the simulated profile obtained using ordinary indicator kriging not accounting for the soft information.

The first approach amounts to simulating the N_v different RFs $Z_i(\mathbf{u})$ simultaneously using sequential indicator cosimulation. Practical implementation of that algorithm suffers from these drawbacks:

- The tedious inference and joint modeling of a matrix of $M(M+1)/2$ direct and cross indicator semivariograms

- The computational cost of solving at each simulated grid node M large and often unstable indicator cokriging systems with $(M \times N)$ equations

where $M = N_v \times K$, and K and N are the numbers of thresholds and conditioning data (original z_i-data plus previously simulated values) retained in the cokriging system. For simplification, we consider here K and N to be the same for all N_v variables.

The modeling effort and CPU time can be reduced substantially by substituting the separate kriging of M indicator principal components for the simultaneous cokriging of the M indicator values (Suro-Pérez, 1993). Such an approximation requires that all N_v variables be recorded at each datum location (equally sampled case) and that the indicator principal components are independent at all lags $\mathbf{h}$, not only at $|\mathbf{h}| = 0$.

An alternative to simulating simultaneously the N_v variables Z_i is to simulate each variable in turn according to a predefined hierarchy; see section 8.3.3 and Almeida and Journel (1994). In the simplest case of two variables ($N_v = 2$), the algorithm would proceed as follows:

- Define a hierarchy of the two variables.

- Transform each primary datum $z_1(\mathbf{u}_{\alpha_1})$ into a vector of hard indicator data of type (7.20), $[i_1(\mathbf{u}_{\alpha_1}; z_{11}), \ldots, i_1(\mathbf{u}_{\alpha_1}; z_{K1})]$. Code the z_2-data similarly.

- Define a random path visiting each node of the grid only once.

- At each node $\mathbf{u}'$:

  1. Use simple IK to determine each of the K ccdf values $F(\mathbf{u}'; z_{k1}|(n))$ for the first variable $Z_1(\mathbf{u}')$. For each threshold z_{k1}, the conditioning information consists of indicator transforms of neighboring z_1-data and previously simulated z_1-values. After correction for any order relation deviation, ccdf values are interpolated/extrapolated, and a simulated value $z_1^{(l)}(\mathbf{u}')$ is drawn from the resulting ccdf model.

  2. Code the simulated value $z_1^{(l)}(\mathbf{u}')$ into a vector of local prior probabilities of type (7.25) for Z_2, $[y_2(\mathbf{u}'; z_{12}), \ldots, y_2(\mathbf{u}'; z_{K2})]$.

  3. Determine the K ccdf values for the second variable Z_2 using the colocated cokriging estimator (7.46). For each threshold z_{k2}, the

conditioning information consists of indicator transforms of neighboring z_2-data and previously simulated z_2-values, plus the colocated soft indicator datum $y_2(\mathbf{u}'; z_{k2})$. After correction for order relation deviation, ccdf values are interpolated/extrapolated, and a simulated value $z_2^{(l)}(\mathbf{u}')$ is drawn from the resulting ccdf model.

4. Add the simulated values $z_1^{(l)}(\mathbf{u}')$ and $z_2^{(l)}(\mathbf{u}')$ to the conditioning data set.

5. Proceed to the next node along the random path, and repeat steps 1 to 4.

Only indicator data at the threshold z_{ki} being considered are retained in the kriging system. As discussed in section 7.3.2, it is generally not worth accounting for information at other thresholds when indicator vectors are complete (equally sampled case).

8.5 The LU Decomposition Algorithm

As with sequential Gaussian simulation, simulation through LU decomposition of the covariance matrix capitalizes on the congenial properties of the multiGaussian RF model. LU simulation should be the preferred Gaussian-based algorithm when many small realizations (few nodes) sparsely conditioned are to be generated (Alabert, 1987a; Davis, 1987; Deutsch and Journel, 1992a, p. 143).

Consider the simulation of the continuous attribute z at N grid nodes $\mathbf{u}'_j$ conditional to the data set $\{z(\mathbf{u}_\alpha), \alpha = 1, \ldots, n\}$ with both N and n small. Like any Gaussian-based technique, the LU decomposition algorithm starts with transforming the z-data into normal score y-data with a standard normal cdf. Next, the normality of the two-point distribution of these normal score data is checked.

Under the assumption that the RF $Y(\mathbf{u})$ is multiGaussian, the simulation proceeds as follows:

1. Build the covariance matrix $\mathbf{C}$ between all $(n + N)$ conditioning data locations and simulated grid nodes:

$$
\mathbf{C} = \begin{bmatrix} \mathbf{C}_{11} & \mathbf{C}_{12} \\ \mathbf{C}_{21} & \mathbf{C}_{22} \end{bmatrix} = \begin{bmatrix} [C_Y(\mathbf{u}_\alpha - \mathbf{u}_\beta)] & [C_Y(\mathbf{u}_\alpha - \mathbf{u}'_j)] \\ [C_Y(\mathbf{u}'_i - \mathbf{u}_\beta)] & [C_Y(\mathbf{u}'_i - \mathbf{u}'_j)] \end{bmatrix}
$$

where $C_Y(\mathbf{h})$ is the covariance function of the standard normal RF $Y(\mathbf{u})$, $\mathbf{C}_{11}$ is the $n \times n$ data-to-data covariance matrix, $\mathbf{C}_{22}$ is the $N \times N$ node-to-node covariance matrix, and $\mathbf{C}_{12} = \mathbf{C}_{21}^T$ is the data-to-node covariance matrix.

2. Decompose the matrix $\mathbf{C}$ into the product of a lower and an upper

triangular matrix:

$$C = L \cdot U = \begin{bmatrix} L_{11} & 0 \\ L_{21} & L_{22} \end{bmatrix} \cdot \begin{bmatrix} U_{11} & U_{12} \\ 0 & U_{22} \end{bmatrix}$$

Clearly this step requires the dimension $(n+N)$ to be small (lesser than 10^3).

3. Generate a conditional realization $\{y^{(l)}(\mathbf{u}'_j), j = 1, \ldots, N\}$ as the linear combination:

$$\mathbf{y}^{(l)} = [y^{(l)}(\mathbf{u}'_j)] = L_{21} \cdot L_{11}^{-1} \cdot \mathbf{y}_\alpha + L_{22} \cdot \boldsymbol{\omega}^{(l)} \qquad (8.12)$$

where $\mathbf{y}_\alpha$ is the vector of the n conditioning normal score data and $\boldsymbol{\omega}^{(l)}$ is a vector of N independent standard normal deviates.

4. Back-transform the simulated normal scores $\{y^{(l)}(\mathbf{u}'_j), j = 1, \ldots, N\}$ into simulated values of the original variable $\{z^{(l)}(\mathbf{u}'_j) = \phi^{-1}(y^{(l)}(\mathbf{u}'_j)), j = 1, \ldots, N\}$.

Other realizations, $\{z^{(l')}(\mathbf{u}'_j), j = 1, \ldots, N\}$, $l' \neq l$, are readily obtained by multiplying the matrix L_{22} in expression (8.12) by other vectors of independent standard normal deviates, $\boldsymbol{\omega}^{(l')}$, $l' \neq l$.

The vector $\mathbf{y}^{(l)}$ of simulated values can be viewed as the sum of a first component $(L_{21} \cdot L_{11}^{-1} \cdot \mathbf{y}_\alpha)$ that accounts for the conditioning data and a residual component $(L_{22} \cdot \boldsymbol{\omega}^{(l)})$ needed for reproducing the covariance model $C_Y(\mathbf{h})$. The first component is but the vector of SK estimates of the variable Y at the N simulated grid nodes.

For each realization, sequential Gaussian simulation calls for solving a series of N small kriging systems that account for only neighboring conditioning data and previously simulated values. In contrast, however many realizations are to be generated, the LU decomposition algorithm requires decomposing only once a single large covariance matrix that accounts for all simulated nodes and data locations simultaneously. In practice, the LU algorithm cannot handle more than a few hundred simulated grid nodes and conditioning data, but it allows one to generate many additional realizations at little additional computational cost.

Application: Determination of a block ccdf

Consider the problem of evaluating the *block* posterior ccdf $F_V(\mathbf{u}; z|(n))$ that models the uncertainty about an average z-value over the block $V(\mathbf{u})$:

$$F_V(\mathbf{u}; z|(n)) = \text{Prob}\{Z_V(\mathbf{u}) \leq z|(n)\} \qquad (8.13)$$

Because of the non-linearity of the indicator transform, the block ccdf cannot be derived simply as a linear combination of point ccdfs:

$$[F_V(\mathbf{u}; z|(n))]^* \neq \frac{1}{J} \sum_{j=1}^{J} [F(\mathbf{u}'_j; z|(n))]^*$$

with the point-ccdf $F(\mathbf{u}'_j; z|(n))$ being defined at J points $\mathbf{u}'_j$ discretizing the block $V(\mathbf{u})$ (recall section 7.3.2).

In the absence of block data $z_V(\mathbf{u}_\alpha)$ and corresponding block statistics and block indicator data, the block ccdf (8.13) can be numerically approximated by the cumulative distribution of many simulated block values $z_V^{(l)}(\mathbf{u})$ (Isaaks, 1990; Gómez-Hernández, 1991; Deutsch and Journel, 1992a, p. 90; Glacken, 1996):

$$[F_V(\mathbf{u}, z|(n))]^* \quad - \quad \frac{1}{L}\sum_{l=1}^{L} i_V^{(l)}(\mathbf{u}, z) \qquad (8.14)$$

with the block indicator value defined as $i_V^{(l)}(\mathbf{u}; z) = 1$ if $z_V^{(l)}(\mathbf{u}) \le z$, and zero otherwise. Each simulated block value $z_V^{(l)}(\mathbf{u})$ is obtained by averaging a set of z-values simulated at J points $\mathbf{u}'_j$ discretizing the block $V(\mathbf{u})$:

$$z_V^{(l)}(\mathbf{u}) \quad = \quad \frac{1}{J}\sum_{j=1}^{J} z^{(l)}(\mathbf{u}'_j) \qquad l = 1,\ldots,L$$

The LU decomposition algorithm is particularly appropriate for such an application because it allows fast generation of many conditional realizations over a few grid nodes.

8.6 The p-field Simulation Algorithm

The quantile algorithm (7.76) allows one to generate a series of L simulated values $z^{(l)}(\mathbf{u})$, $l = 1,\ldots,L$, at any location $\mathbf{u}$ by sampling the corresponding ccdf $F(\mathbf{u}; z|(n))$. If the same procedure were repeated at a neighboring location $\mathbf{u}' \neq \mathbf{u}$ using ccdfs conditioned only to the original n data, the realizations $z^{(l)}(\mathbf{u})$ and $z^{(l)}(\mathbf{u}')$ for any given l would not reproduce the target covariance $C(\mathbf{u} - \mathbf{u}')$ since the ccdf at $\mathbf{u}$ does not account for the previously simulated value $z^{(l)}(\mathbf{u}')$ at a location $\mathbf{u}'$ possibly quite close to $\mathbf{u}$. The basic idea of the p-field simulation approach is to sample the different ccdfs using autocorrelated random numbers (p-values) so that the resulting simulated values reproduce the covariance model (Srivastava, 1992; Froidevaux, 1993).

Consider the simulation of a continuous attribute z at N nodes $\mathbf{u}'_j$ conditional to the data set $\{z(\mathbf{u}_\alpha), \alpha = 1,\ldots,n\}$. The p-field simulation algorithm proceeds as follows:

1. At each location $\mathbf{u}'_j$ being simulated, build the ccdf model $F(\mathbf{u}'_j; z|(n))$ using any appropriate algorithm, e.g., any one of the multiGaussian or indicator algorithms introduced in Chapter 7.

2. Generate a set of autocorrelated p-values, $\{p^{(l)}(\mathbf{u}'_j), j = 1,\ldots,N\}$, called probability field or p-field, that is a realization of the RF $P(\mathbf{u})$

with a uniform marginal cdf and a covariance $C_P(\mathbf{h})$ identified with the covariance $C_X(\mathbf{h})$ of the uniform transform of the original variable Z:

$$
\begin{aligned}
\text{Prob}\,\{P(\mathbf{u}) \leq p\} &= p \qquad\quad \forall\, p \in [0,1]\\
C_P(\mathbf{h}) &= C_X(\mathbf{h})\\
&= \mathrm{E}\,\{X(\mathbf{u}) \cdot X(\mathbf{u}+\mathbf{h})\} - [\mathrm{E}\,\{X(\mathbf{u})\}]^2
\end{aligned}
$$

where $X(\mathbf{u}) = F(Z(\mathbf{u})) \in [0,1]$ is the uniform transform of $Z(\mathbf{u})$.

3. At each location $\mathbf{u}'_j$, draw a simulated value $z^{(l)}(\mathbf{u}'_j)$ from the ccdf $F(\mathbf{u}'_j; z|(n))$ using the colocated simulated p-value $p^{(l)}(\mathbf{u}'_j)$:

$$
z^{(l)}(\mathbf{u}'_j) = F^{-1}(\mathbf{u}'_j; p^{(l)}(\mathbf{u}'_j)|(n)) \qquad j = 1, \ldots, N
$$

The correlation between any two neighboring p-values $p^{(l)}(\mathbf{u}'_j)$ and $p^{(l)}(\mathbf{u}'_k)$ induces correlation on the corresponding simulated z-values $z^{(l)}(\mathbf{u}'_j)$ and $z^{(l)}(\mathbf{u}'_k)$.

Steps 2 and 3 are repeated to generate a different realization $\{z^{(l')}(\mathbf{u}'_j), j = 1, \ldots, N\}$, $l' \neq l$.

Secondary information can be incorporated at the first step of ccdf modeling. For example, secondary data can be coded as local prior probabilities for the primary variable Z, then accounted for in the determination of the ccdfs using any one of the indicator kriging algorithms introduced in section 7.3.3.

The p-field simulation algorithm dissociates the two tasks of (1) conditioning to the original data, a task accounted for by the ccdf models $F(\mathbf{u}'_j; z|(n))$, and (2) imparting spatial correlation to the simulated z-values by imposing spatial correlation on the values used to sample these ccdfs. Once ccdfs at every location $\mathbf{u}'_j$ are modeled, the generation of L realizations $\{z^{(l)}(\mathbf{u}'_j), j = 1, \ldots, N\}$, $l = 1, \ldots, L$, requires only the generation of L different non-conditional p-field realizations $\{p^{(l)}(\mathbf{u}'_j), j = 1, \ldots, N\}$, $l = 1, \ldots, L$, which renders the p-field algorithm computationally fast.

In the sequential simulation approach, the correlation between simulated z-values originates from the conditioning of the ccdfs to all previously simulated values in addition to the n original data. Thus, the ccdf at each location $\mathbf{u}'_j$ must be modeled as many times as there are realizations to be generated.

Reproduction of model statistics

Reproduction of the z-data values is ensured through the ccdfs $F(\mathbf{u}; z|(n))$. Indeed, at any datum location $\mathbf{u}_\alpha$, the ccdf is a unit-step function identifying the datum value $z(\mathbf{u}_\alpha)$. Thus, whatever the simulated p-field value $p^{(l)}(\mathbf{u}_\alpha)$ at that location,

$$
F^{-1}(\mathbf{u}_\alpha; p^{(l)}(\mathbf{u}_\alpha)|(n)) = z^{(l)}(\mathbf{u}_\alpha) = z(\mathbf{u}_\alpha) \qquad \forall\, p^{(l)}(\mathbf{u}_\alpha) \in [0,1]
$$

The one-point and two-point statistics of the simulated z-values are controlled by the statistics of the p-field values. Journel (1995) proved that, under conditions of ergodicity and on average over a large number L of realizations, the z-histogram and the covariance of the z-uniform scores are reproduced; that is,

$$\frac{1}{L}\sum_{l=1}^{L}\int_{\mathcal{A}} i^{(l)}(\mathbf{u};z)d\mathbf{u} \simeq F(z)$$

$$\frac{1}{L}\sum_{l=1}^{r}\int_{\mathcal{A}\cap\mathcal{A}_{-\mathbf{h}}} i^{(l)}(\mathbf{u};z)\cdot i^{(l)}(\mathbf{u}+\mathbf{h};z')d\mathbf{u} \simeq F(\mathbf{h};z,z')$$

with $i^{(l)}(\mathbf{u};z) = 1$ if $z^{(l)}(\mathbf{u}) \leq z$, and zero otherwise; $\mathcal{A}$ is the simulation area, and $\mathcal{A}\cap\mathcal{A}_{-\mathbf{h}}$ is the intersection of area $\mathcal{A}$ with its translation by vector $-\mathbf{h}$.

In words, the one-point and two-point ccdfs of the simulated values $z^{(l)}(\mathbf{u})$ averaged over all possible locations $\mathbf{u} \in \mathcal{A}$ approximate the stationary one-point and two-point z-cdfs.

Example

Consider the conditional simulation of Cd concentrations along the NE-SW transect. Figure 8.11 depicts the different steps of the p-field simulation algorithm, as follows:

1. The conditional cdf is first determined at each simulated location; in this example, probability kriging was used. Figure 8.11 (top graph) shows the NE-SW profiles of ccdf values corresponding to the four thresholds $z_k = 0.8, 1.38, 1.88$, and 2.26 ppm.

2. A **non**-conditional p-field realization is then generated in two-steps: (1) normal score values, $\{y^{(l)}(\mathbf{u}'_j), j = 1,\ldots,N\}$, are generated using sequential Gaussian simulation and the semivariogram of uniform transforms shown in Figure 7.19 (page 303, middle graph), (2) the simulated y-values are then transformed into uniform p-values, $p^{(l)}(\mathbf{u}'_j) = G(y^{(l)}(\mathbf{u}'_j)), j = 1,\ldots,N$, where $G(.)$ is the Gaussian cdf. The p-field realization is shown in Figure 8.11 (second row).

3. The simulated p-values are used to sample the conditional cdfs. For example, the simulated z-values at locations $\mathbf{u}'_1$ and $\mathbf{u}'_2$ are, respectively, 0.46 and 1.79 ppm, corresponding to simulated p-values 0.29 and 0.55 (Figure 8.11, third row). The resulting simulated profile is shown at the bottom of Figure 8.11.

Remark

The sequential Gaussian algorithm used in the previous example to generate the p-field does not capitalize on the fact that the p-field is non-conditional.

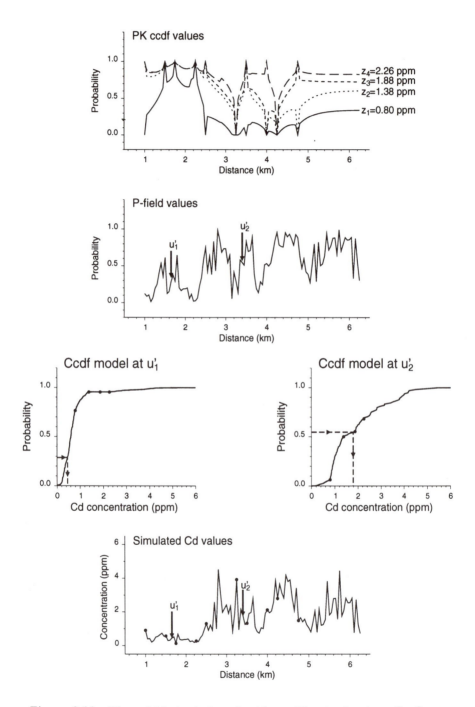

Figure 8.11: The p-field simulation algorithm. The simulated profile (bottom graph) is obtained by sampling ccdf models using a profile of probability values, called p-field (second row). The sampling procedure is illustrated for ccdfs at locations $\mathbf{u}_1'$ and $\mathbf{u}_2'$ (third row).

Much faster simulation algorithms exist for the reproduction of any covariance model as long as there is no data-conditioning, for example, simulations using the spectral decomposition theorem (Borgman et al., 1984; Chu and Journel, 1994), or the non-conditional version of the turning bands algorithm (Journel and Huijbregts, 1978, p. 498), or simulations using moving averages (Journel and Huijbregts, 1978, p. 505; Luster, 1985). This book does not cover the vast field of simulation algorithms that cannot be made directly conditional to local data.

8.7 Simulated Annealing

Simulated annealing is a generic name for a family of optimization algorithms based on the principle of stochastic relaxation (Geman and Geman, 1984; Farmer, 1988). An initial (seed) image is gradually perturbed so as to match constraints such as reproduction of a target histogram and covariance while honoring data values at their locations. For example, the initial random image shown at the top of Figure 8.12 (left graph) is gradually modified by swapping pairs of values so as to achieve an almost perfect reproduction of the target Cd semivariogram model depicted by the solid line on the graphs at the right of Figure 8.12. Unlike previous simulation algorithms, the creation of a stochastic image is here formulated as an optimization problem without reference to a random function model. Because there are usually many (approximate) solutions to that optimization problem, the set of acceptable solutions is taken as the set of simulated realizations, hence the geostatistical understanding of the qualifier *simulated* in simulated annealing (Deutsch, 1994a). The qualifier *annealing* relates to the optimization process mimicking the metallurgical process of slow cooling of a molten alloy.

8.7.1 Simulated annealing paradigm

Consider the simulation of the continuous attribute z at N grid nodes $\mathbf{u}'_j$ conditional to the data $z(\mathbf{u}_\alpha)$, $\alpha = 1, \ldots, n$ and such that the z-semivariogram model $\gamma(\mathbf{h})$ is reproduced over the first S lags. Geostatistical simulated annealing requires an objective function that measures the deviation between the target and current statistics of the realization at each ith perturbation. If the objective is to reproduce the z-semivariogram model, the objective function could be

$$O(i) = \sum_{s=1}^{S} \left[\gamma(\mathbf{h}_s) - \widehat{\gamma}_{(i)}(\mathbf{h}_s) \right]^2$$

where $\gamma(\mathbf{h}_s)$ is the value of the target z-semivariogram model at lag $\mathbf{h}_s$, and $\widehat{\gamma}_{(i)}(\mathbf{h}_s)$ is the corresponding experimental z-semivariogram value of the realization at the ith perturbation, $\{z_{(i)}^{(l)}(\mathbf{u}'_j), j = 1, \ldots, N\}$.

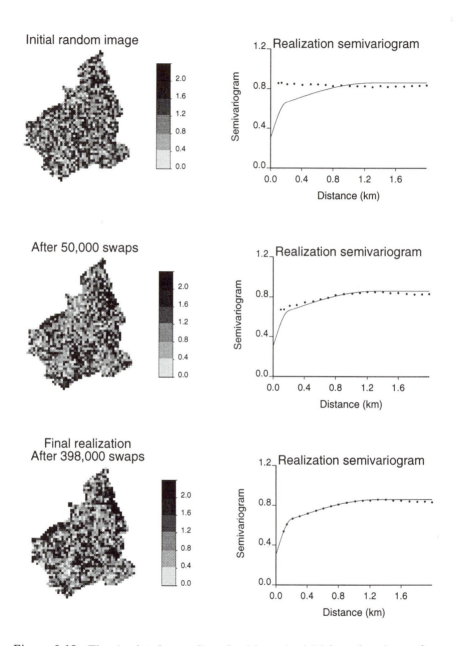

Figure 8.12: The simulated annealing algorithm. An initial random image (top graph) is gradually modified by swapping pairs of values so as to achieve reproduction of the Cd semivariogram model (bottom graph).

Once the objective function has been established, the simulation (actually an optimization) process amounts to systematically modifying an initial realization so as to decrease the value of that objective function, getting the realization acceptably close to the target statistics. The general annealing algorithm proceeds as follows:

1. Create an initial realization, $\{z_{(0)}^{(l)}(\mathbf{u}'_j), j = 1, \ldots, N\}$, that honors data values at their locations and may already approximate some of the target statistics, such as the variance or sill of the target z-semivariogram.

2. Compute the initial value of the objective function corresponding to that initial realization:

$$O(0) \;=\; \sum_{s=1}^{S} \left[\gamma(\mathbf{h}_s) - \widehat{\gamma}_{(0)}(\mathbf{h}_s)\right]^2$$

3. Perturb the realization by some simple mechanism, such as swapping a pair of z-values: $z_{(0)}^{(l)}(\mathbf{u}'_j)$ becomes $z_{(0)}^{(l)}(\mathbf{u}'_i)$ and vice versa.

4. Assess the impact of the perturbation on the reproduction of target statistics by recomputing the objective function, $O_{new}(0)$, accounting for the modification of the initial image.

5. Accept or reject the perturbation on the basis of a specified decision rule; for example, only the perturbations that diminish the objective function, $O_{new}(0) < O(0)$, are accepted.

6. If the perturbation is accepted, update the initial realization into a new image $\{z_{(1)}^{(l)}(\mathbf{u}'_j), j = 1, \ldots, N\}$ with objective function value $O(1) = O_{new}(0)$.

7. Repeat steps 3 to 6 until the target constraints are acceptably reached or until the perturbations do not reduce significantly the objective function further.

Other realizations $\{z^{(l')}(\mathbf{u}'_j), j = 1, \ldots, N\}, l' \neq l$, are generated by repeating the entire process starting from different initial images. Typically, the number of nodes N is so large and the semivariogram is so little constraining that there exist many solutions to the optimization problem. The realizations drawn are samples from that set of approximate solutions.

Simulated annealing is conceptually simple and offers great flexibility to account for various constraints built into the objective function. The optimization process, however, relies on brute force, CPU-intensive, trial and error to gradually achieve reproduction of the target statistics. Implementation tips thus play an essential role for the successful application of simulated annealing; see the next section and Deutsch and Journel (1992b), Deutsch and Cockerham (1994).

8.7.2   Implementation tips

There are many possible implementations of the general simulated annealing paradigm. Variants differ in the way the initial image is generated and then perturbed, in the components that enter the objective function, and in the type of decision rule and convergence criterion that are adopted.

The initial image

Simulated annealing requires the prior determination of an initial image,[1] $\{z_{(0)}(\mathbf{u}_j'), j = 1, \ldots, N\}$. That prior determination should be such that:

- The initial image is easily generated.

- The image already matches simple target constraints (e.g., honoring of data values. reproduction of target histogram) so as to accelerate the subsequent optimization process,

- All initial images are "equally probable" in that each image is equally likely to have been drawn. Beware that using the same image as a starting point for several different runs may lead to artificial similarity between final realizations and hence an understatement of uncertainty.

Typically, the initial image is generated by freezing data values at their locations and assigning to each unsampled grid node a z-value drawn at random from the target cdf $F(z)$. This approach is fast and yields a set of initial images that already honor the conditioning data and match the target histogram.

The initial image could also be a realization generated by any one of the previous simulation algorithms, for example, p-field or sequential simulation algorithms. Simulated annealing is then used as a post-processor with an objective of either a better reproduction of the target statistics or the imposition of additional constraints that cannot be readily incorporated by other simulation algorithms.

The perturbation mechanism

The two mechanisms most commonly used for sequentially modifying the initial image are as follows:

1. Swap the z-values at any two unsampled locations $\mathbf{u}_j'$ and $\mathbf{u}_k'$ chosen at random:

$$\begin{cases} z_{(i)}(\mathbf{u}_j') &= z_{(i-1)}(\mathbf{u}_k') \\ z_{(i)}(\mathbf{u}_k') &= z_{(i-1)}(\mathbf{u}_j') \end{cases}$$

---

[1] To simplify notation, the superscript (l), which refers to a particular realization, is omitted in this section.

Such a perturbation mechanism allows one to keep unchanged the histogram of the initial image. Thus, there is no need to include reproduction of the histogram in the objective function as long as the initial image already matches the target histogram.

2. Select randomly a single location $\mathbf{u}'_j$ and modify the corresponding z-value $z_{(i-1)}(\mathbf{u}'_j)$ according to some mechanism; for example, the target histogram $F(z)$ is sampled anew for a different value:

$$z_{(i)}(\mathbf{u}'_j) \;=\; F^{-1}(p_j)$$

with p_j being a random number in $[0, 1]$. Unlike the swapping mechanism, preserving the initial histogram here requires a specific component in the objective function.

In both cases, conditioning data are never perturbed, thereby ensuring that the final realization honors data values.

The objective function

Simulated annealing allows one to account for various types of information as long as these can be quantified to enter a global objective function. That function appears as a weighted sum of C components O_c, each measuring the deviation of the current realization (at the ith perturbation) from the target statistics:

$$O(i) \;=\; \sum_{c=1}^{C} \omega_c \, O_c(i) \tag{8.15}$$

where the weight ω_c controls the relative importance of the cth component in the objective function.

Defining the components O_c
Different components can be incorporated in the objective function, depending on the target statistics to be reproduced. Six examples follow:

(1) *One-point cdf*
A typical target statistic is the, possibly declustered, one-point cumulative distribution of the z-data, $F(z)$. If the range of variation of z is discretized by a series of K thresholds z_k, the deviation between target and current cdfs can be measured as

$$O_c(i) \;=\; \sum_{k=1}^{K} \Big[F(z_k) - \widehat{F}_{(i)}(z_k) \Big]^2 \tag{8.16}$$

where $\widehat{F}_{(i)}(z_k)$ is the cumulative frequency at threshold z_k calculated from the realization at the ith perturbation.

(2) *Semivariogram model*

Reproduction of the z-semivariogram model $\gamma(\mathbf{h})$ is generally limited to a specified number S of lags. One measure of the deviation between target and current semivariogram values is

$$O_c(i) \;=\; \sum_{s=1}^{S} \frac{\left[\gamma(\mathbf{h}_s) - \widehat{\gamma}_{(i)}(\mathbf{h}_s)\right]^2}{[\gamma(\mathbf{h}_s)]^2} \tag{8.17}$$

where $\widehat{\gamma}_{(i)}(\mathbf{h}_s)$ is the semivariogram value at lag $\mathbf{h}_s$ of the realization at the ith perturbation. The division by the square of the semivariogram model at each lag $\mathbf{h}_s$ gives more weight to reproduction of the z-semivariogram model near the origin.

(3) *Indicator semivariogram models*

As with sequential indicator algorithms, simulated annealing allows one to account for class-specific patterns of spatial continuity as modeled by indicator semivariograms $\gamma_I(\mathbf{h}; z_k)$ defined at K different thresholds z_k. One measure of deviation between target and current indicator semivariogram values is

$$O_c(i) \;=\; \sum_{k=1}^{K}\sum_{s=1}^{S} \frac{\left[\gamma_I(\mathbf{h}_s; z_k) - \widehat{\gamma}_I^{(i)}(\mathbf{h}_s; z_k)\right]^2}{[\gamma(\mathbf{h}_s; z_k)]^2}$$

where $\widehat{\gamma}_I^{(i)}(\mathbf{h}_s; z_k)$ is the indicator semivariogram value at lag $\mathbf{h}_s$ and threshold z_k for the realization at the ith perturbation.

(4) *Multiple-point* statistics

All previous simulation techniques are limited to reproduction of two-point statistics in that (indicator) semivariograms or covariance functions involve only two locations at a time. Two-point statistics, however, are often not enough to characterize complex features, such as curvilinear structures, cross-bedding, and meandering geometries (Guardiano and Srivastava, 1993). Reproduction of such complex spatial features calls for considering more than two locations at a time, say, three or four locations as illustrated by the examples of Figure 8.13.

Consider a J-point configuration defined by J separation vectors $\mathbf{h}_1, \dots, \mathbf{h}_J$, with $\mathbf{h}_1 = 0$ by convention. The probability that the J values $z(\mathbf{u} + \mathbf{h}_1), \dots$, $z(\mathbf{u} + \mathbf{h}_J)$ are jointly no greater than the J threshold values $z_1, \dots, z_J$ is defined as

$$\phi(\mathbf{h}_1, \dots, \mathbf{h}_J; z_1, \dots, z_J) \;=\; \mathrm{E}\left\{\prod_{j=1}^{J} I(\mathbf{u} + \mathbf{h}_j; z_j)\right\} \tag{8.18}$$

The quantity (8.18) can be read as a multiple-point non-centered indicator covariance. For $J = 2$, one retrieves the usual two-point indicator covariance

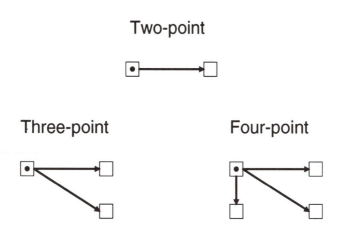

Figure 8.13: Examples of two-point, three-point, and four-point data configurations.

as $C_I(\mathbf{h}; z_1, z_2) = \phi(\mathbf{h}_1, \mathbf{h}; z_1, z_2) - F(z_1) \cdot F(z_2)$. If all J threshold values are equal to the same value z_c and all J separation vectors are multiples of the same vector $\mathbf{h}$, the multiple-point statistic (8.18) is the connectivity function introduced by Journel and Alabert (1988):

$$\phi(J; z_c) = \mathrm{E}\left\{\prod_{j=1}^{J} I(\mathbf{u} + (j-1)\mathbf{h}; z_c)\right\} \qquad (8.19)$$

The quantity (8.19) measures the probability that a string of J values oriented along the direction of $\mathbf{h}$ are jointly no greater than a given threshold value z_c.

Multiple-point statistics of type (8.18) or (8.19) can be reproduced by adding into the objective function a component of type

$$O_c(i) = \left[\phi(\mathbf{h}_1, \ldots, \mathbf{h}_J; z_1, \ldots, z_J) - \widehat{\phi}_{(i)}(\mathbf{h}_1, \ldots, \mathbf{h}_J; z_1, \ldots, z_J)\right]^2$$

where the experimental frequency $\widehat{\phi}_{(i)}(\mathbf{h}_1, \ldots, \mathbf{h}_J; z_1, \ldots, z_J)$ is calculated from the realization at the ith perturbation.

The difficulty resides in the inference of the target multiple-point statistics $\phi(\mathbf{h}_1, \ldots, \mathbf{h}_J; z_1, \ldots, z_J)$. For example, the inference of a three-point statistic $\phi(\mathbf{h}_1, \mathbf{h}_2, \mathbf{h}_3; z_1, z_2, z_3)$ requires the availability of a series of triplet values with the same three-point configuration, say, the same geometric configuration as in Figure 8.13 (left bottom graph). Non-regular gridding and data sparsity generally preclude the computation of such statistics from sample data. In most applications, multiple-point statistics are derived from an array of values referred to as a *training image* or control pattern (Farmer, 1992; Deutsch and Journel, 1992b). Such a training image may also originate from better sampled zones or can be synthetised from ancillary information.

(5) *Correlation coefficient*

Let Y be a better sampled or previously simulated secondary variable related to the primary variable Z. Provided the Z-Y relationship is correctly summarized by the linear correlation coefficient $\rho_{ZY}(0)$, the secondary information can be incorporated by forcing the set of simulated z-values to reproduce the target linear correlation with the colocated secondary y-data. This is done by including in the objective function a component measuring deviation between target and current correlation coefficients:

$$O_c(i) \;=\; \left[\rho_{ZY}(0) - \hat{\rho}^{(i)}_{ZY}(0)\right]^2$$

where $\hat{\rho}^{(i)}_{ZY}(0)$ is the correlation coefficient calculated from pairs of colocated simulated z-values and y-data at the ith perturbation, $(z_{(i)}(\mathbf{u}'_j), y(\mathbf{u}'_j))$.

(6) *Cross semivariogram*

The spatial cross correlation between two variables Z and Y, as modeled by the cross semivariogram $\gamma_{ZY}(\mathbf{h})$, can be reproduced by including in the objective function a component of the type

$$O_c(i) \;=\; \sum_{s=1}^{S} \frac{\left[\gamma_{ZY}(\mathbf{h}_s) - \hat{\gamma}^{(i)}_{ZY}(\mathbf{h}_s)\right]^2}{\left[\gamma_{ZY}(\mathbf{h}_s)\right]^2}$$

where $\hat{\gamma}^{(i)}_{ZY}(\mathbf{h}_s)$ is the cross semivariogram value between simulated z-values and y-data at the ith perturbation.

Unlike RF-based simulation algorithms, simulated annealing does not call for solving any cokriging system. Thus, there is no need for a prior joint modeling of auto and cross covariance functions under the constraint of positive semi-definiteness; this provides more flexibility at the risk of inconsistencies between direct and cross semivariogram values. In such a case, simulated annealing might generate realizations that match only roughly the inconsistent target model.

Updating the components O_c

The optimization process calls for recalculating all components O_c of the objective function (8.15) after each perturbation. Since millions of such perturbations may be needed to approximate the target statistics, it is critical that (1) each component O_c can be quickly updated, and (2) there are not too many components and they are not strongly conflicting. The common perturbation mechanisms typically involve only two locations at a time; consequently, most statistics can be updated locally, using a few nodal z-values, rather than globally considering all N grid nodes.

Consider, for example, that the ith perturbation consists of substituting the new value $z'(\mathbf{u}'_j)$ for the former value $z(\mathbf{u}'_j)$. The semivariogram value at any particular lag $\mathbf{h}$, $\widehat{\gamma}_{(i-1)}(\mathbf{h})$, is updated by subtracting the contribution of the former value $z(\mathbf{u}'_j)$ and adding the contribution of the new value $z'(\mathbf{u}'_j)$:

$$\widehat{\gamma}_{(i)}(\mathbf{h}) \;=\; \widehat{\gamma}_{(i-1)}(\mathbf{h}) \;-\; \left([z(\mathbf{u}'_j) - z(\mathbf{u}'_j + \mathbf{h})]^2 + [z(\mathbf{u}'_j - \mathbf{h}) - z(\mathbf{u}'_j)]^2\right)$$
$$+ \left([z'(\mathbf{u}'_j) - z(\mathbf{u}'_j + \mathbf{h})]^2 + [z(\mathbf{u}'_j - \mathbf{h}) - z'(\mathbf{u}'_j)]^2\right)$$

Whatever the number N of grid nodes, the previous updating of the semivariogram involves only two pairs of z values at each lag $\mathbf{h}$. In general, for any J-point statistic, there are J configurations that change if a single location is perturbed.

Weighting the components O_c
Recall the definition (8.15) of the objective function:

$$O(i) \;=\; \sum_{c=1}^{C} w_c\, O_c(i)$$

The role of the weighting scheme (w_c) is threefold:

1. Account for differences in the units of measurements between components. All components should be standardized to prevent the component with the largest unit from dominating the objective function.

2. Control the relative importance of each component in the objective function. In the presence of conflicting components, one may want to give greater weight to the reproduction of the statistic that is either most reliable or most consequential for the problem at hand. The relative importance of each component may also change during the optimization process; for example, at the beginning of the simulation one may give priority to the reproduction of coarse spatial features by giving more weight to the reproduction of the semivariogram at large lags $\mathbf{h}_s$.

3. Account for the average magnitude of each component change caused by a single perturbation, $|\Delta O_c(i)| = |O_c(i) - O_c(i-1)|$. In practice, the average change, $\overline{|\Delta O_c|}$, can be approximated during a prior calibration run where a certain number M (say, 1,000) of independent perturbations are applied to the same initial image. The M resulting changes are then averaged as

$$\overline{|\Delta O_c|} \;=\; \frac{1}{M} \sum_{m=1}^{M} |O_c^{(m)}(1) - O_c(0)|$$

where $|O_c^{(m)}(1) - O_c(0)|$ measures the impact of the mth perturbation on the initial value of the cth component, $O_c(0)$.

Once all components have been standardized, a possible weighting system is

$$\omega_c \;=\; \frac{\lambda_c}{|\overline{\Delta O_c}|} \qquad c = 1, \ldots, C \qquad \text{with } \sum_{c=1}^{C} \lambda_c = 1$$

The parameter λ_c controls the relative importance of the cth component, whereas the denominator $|\overline{\Delta O_c}|$ ensures that each component contributes equally to the change in the objective function.

The decision rule

Different criteria can be used to decide whether a given perturbation is accepted or rejected during the optimization process.

The maximum a posteriori criterion
One straightforward approach consists of accepting only the perturbations that diminish the objective function, that is,

$$\text{Prob \{Accept } i\text{th pert.\}} \;=\; \begin{cases} 1 & \text{if } O(i) \leq O(i-1) \\ 0 & \text{otherwise} \end{cases}$$

When the number of possible perturbations is small, one could retain the perturbation that reduces the value of the objective function the most. This latter approach is called the Maximum a Posteriori (MAP) model (Geman and Geman, 1984; Doyen et al., 1989). Such steepest descent-type approach tends to quickly "freeze" the realization, which could then be trapped in a sub-optimal situation far from the target statistics. Therefore, the MAP criterion is better suited to situations where a fairly advanced image is post-processed using simulated annealing (see section 8.9.1).

The annealing schedule
A decision rule commonly used in the application of simulated annealing amounts to accepting unfavorable perturbations according to a negative exponential probability distribution:

$$\text{Prob \{Accept } i\text{th pert.\}} \;=\; \begin{cases} 1 & \text{if } O(i) \leq O(i-1) \\ \exp\left(\frac{[O(i-1)-O(i)]}{t(i)}\right) & \text{otherwise} \end{cases}$$

The larger the parameter $t(i)$ of the probability distribution, called *temperature*, the greater the probability that an unfavorable perturbation will be accepted at the ith iteration. The previous criterion corresponds to a constant zero temperature, $t(i) = 0 \;\; \forall \; i$: all unfavorable perturbations are then rejected.

The idea is to start with an initially high temperature $t(0)$, which allows one to accept a large proportion of unfavorable perturbations at the beginning of the simulation. As the simulation proceeds, the temperature is gradually lowered so as to limit discontinuous modification of the stochastic image. The simulation algorithm thus appears analogous to the metallurgical annealing process whereby a very hot melt is slowly cooled so that molecules reorder themselves into a low-energy system.

Two important issues are the timing and magnitude of the temperature reduction, which defines the *annealing schedule*. The temperature is generally lowered by multiplying the initial temperature $t(0)$ by a reduction factor λ ($\lambda < 1$) whenever enough perturbations have been accepted or too many have been tried (Farmer, 1992; Press et al., 1986). The maximum number of accepted or attempted perturbations is chosen as a multiple of the number N of grid nodes. The objective is to find a balance between a temperature reduction that is too slow, thereby unnecessarily increasing the convergence time, and a temperature reduction that is too fast, which may freeze the image at some local minimum of the objective function unacceptably remote from the target statistics.

The convergence criterion

As with any iterative algorithm, a criterion for stopping the optimization process must be defined. Possible stopping criteria follow:

1. The objective function reaches a sufficiently small value O_{min}.

2. The maximum number of perturbations that can be attempted at the same temperature has been exceeded a certain number of times.

3. The proportion of accepted perturbations is smaller than a given threshold value.

Example

The spatial distribution of Cd values over the study area is simulated using simulated annealing and the three annealing schedules given in Table 8.1. All three algorithms start with the same initial random image shown in Figure 8.12 (page 410, left top). That image is perturbed by swapping two values at a time, which ensures that the initial target histogram is not perturbed. The objective function includes a single component of type (8.17) that controls reproduction of the isotropic Cd semivariogram model over the first 20 lags.

Figure 8.14 (first row) shows the value of the objective function versus the number of swaps for the three different annealing schedules. Note the following:

- The value of the objective function increases at the beginning of the simulation when a large proportion of unfavorable perturbations are accepted.

Table 8.1: Parameters of annealing schedules used in the example of Figure 8.14. The inital temperature t_0 is reduced by a factor λ whenever enough perturbations have been attempted ($K_{attempt} \cdot N$) or have been accepted ($K_{accept} \cdot N$). The simulation is stopped when either the target low value O_{min} is reached or the maximum number of attempted perturbations at the same temperature has been reached S times.

Schedule	t_0	λ	$K_{attempt}$	K_{accept}	O_{min}	S
Default	1.0	0.10	100	10	0.001	3
Fast	1.0	0.05	50	5	0.001	3
Very fast	0.5	0.01	10	2	0.001	3

- The value of the objective function decreases sharply when the temperature is lowered. The magnitude of the drop is greatest for the very fast annealing schedule with a small factor λ.

- The very fast annealing schedule is stopped after only 64,000 swaps, whereas the default schedule is stopped after 398,000 swaps. In this example, the three schedules yield a similar, excellent reproduction of the target semivariogram model (Figure 8.14, second row). In other situations, there might be a risk of getting trapped in unacceptable sub-optima when using an annealing schedule that is too fast.

- The three realizations shown at the bottom of Figure 8.14 are quite different, yet they all honor the 259 Cd data at their locations and reproduce closely the target semivariogram model. Such differences between realizations reflect the existence of many approximate solutions to the optimization problem, and these can be used to model spatial uncertainty. In this example, uncertainty may have been understated because the same initial image has been used.

8.8   Simulation of Categorical Variables

Variables, such as the concentration of a metal in the soil, may appear to change abruptly in space at terrace bluffs and across the boundaries between the outcrops of contrasting rocks. In such cases, the phenomenon should be modeled as a mixture of populations each with possibly different patterns of spatial continuity. The simulation would proceed in two steps: (1) the relative geometry of the different populations, say, different rocks, is modeled, and (2) the spatial distribution of the continuous attribute specific to each population is simulated using any of the previously mentioned algorithms. This two-step approach allows for better reproduction of long-range

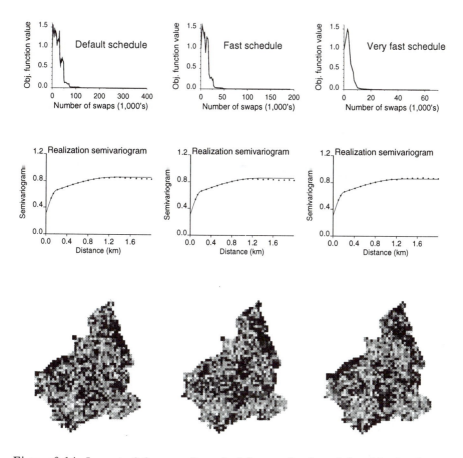

Figure 8.14: Impact of the annealing schedule on reduction of the objective function value versus number of swaps (top graphs) and reproduction of the target Cd semivariogram model (second row). The corresponding final realizations are shown at the bottom of the figure.

heterogeneities as generated by categorical boundaries and, simultaneously, the short-range heterogeneities within each category (Alabert and Massonnat, 1990; Damsleth et al., 1990).

Consider, for example, a stratification of the study area based on the geologic map (Figure 8.15, left top graph). Stratum 1 includes Argovian rocks with the smallest proportion of contaminated locations; the four other geologic formations constitute stratum 2. Figure 8.15 (right top graph) shows the semivariograms of the Cd normal score data within each stratum. Concentrations appear to vary more continuously within the first stratum (smaller relative nugget effect), which relates to the better connectivity of small Cd concentrations revealed by the indicator semivariograms shown in Figure 2.19 (page 45). The spatial distribution of Cd concentration is simulated within

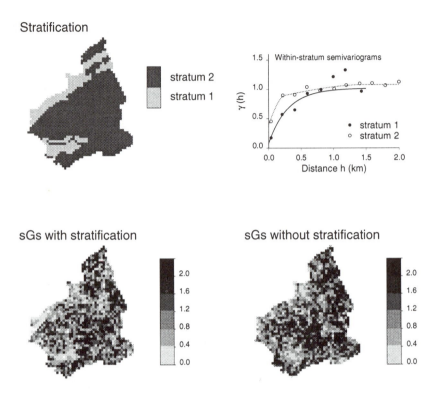

Figure 8.15: A two-step approach for simulating a mixture of populations identified with the different states of a categorical attribute. The study area is first stratified according to rock types, then Cd values are simulated within each stratum using sGs and stratum-specific semivariograms. The right bottom graph shows an sGs realization generated without prior stratification.

each stratum using sequential Gaussian simulation conditioned by data and semivariogram models specific to that stratum. The resulting realization shows a better contrast between low-valued Argovian rocks and the other rocks than if a single population and semivariogram model is used; compare the two sGs realizations at the bottom of Figure 8.15.

In the example of Figure 8.15, the geometry of the two strata is retrieved directly from the geologic map. In many applications, there is no such exhaustive categorical map, and the relative geometry of the different populations should be simulated first. Many algorithms can be used to simulate categorical variables: Boolean and object-oriented algorithms (Ripley, 1987; Haldorsen et al., 1988; Suro-Pérez, 1991); single or multiple truncations of a Gaussian field (Journel and Isaaks, 1984; Matheron et al., 1987; Xu and Journel, 1993); indicator-based algorithms (Journel, 1989; Alabert and Massonnat, 1990; Deutsch and Journel, 1992a); simulated annealing (Deutsch and Journel, 1992b; Farmer, 1992; Goovaerts and Journel, 1996); p-field simulation (Xu, 1995a). The following presentation is limited to the class of indicator simulation algorithms.

Sequential indicator simulation

Consider the simulation of the spatial distribution of K mutually exclusive categories s_k conditional to the data set $\{s(\mathbf{u}_\alpha), \alpha = 1, \ldots, n\}$; see subsequent definition. Sequential indicator simulation of categorical variables follows a procedure similar to that described in section 8.4 for continuous variables:

- Transform each categorical datum $s(\mathbf{u}_\alpha)$ into a vector of K hard indicator data, defined as

$$i(\mathbf{u}_\alpha; s_k) = \begin{cases} 1 & \text{if } s(\mathbf{u}_\alpha) = s_k \\ 0 & \text{otherwise} \end{cases} \qquad k = 1, \ldots, K$$

- Define a random path visiting each node of the grid only once.

- At each node $\mathbf{u}'$:

  1. Determine the conditional probability of occurrence of each category s_k, $[p(\mathbf{u}'; s_k|(n))]^*$, using simple or ordinary indicator (co)kriging. The conditioning information (n) consists of neighboring original indicator data and previously simulated indicator values.

  2. Correct these probabilities for order relation deviations.

  3. Define **any** ordering of the K categories and build a cdf-type function by adding the corresponding probabilities of occurrence; for example,

$$[F(\mathbf{u}'; s_{k'}|(n))]^* = \sum_{k=1}^{k'} [p(\mathbf{u}'; s_k|(n))]^* \qquad k' = 1, \ldots, K$$

  4. Draw a random number p uniformly distributed in $[0, 1]$. The simulated category at location $\mathbf{u}'$ is the one that corresponds to the probability interval that includes p:

$$s^{(l)}(\mathbf{u}') = s_{k'} \quad \text{such as} \quad [F(\mathbf{u}'; s_{k'-1}|(n))]^* < p \leq [F(\mathbf{u}'; s_{k'}|(n))]^*$$

  5. Add that simulated value $s^{(l)}(\mathbf{u}')$ to the conditioning data set.

  6. Proceed to the next node along the random path, and repeat steps 1–5.

Repeat the entire sequential procedure with a different random path to generate another realization $\{s^{(l')}(\mathbf{u}'_j), j = 1, \ldots, N\}$, $l' \neq l$.

Remarks

1. Because the random value p is uniformly distributed in $[0, 1]$, the arbitrary ordering of the K categories affects neither which category is drawn nor their spatial distribution (Alabert and Massonnat, 1990).

2. Because the K indicator RVs $I(\mathbf{u}; s_k)$ are linearly related (they add to 1), there is a risk of numerical instabilities if all K categories are considered together in a single cokriging system (see related discussion in section 7.3.6).

3. Actual data about the prevailing category s_k may be supplemented by soft information, such as the knowledge of absence of one or more categories or a set of prior probabilities of occurrence provided by ancillary data. The indicator algorithms introduced in section 7.3.3 allow one to account for both hard and soft indicator data in the determination of the conditional probabilities at each simulated node.

Example

Sequential indicator simulation (sis) is used to simulate the spatial distribution of the four land uses over the study area conditional to the 50 data shown at the top of Figure 7.42 (page 355). The conditional probabilities are determined using ordinary indicator kriging and the indicator semivariogram models displayed at the bottom of Figure 7.30. At each grid node, the conditioning information consists of the 12 closest land use data and the 12 closest previously simulated values.

Figure 8.16 shows one realization (top graph) and the corresponding direct and cross indicator semivariograms. Note the following:

- The realization departs significantly from the target sample proportions (4, 58, 18, 20%). For correction of such departure, see section 8.9.1.

- The direct indicator semivariograms are reasonably well reproduced.

- The indicator cross semivariograms, which measure the transition frequencies between two different categories, are poorly reproduced since they were not accounted for in the indicator algorithm, in this case ordinary indicator kriging.

8.9   Miscellaneous Aspects of Simulation

Algorithms for simulating continuous and categorical variables were introduced in sections 8.2–8.8. First, this section shows how a post-processing of realizations allows one to improve reproduction of model statistics. Next, different ways of summarizing the spatial uncertainty represented by the series of alternative realizations are reviewed. Last, guidelines for selecting from

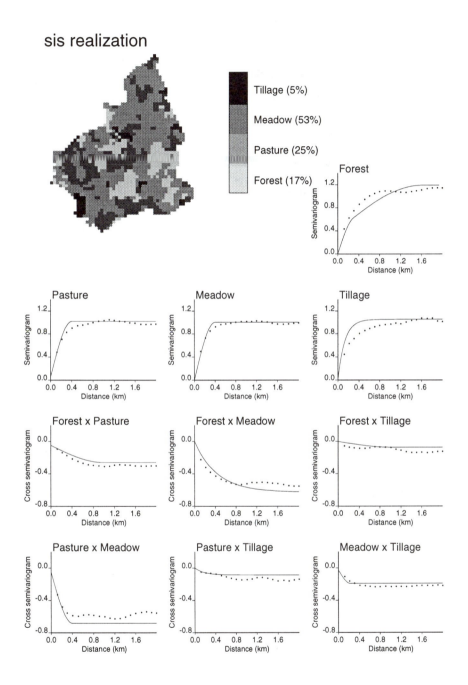

Figure 8.16: Sequential indicator realization of the spatial distribution of four land uses over the study area, and the corresponding direct and cross indicator semivariograms. Solid lines depict the target semivariogram models.

the toolbox of simulation algorithms the one best suited to the problem at hand are offered.

8.9.1   Reproduction of model statistics

Stochastic realizations rarely match model statistics exactly, nor should they. Consider, for example, the three sGs realizations shown in Figure 8.2 (page 373). The three corresponding realization semivariograms $\gamma^{(l)}(\mathbf{h})$, $l = 1$, 2, and 3, fluctuate around the model $\gamma(\mathbf{h})$ depicted by the solid line in Figure 8.17 (left graph). Similar fluctuations are observed for the marginal cdfs (Figure 8.17, right graph). Such discrepancies between realization and model statistics are referred to as *ergodic fluctuations*.

Several factors control the importance of ergodic fluctuations displayed by a realization (Deutsch and Journel, 1992a, p. 127–129):

1. The algorithm used to generate the set of realizations.

   Unlike simulated annealing, Gaussian- and indicator-based simulation algorithms reproduce the semivariogram model(s) only in expected value, that is, on average over many realizations. Consequently, larger fluctuations of realization semivariograms are expected when using sequential simulation algorithms.

2. The density of conditioning data.

   As more data are used to condition the realizations, the realization statistics become increasingly similar and closer to the target statistics, if these were modeled from the same data.

3. The semivariogram parameters and the size of the simulation grid.

   Ergodic fluctuations of realization semivariograms are generally important when the range of the semivariogram model is large with respect to the size of the simulated area, particularly if the relative nugget effect is small.

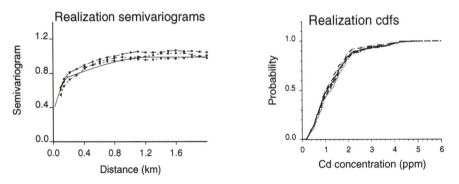

Figure 8.17: Fluctuations in the reproduction of the target semivariogram model and target cdf (solid line) by the three realizations of Figure 8.2.

When departures from model statistics are deemed too important, the realization can be discarded and another realization generated. Computation time may prevent one from creating many realizations and selecting only those with desirable statistics. An alternative is to post-process the few realizations available so as to better reproduce or even identify these target statistics.

Beware that model statistics are inferred from sample information, which necessarily departs from the population parameters, particularly when data are sparse or in the presence of preferential sampling. Ergodic fluctuations allow one to account indirectly for the uncertainty about sample statistics. Reduction or removal of such fluctuations may lead to a false sense of certainty about the simulated features; the selected realizations are artificially made to look alike through too stringent constraints of reproduction of target statistics. A more rigorous approach (Journel, 1994b) consists of a formal randomizing of the semivariogram models and then accounting for such variation in the simulation algorithms.

Data that are subject to measurement error should not be exactly honored by the realization. Moreover, the part of the nugget effect arising from these errors should not be reproduced by the realization semivariogram. To filter the noise due to measurement error from the realization, Marcotte (1995) proposed to post-process it using a modified version of kriging analysis introduced in section 5.6. The filtered image does not honor noisy data at their locations, and the measurement error variance is removed from the realization semivariogram.

Posterior identification of histograms

Consider a realization $\{z^{(l)}(\mathbf{u}), \mathbf{u} \in \mathcal{A}\}$ of the RF $Z(\mathbf{u})$ conditional to the data $z(\mathbf{u}_\alpha), \alpha = 1, \ldots, n$. The cdf of the simulated values, denoted $F^{(l)}(z)$, is deemed too different from the target cdf $F(z)$. Journel and Xu (1994) have proposed an algorithm that allows improving reproduction of the target cdf while still honoring the conditioning data and without significant modification of the spatial correlation patterns in the original realization.

A transform that allows exact identification of the target cdf is a generalization of the normal score transform (7.8) and is written

$$z_c^{(l)}(\mathbf{u}) \;=\; F^{-1}\left[F^{(l)}(z^{(l)}(\mathbf{u}))\right] \qquad \forall\, \mathbf{u} \in \mathcal{A} \qquad (8.20)$$

where $z_c^{(l)}(\mathbf{u})$ is the value corrected from the original simulated value $z^{(l)}(\mathbf{u})$. The set of corrected values $\{z_c^{(l)}(\mathbf{u}), \mathbf{u} \in \mathcal{A}\}$ identifies (in expected value) the target cdf:

$$\text{Prob}\{Z_c^{(l)}(\mathbf{u}) \leq z\} \;=\; \text{Prob}\{F^{(l)}(Z^{(l)}(\mathbf{u})) \leq F(z)\} \;=\; F(z)$$

since the uniform score $F^{(l)}(Z^{(l)}(\mathbf{u}))$ is, by definition, uniformly distributed.

Relation (8.20) amounts to identification of the p-quantiles of the two distributions of $z_c^{(l)}(\mathbf{u})$ and $z^{(l)}(\mathbf{u})$:

$$F(z_c^{(l)}(\mathbf{u})) = F^{(l)}(z^{(l)}(\mathbf{u})) = p$$

Such quantile identification preserves by definition the ranks of the original values $z^{(l)}(\mathbf{u})$, and hence the structures seen on the original realization are unchanged.

The correction (8.20), however, also affects data locations, hence

$$z_c^{(l)}(\mathbf{u}_\alpha) \neq z^{(l)}(\mathbf{u}_\alpha) = z(\mathbf{u}_\alpha) \qquad \alpha = 1,\ldots, n$$

The exactitude property of the original realization could be preserved by correcting only the unsampled locations, but such selective correction may create artifact discontinuities next to the data locations. The solution consists of applying the correction (8.20) progressively as the location $\mathbf{u}$ gets farther away from the data locations:

$$z_{cc}^{(l)}(\mathbf{u}) = z^{(l)}(\mathbf{u}) + \lambda(\mathbf{u}) \left[z_c^{(l)}(\mathbf{u}) - z^{(l)}(\mathbf{u}) \right] \qquad (8.21)$$

with the relative correction factor $\lambda(\mathbf{u})$ defined as

$$\lambda(\mathbf{u}) = [\sigma_K(\mathbf{u})/\sigma_{max}]^\omega$$

where $\omega > 0$ is the correction parameter, $\sigma_K^2(\mathbf{u})$ is a kriging variance calculated using only the n data $z(\mathbf{u}_\alpha)$, and σ_{max} is the maximum kriging variance observed over the study area $\mathcal{A}$. The following are noteworthy:

- At a datum location $\mathbf{u}_\alpha$: $\sigma_K^2(\mathbf{u}_\alpha) = 0$ and $\lambda(\mathbf{u}_\alpha) = 0$, $\forall\, \omega$, hence the exactitude property of the original realization is preserved:

$$z_{cc}^{(l)}(\mathbf{u}_\alpha) = z^{(l)}(\mathbf{u}_\alpha) = z(\mathbf{u}_\alpha) \qquad \alpha = 1,\ldots, n$$

- As the location $\mathbf{u}$ gets farther away from data locations, the kriging variance $\sigma_K^2(\mathbf{u})$, and hence the intensity of the correction controlled by the factor $\lambda(\mathbf{u})$, increases.

- The ratio $\sigma_K(\mathbf{u})/\sigma_{max}$ is smaller than 1, so the correction factor $\lambda(\mathbf{u})$ increases as ω decreases. The parameter ω allows a balance between the two extreme cases of correction at all non-data locations (case $\omega = 0$) and no correction (case $\omega = \infty$).

A similar algorithm has been developed for post-processing realizations of categorical variables. The objective is to reproduce the target proportion of each category while honoring conditioning data without significant modification of class-indicator semivariograms.

Figure 8.18 (left top graph) shows a realization generated by sequential indicator simulation using only five threshold values. The Q-Q plot shown in

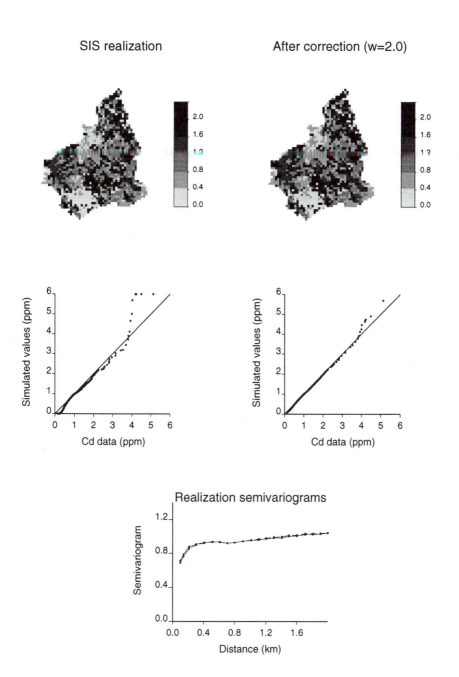

Figure 8.18: Sequential indicator realization that departs from the target cdf before (left column) and after correction (right column) using $\omega = 2.0$. The correction improves the reproduction of the target cdf while preserving the shape of the semivariogram.

Figure 8.18 (left middle graph) indicates a poor reproduction of the target histogram by the realization because of the coarse discretization. The realization is post-processed using the correction in equation (8.21) with $\omega = 2$ (Figure 8.18, right column). The correction algorithm improves the reproduction of the cdf model depicted by the solid line while preserving the shape of the semivariogram.

Post-processing with simulated annealing

Better reproduction of model statistics, not limited to the marginal distribution, can also be achieved by post-processing the realization with unacceptable fluctuations using simulated annealing introduced in section 8.7. Actually, simulated annealing can be used not only to improve reproduction of current statistics but also to impart new properties to the realization, such as multiple-point statistics (Deutsch and Journel, 1992b; Murray, 1992; Goovaerts, 1996).

Consider, for example, post-processing of the land use simulated map shown at the top of Figure 8.16 (page 425) with two objectives: (1) to improve reproduction of the K target class proportions p_k and direct semivariograms $\gamma_I(\mathbf{h}; s_k)$, and (2) to improve reproduction of the cross semivariograms $\gamma_I(\mathbf{h}; s_k, s_{k'})$ not accounted for by the original sequential indicator algorithm. The corresponding two objective function components are:

$$O_1(i) \quad = \quad \sum_{k=1}^{K} \left[p_k - \widehat{p}_k^{(i)} \right]^2$$

$$O_2(i) \quad = \quad \sum_{s=1}^{S} \left(\sum_{k=1}^{K} \sum_{k'=1}^{K} \frac{\left[\gamma_I(\mathbf{h}_s; s_k, s_{k'}) - \widehat{\gamma}_I^{(i)}(\mathbf{h}_s; s_k, s_{k'}) \right]^2}{\left[\gamma_I(\mathbf{h}_s; s_k, s_{k'}) \right]^2} \right)$$

where $\widehat{p}_k^{(i)}$ is the global proportion of category s_k at the ith perturbation, and $\widehat{\gamma}_I^{(i)}(\mathbf{h}_s; s_k, s_{k'})$ is the indicator cross semivariogram value between categories s_k and $s_{k'}$ at lag $\mathbf{h}_s$ calculated from the realization at the ith perturbation.

Because the realization shown in Figure 8.16 already reproduces fairly well global proportions and indicator semivariograms, it should not be completely randomized by accepting too many unfavorable perturbations at the beginning of the post-processing. Thus, the realization was post-processed using the following MAP algorithm that retains the perturbation that diminishes most the value of the objective function:

1. Compute the value of the objective function for the initial realization.

2. For a specified number of iterations, follow these steps:

   - Define a random path that visits all non-conditioned grid nodes.
   - At each node $\mathbf{u}$, consider all K possible categories and compute the values of the corresponding K objective functions.

- Allocate the node **u** to the category s_k associated with the smallest value of the objective function.
- Proceed to the next node along the random path.

The post-processing was stopped after 10 iterations when the proportion of changes per iteration was found to be less than 1% of the total number of grid nodes.

Figure 8.19 shows the post-processed realization and the corresponding direct and cross indicator semivariograms. Simulated annealing improves the reproduction of the target global proportions (4, 58, 18, 20%) and indicator semivariogram models. Significant discrepancies still remain between model and realization cross semivariograms, for example, between pasture and meadow. Such departures may be due to an inconsistent target coregionalization model. Recall that the direct and cross semivariograms were modeled independently without checking for the positive semi-definiteness of the full matrix of cross covariance functions.

8.9.2 Visualization of spatial uncertainty

The set of alternative realizations generated by stochastic simulation provides a measure of uncertainty about the spatial distribution of attribute values. To depict visually that uncertainty, several authors (Srivastava, 1994; Wang, 1994) have developed algorithms that show the realizations one at a time in rapid succession, like the frames of an animated cartoon, say, eight realizations per second. Like an animated cartoon, successive realizations must be similar enough to allow the eye to catch gradual changes. Such a similarity can be achieved by ranking the realizations appropriately (Wang, 1994) or by using a simulation algorithm that generates realizations that are incrementally different (Srivastava, 1994). The animated display of realizations allows one to distinguish areas that remain stable over all realizations (low uncertainty) from those where large fluctuations occur between realizations (high uncertainty).

The series of L realizations can also be post-processed and the uncertainty information summarized using different types of displays, as follows:

1. Probability maps.
   At each simulated grid node $\mathbf{u}'_j$, the probability of exceeding a given threshold z_k is evaluated as the proportion of the L simulated values $z^{(l)}(\mathbf{u}'_j)$ that exceed that threshold. The map of such probabilities is referred to as a probability map. For example, Figure 8.20 (top graph) shows the probability map for exceeding the critical Cd concentration 0.8 ppm. That map was determined from 100 sGs realizations. The two low-valued zones correspond to Argovian rocks where smaller Cd concentrations were measured.

2. Quantile maps.
   Rather than displaying the probability of exceeding a particular thresh-

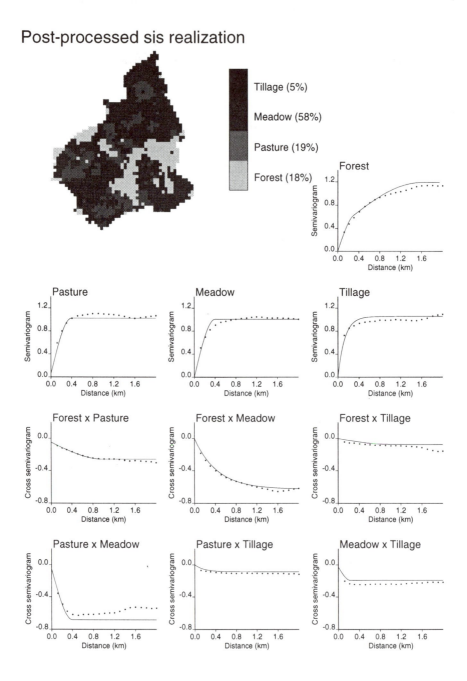

Figure 8.19: Post-processing of the sequential indicator realization of Figure 8.16 using simulated annealing. Note the better reproduction of target semivariogram models.

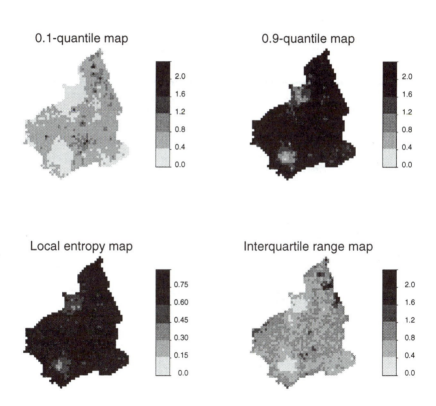

Figure 8.20: Different displays summarizing the spatial uncertainty model provided by a set of 100 sGs realizations of the spatial distribution of Cd values over the study area.

old z_k, one may map the p-quantile values corresponding to any given probability p. For example, the 0.1-quantile map in Figure 8.20 (left middle graph) shows at each grid node the Cd value that is exceeded by 90% of the simulated values there; high-valued areas thus indicate that the unknown Cd values are certainly large. Conversely, the 0.9-quantile map depicts the Cd values that are exceeded by only 10% of simulated valuese (Figure 8.20, right middle graph). Low-valued areas in a 0.9-quantile map indicate unknown Cd values that are almost certainly small (Argovian rocks).

3. Maps of spread.
   Local differences between realizations can be depicted by mapping some measure of the spread of the distribution of the L simulated values at each simulated grid node; for example, the local entropy, the variance, or the interquartile range of the local ccdf, as defined in section 7.4.1. For example, Figure 8.20 (bottom maps) shows the maps of local entropy and interquartile range determined from the 100 sGs realizations. Both maps indicate greater certainty (lighter grey) on Argovian rocks where Cd concentrations are consistently small. The uncertainty is greater where large and medium Cd concentrations are intermingled or in the west part of the study area where data are sparse (see Figure 1.1, page 5).

8.9.3   Choosing a simulation algorithm

The practitioner may get confused in the face of an ever-growing palette of simulation algorithms available. There is no simulation algorithm that is best for all cases but rather a toolbox of alternative algorithms from which to choose or to build the algorithm best suited for the problem at hand. Building from Deutsch (1994a), four criteria can be used to select an appropriate simulation algorithm:

1. The algorithm that best fits the goals of the study

2. The human and CPU time required for generating a given set of realizations

3. The amount of relevant information accounted for (conditioning data)

4. The precision and accuracy of probabilistic prediction, that is, of the distribution of outcomes resulting from the application of a given transfer function (flow simulator, remediation process) to the set of realizations

As shown in previous sections, the p-field and the various sequential Gaussian simulation algorithms are less demanding than either sequential indicator simulation or simulated annealing in terms of inference and computational effort. The rapid growth of computational capabilities, however, tends to attenuate that difference.

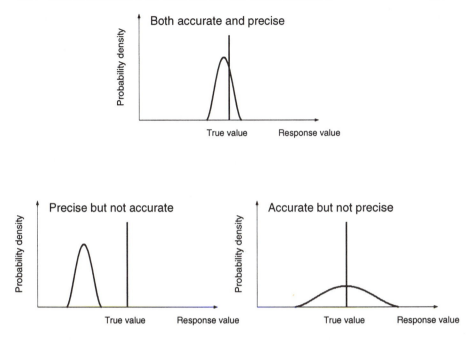

Figure 8.21: Illustration of the accuracy and precision of an output distribution generated by any transfer function, such as the cost function of Figure 8.3.

The greater CPU cost of indicator-based and simulated annealing algorithms is balanced by their greater flexibility in incorporating various types of information, such as class-specific patterns of spatial continuity, soft data, and multiple-point statistics.

The fourth criterion relates to the distribution of response values obtained by processing the set of realizations, for example, the distribution of costs shown at the top of Figure 8.3 (page 375). Recall that this distribution provides an assessment of the risk associated with declaring the study area safe with respect to Cd. A response distribution is accurate if some fixed probability interval (e.g., the symmetric 50% probability interval or interquartile range) contains the true response, in this case the actual cost if no remedial measure is taken. The precision of the response distribution is measured by its narrowness. Note that accuracy can be evaluated only if the actual true value is known, for example, during calibration exercises. A good simulation algorithm should generate an output distribution that is both accurate and precise (Figure 8.21, top graph). The worst situation is an output distribution that is precise but not accurate, because it gives a false sense of confidence to a prediction that is actually wrong (Figure 8.21, left bottom graph). Accuracy can be achieved at the expense of precision through an output distribution with a very large spread (Figure 8.21, right bottom graph).

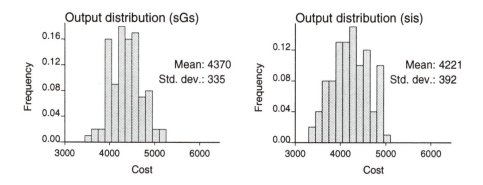

Figure 8.22: Two distributions of costs associated with a wrong decision to declare the study area safe with respect to Cd. These distributions are obtained by post-processing 100 realizations of the spatial distribution of Cd values generated using either the sequential Gaussian or the sequential indicator simulation algorithm.

The relative precision resulting from each simulation algorithm is readily assessed by measuring the spread of the output distribution, which is often referred to as the *space of uncertainty*. The extent of this space depends on the particular algorithm used to generate the realizations. For example, Figure 8.22 shows the probability distributions of the cost (8.2) computed from two sets of 100 realizations generated by two different algorithms: sequential Gaussian and sequential indicator simulation. Precision also depends on the transfer function being applied to the realizations. For the same data set, depending on the transfer function and the specific response value retained, different algorithms can be best. Note that conditioning to more information does not necessarily reduce the space of uncertainty. Incorporating additional information that conflicts with current data may increase the uncertainty in the response variable.

Chapter 9

Summary

Chapters 2–8 covered the sequence typical of a geostatistical study, beginning with exploratory data analysis, then quantitative modeling of spatial continuity, prediction of attribute values at unvisited locations, and, last, assessment of local and spatial uncertainty about unsampled values. This final chapter provides a synopsis of previous chapters and points out topics that deserve further attention.

Data collection

Characterization of any region requires a prior collection of data at specific locations within the region. In this book, one considered the situation where data have already been collected and the main challenge is to infer statistics representative of the study area from data that might have been preferentially sampled. The problem of designing sampling schemes has not been addressed. Information about how geostatistics can be used in sampling design is given in Webster and Oliver (1990, p. 272–290).

The number and spatial configuration of data are controlled by several factors, such as the available technology and resources, the accessibility of some areas, the need for sampling more densely areas deemed critical, the combination of different measurement devices used, etc. The sampling strategy should always be clearly documented because it may guide or skew the results of the exploratory data analysis.

When dealing with attributes that vary over time, such as pollutant concentrations, spatial data should be collected within a short time to avoid mixing spatial and temporal fluctuations. If the study pursues different objectives that call for different sampling strategies, the sampling scheme should be designed to simultaneously achieve these objectives. For example, the Jura data were collected using a combination of a nested and regular sampling schemes, which allowed characterization of short-range variability of metal concentrations while providing uniform coverage for subsequent mapping.

Exploratory data analysis

The second step consists of getting familiar with the data using descriptive tools, such as a location map, histogram, scattergram, or h-scattergram as introduced in Chapter 2. Exploratory data analysis may indicate the existence of several populations with significantly different features. In such a case, one should consider splitting the data into more homogeneous subsets prior to statistical analysis. Such subdivision might not be possible because of the lack of data or the inability to delineate the different populations in the field.

Prior information about the range of attribute values, their cross relations and distributions in space should be used with the data to build summary statistics, such as means, variances, correlation coefficients, and semivariograms. Beware that sample statistics are affected by spatial clusters and extreme values. If there is no strong physical reason for discarding extreme values or possibility to treat them separately, their influence could be reduced by an appropriate transformation of data or by using more robust statistics. When specific subareas are preferentially sampled, the global representativity of sample statistics should be questioned, and the declustering techniques introduced in section 4.1.1 should be considered.

Probabilistic models

To go beyond the data and estimate attribute values at unvisited locations, a model of variability in space must be established. Our imperfect knowledge of how physical processes operate and interact over the study area usually precludes any deterministic modeling whereby a single and deemed exact value is associated with each unsampled location. As an alternative until deterministic models become available, the random function (RF) model is introduced in Chapter 3. The random function allows modeling of spatial variability in that it yields at each unsampled location a probability distribution of possible values rather than a single value.

One important aspect of the RF approach is the concept of stationarity, which allows inference of statistics, such as the mean, covariance, or semivariogram, by pooling data over areas deemed homogeneous. It is worth recalling that stationarity is not a characteristic of the physical phenomenon under study, rather it is a modeling decision needed to specify the parameters of the RF model used. Because such stationarity cannot be proven or refuted from the data alone, however, the decision of stationarity, like any model decision, can be deemed inappropriate if its consequences do not allow one to reach the goals of the study.

Structural analysis

One major step in any geostatistical analysis is to build a licit semivariogram (covariance) model that captures the relevant spatial features of the attribute under study. Subsequent applications, such as prediction or assessment of uncertainty, rely on this particular model of spatial variability.

As emphasized in Chapter 4, semivariogram modeling is not simply an exercise in fitting curves to experimental values. Important decisions regarding the number, type, and anisotropy of basic semivariogram models must be made by the user rather than left to some automatic-fitting algorithm. The user's expertise is particularly critical whenever sparse data and measurement errors lead to noisy sample semivariograms.

Modeling a coregionalization is made difficult because the different direct and cross semivariograms cannot be modeled independently of one another. The commonly used linear model of coregionalization is reasonably flexible and allows an easy check of the positive semi-definiteness condition. The consistency constraints become very cumbersome when many attributes are considered altogether. For more than two attributes, interactive graphical programs including semi-automatic fitting algorithms should be considered (see Appendix A).

Local estimation

In Chapter 5, three variants of the linear regression algorithm (simple kriging, ordinary kriging, and kriging with a trend) are introduced for predicting values of a single attribute at unsampled locations. Explicit modeling of local trends is typically needed only when the location being estimated is outside the geographic range of data (extrapolation situation). In interpolation situations, ordinary kriging implicitly rescales the mean value within each local search neighborhood.

Block kriging allows for estimation of attribute values linearly averaged over supports (segment, area, volume) much larger than data supports. Beware that arithmetic averages of data are not the same as physical averages if the attribute does not average linearly in space, for example permeability or pH.

All interpolation algorithms tend to smooth out local details of the spatial variability of the attribute, leading to overestimation of small values and underestimation of large ones. Such smoothing depends on the local data configuration: high-frequency components are progressively filtered as the location being estimated gets farther away from data locations. Maps of kriging estimates appear artificially more variable in densely sampled areas than they do where data are sparse. Building on that limitation, kriging analysis allows one to filter one or more spatial components of the semivariogram or covariance model, resulting in maps of low- or high-frequency components of spatial variability. Such maps can be used as exploratory tools to detect areas that depart substantially from a regional background.

Accounting for secondary information

When secondary information is available at all locations, the study area can be stratified according to secondary data. Then the primary attribute is estimated within each stratum using primary data and covariance model specific to that stratum. An alternative is to use secondary data to inform on the spatial trend of the primary attribute.

The cokriging algorithm, as introduced in Chapter 6, allows one to account for non-exhaustive secondary information, capitalizing on the spatial cross correlations between primary and secondary variables. Recurrent practice has shown that cokriging yields better re-estimation scores than kriging only when the primary variable is undersampled with respect to the secondary variables. In other situations, the theoretical increase in precision is not worth the additional CPU cost and modeling effort required by cokriging relative to kriging.

The computational cost of cokriging can be alleviated by retaining only the secondary data colocated with or nearest to the location being estimated. Provided it is not invalidated, a Markov-type model allows a further reduction of inference and modeling effort. Analysis of re-estimation scores at test locations has shown that colocated cokriging with cross correlograms instead of cross covariances reduces the mean absolute prediction errors and the proportion of misclassified locations.

Cokriging can also be used to incorporate soft information about the primary attribute, such as constraint interval or prior probability distributions as derived from secondary information.

Assessment of local uncertainty

Chapter 7 is devoted to modeling uncertainty at unsampled locations, then using these uncertainty models for risk analysis and decision making.

Analysis of re-estimation scores at test locations has shown that the error variance provided by kriging algorithms is poorly correlated with actual estimation error, hence in general the kriging variance cannot be used alone as a measure of local uncertainty. The major shortcoming of the kriging variance is that it does not depend on data values. A better measure of local uncertainty is provided by a model of the probability distribution of the unknown made conditional to neighboring data values. This approach is more rigorous because local uncertainty is modeled independently of the derivation of an optimal estimate. The easiest approach for modeling conditional distributions (ccdfs) consists of transforming all original data into normal scores, then adopting a multiGaussian RF model for these normal scores. Under this particular model, all conditional distributions at any location are Gaussian with means and variances identifying the corresponding simple kriging estimates and simple kriging variances. One limitation of the multiGaussian model is that it does not allow for any significant and specific spatial correlation between extreme values, whether large or small.

Class-specific patterns of spatial correlation can be accounted for using an indicator approach that does not assume any particular shape or analytical expression for the conditional distributions. Ccdfs are modeled through a series of threshold values discretizing the range of variation of the attribute. The greater computational and inference cost of indicator algorithms, compared to multiGaussian algorithms, is balanced by their greater flexibility in incorporating various types of information, such as constraint intervals or indirect (soft) categorical or continuous data. The key step is to code each piece of information (hard or soft data) into local prior probabilities. These prior probabilities are then processed together using kriging algorithms, resulting in posterior or updated distributions.

Two major issues of the indicator approach are the correction of order relation deviations and the choice of interpolation/extrapolation models to increase the resolution of the discrete ccdf models. Models for extrapolating lower and upper tails of the ccdf are particularly critical to the resulting statistics. Note that the multiGaussian approach also requires increasing the resolution of the sample cdf.

The uncertainty at an unsampled location can be assessed through spread measures derived from the corresponding ccdf model, for example, the local entropy, the conditional variance, or the interquartile range. Different estimates of the unknown attribute value can be retrieved from the ccdf, depending on the optimality criterion retained. Instead of adopting the arbitrary least-squares criterion that equally penalizes overestimation and underestimation, criteria should be customized to the specific problem at hand.

There are many ways to account for local uncertainty in a decision-making process such as that involving cleaning of hazardous areas. In particular, ccdfs modeled by the geostatistician can be combined with relevant economic functions to come up with expected costs for the different options considered.

Assessment of spatial uncertainty

Chapter 8 addresses the assessment of spatial or joint uncertainty rather than uncertainty at each single unsampled location. Stochastic simulation provides multiple realizations of the spatial distribution of the attribute values, each reproducing statistics deemed consequential for the problem at hand. Typically, simulated realizations do not show the smoothing effect characteristic of interpolated maps. Among the ever-growing repertory of simulation algorithms, this book focuses on four classes of algorithms including but not limited to the Gaussian model: sequential simulation, LU decomposition, p-field approach, and simulated annealing.

Sequential indicator simulation and simulated annealing offer greater flexibility in incorporating diverse types of information (class-specific patterns of spatial continuity, soft data, and multiple-point statistics) at the cost of larger CPU and inference requirements. Different algorithms are often used in conjunction. For example, the relative geometry of different categories (e.g. soil types) can be modeled using sequential indicator simulation, then the spatial distribution of the continuous attribute (e.g. metal concentration) specific to each category can be simulated using a Gaussian-based algorithm or p-field approach. Last, simulated annealing can be used to post-process a simulated map to ensure better reproduction of the target statistics and/or to impose additional constraints that cannot be readily incorporated by other simulation algorithms.

Unlike the ccdf model that provides only a *location-specific* measure of uncertainty, the set of alternative realizations generated by stochastic simulation provides a measure of uncertainty about the *spatial* distribution of attribute values. Such uncertainty can be visualized through an animated display of realizations. Spatial uncertainty can also be displayed using probability maps, quantile maps, or conditional variance maps.

Simulated maps often serve as input to complex transfer functions, such as flow simulators in reservoir engineering or remediation processes in pollution control. The final objective is to model the uncertainty about some response value, such as travel time or remediation efficiency.

As a last remark, beware that uncertainty is not intrinsic to the phenomenon under study: rather it arises from our imperfect knowledge of that phenomenon, it is data-dependent and most importantly model-dependent, that model specifying our prior concept (decisions) about the phenomenon. No model, hence no uncertainty measure, can ever be objective: the point is to accept that limitation and document clearly all aspects of the model.

Appendix A

Fitting an LMC

Goulard (1989) proposed an iterative procedure to fit a linear model of coregionalization (LMC) under the constraint of positive semi-definiteness of the coregionalization matrices $\mathbf{B}_l$. The algorithm aims at minimizing a weighted sum of squares of differences between the experimental and model (cross) semivariogram values:

$$\text{WSS} = \sum_{k=1}^{K} \sum_{i=1}^{N_v} \sum_{j=1}^{N_v} \omega(\mathbf{h}_k) \cdot \frac{[\widehat{\gamma}_{ij}(\mathbf{h}_k) - \gamma_{ij}(\mathbf{h}_k)]^2}{\widehat{\sigma}_i \cdot \widehat{\sigma}_j} \tag{A.1}$$

where $\omega(\mathbf{h}_k)$ is the weight of the kth lag. The weights $\omega(\mathbf{h}_k)$ can be chosen in several ways:

- proportional to the number $N(\mathbf{h}_k)$ of pairs used in the estimate

- proportional to the quantity $N(\mathbf{h}_k)/[\gamma_{ij}(\mathbf{h}_k)]^2$ thereby giving more weight to the first lags, which are of greater interest

- empirically, for example, to ensure a fit that is pleasing to the eye

To prevent the variable with the largest variance from dominating the criterion WSS, each residual $[\widehat{\gamma}_{ij}(\mathbf{h}_k) - \gamma_{ij}(\mathbf{h}_k)]$ is standardized by the product of standard deviations $\widehat{\sigma}_i$ and $\widehat{\sigma}_j$.

Using matrix notation, the criterion (A.1) is rewritten:

$$\text{WSS} = \sum_{k=1}^{K} \omega(\mathbf{h}_k) \cdot \text{tr}\left([\mathbf{V}\{\widehat{\mathbf{\Gamma}}(\mathbf{h}_k) - \mathbf{\Gamma}(\mathbf{h}_k)\}]^2\right) \tag{A.2}$$

where the trace "tr" is the sum of the diagonal elements of the matrix, and $\mathbf{V}$ is the diagonal matrix of inverse standard deviations. The matrices $\widehat{\mathbf{\Gamma}}(\mathbf{h}_k) = [\widehat{\gamma}_{ij}(\mathbf{h}_k)]$ and $\mathbf{\Gamma}(\mathbf{h}_k) = [\gamma_{ij}(\mathbf{h}_k)]$ are the experimental and model semivariogram matrices. The linear model of coregionalization $\mathbf{\Gamma}(\mathbf{h}_k)$ is defined as

$$\mathbf{\Gamma}(\mathbf{h}_k) = \sum_{l=0}^{L} \mathbf{B}_l \, g_l(\mathbf{h}_k) \qquad k = 1, \dots, K \qquad (A.3)$$

where the number and type of basic semivariogram models $g_l(.)$ are specified by the user.

Algorithm

The least-squares fit of a linear model of coregionalization is more demanding than in the univariate case. The difficulty lies in the constraint that the matrices of estimated coefficients $\widehat{\mathbf{B}}_l = [\hat{b}_{ij}^l]$ must be positive semi-definite. The idea is to start with a set of arbitrary coregionalization matrices $\mathbf{B}_l$ and to modify one matrix at a time iteratively so as to minimize the criterion WSS under the constraint of positive semi-definiteness of that matrix. The algorithm proceeds as follows:

1. Choose initial values for the $L + 1$ coregionalization matrices $\mathbf{B}_l$.

2. Remove any one of the $L + 1$ basic semivariogram models, say, $g_{l_0}(\mathbf{h})$, and compute the difference between each experimental semivariogram matrix $\widehat{\mathbf{\Gamma}}(\mathbf{h}_k)$ and the linear model of coregionalization including the L remaining basic structures:

$$\Delta_{l_0}\mathbf{\Gamma}(\mathbf{h}_k) = \widehat{\mathbf{\Gamma}}(\mathbf{h}_k) - \sum_{\substack{l=0 \\ l \neq l_0}}^{L} \widehat{\mathbf{B}}_l \, g_l(\mathbf{h}_k) \qquad (A.4)$$

3. Compute the symmetric matrix $\mathbf{G}_{l_0}$:

$$\mathbf{G}_{l_0} = \sum_{k=1}^{K} \Delta_{l_0}\mathbf{\Gamma}(\mathbf{h}_k) \, g_{l_0}(\mathbf{h}_k) \qquad (A.5)$$

4. Perform the spectral decomposition[1] of the matrix $\mathbf{G}_{l_0}$:

$$\mathbf{G}_{l_0} = \mathbf{Q}_{l_0} \, \mathbf{\Lambda}_{l_0} \, \mathbf{Q}_{l_0}^T \qquad (A.6)$$

where $\mathbf{Q}_{l_0}$ is the $N_v \times N_v$ orthogonal matrix of eigenvectors, and $\mathbf{\Lambda}_{l_0} = [\lambda_{l_0}]$ is the diagonal matrix of eigenvalues. The matrix $\mathbf{G}_{l_0}$ is not necessarily positive semi-definite in that some eigenvalues may be negative.

---

[1] For the reader not familiar with matrix algebra, a brief account of the determination of eigenvalues and eigenvectors is given in Davis (1986), p. 107–148 and Webster and Oliver (1990), p. 291–298.

The positive semi-definite matrix closest (in the least-squares sense) to the matrix $\mathbf{G}_{l_0}$ is

$$\mathbf{G}_{l_0}^+ = \mathbf{Q}_{l_0} \, \mathbf{\Lambda}_{l_0}^+ \, \mathbf{Q}_{l_0}^T \tag{A.7}$$

where $\mathbf{\Lambda}_{l_0}^+$ is the matrix $\mathbf{\Lambda}_{l_0}$ where all negative eigenvalues are reset to zero.

5. The coregionalization matrix $\widehat{\mathbf{B}}_{l_0}$ that minimizes the criterion WSS under the constraint of positive semi-definiteness is

$$\widehat{\mathbf{B}}_{l_0} = \frac{\mathbf{G}_{l_0}^l}{\displaystyle\sum_{k=1}^{K} [g_{l_0}(\mathbf{h}_k)]^2} \tag{A.8}$$

6. Define the new index number $l_0 = l_0 + 1$ (l_0 is reset to zero if $l_0 > L$), and repeat steps 2 to 5 until the criterion WSS is smaller than a threshold value specified by the user.

Remarks

1. Theoretically, the procedure does not necessarily converge nor, if it does, is a unique solution assured. Experience shows, however, that the algorithm almost always converges and leads to similar results whatever the initial values of the coregionalization matrices.

2. The number of variables and basic structures are not limited. An example of application to the joint modeling of 12 variables and 3 structures is given in Goulard and Voltz (1992).

3. Beware that the number of parameters b_{ij}^l to be estimated rapidly increases with the number of variables and structures. For example, this number would be 165 for ten variables and three structures. Too many parameters may cause instability in the estimation results and reduce the overall quality of the fit. One would then retain only the variables that contribute most to the objective pursued (see related discussion in section 4.3.3).

4. Because the algorithm is fast, several combinations of basic semivariogram models $g_l(\mathbf{h})$ can be tried. The problem is then to select one of the resulting linear models. One may choose the model that yields the smallest WSS value. However, this approach suffers from the following limitations:

   - As in the univariate case, the value of the weighted least-squares criterion depends on the number of lags considered and the weighting system adopted by the user. The linear model that yields the smallest WSS value need not be the same for different choices of parameters K and $\omega(\mathbf{h}_k)$.

- The criterion WSS is but a measure of the overall quality of the fit. Beware that a small WSS value can be achieved at the expense of poor fits for the N_v direct semivariograms relative to the $N_v(N_v - 1)/2$ cross semivariogram models. Thus, it is crucial to check visually that the direct semivariograms, in particular that of the primary variable, are correctly modeled.

The least-squares criterion is convenient for the automatic fitting, but it should not be the sole criterion for selecting the final linear model of coregionalization.

Appendix B

List of Acronyms and Notation

B.1 Acronyms

ccdf: conditional cumulative distribution function

cdf: cumulative distribution function

CKT: cokriging with trend model, also known as "universal" cokriging

E-type: conditional expectation estimate

FKA: factorial kriging analysis

IK: indicator kriging

IRF-k: intrinsic random functions of order k

KED: kriging with an external drift

KT: kriging with a trend model, also known as "universal" kriging

KWS: kriging within strata

LMC: linear model of coregionalization

LS: least squares

MAP: maximum a posteriori model

MG: multiGaussian (algorithm or model)

mIK: median indicator kriging

M-type: conditional median estimate

OCK: ordinary cokriging

oICK: ordinary indicator cokriging

oIK: ordinary indicator kriging

OK: ordinary kriging

PCK: principal component kriging

PK: probability kriging

pdf: probability density function

P-P plot: probability-probability plot

Q-Q plot: quantile-quantile plot

RF: random function

RV: random variable

SCK: simple cokriging

sGs: sequential Gaussian simulation

sIK: simple indicator kriging

sis: sequential indicator simulation

SK: simple kriging

SKlm: simple kriging with varying local means

B.2 Common notation

$\forall$: whatever

$\mathcal{A}$: study area

a: range parameter

a_k: coefficient of the kth component of the trend model $m(\mathbf{u})$

$B(z)$: Markov–Bayes calibration parameter

b^l: coefficient of the basic covariance model $c_l(\mathbf{h})$ or semivariogram model $g_l(\mathbf{h})$ in the linear model of regionalization for variable $Z(\mathbf{u})$

b^l_{ij}: coefficient of the basic covariance model $c_l(\mathbf{h})$ or semivariogram model $g_l(\mathbf{h})$ in the linear model of coregionalization between variable $Z_i(\mathbf{u})$ and $Z_j(\mathbf{u})$

$\mathbf{B}_l$: coregionalization matrix including the coefficients b_{ij}^l of the basic covariance model $c_l(\mathbf{h})$ in the linear model of coregionalization $\mathbf{C}(\mathbf{h}) = \sum_{l=0}^{L} \mathbf{B}_l c_l(\mathbf{h})$

$C(0)$: covariance value at separation distance $|\mathbf{h}|=0$. It is also the stationary variance of the RV $Z(\mathbf{u})$

$C(\mathbf{u}, \mathbf{u}')$: non-stationary covariance between the RVs $Z(\mathbf{u})$ and $Z(\mathbf{u}')$

$C(\mathbf{h})$: stationary covariance of the RF $Z(\mathbf{u})$ for lag vector $\mathbf{h}$

$\mathbf{C}(\mathbf{h})$: covariance function matrix (dimension $N_v \times N_v$)

$\mathbf{C} = \mathbf{C}(0)$: variance–covariance matrix

$C(\mathbf{h}) = \sum_{l=0}^{L} b^l c_l(\mathbf{h})$: linear model of regionalization

$C_{ij}(\mathbf{h})$: stationary cross covariance between the two RFs $Z_i(\mathbf{u})$ and $Z_j(\mathbf{u})$ for lag vector $\mathbf{h}$

$C_{ij}(\mathbf{h}) = \sum_{l=0}^{L} b_{ij}^l c_l(\mathbf{h})$: linear model of coregionalization

$C_I(\mathbf{h}; z_k)$: stationary indicator covariance for lag vector $\mathbf{h}$ and threshold z_k of the RF $Z(\mathbf{u})$; it is the covariance of the binary indicator RF $I(\mathbf{u}; z_k)$.

$C_I(\mathbf{h}; z_k, z_{k'})$: stationary indicator cross covariance for lag vector $\mathbf{h}$ and thresholds z_k and $z_{k'}$ of the RF $Z(\mathbf{u})$; it is the cross covariance between the two indicator RFs $I(\mathbf{u}; z_k)$ and $I(\mathbf{u}; z_{k'})$.

$C_{ij}^I(\mathbf{h}; z_{ik}, z_{jk'})$: stationary indicator cross covariance for lag vector $\mathbf{h}$ and thresholds z_{ik} and $z_{jk'}$ of the RFs $Z_i(\mathbf{u})$ and $Z_j(\mathbf{u})$; it is the cross covariance between the two indicator RFs $I(\mathbf{u}; z_{ik})$ and $I(\mathbf{u}; z_{jk'})$.

$C_R(\mathbf{h})$: stationary covariance of the residual component model

$\text{Cov}\{\cdot\}$: covariance

$\text{Circ}(\mathbf{h})$: circular semivariogram model, a function of the separation vector $\mathbf{h}$

$c_l(\mathbf{h})$: lth basic covariance model in the linear model of (co)regionalization

$E\{\cdot\}$: expected value

$E\{Z(\mathbf{u}); z|(n)\}$: conditional expectation of the RV $Z(\mathbf{u})$ given the realizations of n other neighboring RVs (called data)

$\text{Exp}(\frac{\mathbf{h}}{a})$: exponential semivariogram model of practical range a, a function of the separation vector $\mathbf{h}$

$F(\mathbf{u}; z)$: non-stationary cumulative distribution function of the RV $Z(\mathbf{u})$

$F(\mathbf{u}; z|(n))$: non-stationary conditional cumulative distribution function of the continuous RV $Z(\mathbf{u})$ given neighboring information, such as realizations of n other RVs (called data)

$F(\mathbf{u}; s_k|(n))$: non-stationary conditional cumulative distribution function of the categorical RV $S(\mathbf{u})$ given the realizations of n other neighboring RVs (called data)

$F(\mathbf{u}_1, \ldots, \mathbf{u}_N; z_1, \ldots, z_N)$: N-variate cumulative distribution function of the N RVs $Z(\mathbf{u}_1), \ldots, Z(\mathbf{u}_N)$

$F(\mathbf{u}, \mathbf{u}'; z, z')$: non-stationary "two-point" cumulative distribution function of the RF $Z(\mathbf{u})$

$F(\mathbf{h}; z, z')$: stationary "two-point" cumulative distribution function of the RF $Z(\mathbf{u})$

$F(z)$: cumulative distribution function of a RV Z, or stationary cumulative distribution function of a RF $Z(\mathbf{u})$

$F^{-1}(p)$: inverse cumulative distribution function or quantile function for the probability value $p \in [0, 1]$

$F_{ij}(\mathbf{h}; z_i, z_j)$: stationary "two-point" joint cumulative distribution function of RFs $Z_i(\mathbf{u})$ and $Z_j(\mathbf{u})$

$F_{ij}(z_i, z_j)$: joint cumulative distribution function of the two RVs Z_i and Z_j, or stationary joint cumulative distribution function of the two RFs $Z_i(\mathbf{u})$ and $Z_j(\mathbf{u})$

$F_V(\mathbf{u}; z|(n))$: non-stationary conditional cumulative distribution function of the continuous RV $Z_V(\mathbf{u})$ defined over the block $V(\mathbf{u})$

$f_k(\mathbf{u})$: function of the coordinates used in a trend model $m(\mathbf{u})$

$G(y)$: standard normal cumulative distribution function

$G^{-1}(y)$: standard normal quantile function such that $G(G^{-1}(p)) = p \in [0, 1]$

$G(\mathbf{u}; z|(n))$: non-stationary conditional cumulative distribution function of the standard normal RV $Y(\mathbf{u})$ given the information (n) available, for example, realizations of n other neighboring RVs $Y(\mathbf{u}_\alpha)$

$G(\mathbf{h}; z, z')$: stationary "two-point" cdf of the multivariate Gaussian RF $Y(\mathbf{u})$

$\boldsymbol{\Gamma}(\mathbf{h})$: semivariogram matrix (dimension $N_v \times N_v$)

$g_l(\mathbf{h})$: lth basic semivariogram model in the linear model of (co)regionalization

$\gamma(\mathbf{u}, \mathbf{u}')$: non-stationary semivariogram between the two RVs $Z(\mathbf{u})$ and $Z(\mathbf{u}')$

$\gamma(\mathbf{h})$: stationary semivariogram of the RF $Z(\mathbf{u})$ for lag vector $\mathbf{h}$

$\gamma(\mathbf{h}) = \sum_{l=0}^{L} b^l g_l(\mathbf{h})$: linear model of regionalization

$\gamma_{ij}(\mathbf{h})$: stationary cross semivariogram between the two RFs $Z_i(\mathbf{u})$ and $Z_j(\mathbf{u})$ for lag vector $\mathbf{h}$

$\gamma_{ij}(\mathbf{h}) = \sum_{l=0}^{L} b_{ij}^l g_l(\mathbf{h})$: linear model of coregionalization

$\gamma_I(\mathbf{h}; z_k)$: stationary indicator semivariogram for lag vector $\mathbf{h}$ and threshold z_k of the RF $Z(\mathbf{u})$; it is the semivariogram of the binary indicator RF $I(\mathbf{u}; z_k)$.

$\gamma_{ij}^I(\mathbf{h}; z_{ik}, z_{jk'})$: stationary indicator cross semivariogram for lag vector $\mathbf{h}$ and thresholds z_{ik} and $z_{jk'}$ of the RFs $Z_i(\mathbf{u})$ and $Z_j(\mathbf{u})$; it is the cross semivariogram between the two indicator RFs $I(\mathbf{u}; z_{ik})$ and $I(\mathbf{u}; z_{jk'})$.

$H(\mathbf{u})$: entropy of the probability density function at location $\mathbf{u}$

$h = |\mathbf{h}|$: separation distance or lag

h_k: lag number

$\mathbf{h}$: separation vector

$I(\mathbf{u}; z)$: binary indicator RF at location $\mathbf{u}$ and for threshold z

$I(\mathbf{u}; s_k)$: binary indicator RF at location $\mathbf{u}$ and for threshold s_k

$i(\mathbf{u}; z)$: binary indicator value at location $\mathbf{u}$ and for threshold z

$i(\mathbf{u}; s_k)$: binary indicator value at location $\mathbf{u}$ and for category s_k

$[i(\mathbf{u}; z_k)]_{SK}^*$: simple kriging estimate of indicator value $i(\mathbf{u}; z_k)$. The same type of notation applies to other indicator algorithms, for example, ordinary IK.

K: number of threshold values z_k

$K(\mathbf{h})$: non-centered stationary covariance of the RF $Z(\mathbf{u})$ for lag vector $\mathbf{h}$

$L+1$: number of basic models $c_l(\mathbf{h})$ or $g_l(\mathbf{h})$ in the linear model of (co)regionalization

$\lambda_\alpha(\mathbf{u})$: kriging weight associated to z-datum at location $\mathbf{u}_\alpha$ for estimation of the attribute z at location $\mathbf{u}$

$\lambda_{\alpha_i}(\mathbf{u})$: cokriging weight associated to z_i-datum at location $\mathbf{u}_\alpha$ for estimation of the attribute z at location $\mathbf{u}$

$\lambda_\alpha^{SK}(\mathbf{u})$: simple kriging weight associated to z-datum at location $\mathbf{u}_\alpha$ for estimation of the attribute z at location $\mathbf{u}$. The same type of notation applies to other algorithms, for example, OK, KT.

$\lambda_{\alpha_i}^{SCK}(\mathbf{u})$: simple cokriging weight associated to z_i-datum at location $\mathbf{u}_{\alpha_i}$ for estimation of the primary attribute z_1 at location $\mathbf{u}$. The same type of notation applies to other algorithms, for example, OCK.

$\lambda_{\alpha m}^{OK}(\mathbf{u})$: OK weight associated to z-datum at location $\mathbf{u}_\alpha$ for estimation of the trend component $m_{OK}(\mathbf{u})$. The same type of notation applies to KT and KED.

M: median of the cumulative distribution function $F(z)$

m: stationary mean of the RF $Z(\mathbf{u})$

$\mathbf{m}$: vector of stationary means (dimension N_v)

$m(\mathbf{u})$: expected value of the RV $Z(\mathbf{u})$; trend component model in the decomposition $Z(\mathbf{u}) = R(\mathbf{u}) + m(\mathbf{u})$, where $R(\mathbf{u})$ represents the residual component model

$m(\mathbf{u}) = \sum_{k=0}^{K} a_k(\mathbf{u}) f_k(\mathbf{u})$: trend component model

$m_{OK}^*(\mathbf{u})$: OK estimate of the trend component at location $\mathbf{u}$. The same type of notation applies to KT and KED.

$\mu_{OK}(\mathbf{u})$: Lagrange parameter for ordinary kriging at location $\mathbf{u}$. The same type of notation applies to other algorithms, for example, KT and KED.

N: number of interpolation or simulation grid nodes

$N(\mathbf{h})$: number of pairs of data values available at lag vector $\mathbf{h}$

N_v: number of variables Z_i

n: number of data values $s(\mathbf{u}_\alpha)$ or $z(\mathbf{u}_\alpha)$ available over the area $\mathcal{A}$

$n(\mathbf{u})$: number of data values $z(\mathbf{u}_\alpha)$ used for estimation of the attribute z at location $\mathbf{u}$

$n_i(\mathbf{u})$: number of data values $z_i(\mathbf{u}_\alpha)$ used for estimation of the attribute z at location $\mathbf{u}$

$\nu_{ij}(\mathbf{h})$: stationary codispersion coefficient between variables $Z_i(\mathbf{u})$ and $Z_j(\mathbf{u} + \mathbf{h})$ separated by lag vector $\mathbf{h}$

$\Pi_{i=1}^{n} y_i = y_1 \cdot y_2 \ldots y_n$: product

$p(\mathbf{u}; s_k)$: probability for the category s_k to prevail at location $\mathbf{u}$

$p(\mathbf{u}; s_k|(n))$: conditional probability for the category s_k to prevail at location $\mathbf{u}$ given the neighboring information (n)

p_k: global proportion of category s_k within the area $\mathcal{A}$

q_p: p-quantile of the cumulative distribution function $F(z)$, $q_p = F^{-1}(p)$

$\mathbf{R}(\mathbf{h})$: correlogram matrix (dimension $N_v \times N_v$)

$\mathbf{R} = \mathbf{R}(0)$: correlation matrix

$R(\mathbf{u})$: residual component model in the decomposition $Z(\mathbf{u}) = R(\mathbf{u}) + m(\mathbf{u})$, where $m(\mathbf{u})$ represents the trend component model

$R(\mathbf{u}) = \sum_{l=0}^{L} Z^l(\mathbf{u})$: decomposition of the residual component model into $(L+1)$ spatial components $Z^l(\mathbf{u})$ corresponding to the decomposition of the residual covariance $C_R(\mathbf{h}) = \sum_{l=0}^{L} b^l c_l(\mathbf{h})$

$\rho(\mathbf{u}, \mathbf{u}')$: non-stationary correlogram between the RVs $Z(\mathbf{u})$ and $Z(\mathbf{u}')$

$\rho(\mathbf{h})$: stationary correlogram of the RF $Z(\mathbf{u})$ for lag vector $\mathbf{h}$

$\rho_{ij}(\mathbf{h})$: stationary cross correlogram between the RFs $Z_i(\mathbf{u})$ and $Z_j(\mathbf{u})$ for lag vector $\mathbf{h}$

$\rho_{ij} = \rho_{ij}(0)$: linear correlation coefficient between variables Z_i and Z_j

ρ_{ij}^R: rank correlation coefficient between variables Z_i and Z_j

$\rho_I(\mathbf{h}; z_k)$: stationary indicator correlogram for lag vector $\mathbf{h}$ and threshold z_k of the RF $Z(\mathbf{u})$; it is the correlogram of the binary indicator RF $I(\mathbf{u}; z_k)$.

$\rho_I(\mathbf{h}; z_k, z_{k'})$: stationary indicator cross correlogram for lag vector $\mathbf{h}$ and thresholds z_k and $z_{k'}$ of the RF $Z(\mathbf{u})$; it is the cross correlogram between the two indicator RFs $I(\mathbf{u}; z_k)$ and $I(\mathbf{u}; z_{k'})$.

$\rho_{ij}^I(\mathbf{h}; z_{ik}, z_{jk'})$: stationary indicator cross correlogram for lag vector $\mathbf{h}$ and thresholds z_{ik} and $z_{jk'}$ of the RFs $Z_i(\mathbf{u})$ and $Z_j(\mathbf{u})$; it is the cross correlogram between the two indicator RFs $I(\mathbf{u}; z_{ik})$ and $I(\mathbf{u}; z_{jk'})$.

$S(\mathbf{u})$: generic categorical RV at location $\mathbf{u}$, or a generic categorical RF at location $\mathbf{u}$.

$\mathrm{Sph}(\frac{\mathbf{h}}{a})$: spherical semivariogram model of range a, a function of the separation vector $\mathbf{h}$

$\sum_{i=1}^{n} y_i = y_1 + y_2 + \ldots + y_n$: summation

s: categorical attribute

$s(\mathbf{u}_\alpha)$: s-datum value at location $\mathbf{u}_\alpha$

σ^2: variance of the RV Z

$\sigma^2_{SK}(\mathbf{u})$: simple kriging variance associated with the simple kriging estimate $Z^*_{SK}(\mathbf{u})$ at location $\mathbf{u}$. The same type of notation applies to other algorithms, for example, OK and KT.

$\mathbf{u}$: coordinate vector

$\mathbf{u}_\alpha$: datum location

Var$\{\cdot\}$: variance

$Y = \varphi(Z)$: transform function $\varphi(.)$ relating two RVs Y and Z

$Y^l_k(\mathbf{u})$: kth regionalized factor corresponding to the $l+1$ basic covariance model in the linear model of coregionalization $C_{ij}(\mathbf{h}) = \sum_{l=0}^L b^l_{ij} c_l(\mathbf{h})$

$Z = \varphi^{-1}(Y)$: inverse transform function $\varphi(.)$ relating two RVs Z and Y

$Z(\mathbf{u})$: generic continuous RV at location $\mathbf{u}$, or a generic continuous RF at location $\mathbf{u}$

$\mathbf{Z}(\mathbf{u})$: multivariate RF or a vector RF (dimension N_v)

$Z^l(\mathbf{u})$: lth spatial component in the decomposition of the residual component $R(\mathbf{u}) = \sum_{l=0}^L Z^l(\mathbf{u})$. Each spatial component $Z^l(\mathbf{u})$ has a covariance $b^l c_l(\mathbf{h})$ with $C_R(\mathbf{h}) = \sum_{l=0}^L b^l c_l(\mathbf{h})$.

$Z^*_{SK}(\mathbf{u})$: simple kriging estimator of $Z(\mathbf{u})$. The same type of notation applies to other algorithms, for example, OK and KT.

$Z^{(1)^*}_{SCK}(\mathbf{u})$: simple cokriging estimator of the primary attribute z_1 at location $\mathbf{u}$. The same type of notation applies to other multivariate algorithms, for example, OCK.

$\{Z(\mathbf{u}), \mathbf{u} \in \mathcal{A}\}$: set of random variables $Z(\mathbf{u})$ defined at each location $\mathbf{u}$ of the area $\mathcal{A}$

z: continuous attribute

$z(\mathbf{u})$: true value at unsampled location $\mathbf{u}$

$z(\mathbf{u}_\alpha)$: z-datum value at location $\mathbf{u}_\alpha$

$z^{(l)}(\mathbf{u})$: lth realization of the RF $Z(\mathbf{u})$ at location $\mathbf{u}$

$z_i(\mathbf{u}_{\alpha_i})$: z_i-datum value at location $\mathbf{u}_{\alpha_i}$

z_k: kth threshold value for the continuous attribute z

z_{ik}: kth threshold value for the continuous attribute z_i

$z_v(\mathbf{u})$: average value of attribute z over a block V centered at $\mathbf{u}$

$z^*(\mathbf{u})$: an estimate of value $z(\mathbf{u})$

$z^*_{SK}(\mathbf{u})$: simple kriging estimate of value $z(\mathbf{u})$. The same type of notation applies to other algorithms, for example, OK and KT.

$z^{(1)^*}_{SCK}(\mathbf{u})$: simple cokriging estimate of the primary attribute z_1 at location $\mathbf{u}$. The same type of notation applies to other multivariate algorithms, for example, OCK.

$z^*_E(\mathbf{u})$: E-type estimate of value $z(\mathbf{u})$, obtained as an arithmetic average of multiple simulated realizations $z^{(l)}(\mathbf{u})$ of the RF $Z(\mathbf{u})$

Appendix C

The Jura data

Data set provided by J.-P. Dubois, IATE-Pédologie, Ecole Polytechnique Fédérale de Lausanne, 1015 Lausanne, Switzerland. Used with permission.

C.1   Prediction and validation sets

Spatial coordinates and values of categorical and continuous attributes at the 359 sampled sites. The 100 test locations are denoted with a star.
Rock types: 1: Argovian, 2: Kimmeridgian, 3: Sequanian, 4: Portlandian, 5: Quaternary. Land uses: 1: forest; 2: pasture, 3: meadow, 4: tillage.

X	Y	Rock	Land	Cd	Cu	Pb	Co	Cr	Ni	Zn
km	km	type	use	ppm	ppm	ppm	ppm	ppm	ppm	ppm
2.386	3.077	3	3	1.740	25.72	77.36	9.32	38.32	21.32	92.56
2.544	1.972	2	2	1.335	24.76	77.88	10.00	40.20	29.72	73.56
2.807	3.347	3	2	1.610	8.88	30.80	10.60	47.00	21.40	64.80
4.308	1.933	2	3	2.150	22.70	56.40	11.92	43.52	29.72	90.00
4.383	1.081	5	3	1.565	34.32	66.40	16.32	38.52	26.20	88.40
3.244	4.519	5	3	1.145	31.28	72.40	3.50	40.40	22.04	75.20
3.925	3.785	5	3	0.894	27.44	60.00	15.08	30.52	21.76	72.40
2.116	3.498	1	3	0.525	66.12	141.00	4.20	25.40	9.72	72.08
1.842	0.989	1	3	0.240	22.32	52.40	4.52	27.96	11.32	56.40
1.709	1.843	3	3	0.625	18.72	41.60	12.08	33.32	16.88	75.60
3.800	4.578	1	3	3.873	22.24	46.00	9.84	46.00	21.64	143.20
2.699	1.199	1	1	1.425	8.76	56.40	7.56	40.80	15.72	66.80
3.033	4.384	5	3	0.455	21.36	38.36	2.36	19.24	7.08	37.88
4.232	1.588	2	1	1.025	5.88	52.80	8.84	28.32	17.40	47.60
4.750	1.369	2	3	0.775	17.60	32.92	11.96	35.60	23.76	70.80
2.341	4.135	1	3	0.620	18.52	42.00	5.32	31.00	10.80	46.80
2.827	3.667	5	3	0.400	3.96	21.48	4.68	20.52	6.68	25.20
3.043	4.692	1	3	0.625	8.24	29.92	3.53	25.20	11.32	31.32
1.687	1.174	1	3	0.355	21.56	46.80	3.64	18.40	6.64	39.56
2.235	2.386	5	3	1.585	40.88	88.68	12.44	40.56	25.08	114.28

(continued)

X	Y	Rock	Land	Cd	Cu	Pb	Co	Cr	Ni	Zn
km	km	type	use	ppm	ppm	ppm	ppm	ppm	ppm	ppm
3.122	1.695	2	3	1.885	35.92	72.88	12.92	44.00	34.72	120.28
2.219	2.774	2	2	1.300	8.80	56.52	13.04	32.48	31.52	71.76
3.185	2.092	2	3	1.575	22.12	43.68	11.20	35.72	35.00	90.68
3.514	5.119	1	2	4.191	15.48	33.36	9.28	52.00	21.76	145.60
4.459	2.624	2	1	1.485	11.16	34.64	6.84	19.00	15.84	82.80
4.339	1.473	5	3	0.915	39.80	64.80	10.80	32.72	18.60	86.00
4.920	1.417	2	3	0.705	15.76	39.52	8.56	30.28	21.96	66.40
3.648	3.780	2	3	0.620	48.80	63.20	14.16	28.56	24.44	78.00
2.916	4.916	1	1	0.885	4.52	37.16	6.28	27.92	13.88	46.80
3.127	1.691	2	3	1.960	47.12	93.92	13.40	45.20	35.12	123.84
3.441	4.719	5	2	3.155	48.80	55.60	9.28	62.40	24.28	151.60
3.336	4.613	5	3	1.280	24.96	49.20	5.92	38.52	23.28	81.60
3.649	2.674	2	1	1.785	9.65	58.40	7.44	23.72	17.72	58.00
3.875	4.924	5	3	1.895	25.28	35.60	11.48	52.00	26.72	141.60
2.612	4.114	1	3	0.570	33.80	58.40	4.88	27.32	7.44	47.60
2.340	4.141	1	3	0.540	17.40	36.88	5.48	35.72	10.60	53.20
3.811	1.318	2	3	0.750	16.40	39.80	12.60	36.20	26.60	80.80
3.661	3.788	2	3	0.875	65.20	69.60	14.64	28.88	20.80	76.40
4.340	3.736	3	2	2.818	29.12	46.00	11.24	35.04	21.96	80.80
2.972	0.988	3	2	3.805	112.04	229.56	9.16	67.04	28.88	219.32
2.286	3.068	3	3	0.605	28.80	56.60	9.68	27.32	12.80	67.16
1.813	2.116	3	3	1.003	21.20	48.80	11.32	33.48	20.56	110.00
3.379	4.308	5	3	1.900	10.28	36.44	8.52	46.40	24.80	83.60
1.965	2.807	3	3	0.895	60.72	89.68	12.60	41.60	24.16	102.44
3.559	4.009	3	2	4.020	21.08	52.00	6.16	60.80	21.96	101.60
2.267	4.189	1	3	0.560	12.72	39.64	6.64	35.12	11.12	57.20
2.965	1.002	3	2	1.390	57.20	129.20	10.36	45.60	24.60	149.20
3.120	1.469	3	1	1.910	11.60	69.16	2.12	31.12	8.00	50.40
3.454	4.654	5	2	1.070	10.88	29.88	7.80	38.24	19.92	70.80
1.071	1.567	3	2	0.795	9.92	37.48	9.96	29.08	15.28	65.60
3.573	3.542	5	3	0.550	50.40	77.20	9.56	26.60	16.00	76.00
3.331	4.609	5	3	1.255	49.60	80.40	7.00	42.80	25.44	101.60
3.541	1.739	3	4	2.055	10.80	43.60	6.32	29.80	20.80	59.36
3.303	3.963	3	3	1.220	51.60	80.80	11.00	34.04	20.32	90.00
3.316	2.987	5	3	0.975	30.10	48.80	14.60	33.00	23.60	83.20
1.879	1.171	1	3	0.465	23.32	54.80	4.00	24.64	9.16	44.40
1.122	2.267	2	3	1.145	14.80	47.60	11.08	37.08	31.56	74.40
4.423	1.078	5	3	1.525	22.00	46.80	16.92	38.60	26.80	87.60
1.241	1.155	1	1	1.455	7.28	36.52	4.52	20.20	12.68	39.60
1.706	1.848	3	3	1.120	20.72	41.20	11.92	34.84	19.96	83.20
3.260	5.420	2	3	1.819	17.60	35.80	11.72	38.04	28.60	82.40
4.081	0.897	5	3	0.495	14.32	32.76	11.20	22.40	10.92	50.40
3.077	2.926	2	2	2.120	12.32	42.40	10.12	35.32	21.52	64.80
3.406	4.741	5	2	2.150	28.96	42.80	7.32	48.00	19.72	120.80
3.699	3.800	2	3	0.650	35.56	80.80	11.60	27.44	19.28	74.00
1.619	2.883	3	3	0.840	22.84	50.40	6.80	30.56	16.08	58.80
3.196	1.814	2	1	3.925	47.40	84.16	12.92	25.60	26.12	92.36
3.447	4.733	5	2	1.980	27.12	39.92	8.40	40.80	17.68	108.00
3.009	4.738	1	3	0.450	12.44	34.00	3.53	25.64	10.00	39.40
3.452	4.736	5	2	2.960	38.96	70.40	7.20	45.20	15.00	148.40
2.908	1.263	1	3	1.685	9.40	34.56	9.84	31.28	14.72	84.32
4.578	1.512	2	3	1.305	17.50	38.96	13.52	45.52	27.20	77.60
3.512	5.124	1	2	4.227	16.88	34.64	9.64	48.40	22.92	136.80
3.312	2.991	5	3	0.980	22.70	41.60	12.52	32.60	19.40	72.80
2.159	2.041	2	3	1.570	17.44	49.60	10.44	37.72	28.32	81.60
4.742	1.381	2	3	0.860	15.12	32.84	11.88	32.72	23.48	63.60
3.592	4.027	3	2	2.170	12.52	37.68	10.84	55.20	29.48	74.40
3.152	3.272	2	3	1.560	34.00	76.40	12.80	42.60	23.52	94.00
4.781	1.375	2	3	0.810	12.08	36.28	10.88	27.44	20.84	58.80
2.319	2.405	5	3	1.710	56.24	104.68	12.84	43.40	26.96	125.40

X km	Y km	Rock type	Land use	Cd ppm	Cu ppm	Pb ppm	Co ppm	Cr ppm	Ni ppm	Zn ppm
3.617	2.084	2	3	0.845	27.92	78.80	12.80	33.72	26.20	79.60
2.278	0.929	1	3	0.357	13.24	35.40	4.08	19.36	6.56	46.80
3.792	2.695	2	1	1.535	5.88	55.20	6.76	20.32	13.40	52.40
4.113	2.700	2	1	1.285	13.64	84.40	1.55	8.72	7.16	65.20
3.043	4.698	1	3	1.820	7.60	27.04	3.92	27.36	10.44	39.80
3.271	2.160	5	3	0.945	37.40	59.04	13.32	37.80	32.60	104.36
1.433	1.525	5	3	1.605	13.48	49.60	7.36	67.60	16.20	75.20
3.184	2.098	2	3	2.230	24.60	44.88	11.92	37.80	37.72	96.60
3.675	4.293	3	2	0.890	4.40	26.76	8.60	37.68	22.96	66.00
3.681	5.690	2	2	1.898	15.16	59.20	12.64	54.00	28.36	86.00
2.461	3.423	5	4	0.370	15.72	29.36	3.14	22.60	6.08	47.20
4.576	1.411	2	3	1.380	15.64	34.44	11.28	37.60	25.76	71.60
2.223	0.580	2	3	0.285	14.16	44.00	9.60	38.00	24.52	76.00
3.031	1.201	3	3	0.610	16.40	52.56	6.44	21.84	10.36	51.24
3.347	2.505	2	1	0.625	5.36	41.20	6.96	33.20	19.20	43.60
2.465	4.003	1	3	0.325	9.72	24.64	3.19	21.40	4.72	35.08
3.175	2.102	2	3	1.985	18.92	42.68	12.20	38.40	37.20	89.84
3.733	0.974	2	3	0.950	15.40	36.60	8.92	40.80	26.52	74.00
3.085	1.677	2	3	2.040	26.20	63.92	11.80	40.40	29.80	107.20
2.595	3.638	5	3	0.325	17.52	44.40	3.71	14.52	4.76	40.80
4.279	3.283	2	3	0.830	19.20	36.36	13.36	33.60	24.64	84.40
2.933	0.978	3	2	0.800	166.40	146.80	12.40	31.44	16.00	100.40
3.590	4.033	3	2	2.080	11.52	37.56	7.92	56.40	24.84	79.60
2.883	1.237	1	3	2.550	12.12	36.56	8.52	31.24	16.12	96.96
2.429	1.620	3	3	2.535	31.96	77.60	11.84	52.40	26.64	126.80
2.333	2.417	5	3	1.435	53.20	107.60	9.12	37.44	21.32	110.32
0.626	1.652	2	1	0.375	4.20	37.12	7.28	33.32	14.40	46.40
2.969	0.778	2	2	1.930	77.32	157.28	12.44	43.04	33.76	166.28
0.971	1.576	3	3	1.215	9.80	35.64	10.08	29.28	15.20	64.00
1.392	1.846	3	3	1.050	7.44	37.68	10.60	39.24	20.00	48.40
2.540	3.320	3	3	1.260	17.20	53.60	9.76	45.32	17.60	62.80
1.738	1.771	3	3	0.685	19.64	41.60	11.92	33.64	16.24	76.00
2.243	3.031	3	3	0.705	19.20	34.00	11.80	29.92	14.60	59.88
3.656	2.630	2	1	1.385	9.23	80.40	2.07	9.72	5.36	49.20
2.325	4.143	1	3	0.505	14.00	33.24	5.48	34.00	10.92	54.40
4.550	1.416	2	3	1.740	18.10	41.60	12.40	44.32	27.72	73.60
4.325	1.436	5	3	0.560	19.40	36.76	11.60	26.92	13.72	57.20
3.669	4.291	3	2	1.917	8.40	35.12	8.44	33.48	18.72	59.60
1.877	1.186	1	3	0.320	26.16	45.20	3.80	18.20	6.44	39.88
3.044	1.123	3	3	0.630	18.04	38.24	12.00	31.28	18.68	80.76
3.001	2.581	2	2	4.180	8.20	42.04	9.60	45.32	22.20	81.72
3.136	2.091	2	3	1.110	17.92	38.72	14.20	34.80	30.32	79.00
2.633	3.650	5	3	0.425	16.20	50.40	3.09	22.12	5.04	61.60
3.721	4.239	3	3	0.280	18.00	41.20	8.00	25.96	12.28	46.40
2.333	2.367	5	3	1.390	59.16	101.92	12.36	42.12	28.20	135.88
4.038	2.354	4	1	2.805	10.50	44.00	12.20	39.20	32.92	65.60
3.117	1.702	2	3	1.710	30.00	58.60	14.12	42.80	35.00	106.36
1.824	0.999	1	3	0.165	23.64	56.40	3.75	18.60	7.08	44.00
2.850	1.890	2	1	1.360	8.68	53.60	2.81	10.76	7.08	28.88
2.502	3.988	1	3	0.240	20.72	22.36	3.76	18.92	5.20	32.24
1.443	1.509	5	3	1.445	15.40	50.80	7.76	61.20	20.68	75.60
2.008	1.350	1	3	0.515	16.04	35.04	5.00	32.56	8.96	50.40
2.978	4.009	5	3	0.560	8.35	25.40	6.52	24.40	12.40	49.20
1.662	1.425	3	3	0.905	44.40	82.00	11.24	37.40	16.56	78.40
1.587	1.080	1	3	0.605	26.24	54.00	5.16	29.28	11.44	54.00
3.994	3.814	5	3	2.423	32.52	80.40	14.48	45.60	28.32	84.00
4.383	2.279	2	1	0.785	6.72	43.60	7.84	35.92	22.72	60.80
2.901	4.919	1	1	0.810	5.48	28.80	4.64	28.60	14.92	42.00
4.156	1.242	4	2	1.100	5.72	32.48	7.84	32.20	17.92	44.00
4.475	3.526	2	4	0.951	38.96	78.80	15.28	29.24	20.36	70.80

X	Y	Rock	Land	Cd	Cu	Pb	Co	Cr	Ni	Zn
km	km	type	use	ppm	ppm	ppm	ppm	ppm	ppm	ppm
2.580	3.654	5	3	0.275	7.48	20.20	3.55	16.20	4.64	33.44
0.701	1.997	3	1	1.580	9.84	64.80	12.60	60.80	32.20	86.00
3.649	3.887	2	3	1.060	26.44	51.20	11.48	41.60	29.24	80.80
2.623	0.853	3	3	2.580	29.24	59.20	11.64	45.20	27.36	106.40
1.315	1.975	3	2	3.680	14.44	56.40	11.12	52.00	28.72	89.60
3.656	2.670	2	1	1.875	9.52	70.00	11.12	36.12	25.40	65.20
4.426	0.821	2	2	0.805	8.76	27.20	13.00	48.32	26.00	61.60
2.882	3.693	5	3	0.535	8.76	27.04	4.60	20.80	5.40	32.92
1.108	1.535	3	2	2.055	60.00	195.60	8.08	50.40	23.16	108.00
4.840	1.089	2	2	0.555	10.80	36.52	10.28	27.64	19.60	64.80
1.468	2.192	3	2	1.510	10.88	44.40	10.92	57.60	30.68	80.40
2.319	4.103	1	3	0.580	20.52	34.88	5.48	36.00	10.92	50.40
3.585	4.040	3	2	1.970	24.64	66.00	8.00	51.20	25.76	99.20
3.647	3.786	2	3	0.765	42.00	70.80	13.84	29.72	20.96	90.80
2.310	2.732	2	2	0.735	17.68	39.20	12.72	39.00	23.24	79.04
3.962	2.009	2	3	1.445	42.92	78.40	12.40	43.52	34.40	104.40
3.470	5.555	2	2	2.234	12.52	48.40	13.48	45.20	29.68	64.00
1.839	0.995	1	3	0.225	27.72	55.60	4.24	24.88	10.12	56.00
1.706	1.165	1	3	0.740	25.52	60.80	4.36	20.40	8.68	62.40
2.774	1.544	3	2	1.915	19.32	45.72	8.44	42.32	20.64	90.24
3.493	5.092	1	2	3.421	14.80	34.32	9.80	52.40	20.44	134.40
1.695	3.227	1	1	1.835	10.88	56.68	5.88	35.08	21.44	97.48
4.409	1.084	5	3	1.770	18.92	48.80	17.32	37.72	27.00	89.20
2.596	3.212	3	2	3.995	15.92	45.16	8.64	49.72	22.00	118.24
1.897	1.220	1	3	0.260	9.24	27.00	3.48	16.24	4.76	27.20
2.083	1.695	2	1	0.440	11.28	56.40	6.00	29.28	16.68	54.80
3.673	4.278	3	2	1.015	7.36	34.00	9.28	44.00	19.28	70.40
2.731	3.002	2	1	0.485	6.60	32.44	9.00	27.32	20.92	57.96
4.264	3.391	2	3	2.828	79.20	135.20	11.16	32.64	24.96	132.80
3.792	2.707	2	1	1.380	6.44	45.60	8.16	26.92	17.12	55.20
3.390	1.048	2	2	1.660	24.60	57.28	10.12	42.92	29.32	123.32
3.184	5.075	3	3	0.685	11.36	23.68	7.08	33.76	16.32	53.60
3.644	2.677	2	1	0.930	6.77	51.20	9.56	37.40	23.12	56.40
3.504	5.130	1	2	5.129	14.76	36.80	8.60	39.24	18.28	135.60
2.242	3.036	3	3	0.725	17.00	34.08	10.32	26.32	12.72	56.72
3.332	4.628	5	2	0.740	96.00	59.20	4.52	35.92	17.20	83.20
3.498	3.196	2	2	0.990	14.50	36.24	15.12	32.80	22.92	65.60
1.883	1.176	1	3	0.195	25.12	49.60	3.92	21.80	7.52	43.20
3.322	3.001	5	3	1.340	25.30	44.40	14.32	35.00	23.12	81.20
1.047	1.922	3	2	1.240	18.32	52.00	14.20	43.60	22.52	83.20
3.843	3.121	5	3	1.840	33.32	44.40	10.80	39.52	35.28	86.00
1.301	0.599	3	4	0.910	21.20	64.80	13.24	32.84	18.00	75.20
1.106	1.548	3	2	1.270	19.12	65.20	9.16	52.80	19.44	65.20
1.114	1.535	3	2	1.480	106.80	226.40	9.12	58.80	26.16	140.40
1.511	0.734	1	2	1.245	22.64	64.40	7.84	47.20	21.96	71.20
1.317	1.501	5	3	1.385	120.80	66.40	6.44	42.80	19.64	94.80
1.439	1.523	5	3	1.720	16.40	52.40	7.24	54.80	16.52	75.60
1.651	1.157	1	4	0.525	26.28	46.40	3.88	20.84	8.36	49.60
1.701	1.169	1	3	0.690	17.04	52.40	4.24	23.48	9.04	45.60
1.857	0.659	3	2	3.530	66.00	118.00	10.08	52.80	24.44	124.80
1.346	1.954	3	2	0.930	9.00	38.12	10.56	39.04	20.24	69.20
1.354	1.966	3	2	1.165	8.64	38.32	10.72	43.60	24.00	59.60
1.350	1.970	3	2	3.810	17.92	64.80	10.12	60.80	27.08	102.00
1.932	1.004	1	3	0.135	10.44	37.00	4.52	15.08	5.24	32.56
1.837	1.037	1	3	0.215	23.44	46.80	3.88	23.00	8.16	44.00
1.690	1.858	3	3	1.775	29.76	54.80	12.96	38.72	22.36	91.60
1.720	1.833	3	3	0.585	24.36	50.00	11.60	31.68	15.96	76.80
1.543	2.537	2	3	1.563	26.72	52.40	10.40	45.20	28.72	104.00
2.353	1.274	1	3	0.275	8.24	32.56	4.32	15.40	4.20	37.88
1.889	2.462	2	1	2.505	12.16	65.20	10.20	48.40	31.04	78.40

X	Y	Rock	Land	Cd	Cu	Pb	Co	Cr	Ni	Zn
km	km	type	use	ppm	ppm	ppm	ppm	ppm	ppm	ppm
2.327	2.417	5	3	1.385	58.72	116.48	9.68	39.72	22.60	123.00
2.504	1.965	2	2	1.790	19.92	62.36	10.44	41.84	25.64	80.32
2.599	1.996	2	3	2.115	61.32	172.12	13.16	43.08	53.20	133.88
2.561	1.984	2	2	1.630	18.08	63.16	13.56	51.36	42.92	77.40
2.898	1.274	1	3	2.700	15.56	38.28	6.52	32.76	15.00	99.60
2.902	1.263	1	3	4.495	10.24	32.92	10.32	39.16	22.64	101.68
3.085	1.214	3	3	0.765	37.60	68.40	7.68	18.52	11.28	86.20
3.046	1.205	3	3	0.400	9.60	32.76	11.16	21.64	13.36	44.00
3.025	1.201	3	3	0.320	12.92	40.20	5.60	18.52	7.20	43.72
2.040	3.153	1	3	0.655	17.20	41.88	5.00	16.56	8.24	51.24
2.234	2.737	2	2	1.110	10.28	47.80	14.60	39.28	24.44	71.12
2.220	2.732	2	2	0.770	16.00	44.40	12.28	43.48	23.88	62.28
2.224	2.728	2	2	1.205	11.36	41.04	13.80	46.84	27.44	85.52
2.580	2.311	2	2	1.555	15.40	51.24	8.84	40.16	23.92	77.56
2.254	3.045	3	3	0.575	31.80	54.28	9.68	27.12	13.20	66.92
2.656	2.656	2	3	0.525	79.20	138.56	12.52	26.12	18.00	91.48
3.465	1.393	3	3	0.705	48.72	86.68	12.52	28.72	17.60	96.60
2.553	3.314	3	3	0.990	9.28	38.00	8.80	38.72	15.72	60.12
2.538	3.314	3	3	1.655	31.52	90.36	8.96	50.60	17.32	81.60
2.402	3.979	1	3	0.515	13.12	26.96	4.08	30.12	8.20	43.20
2.456	3.990	1	3	0.350	11.60	27.68	3.27	26.32	4.92	37.16
2.537	3.768	1	3	0.230	8.72	26.84	3.79	18.20	5.68	32.08
2.584	3.649	5	3	0.305	17.12	48.40	3.72	18.00	5.60	44.80
3.887	1.663	2	3	1.535	29.60	56.80	15.20	40.72	32.40	85.60
4.412	1.088	5	3	1.530	17.72	47.60	17.72	39.52	26.40	80.80
2.795	3.645	5	3	0.300	13.40	37.76	4.24	17.32	6.32	27.12
2.837	3.679	5	3	0.425	4.68	22.56	5.08	22.72	8.00	31.48
2.832	3.676	5	3	0.360	4.44	18.96	6.68	26.12	11.80	28.92
3.308	2.963	5	3	1.060	28.00	50.00	17.32	36.00	23.92	78.40
3.422	2.851	3	3	0.900	49.00	70.40	9.52	38.72	26.20	94.80
3.692	2.430	2	1	1.610	9.44	42.80	5.00	19.40	8.76	31.16
4.331	1.449	5	3	0.950	19.60	37.88	12.00	29.20	18.72	61.20
4.333	1.455	5	3	0.820	32.10	63.20	10.32	29.80	16.72	67.60
4.502	1.167	5	3	0.860	12.70	37.08	12.12	28.40	20.12	61.60
3.228	3.617	2	3	1.835	45.92	80.40	13.32	47.40	24.00	90.00
3.768	2.775	2	1	0.220	6.67	31.88	10.80	26.32	17.52	45.20
3.768	2.675	2	1	1.095	6.20	46.80	12.12	41.32	28.80	56.80
3.798	2.694	2	1	0.995	5.38	51.20	9.68	32.80	19.12	54.00
4.590	1.414	2	3	0.960	16.40	30.76	11.72	35.20	21.92	64.80
4.571	1.407	2	3	1.400	15.24	38.40	10.72	38.64	26.40	70.80
4.745	1.372	2	3	0.690	16.44	34.12	11.96	35.64	23.68	66.40
4.788	1.647	2	3	0.720	11.40	32.68	8.80	34.08	26.72	63.20
2.974	4.937	1	2	2.260	12.64	36.28	5.88	36.44	21.52	77.60
2.906	4.922	1	1	1.130	5.92	21.60	4.32	20.32	13.20	32.84
3.109	4.729	1	3	0.645	8.40	28.56	3.12	22.68	8.88	42.40
3.040	4.713	1	3	0.630	7.24	26.00	3.66	23.96	9.36	36.64
3.919	3.466	5	3	0.860	12.80	28.60	9.76	28.40	17.48	58.80
4.189	3.045	4	3	1.645	35.60	67.60	8.08	46.40	20.32	90.40
3.652	4.312	3	1	3.149	8.24	56.00	10.72	51.20	27.72	98.00
3.894	3.805	5	3	0.552	50.00	91.20	14.32	29.00	20.12	79.20
3.917	3.773	5	3	1.568	18.40	43.60	13.56	30.36	22.28	71.60
3.925	3.779	5	3	0.855	11.40	29.68	14.24	29.20	21.00	59.60
4.252	3.292	2	3	0.733	39.20	65.60	11.80	26.00	22.24	74.00
4.292	3.290	2	3	2.080	71.60	52.80	11.92	31.68	28.64	110.80
4.283	3.278	2	3	0.844	21.04	34.68	13.88	37.12	26.48	90.00
3.530	4.999	1	1	0.490	4.64	31.64	7.44	34.52	16.76	56.80
4.070	4.157	3	3	1.916	45.60	50.00	14.36	47.60	28.44	90.80
4.610	3.315	2	1	1.384	7.48	48.00	12.68	30.76	20.72	46.80
3.605	5.345	3	3	0.849	18.40	35.00	9.04	29.48	17.56	58.00

X	Y	Rock	Land	Cd	Cu	Pb	Co	Cr	Ni	Zn
km	km	type	use	ppm	ppm	ppm	ppm	ppm	ppm	ppm
⋆2.672	3.558	5	3	1.570	18.60	38.20	8.28	37.12	18.60	65.20
⋆3.589	4.443	1	3	2.045	11.48	33.36	10.80	40.80	21.52	112.80
⋆4.010	4.713	1	2	1.203	13.04	26.56	12.00	53.20	23.92	91.60
⋆2.942	3.137	5	2	0.490	5.64	25.88	10.92	23.40	14.60	41.20
⋆1.409	2.748	3	3	0.692	10.32	31.16	8.12	27.16	14.64	50.40
⋆3.978	2.910	2	1	1.750	8.36	37.72	9.12	35.48	26.40	63.20
⋆2.715	2.100	2	1	0.415	4.44	41.00	9.12	30.32	24.24	53.16
⋆3.870	0.762	3	2	0.685	10.92	30.84	11.72	31.92	13.12	49.28
⋆2.445	2.521	2	3	0.920	30.28	68.12	10.56	49.04	31.52	102.72
⋆3.827	2.219	2	1	2.120	7.35	54.40	6.36	23.00	14.52	72.40
⋆2.488	1.064	1	3	0.495	17.08	46.80	8.52	31.40	16.12	57.60
⋆1.646	0.524	2	2	1.060	18.88	55.20	14.44	37.28	25.24	76.40
⋆3.136	2.370	4	1	0.790	3.98	35.28	5.20	33.12	14.92	32.60
⋆3.740	5.134	1	1	0.772	8.16	30.16	10.84	39.04	24.24	56.80
⋆1.678	2.327	3	3	1.188	31.36	72.40	9.68	37.44	22.36	108.40
⋆4.399	3.180	4	3	1.615	91.20	108.80	11.96	34.56	30.20	156.80
⋆3.935	4.368	3	3	3.023	19.72	45.60	12.76	40.80	25.36	105.20
⋆3.330	1.604	3	3	2.315	113.12	144.36	9.92	56.20	25.40	147.40
⋆3.557	2.640	2	1	2.650	12.30	54.40	8.84	25.72	18.92	58.40
⋆1.376	0.945	1	3	3.780	32.76	94.40	9.68	42.80	23.52	175.20
⋆2.024	2.251	3	3	1.805	55.60	142.00	12.60	38.00	23.12	124.80
⋆3.310	4.594	5	3	1.580	56.40	93.60	5.80	40.40	22.52	108.80
⋆4.097	1.798	2	4	1.930	19.30	46.40	13.80	45.00	35.72	90.00
⋆2.326	3.633	1	3	0.415	18.32	31.92	2.36	20.12	6.28	40.72
⋆3.514	4.098	3	2	0.675	9.24	27.24	6.56	26.00	11.64	50.00
⋆3.168	4.173	1	3	0.745	8.52	30.28	3.94	26.56	15.40	52.40
⋆4.292	1.039	5	3	1.420	18.80	36.48	11.12	27.52	20.60	63.20
⋆2.834	0.988	3	2	1.425	13.32	53.60	15.24	49.20	25.56	108.80
⋆4.443	1.723	5	2	1.310	17.70	48.40	12.72	34.80	19.60	80.40
⋆3.482	2.295	2	2	1.765	127.00	300.00	10.32	40.52	30.80	192.00
⋆3.665	4.789	1	3	0.647	9.76	21.76	8.72	31.12	13.88	60.80
⋆4.173	2.144	2	2	1.810	17.90	48.80	13.20	44.12	29.60	92.80
⋆2.143	1.139	1	3	0.394	39.56	105.60	4.44	21.64	8.92	72.40
⋆3.061	2.025	2	3	1.600	20.60	35.72	10.92	37.72	25.32	92.64
⋆2.985	1.679	2	3	1.675	22.92	61.44	14.40	46.28	43.68	111.08
⋆3.438	3.752	2	2	1.280	13.04	43.60	6.36	31.00	18.60	65.20
⋆4.637	0.956	2	3	0.870	16.50	35.32	12.80	33.52	22.52	78.40
⋆3.049	5.285	2	1	0.800	7.68	48.40	12.40	49.20	27.16	60.40
⋆1.106	1.366	1	3	0.475	22.68	55.20	3.96	22.16	7.92	46.80
⋆3.633	2.986	2	3	0.855	25.50	51.60	12.40	32.92	22.32	80.40
⋆3.708	3.331	2	2	1.340	10.32	41.20	9.32	28.48	23.28	58.00
⋆2.909	1.334	1	3	1.805	28.92	74.36	6.12	24.80	11.24	111.32
⋆2.102	2.597	2	3	0.825	31.20	70.40	15.32	36.52	25.36	75.44
⋆1.797	1.215	1	3	0.545	33.08	58.00	4.24	21.80	8.88	47.20
⋆3.255	1.253	3	3	1.780	154.60	239.96	11.40	41.00	24.52	259.84
⋆4.129	3.601	2	1	2.566	7.88	55.20	10.60	32.84	23.60	59.20
⋆3.363	3.407	2	2	0.610	7.86	34.84	14.80	28.52	19.32	53.20
⋆1.452	1.290	1	3	0.585	15.16	56.40	5.80	39.88	13.20	51.20
⋆4.745	3.105	1	1	1.436	14.20	32.68	5.24	19.44	9.52	44.80
⋆0.491	1.862	3	1	2.415	21.32	48.40	7.88	32.72	18.16	49.20
⋆2.369	2.176	5	3	0.750	73.12	139.16	15.60	29.76	20.20	95.68
⋆3.676	1.528	3	3	0.805	38.72	88.32	11.12	32.00	19.52	88.16
⋆1.257	2.057	3	2	0.650	12.80	41.60	11.44	37.08	19.64	52.00
⋆3.903	2.565	2	1	0.510	5.96	62.40	1.65	3.32	1.98	60.40
⋆1.603	1.981	2	4	0.705	16.04	38.24	13.04	29.84	17.92	70.40
⋆3.395	5.210	3	3	1.112	13.12	25.56	9.80	36.56	19.48	65.60
⋆2.564	1.409	1	4	1.310	117.60	152.80	8.44	41.60	20.40	145.20
⋆0.912	2.132	2	2	2.200	11.88	51.60	10.48	60.40	30.80	78.40
⋆3.093	3.828	3	3	0.450	11.50	30.92	7.96	23.20	9.36	43.20
⋆0.836	1.787	3	3	0.825	12.96	50.00	14.32	40.80	22.84	80.40
⋆2.758	0.643	2	1	1.245	13.08	88.00	8.08	39.56	18.64	86.80

X	Y	Rock	Land	Cd	Cu	Pb	Co	Cr	Ni	Zn
km	km	type	use	ppm	ppm	ppm	ppm	ppm	ppm	ppm
⋆3.017	3.482	2	2	1.090	8.80	26.32	12.80	38.72	18.80	59.20
⋆3.212	2.716	3	3	0.780	29.80	60.40	12.52	35.52	24.00	75.60
⋆4.022	1.453	2	2	1.950	22.60	52.00	10.60	47.00	36.72	100.00
⋆4.054	3.256	2	3	1.310	21.16	52.80	8.92	35.04	20.08	78.80
⋆2.521	2.867	2	2	1.585	11.40	39.36	13.20	42.00	33.12	75.64
⋆3.319	4.864	1	2	0.520	6.04	21.12	5.08	32.68	15.40	37.52
⋆2.823	4.249	1	1	1.005	7.08	35.40	5.36	18.12	8.20	51.60
⋆1.722	0.869	1	3	0.570	21.36	67.20	4.08	24.88	9.68	56.80
⋆2.251	3.288	5	3	0.330	5.72	18.68	1.92	14.92	4.68	26.80
⋆1.182	1.711	3	2	2.535	8.72	55.60	12.56	70.00	26.24	71.60
⋆1.830	3.018	3	3	2.685	32.68	69.32	10.12	43.52	27.88	111.76
⋆3.752	1.874	2	2	0.690	20.80	55.60	13.60	37.60	28.40	80.40
⋆2.218	1.485	1	2	0.375	19.36	45.20	12.04	34.08	16.36	70.00
⋆1.873	1.560	3	3	1.955	26.08	60.40	9.68	37.88	20.88	110.80
⋆3.600	1.183	2	3	1.185	22.72	49.72	10.00	34.20	27.40	89.88
⋆2.887	4.889	1	1	1.325	5.40	37.52	3.74	27.64	14.44	46.40
⋆1.527	1.636	3	3	0.795	42.00	84.80	7.92	27.40	13.20	78.80
⋆1.333	2.407	2	3	1.630	20.56	46.00	13.40	43.60	37.52	98.40
⋆2.413	0.718	3	3	0.677	11.92	38.16	12.48	34.52	23.92	69.20
⋆1.948	1.906	3	3	1.298	22.24	45.60	12.00	49.60	26.04	96.40
⋆1.754	2.672	2	3	1.530	16.24	43.60	10.36	46.40	29.80	77.20
⋆2.294	1.830	2	2	0.670	8.72	32.76	9.72	46.80	23.24	81.60
⋆2.639	1.755	3	1	1.010	5.96	67.36	9.96	28.68	17.44	52.48
⋆1.905	3.363	1	3	0.500	16.68	39.76	4.80	23.12	9.36	46.40
⋆2.175	2.942	3	2	1.665	37.00	43.80	14.80	40.28	30.08	87.84
⋆2.791	2.446	2	3	1.520	16.32	57.52	9.16	36.20	24.12	84.40
⋆2.866	2.791	2	3	1.315	14.00	45.96	11.20	30.92	24.32	79.08
⋆3.407	1.953	2	2	1.325	9.92	38.40	9.12	37.32	24.60	84.08
⋆3.946	1.107	2	2	0.495	7.52	33.00	8.28	33.00	20.60	48.80
⋆2.452	3.995	1	3	0.395	12.72	28.24	3.56	24.00	5.60	39.64
⋆2.747	3.903	1	2	0.380	3.55	21.12	5.44	25.52	11.40	25.00
⋆3.287	3.061	5	3	2.085	39.00	52.40	13.20	45.92	26.40	104.00
⋆4.367	1.377	5	3	2.610	24.00	47.20	20.60	37.20	29.40	86.40
⋆4.713	1.302	2	3	0.845	10.28	33.64	9.68	29.92	23.80	65.60
⋆4.248	2.489	4	1	1.220	5.52	48.80	5.24	27.04	21.04	46.40
⋆3.784	3.677	2	1	0.640	6.68	34.32	13.92	28.32	18.00	55.20
⋆4.324	2.835	2	1	1.650	8.88	60.00	8.72	33.36	22.72	80.00
⋆3.859	4.022	3	3	1.433	43.60	60.80	13.32	47.60	29.12	87.20
⋆2.593	3.312	3	3	0.325	8.08	26.20	10.60	30.00	14.00	54.96

C.2 Transect data set

Spatial coordinates and values of primary and secondary attributes measured along the NE-SW transect. Rock types: 1: Argovian, 2: Kimmeridgian, 3: Sequanian, 4: Portlandian, 5: Quaternary. A dash denotes missing values.

X	Cd	Ni	Rock	Block Ni	X	Cd	Ni	Rock	Block Ni
km	ppm	ppm	type	ppm	km	ppm	ppm	type	ppm
1.00	0.910	18.00	3	18.814	3.65	-	-	2	25.000
1.05	-	-	3	19.091	3.70	-	-	2	24.883
1.10	-	-	3	19.353	3.75	-	26.20	2	24.611
1.15	-	-	3	19.533	3.80	-	-	2	23.511
1.20	-	-	3	19.634	3.85	-	-	2	22.266
1.25	-	21.96	1	19.470	3.90	-	-	2	21.023
1.30	-	-	1	18.008	3.95	-	-	2	19.784
1.35	-	-	1	16.313	4.00	2.120	14.52	2	18.930
1.40	-	-	1	14.600	4.05	-	-	2	20.344
1.45	-	-	1	12.873	4.10	-	-	2	22.188
1.50	0.570	9.68	1	11.233	4.15	-	-	2	24.129
1.55	-	-	1	10.307	4.20	-	-	4	26.166
1.60	-	-	1	9.439	4.25	2.805	32.92	4	27.869
1.65	-	-	1	8.501	4.30	-	-	4	27.083
1.70	-	-	1	7.500	4.35	-	-	4	25.880
1.75	0.135	5.24	1	6.644	4.40	-	-	4	24.609
1.80	-	-	1	6.610	4.45	-	-	4	23.269
1.85	-	-	1	6.803	4.50	-	21.04	4	21.938
1.90	-	-	1	7.099	4.55	-	-	4	21.003
1.95	-	-	1	7.491	4.60	-	-	4	20.105
2.00	-	8.92	1	7.860	4.65	-	-	4	19.194
2.05	-	-	1	7.727	4.70	-	-	4	18.273
2.10	-	-	1	7.542	4.75	1.485	15.84	4	17.484
2.15	-	-	1	7.394	4.80	-	-	2	17.529
2.20	-	-	1	7.282	4.85	-	-	2	17.716
2.25	0.275	4.20	1	7.488	4.90	-	-	2	17.914
2.30	-	-	1	9.453	4.95	-	-	2	18.124
2.35	-	-	1	11.720	5.00	-	-	2	18.343
2.40	-	-	1	14.016	5.05	-	-	2	18.572
2.45	-	-	1	16.332	5.10	-	-	2	18.789
2.50	1.310	20.40	1	18.502	5.15	-	-	2	18.974
2.55	-	-	1	19.599	5.20	-	-	2	19.123
2.60	-	-	1	20.545	5.25	-	-	2	19.235
2.65	-	-	1	21.486	5.30	-	-	2	19.309
2.70	-	-	3	22.416	5.35	-	-	2	19.401
2.75	-	20.64	3	23.583	5.40	-	-	2	19.567
2.80	-	-	3	26.263	5.45	-	-	2	19.813
2.85	-	-	3	29.179	5.50	-	-	2	20.140
2.90	-	-	3	32.048	5.55	-	-	2	20.549
2.95	-	-	2	34.862	5.60	-	-	2	20.978
3.00	-	43.68	2	37.047	5.65	-	-	2	21.361
3.05	-	-	2	35.837	5.70	-	-	2	21.694
3.10	-	-	2	33.941	5.75	-	-	2	21.973
3.15	-	-	2	32.014	5.80	-	-	2	22.197
3.20	-	-	2	30.059	5.85	-	-	2	22.374
3.25	3.925	26.12	2	28.260	5.90	-	-	2	22.515
3.30	-	-	2	27.550	5.95	-	-	2	22.617
3.35	-	-	2	27.014	6.00	-	-	2	22.680
3.40	-	-	2	26.443	6.05	-	-	2	22.702
3.45	-	-	2	25.837	6.10	-	-	2	22.703
3.50	1.325	24.60	2	25.289	6.15	-	-	2	22.703
3.55	-	-	2	25.174	6.20	-	-	2	22.703
3.60	-	-	2	25.097	6.25	-	-	2	22.703

Bibliography

Abramovitz, M. and I. A. Stegun, editors. 1972. *Handbook of Mathematical Functions: with Formulas, Graphs, and Mathematical Tables*, 9th (revised) printing. Dover, New York. 1046 p.

Alabert, F. G. 1987a. The practice of fast conditional simulations through the LU decomposition of the covariance matrix. *Mathematical Geology*, 19(5):369–386.

Alabert, F. G. 1987b. *Stochastic Imaging of Spatial Distributions Using Hard and Soft Information*. Master's thesis, Stanford University, Stanford, CA.

Alabert, F. G. and G. J. Massonnat. 1990. Heterogeneity in a complex turbiditic reservoir: Stochastic modelling of facies and petrophysical variability. In *65th Annual Technical Conference and Exhibition*, number 20604, pages 775–790. Society of Petroleum Engineers.

Almeida, A. S. 1993. *Joint Simulation of Multiple Variables with a Markov-type Coregionalization Model*. Doctoral dissertation, Stanford University, Stanford, CA.

Almeida, A. S. and A. G. Journel. 1994. Joint simulation of multiple variables with a Markov-type coregionalization model. *Mathematical Geology*, 26(5):565–588.

Anderson, T. 1958. *An Introduction to Multivariate Statistical Analysis*. John Wiley & Sons, New York.

Atteia, O., J.-P. Dubois, and R. Webster. 1994. Geostatistical analysis of soil contamination in the Swiss Jura. *Environmental Pollution*, 86:315–327.

Barnes, R. J. 1991. The variogram sill and the sample variance. *Mathematical Geology*, 23:673–678.

Barnes, R. J. and T. Johnson. 1984. Positive kriging. In G. Verly, M. David, A. G. Journel, and A. Maréchal, editors, *Geostatistics for Natural Resources Characterization*, volume 1, pages 231–244. Reidel, Dordrecht.

Borgman, L., M. Taheri, and R. Hagan. 1984. Three-dimensional frequency-domain simulations of geological variables. In G. Verly, M. David,

A. G. Journel, and A. Maréchal, editors, *Geostatistics for Natural Resources Characterization*, volume 2, pages 517–541. Reidel, Dordrecht.

Bourgault, G. 1994. Robustness of noise filtering by kriging analysis. *Mathematical Geology*, 26(6):733–752.

Bourgault, G. and A. G. Journel. 1994. Gaussian or indicator-based simulation? which variogram is more relevant? In *International Association for Mathematical Geology Annual Conference*, pages 32–37.

Bourgault, G., A. G. Journel, S. M. Lesh, J. D. Rhoades, and D. L. Corwin. 1995. Geostatistical analysis of a soil salinity data set. In *Application of GIS to the Modeling of Non-point Source Pollutants in the Vadose Zone*, pages 53–114, ASA-CSSA-SSSA Bouyoucos conference, Mission Inn, Riverside, CA.

Bourgault, G. and D. Marcotte. 1991. Multivariable variogram and its application to the linear model of coregionalization. *Mathematical Geology*, 23(7):899–928.

Chilès, J. and A. Guillen. 1984. Variogrammes et krigeages pour la gravimétrie et le magnétisme. In J.-J. Royer, editor, *Computers in Earth Sciences for Natural Resources Characterization*, volume 20, pages 455–468. Sciences de la Terre, Nancy, France.

Christakos, G. 1984. On the problem of permissible covariance and variogram models. *Water Resources Research*, 20(2):251–265.

Christakos, G. 1990. A Bayesian/maximum-entropy view to the spatial estimation problem. *Mathematical Geology*, 22(7):763–777.

Chu, J. 1993. Xgam: A 3-D interactive graphic software for modeling variograms and crossvariograms under conditions of positive definiteness. In *Report 6, Stanford Center for Reservoir Forecasting*, Stanford, CA.

Chu, J. 1996. Fast sequential indicator simulation: Beyond reproduction of indicator variograms. *Mathematical Geology*, 28(7):923–936.

Chu, J. and A. G. Journel. 1994. Conditional fBm simulation with dual kriging. In R. Dimitrakopoulos, editor, *Geostatistics for the Next Century*, pages 407–421. Kluwer, Dordrecht.

Chu, J., W. Xu, H. Zhu, and A. G. Journel. 1991. The Amoco case study. In *Report 4, Stanford Center for Reservoir Forecasting*, Stanford, CA.

Clark, I., K. L. Basinger, and W. V. Harper. 1989. MUCK—A novel approach to cokriging. In B. E. Buxton, editor, *Proceedings of the Conference on Geostatistical, Sensitivity, and Uncertainty Methods for Ground-water Flow and Radionuclide Transport Modeling*, pages 473–493. Battelle Press.

Cressie, N. 1985. Fitting variogram models by weighted least squares. *Mathematical Geology*, 17(5):563–586.

Daly, C., C. Lajaunie, and D. Jeulin. 1989. Application of multivariate kriging to the processing of noisy images. In M. Armstrong, editor, *Geostatistics*, volume 2, pages 749–760. Kluwer, Dordrecht.

Damsleth, E., C. B. Tjolsen, K. H. Omre, and H. H. Haldorsen. 1990. A two-stage stochastic model applied to a North Sea reservoir. In *65th Annual Technical Conference and Exhibition*, pages 791–802. Society of Petroleum Engineers.

David, M. 1977. *Geostatistical Ore Reserve Estimation*. Elsevier, Amsterdam, 001 p.

Davis, B. M. 1987. Uses and abuses of cross validation in geostatistics. *Mathematical Geology*, 19(3):241–248.

Davis, B. M. and K. A. Greenes. 1983. Estimating using spatially distributed multivariate data: An example with coal quality. *Mathematical Geology*, 15(2):287–300.

Davis, J. C. 1986. *Statistics and Data Analysis in Geology*, 2nd edition. John Wiley & Sons, New York, 646 p.

Davis, M. 1987. Production of conditional simulations via the LU decomposition of the covariance matrix. *Mathematical Geology*, 19(2):91–98.

Delfiner, P. 1976. Linear estimation of non-stationary spatial phenomena. In M. Guarascio, M. David, and C. J. Huijbregts, editors, *Advanced Geostatistics in the Mining Industry*, pages 49–68. Reidel, Dordrecht.

Deutsch, C. V. 1989. DECLUS: A Fortran 77 program for determining optimum spatial declustering weights. *Computers & Geosciences*, 15(3):325–332.

Deutsch, C. V. 1994a. Algorithmically-defined random function models. In R. Dimitrakopoulos, editor, *Geostatistics for the Next Century*, pages 422–435. Kluwer, Dordrecht.

Deutsch, C. V. 1994b. Constrained modeling of histograms and cross plots with simulated annealing. In *Report 7, Stanford Center for Reservoir Forecasting*, Stanford, CA.

Deutsch, C. V. and P. W. Cockerham. 1994. Practical considerations in the application of simulated annealing to stochastic simulation. *Mathematical Geology*, 26(1):67–82.

Deutsch, C. V. and A. G. Journel. 1992a. *GSLIB: Geostatistical Software Library and User's Guide*. Oxford University Press, New York, 340 p.

Deutsch, C. V. and A. G. Journel. 1992b. Annealing techniques applied to the integration of geological and engineering data. In *Report 5, Stanford Center for Reservoir Forecasting*, Stanford, CA.

Doyen, P. M., T. M. Guidish, and M. de Buyl. 1989. Seismic discrimination of lithology in sand/shale reservoirs: A Bayesian approach. In *SEG 59th Annual Meeting*, Dallas, TX.

Dubrule, O. 1983. Two methods with different objectives: Splines and kriging. *Mathematical Geology*, 15(2):245–257.

Edwards, C. and D. Penney. 1982. *Calculus and Analytical Geometry*. Prentice-Hall, Englewood Cliffs, N.J.

Englund, E. and A. Sparks. 1991. *Geo-EAS 1.2.1 Geostatistical Environmental Assessment Software, User's Guide, EPA Report # 60018-91/008*. U.S. EPA, EMS Lab, Las Vegas, NV.

Farmer, C. 1988. The generation of stochastic fields of reservoir parameters with specified geostatistical distributions. In S. Edwards and P. R. King, editors, *Mathematics in Oil Production*, pages 235–252. Clarendon Press, Oxford.

Farmer, C. 1992. Numerical rocks. In P. R. King, editor, *The Mathematical Generation of Reservoir Geology*. Clarendon Press, Oxford.

FOEFL (Swiss Federal Office of Environment and Landscape). 1987. *Commentary on the Ordinance Relating to Pollutants in Soil (VSBo of June 9, 1986)*. FOEFL, Bern.

Froidevaux, R. 1990. *Geostatistical Toolbox Primer, version 1.30*. FSS International, Troinex, Switzerland.

Froidevaux, R. 1993. Probability field simulation. In A. Soares, editor, *Geostatistics Tróia '92*, volume 1, pages 73–84. Kluwer Academic Publishers, Dordrecht.

Galli, A., F. Gerdil-Neuillet, and C. Dadou. 1984. Factorial kriging analysis: A substitute to spectral analysis of magnetic data. In G. Verly, M. David, A. G. Journel, and A. Maréchal, editors, *Geostatistics for Natural Resources Characterization*, volume 2, pages 543–557. Reidel, Dordrecht.

Galli, A. and G. Meunier. 1987. Study of a gas reservoir using the external drift method. In G. Matheron and M. Armstrong, editors, *Geostatistical Case Studies*, pages 105–119. Reidel, Dordrecht.

Geman, S. and D. Geman. 1984. Stochastic relaxation, Gibbs distributions, and the Bayesian restoration of images. *IEEE Transactions on Pattern Analysis and Machine Intelligence*, PAMI-6(6):721–741.

Glacken, I. 1996. *Change of support by direct conditional block simulation*. Master's thesis, Stanford University, Stanford, CA.

Gómez-Hernández, J. 1991. *A Stochastic Approach to the Simulation of Block Conductivity Fields Conditioned upon Data Measured at a Smaller Scale*. Doctoral dissertation, Stanford University, Stanford, CA.

Gómez-Hernández, J. and E. Cassiraga. 1994. Theory and practice of sequential simulation. In M. Armstrong and P. A. Dowd, editors, *Geostatistical Simulations*, pages 111–124. Kluwer Academic Publishers, Dordrecht.

Gómez-Hernández, J. and A. G. Journel. 1993. Joint sequential simulation of multiGaussian fields. In A. Soares, editor, *Geostatistics Tróia '92*, volume 1, pages 85–94. Kluwer Academic Publishers, Dordrecht.

Gómez-Hernández, J. and R. M. Srivastava. 1990. ISIM3D: An ANSI-C three-dimensional multiple indicator conditional simulation program. *Computers & Geosciences*, 16(4):395–440.

Gómez-Hernández, J. and X. Wen. 1994. To be or not to be multiGaussian? That is the question. In *Report 7, Stanford Center for Reservoir Forecasting, Stanford, CA*.

Goovaerts, P. 1992. Factorial kriging analysis: A useful tool for exploring the structure of multivariate spatial soil information. *Journal of Soil Science*, 43(4):597–619.

Goovaerts, P. 1993. Spatial orthogonality of the principal components computed from coregionalized variables. *Mathematical Geology*, 25(3):281–302.

Goovaerts, P. 1994a. Comparative performance of indicator algorithms for modeling conditional probability distribution functions. *Mathematical Geology*, 26(3):389–411.

Goovaerts, P. 1994b. Comparison of CoIK, IK and mIK performances for modeling conditional probabilities of categorical variables. In R. Dimitrakopoulos, editor, *Geostatistics for the Next Century*, pages 18–29. Kluwer, Dordrecht.

Goovaerts, P. 1994c. On a controversial method for modeling a coregionalization. *Mathematical Geology*, 26(2):197–204.

Goovaerts, P. 1994d. Study of spatial relationships between two sets of variables using multivariate geostatistics. *Geoderma*, 62:93–107.

Goovaerts, P. 1996. Stochastic simulation of categorical variables using a classification algorithm and simulated annealing. *Mathematical Geology*, 28(7):909–921.

Goovaerts, P. 1997. Ordinary cokriging revisited. *Mathematical Geology*, 29, in press.

Goovaerts, P. and C. Chiang. 1993. Temporal persistence of spatial patterns for mineralizable nitrogen and selected soil properties. *Soil Science Society of America Journal*, 57(2):372–381.

Goovaerts, P. and A. G. Journel. 1995. Integrating soil map information in modelling the spatial variation of continuous soil properties. *European Journal of Soil Science*, 46(3):397–414.

Goovaerts, P. and A. G. Journel. 1996. Accounting for local probabilities in stochastic modeling of facies data. *SPE Journal*, 1(1):21–29.

Goovaerts, P. and Ph. Sonnet. 1993. Study of spatial and temporal variations of hydrogeochemical variables using factorial kriging analysis. In

A. Soares, editor, *Geostatistics Tróia '92*, volume 2, pages 745–756. Kluwer Academic Publishers, Dordrecht.

Goovaerts, P., Ph. Sonnet, and A. Navarre. 1993. Factorial kriging analysis of springwater contents in the Dyle river basin, Belgium. *Water Resources Research*, 29(7):2115–2125.

Goovaerts, P. and R. Webster. 1994. Scale-dependent correlation between topsoil copper and cobalt concentrations in Scotland. *European Journal of Soil Science*, 45(1):79–95.

Goulard, M. 1989. Inference in a coregionalization model. In M. Armstrong, editor, *Geostatistics*, volume 1, pages 397–408. Kluwer, Dordrecht.

Goulard, M. and M. Voltz. 1992. Linear coregionalization model: Tools for estimation and choice of cross-variogram matrix. *Mathematical Geology*, 24(3):269–286.

Grzebyk, M. 1993. *Ajustement d'une Corégionalisation_Stationnaire*. Doctoral dissertation, Ecole Nationale Supérieure des Mines de Paris.

Guardiano, F. and R. M. Srivastava. 1993. Multivariate geostatistics: Beyond bivariate moments. In A. Soares, editor, *Geostatistics Tróia '92*, volume 1, pages 133–144. Kluwer Academic Publishers, Dordrecht.

Haldorsen, H. H., P. J. Brand, and C. J. Macdonald. 1988. Review of the stochastic nature of reservoirs. In S. Edwards and P. R. King, editors, *Mathematics in Oil Production*, pages 109–209. Clarendon Press, Oxford.

Haslett, J., R. Bradley, P. S. Craig, G. Wills, and A. R. Unwin. 1991. Dynamic graphics for exploring spatial data, with application to locating global and local anomalies. *American Statistician*, 45:234–242.

Hudson, G. 1993. Kriging temperature in Scotland using the external drift method. In A. Soares, editor, *Geostatistics Tróia '92*, volume 2, pages 577–588. Kluwer Academic Publishers, Dordrecht.

Isaaks, E. H. 1984. *Risk Qualified Mappings for Hazardous Waste Sites: A Case Study in Distribution-free Geostatistics*. Master's thesis, Stanford University, Stanford, CA.

Isaaks, E. H. 1990. *The Application of Monte Carlo Methods to the Analysis of Spatially Correlated Data*. Doctoral dissertation, Stanford University, Stanford, CA.

Isaaks, E. H. and R. M. Srivastava. 1989. *An Introduction to Applied Geostatistics*. Oxford University Press, New York, 561 p.

Journel, A. G. 1980. The lognormal approach to predicting local distributions of selective mining unit grades. *Mathematical Geology*, 12(4):285–303.

Journel, A. G. 1983. Non-parametric estimation of spatial distributions. *Mathematical Geology*, 15(3):445–468.

Journel, A. G. 1984a. Mad and conditional quantile estimators. In G. Verly, M. David, A. G. Journel, and A. Maréchal, editors, *Geostatistics for Natural Resources Characterization*, volume 2, pages 261–270. Reidel, Dordrecht.

Journel, A. G. 1984b. The place of non-parametric geostatistics. In G. Verly, M. David, A. G. Journel, and A. Maréchal, editors, *Geostatistics for Natural Resources Characterization*, volume 1, pages 307–355. Reidel, Dordrecht.

Journel, A. G. 1986a. Geostatistics: Models and tools for the earth sciences. *Mathematical Geology*, 18(1):119–140.

Journel, A. G. 1986b. Constrained interpolation and qualitative information. *Mathematical Geology*, 18(3):269–286.

Journel, A. G. 1987. Geostatistics for the environmental sciences, EPA project no. cr 811893. Technical report, U.S. EPA, EMS Lab, Las Vegas, NV.

Journel, A. G. 1989. *Fundamentals of Geostatistics in Five Lessons.* Volume 8 Short Course in Geology. American Geophysical Union, Washington, D.C.

Journel, A. G. 1994a. Modeling uncertainty: Some conceptual thoughts. In R. Dimitrakopoulos, editor, *Geostatistics for the Next Century*, pages 30–43. Kluwer, Dordrecht.

Journel, A. G. 1994b. Resampling from stochastic simulations. *Environmental and Ecological Statistics*, 1:63–84.

Journel, A. G. 1995. Probability fields: Another look and a proof. In *Report 8, Stanford Center for Reservoir Forecasting*, Stanford, CA.

Journel, A. G. and F. G. Alabert. 1988. Focusing on spatial connectivity of extreme valued attributes: Stochastic indicator models of reservoir heterogeneities. SPE paper # 18324.

Journel, A. G. and C. V. Deutsch. 1993. Entropy and spatial disorder. *Mathematical Geology*, 25(3):329–355.

Journel, A. G. and C. J. Huijbregts. 1978. *Mining Geostatistics.* Academic Press, New York, 600 p.

Journel, A. G. and E. H. Isaaks. 1984. Conditional indicator simulation: Application to a Saskatchewan uranium deposit. *Mathematical Geology*, 16(7):685–718.

Journel, A. G. and D. Posa. 1990. Characteristic behavior and order relations for indicator variograms. *Mathematical Geology*, 22(8):1011–1025.

Journel, A. G. and S. E. Rao. 1996. Deriving conditional distributions from ordinary kriging. In *Report 9, Stanford Center for Reservoir Forecasting*, Stanford, CA.

Journel, A. G. and M. E. Rossi. 1989. When do we need a trend model in kriging? *Mathematical Geology*, 21(7):715–739.

Journel, A. G. and W. Xu. 1994. Posterior identification of histograms conditional to local data. *Mathematical Geology*, 26(3):323–359.

Krige, D. G. 1951. *A Statistical Approach to Some Mine Valuations and Allied Problems at the Witwatersrand*. Master's thesis, University of Witwatersrand.

Luenberger, D. G. 1969. *Optimization by Vector Space Methods*. John Wiley & Sons, New York.

Luster, G. R. 1985. *Raw Materials for Portland Cement: Applications of Conditional Simulation of Coregionalization*. Doctoral dissertation, Stanford University, Stanford, CA.

Ma, Y. Z. and J. J. Royer. 1988. Local geostatistical filtering: Application to remote sensing. *Sciences de la Terre*, Série Informatique, 27:17–36.

Mallet, J. L. 1980. Régression sous contraintes linéaires: Application au codage des variables aléatoires. *Revue de Statistique Appliquée*, 28(1):57–68.

Marcotte, D. 1995. Conditional simulation with data subject to measurement error: Post-simulation filtering with modified factorial kriging. *Mathematical Geology*, 27(6):749–762.

Maréchal, A. 1984. Kriging seismic data in presence of faults. In G. Verly, M. David, A. G. Journel, and A. Maréchal, editors, *Geostatistics for Natural Resources Characterization*, pages 271–294. Reidel, Dordrecht.

Matheron, G. 1970. *La Théorie des Variables Régionalisées et ses Applications*. Fascicule 5, Les Cahiers du Centre de Morphologie Mathématique, Ecole des Mines de Paris, Fontainebleau, 212 p.

Matheron, G. 1979. Recherche de simplification dans un problème de cokrigeage. Internal note N-628, Centre de Géostatistique, Fontainebleau.

Matheron, G. 1982. Pour une analyse krigeante de données régionalisées. Internal note N-732, Centre de Géostatistique, Fontainebleau.

Matheron, G. 1989. *Estimating and Choosing*. Springer-Verlag, Berlin, 141 p.

Matheron, G., H. Beucher, C. de Fouquet, A. Galli, D. Guerillot, and C. Ravenne. 1987. Conditional simulation of the geometry of fluvio-deltaic reservoirs. SPE paper # 16753.

Murray, C. 1992. *Geostatistical Applications in Petroleum Geology and Sedimentary Geology*. Doctoral dissertation, Stanford University, Stanford, CA.

Myers, D. E. 1982. Matrix formulation of co-kriging. *Mathematical Geology*, 14(3):249–257.

Myers, D. E. 1991. Pseudo-cross variograms, positive-definiteness, and cokriging. *Mathematical Geology*, 23(6):805–816.

Papritz, A. and H. Flühler. 1994. Temporal change of spatially autocorrelated soil properties: Optimal estimation by cokriging. *Geoderma*, 25:1015–1026.

Papritz, A., H. R. Kunsch, and R. Webster. 1993. On the pseudo cross variogram. *Mathematical Geology*, 25(8):1015–1026.

Posa, D. 1989. Conditioning of the stationary kriging matrices for some well-known covariance models. *Mathematical Geology*, 21(7):755–765

Press, W. H., B. P. Flannery, S. A. Teukolsky, and W. T. Vetterling. 1986. *Numerical Recipes*. Cambridge University Press, New York.

Ripley, B. D. 1981. *Spatial Statistics*. John Wiley & Sons, New York, 252 p.

Ripley, B. D. 1987. *Stochastic Simulation*. John Wiley & Sons, New York, 237 p.

Rouhani, S. and H. Wackernagel. 1990. Multivariate geostatistical approach to space-time data analysis. *Water Resources Research*, 26(4):585–591.

Sandjivy, L. 1984. The factorial kriging analysis of regionalized data. Its application to geochemical prospecting. In G. Verly, M. David, A. G. Journel, and A. Maréchal, editors, *Geostatistics for Natural Resources Characterization*, pages 559–571. Reidel, Dordrecht.

Séguret, S. and P. Huchon. 1990. Trigonometric kriging: A new method for removing the diurnal variation from geomagnetic data. *Journal of Geophysical Research*, 95(13):21,383–21,397.

Shannon, C. E. 1948. A mathematical theory of communication. *Bell System Technical Journal*, 27:379–623.

Silverman, B. W. 1986. *Density Estimation for Statistics and Data Analysis*. Chapman and Hall, New York.

Soares, A. 1992. Geostatistical estimation of multi-phase structures. *Mathematical Geology*, 24(2):149–160.

Sousa, A. J. 1989. Geostatistical data analysis—An application to ore typology. In M. Armstrong, editor, *Geostatistics*, volume 2, pages 851–860. Kluwer, Dordrecht.

Srivastava, R. M. 1987a. Minimum variance or maximum profitability? *CIM Bulletin*, 80(901):63–68.

Srivastava, R. M. 1987b. *A Non-ergodic Framework For Variogram and Covariance Functions*. Master's thesis, Stanford University, Stanford, CA.

Srivastava, R. M. 1992. Reservoir characterization with probability field simulation. In *SPE Annual Conference and Exhibition, Washington, D.C.*, number 24753, pages 927–938, Washington, D.C., Society of Petroleum Engineers.

Srivastava, R. M. 1994. The visualization of spatial uncertainty. In J. M. Yarus and R. L. Chambers, editors, *Stochastic Modeling and Geostatistics. Principles, Methods, and Case Studies*, volume 3 of *AAPG Computer Applications in Geology*, pages 339–345.

Srivastava, R. M. and H. M. Parker. 1989. Robust measures of spatial continuity. In M. Armstrong, editor, *Geostatistics*, pages 295–308. Kluwer, Dordrecht.

Stein, A., M. Hoogerwerf, and J. Bouma. 1988. Use of soil-map delineations to improve (co)kriging of point data on moisture deficits. *Geoderma*, 43:163–177.

Sullivan, J. 1984. Conditional recovery estimation through probability kriging: Theory and practice. In G. Verly, M. David, A. G. Journel, and A. Maréchal, editors, *Geostatistics for Natural Resources Characterization*, pages 365–384. Reidel, Dordrecht.

Sullivan, J. 1985. *Non-parametric Estimation of Spatial Distributions*. Doctoral dissertation, Stanford University, Stanford, CA.

Suro-Pérez, V. 1991. Generation of a turbiditic reservoir: The Boolean alternative. In *Report 4, Stanford Center for Reservoir Forecasting*, Stanford, CA.

Suro-Pérez, V. 1993. Indicator principal component kriging: The multivariate case. In A. Soares, editor, *Geostatistics Tróia '92*, volume 1, pages 441–454. Kluwer Academic Publishers, Dordrecht.

Suro-Pérez, V. and A. G. Journel. 1991. Indicator principal component kriging. *Mathematical Geology*, 23(5):759–788.

Tran, T. 1994. Improving variogram reproduction on dense simulation grids. *Computers & Geosciences*, 20(7):1161–1168.

Van Meirvenne, M., K. Scheldeman, G. Baert, and G. Hofman. 1994. Quantification of soil textural fractions of Bas-Zaire using soil map polygons and/or point observations. *Geoderma*, 62:69–82.

Verly, G. 1986. MultiGaussian kriging—A complete case study. In R. V. Ramani, editor, *Proceedings of the 19th International APCOM Symposium*, pages 283–298, Littleton, CO. Society of Mining Engineers.

Verly, G. 1993. Sequential Gaussian co-simulation: A simulation method integrating several types of information. In A. Soares, editor, *Geostatistics Tróia '92*, volume 1, pages 543–554. Kluwer Academic Publishers, Dordrecht.

Voltz, M. and R. Webster. 1990. A comparison of kriging, cubic splines and classification for predicting soil properties from sample information. *Journal of Soil Science*, 41(3):473–490.

Wackernagel, H. 1988. Geostatistical techniques for interpreting multivariate spatial information. In C. F. Chung, A. G. Fabbri, and R. Sinding-

Larsen, editors, *Quantitative Analysis of Mineral and Energy Resources*, pages 393–409. Reidel, Dordrecht.

Wackernagel, H. 1994. Cokriging versus kriging in regionalized multivariate data analysis. *Geoderma*, 62:83–92.

Wackernagel, H. 1995. *Multivariate Geostatistics: An introduction with applications*. Springer-Verlag, Berlin, 256 p.

Wackernagel, H., P. Petitgas, and Y. Touffait. 1989. Overview of methods for coregionalization analysis. In M. Armstrong, editor, *Geostatistics*, volume 1, pages 409–420. Kluwer, Dordrecht.

Wackernagel, H. and H. Sanguinetti. 1993. Gold prospecting with factorial cokriging in the Limousin, France. In J. C. Davis and U. C. Herzfeld, editors, *Computers in Geology: 25 years of progress. Studies in Mathematical Geology*, volume 5, pages 33–43. Kluwer, Dordrecht.

Wackernagel, H., R. Webster, and M. A. Oliver. 1988. A geostatistical method for segmenting multivariate sequences of soil data. In H. H. Bock, editor, *Classification and Related Methods of Data Analysis*, pages 641–650. Elsevier North Holland, Amsterdam.

Wang, L. 1994. Animated visualization of multiple simulated realizations and assessment of uncertainty. In *Report 7, Stanford Center for Reservoir Forecasting*, Stanford, CA.

Webster, R., O. Atteia, and J.-P. Dubois. 1994. Coregionalization of trace metals in the soil in the Swiss Jura. *European Journal of Soil Science*, 45:205–218.

Webster, R. and M. A. Oliver. 1990. *Statistical Methods in Soil and Land Resource Survey*. Oxford University Press, New York, 316 p.

Xiao, H. 1985. *A Description of the Behavior of Indicator Variograms for a Bivariate Normal Distribution*. Master's thesis, Stanford University, Stanford, CA.

Xu, W. 1994. Convex kriging. In *Report 7, Stanford Center for Reservoir Forecasting*, Stanford, CA.

Xu, W. 1995a. Stochastic modeling of lithofacies: Alternatives. In *Report 8, Stanford Center for Reservoir Forecasting*, Stanford, CA.

Xu, W. 1995b. *Stochastic Modeling of Reservoir Lithofacies and Petrophysical Properties*. Doctoral dissertation, Stanford University, Stanford, CA.

Xu, W. and A. G. Journel. 1993. Gtsim: Gaussian truncated simulation of lithofacies. In *Report 6, Stanford Center for Reservoir Forecasting*, Stanford, CA.

Xu, W. and A. G. Journel. 1994. Dssim: A general sequential simulation algorithm. In *Report 7, Stanford Center for Reservoir Forecasting*, Stanford, CA.

Xu, W. and A. G. Journel. 1995. Histogram and scattergram smoothing using convex quadratic programming. *Mathematical Geology*, 27(1):83–103.

Xu, W., T. Tran, R. M. Srivastava, and A. G. Journel. 1992. Integrating seismic data in reservoir modeling: The collocated cokriging alternative. SPE paper # 24742.

Zhu, H. 1992. Dual kriging. *Geostat Newsletter*, 4:4–5.

Zhu, H. and A. G. Journel. 1989. Indicator conditioned estimator. *Transactions, Society for Mining, Metallurgy and Exploration, Inc.*, 286:1880–1886.

Zhu, H. and A. G. Journel. 1993. Formatting and integrating soft data: Stochastic imaging via the Markov–Bayes algorithm. In A. Soares, editor, *Geostatistics Tróia '92*, volume 1, pages 1–12. Kluwer Academic Publishers, Dordrecht.

Index